Lecture Notes in Computer Science 6735

Commenced Publication in 1973
Founding and Former Series Editors:
Gerhard Goos, Juris Hartmanis, and Jan van Leeuwen

Benedikt Löwe
Dag Normann
Ivan Soskov
Alexandra Soskova (Eds.)

Models of Computation in Context

7th Conference on Computability in Europe, CiE 2011
Sofia, Bulgaria, June 27 - July 2, 2011
Proceedings

 Springer

Volume Editors

Benedikt Löwe
University of Amsterdam
Institute for Logic, Language and Computation
P.O. Box 94242
1090 GE Amsterdam, The Netherlands
E-mail: b.loewe@uva.nl

Dag Normann
University of Oslo
Department of Mathematics
P.O. Box 1053 Blindern
0316 Oslo, Norway
E-mail: dnormann@math.uio.no

Ivan Soskov
Sofia University
Faculty of Mathematics and Informatics
Department of Mathematical Logic and Applications
5 James Bourchier blvd.
1164 Sofia, Bulgaria
E-mail: soskov@fmi.uni-sofia.bg

Alexandra Soskova
Sofia University
Faculty of Mathematics and Informatics
Department of Mathematical Logic and Applications
5 James Bourchier blvd.
1164 Sofia, Bulgaria
E-mail: asoskova@fmi.uni-sofia.bg

ISSN 0302-9743   e-ISSN 1611-3349
ISBN 978-3-642-21874-3   e-ISBN 978-3-642-21875-0
DOI 10.1007/978-3-642-21875-0
Springer Heidelberg Dordrecht London New York

Library of Congress Control Number: 2011930187
CR Subject Classification (1998): F.2, F.1, G.2, I.1
LNCS Sublibrary: SL 1 – Theoretical Computer Science and General Issues

Typesetting: Camera-ready by author, data conversion by Scientific Publishing Services, Chennai, India

Printed on acid-free paper

Springer is part of Springer Science+Business Media (www.springer.com)

Preface

CiE 2011: Models of Computation in Context
Sofia, Bulgaria, June 27 – July 2, 2011

CiE 2011 was the seventh meeting in the conference series "Computability in Europe" and the third meeting after the foundation of the Association CiE in Athens 2008. Both the conference series and the association promote the development of computability-related science, ranging over mathematics, computer science and applications in various natural and engineering sciences such as physics and biology, and also including the promotion of related non-scientific fields such as philosophy and history of computing. The title and topic of the seventh conference, *Models of Computation in Context*, highlights the two pillars on which the CiE movement rests: the mathematical representation of computation as a tool for understanding computability, and its application and adaptation to various real-world contexts. This conference was held at Sofia University in Bulgaria.

The first six of the CiE conferences were held at the University of Amsterdam in 2005, at the University of Wales Swansea in 2006, at the University of Siena in 2007, at the University of Athens in 2008, at the University of Heidelberg in 2009 and at the University of the Azores in 2010. The proceedings of these meetings, edited in 2005 by S. Barry Cooper, Benedikt Löwe and Leen Torenvliet, in 2006 by Arnold Beckmann, Ulrich Berger, Benedikt Löwe and John V. Tucker, in 2007 by S. Barry Cooper, Benedikt Löwe and Andrea Sorbi, in 2008 by Arnold Beckmann, Costas Dimitracopoulos and Benedikt Löwe, in 2009 by Klaus Ambos-Spies, Benedikt Löwe and Wolfgang Merkle and in 2010 by Fernando Ferreira, Benedikt Löwe, Elvira Mayordomo and Luís Mendes Gomes, were published as *Springer Lecture Notes in Computer Science*, Volumes 3526, 3988, 4497, 5028, 5365 and 6158, respectively.

CiE and its conferences have changed our perceptions of computability and its interface with other areas of knowledge. The large number of mathematicians and computer scientists attending those conference had their view of computability theory enlarged and transformed: They discovered that its foundations were deeper and more mysterious, its technical development more vigorous, its applications wider and more challenging than they had known. The annual CiE conference has become a major event, and is the largest international meeting focused on computability theoretic issues. Future meetings are in the planning. The meeting in 2012 will be in Cambridge, UK, and will be a part of the Turing Centenary Celebrations. The series is coordinated by the CiE Conference Series Steering Committee consisting of Luís Antunes (Porto, Secretary), Arnold Beckmann (Swansea), Paola Bonizzoni (Milano), S. Barry Cooper (Leeds), Viv Kendon (Leeds), Benedikt Löwe (Amsterdam, Chair), Dag Normann (Oslo), and Peter van Emde Boas (Amsterdam).

The conference was based on invited tutorials and lectures, and a set of special sessions on a range of subjects; there were also many contributed papers and informal presentations. This volume contains 11 of the invited lectures and under 40% of the submitted contributed papers, all of which have been refereed. The *Best Student Paper Award* was given to Stefan Vatev for his paper "Conservative Extensions of Abstract Structures." We thank Springer for sponsoring this award.

Jack Lutz (Iowa State University) and Geoffrey Pullum (University of Edinburgh) were the tutorial speakers. The conference had the following invited speakers: Christel Baier (Dresden), Michiel van Lambalgen (Amsterdam), Antonio Montalban (Chicago), Alexandra Shlapentokh (Greenville), Theodore Slaman (Berkeley), Janet Thornton (Cambridge) and Alasdair Urquhart (Toronto).

CiE 2011 had six special sessions:

Computability in Analysis, Algebra, and Geometry.
Organizers. Alexandra Shlapentokh (Greenville) and Dieter Spreen (Siegen).
Speakers. Ulrich Berger, Vasco Brattka, Valentina Harizanov and Russell Miller.
Classical Computability Theory.
Organizers. Douglas Cenzer (Gainesville) and Bjørn Kjos-Hanssen (Honolulu).
Speakers. Mingzhong Cai, Rachel Epstein, Charles Harris and Guohua Wu.
Natural Computing.
Organizers. Erzébet Csuhaj-Varjú (Budapest) and Ion Petre (Turku).
Speakers. Natalio Krasnogor, Martin Kutrib, Victor Mitrana and Agustín Riscos-Núñez.
Relations Between the Physical World and Formal Models of Computability.
Organizers. Viv Kendon (Leeds) and Sonja Smets (Groningen).
Speakers. Pablo Arrighi, Časlav Brukner, Elham Kashefi and Prakash Panangaden.
Theory of Transfinite Computations.
Organizers. Peter Koepke (Bonn) and Chi Tat Chong (Singapore).
Speakers. Merlin Carl, Sy D. Friedman, Wei Wang and Philip Welch.
Computational Linguistics.
Organizers. Tejaswini Deoskar (Amsterdam) and Tinko Tinchev (Sofia).
Speakers. Klaus U. Schulz, Ian Pratt-Hartmann and Robin Cooper.

The CiE 2011 conference was organised by Douglas Cenzer (Gainesville), Angel Dichev (Sofia), Damir Dzhafarov (Chicago), Hristo Ganchev (Sofia), Dimitar Guelev (Sofia), Vladislav Nenchev (Sofia), Stela Nikolova (Sofia), Dimitar Shiachki (Sofia), Alexandra Soskova (Chair, Sofia), Mariya Soskova (Sofia), Stefan Vatev (Sofia), Mitko Yanchev (Sofia) and Anton Zinoviev (Sofia).

We are happy and thankful to acknowledge the generous support of the following sponsors: The National Science Foundation (USA), the Association for Symbolic Logic, the European Mathematical Society, the European Social Fund, the Bulgarian National Science Fund, the Bulgarian Ministry of Education, Youth and Science, the company *Partners 1993*, the publisher Elsevier B.V., and the company *Opencode Mobile Network Systems*. Special thanks are due to the following sponsors who supported this volume: The *Haemimont Foundation* (a corporation engaged in not-for-profit activities for the public good, especically in the area of education), and *Astea Solutions* (a custom software development company which—in collaboration with The Andrew W. Mellon Foundation and the University of Southern California—has developed *Sophie*, an open source, electronic authoring, reading, and publishing set of tools).

The high scientific quality of the conference was possible through the conscientious work of the Programme Committee, the special session organizers and the referees. We are grateful to all members of the Programme Committee for their efficient evaluations and extensive debates, which established the final programme, and we also thank the referees.

We thank Andrej Voronkov for his EasyChair system which facilitated the work of the Programme Committee and the editors considerably.

April 2011

Benedikt Löwe
Dag Normann
Ivan Soskov
Alexandra Soskova

Organization

Programme Committee

Selim Akl, Kingston
Albert Atserias, Barcelona
Anthony Beavers, Evansville
Arnold Beckmann, Swansea
Paola Bonizzoni, Milan
Anne Condon, Vancouver
Thierry Coquand, Gothenburg
Anuj Dawar, Cambridge
Fernando Ferreira, Lisbon
Denis Hirschfeldt, Chicago
Radha Jagadeesan, Chicago
Neil Jones, Copenhagen
Natasha Jonoska, New York
Achim Jung, Birmingham
Viv Kendon, Leeds
Julia Knight, Notre Dame

Phokion Kolaitis, Santa Cruz
Benedikt Löwe, Amsterdam
Elvira Mayordomo, Zaragoza
Dag Normann(Chair), Oslo
Jan-Willem Romeijn, Groningen
Marie-France Sagot, Lyon
Dirk Schlimm, Montreal
Tony Seda, Cork
Nir Shavit, Tel Aviv
Ivan Soskov(Chair), Sofia
Alexandra Soskova, Sofia
Sarah Teichmann, Cambridge
Peter Van Emde Boas, Amsterdam
Jan Van Leeuwen, Utrecht
Klaus Wagner, Würzburg
Andreas Weiermann, Ghent

List of Referees

Allender, Eric
Baixeries, Jaume
Bernardinello, Luca
Brattka, Vasco
Buhrman, Harry
Busic, Ana
Buss, Sam
Carpi, Arturo
Dassow, Jürgen
De Miguel Casado, Gregorio
Deonarine, Andrew
Deoskar, Tejaswini
Fermüller, Chris
Fernandes, António
Franco, Giuditta
Fuhr, Norbert
Gierasimczuk, Nina
Kalimullin, Iskander
Kötzing, Timo
Lempp, Steffen
Lewis, Andrew

Lubarsky, Robert
Miller, Joseph
Montalban, Antonio
Murlak, Filip
Ng, Keng Meng
Nguyen, Phuong
Nordstrom, Bengt
Pauly, Arno
Polyzotis, Neoklis
Rettinger, Robert
Schlage-Puchta, Jan-Christoph
Solomon, Reed
Soskova, Mariya
Sterkenburg, Tom
Stukachev, Alexey
ten Cate, Balder
Van Gasse, Bart
Vereshchagin, Nikolai
Weller, Daniel
Yunes, Jean-Baptiste
Zandron, Claudio

Table of Contents

Applying Causality Principles to the Axiomatization of Probabilistic Cellular Automata

Pablo Arrighi[1,2], Renan Fargetton[2], and Vincent Nesme[3],
and Eric Thierry[1]

[1] École normale supérieure de Lyon, LIP,
46 allée d'Italie, 69008 Lyon, France
{parrighi,ethierry}@ens-lyon.fr
[2] Université de Grenoble, LIG,
220 rue de la chimie, 38400 SMH, France
renan.fargetton@imag.fr
[3] Universität Potsdam, Karl-Liebknecht-Str. 24/25,
14476 Potsdam, Germany
nesme@qipc.org

Abstract. Cellular automata (CAs) consist of an bi-infinite array of identical cells, each of which may take one of a finite number of possible states. The entire array evolves in discrete time steps by iterating a global evolution G. Further, this global evolution G is required to be shift-invariant (it acts the same everywhere) and causal (information cannot be transmitted faster than some fixed number of cells per time step). At least in the classical [7], reversible [11] and quantum cases [1], these two top-down axiomatic conditions are sufficient to entail more bottom-up, operational descriptions of G. We investigate whether the same is true in the probabilistic case.

1 Introduction

Due to their built-in symmetries, CAs constitute a clearly physics-like model of computation. They model spatially distributed computation in space as we know it, and therefore they constitute a framework for studying and proving properties about such systems – by far the most established. Conceived amongst theoretical physicists such as Ulam and Von Neumanns, CAs were soon considered as a possible model for particle physics, fluids, and the related differential equations.

Each of the contexts in which CAs are studied brings its own set of reasons to study probabilistic extensions of CAs. Usually, papers studying this model begin with a definition of PCAs, and these definitions are all variations of the same concept. Namely, these are random maps having the property that they decompose in two phases: first, a classical CA is applied, and second, a model of noise is applied, namely independent random maps on each individual cell. We refer to this class as Standard-PCAs and now explain its drawbacks.

B. Löwe et al. (Eds.): CiE 2011, LNCS 6735, pp. 1–10, 2011.

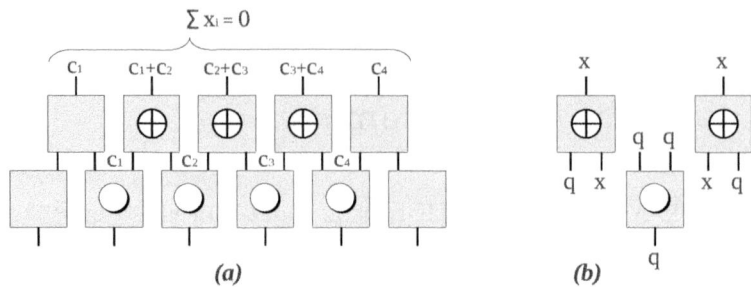

Fig. 1. *(a) Parity.* Time flows upwards. Wires are bits. Boxes are stochastic matrices. Boxes marked with a circle are coin tosses. Boxes marked $\oplus$ perform addition modulo two. Input cells $0 \ldots n$ are ignored. Output random variables $(X_i)_{i=1\ldots n}$ have the property that for strict subset I of $1 \ldots n$, X_I is uniformly distributed, yet their global parity is always 0. *(b)* As such *Parity* is a localized random map, but it can be modified into a translation-invariant random map over entire configurations of the form $\ldots qq01001qq\ldots$, by ensuring that for any $x \in \{q, 0, 1\}$, $qx \mapsto x$, $xq \mapsto x$, $qq \mapsto q$, and hence making the coin tosses conditional upon their input cells not being q. *Parity* does not pertain to the class of Standard-PCAs, but it ought to be considered a valid PCA.

First, the composition of two Standard-PCAs is not necessarily a Standard-PCA. Indeed, applying a Standard-PCA only once on a deterministic configuration cannot create correlations between cells, as opposed to *Parity* (see Figure 1), which is the composition of two Standard-PCAs. Also, the intuition behind this class is simple but ad-hoc; i.e. not founded upon some meaningful high-level principles. We want to define PCAs in a more robust manner, which is a long-standing problem. Criteria for a good axiomatic definition include: being composable, and being based on high-level principles, while entailing an operational description (i.e. implementation by local mechanisms).

As the crucial problem will be to find the good definition of locality, we will tend not to worry too much about translational invariance and neighborhoods. Some of our examples will be presented in a form that is not directly translation invariant but can be made so using the technique explained in the caption of Figure 1.(b). As for the neighborhood, all our definitions will be given *modulo grouping* and shifts, which means you can always compose with a shift and group cells into supercells before applying definitions. For instance, with this perspective, it would be correct to assume that the neighborhood of a common deterministic CA is $\{0; 1\}$.

Why not generalize Hedlund's characterization of CAs to PCAs? We will begin by investigating precisely that route and show that continuity arguments fail to characterize PCAs (Section 2), as they do not forbid spontaneous generation of statistical correlation in arbitrarily distant places (Section 3). Counter-examples are necessarily non-signalling, non-local random maps. That such objects can exist is now a well-known fact in quantum theory, where non-signalling non-locality arises from entanglement in Bell's test experiment [3]. Recently their study has been abstracted away from quantum theory, for instance via the rather topical non-local box (NL box) [2,12].

This points out the weakness of the non-signalling condition in the probabilistic/stochastic setting, which is a well-known issue in foundations of physics. In this context, more robust causality principles have been considered. In fact, Bell's test experiments are motivated by a 'local causality', of which there exist several variants, all of them stemming from Reichenbach's 'principle of common cause'. Now, since PCAs are nothing but simplifications of those spaces used in physics, this principle of common cause, if it is at all robust, should therefore lead to the axiomatization of PCAs. We investigate this intriguing question (Section 4) and answer it in an unexpected fashion (Section 5). But let us start with a few definitions and notations.

We will only consider CAs over finite configurations. A *finite configuration* over a finite alphabet Σ with a distinguished *quiescent* state q is a function $c : \mathbb{Z} \longrightarrow \Sigma$ that sends every integer, except a finite number of them, on q. The (countable) set of finite configurations is denoted $\mathcal{C}$. Countability is not crucial to the arguments developed in this paper, but it certainly makes the following definitions much easier.

Random variables denoted by X range over $\mathcal{C}$, i.e. entire configurations. Random variables denoted by X_i range over Σ, i.e. cells. Random variables denoted by X_I range over Σ^I, i.e. sets of cells. Random variables corresponding to time t, are denoted X^t, X_I^t. We denote $\rho_Y : \text{range}(Y) \to [0, 1]$ the probability distribution of Y. For convenience in the particular case when Y is fully determined on y, i.e. such that $\rho_Y(x) = \delta_{xy}$, ρ_Y will be noted $\tilde{y}$. Moreover for convenience $\rho_{X_I^t}$ will be denoted ρ_I^t, and referred to as the *distribution of cells I at time t*. Whenever I is finite, we can make ρ_I^t explicit as the vector $\left(\rho_I^t(w)\right)_{w \in \Sigma^I}$ with the $|\Sigma|^{|I|}$ entries listed in the lexicographic order.

The distribution of a random map S from Σ^I to Σ^J can be represented by the (potentially infinite) stochastic matrix $m_S = \left(\rho_{S\tilde{x}}(y)\right)_{y \in \text{range}(Y), x \in \text{range}(X)}$. We assume that the random variables over configurations $(X^t)_{t \in \mathbb{N}}$ follow a stochastic process, i.e. that they form a Markov chain $Pr(X^{n+1} = x^{n+1} | X^n = x^n) = Pr(X^{n+1} = x^{n+1} | X^n = x^n, \ldots, X^0 = x^0)$ obeying the recurrence relation $\rho^t = S^t \rho$ for some random map S over configurations.

Our problem is to determine what it means for a random map G over configurations to be *causal*, meaning that arbitrarily remote regions I and J do not influence each other *by any means*. Then PCAs will just be causal, shift-invariant random maps. Unless stated otherwise, we are thinking of each of those random maps as an autonomous block, therefore all random maps mentioned are independent; hence we can assimilate them with the stochastic matrix representing their distribution.

2   Continuity

In the deterministic case, CAs can be given a simple operational definition by means of the local transition function. A local transition function is a function

$\delta : \Sigma^2 \to \Sigma$, and the corresponding CA is then defined as $F : \mathcal{C} \to \mathcal{C}$ by $F(c)_x = \delta(c_x, c_{x+1})$; that amounts to supposing the neighborhood is $\{0; 1\}$, which is true modulo shifts and grouping.

Deterministic case CAs were axiomatized by the celebrated Curtis-Lyndon-Hedlund Theorem[7], which we now recall. First the space of configurations is endowed with a metric $d(.,.) : \mathcal{C} \times \mathcal{C} \longrightarrow \mathbb{R}^+$ such that $d(c, c') = 0$ if $c = c'$ and $d(c, c') = 1/2^k$ with $k = \min\{i \in \mathbb{N} \mid c_{-i...i} \neq c'_{-i...i}\}$. With respect to this metric, a function $F : \mathcal{C} \longrightarrow \mathcal{C}$ is uniformly continuous if and only if for all $n \in \mathbb{N}$, there exists $m \in \mathbb{N}$ such that for all $c, c' \in \mathcal{C}_\infty$, $c_{-m...m} = c'_{-m...m}$ implies $F(c)_{-n...n} = F(c')_{-n...n}$. Notice that rephrased in this manner, uniform continuity is a synonym for non-signalling, i.e. the fact that information does not propagate faster than a fixed speed bound. Continuity on the other hand expresses a somewhat strange form of relaxed non-signalling, where information does not propagate faster than a certain speed bound, but this speed bound depends upon the input. However, it so happens that the two notions coincide for compact spaces. Moreover, classical CAs are easily defined upon infinite configurations $\mathcal{C}_\infty : \mathbb{Z} \to \Sigma$, for which the same $d(.,.)$ happens to be a compact metric. This yields:

Theorem 1 (Curtis, Lyndon, Hedlund)
A function $F : \mathcal{C}_\infty \longrightarrow \mathcal{C}_\infty$ is continuous and shift-invariant if and only if it is the global evolution of a cellular automaton over $\mathcal{C}_\infty$.

However, $\mathcal{C}$ is not compact for d. In this case we must assume the stronger, uniform continuity for the theorem to work. Generally speaking, it is rather difficult to find a relevant compact metric for probabilistic extensions of $\mathcal{C}_\infty$ — and not worth the effort for the sole purpose of axiomatizing PCAs. Indeed, let us directly assume the probabilistic counterpart of non-signalling (a.k.a uniform continuity for some extended metric):

Definition 1 (Non-signalling). *A random map G over configurations is non-signalling if and only if for any random configurations X, Y, and for any cell i, we have:*

$$\rho_{X_{i-1,i}} = \rho_{Y_{i-1,i}} \quad \Rightarrow \quad \rho_{(GX)_i} = \rho_{(GY)_i}.$$

For example, *Parity* (see Figure 1) is non-signalling by construction. Is it reasonable to say, à la Hedlund, that PCAs are the non-signalling, shift-invariant random maps? Surprisingly, this is not the case. Imagine that Alice in Paris (cell 0) tosses a fair coin, whilst Bob in New York (cell $n + 1$) tosses another. Imagine that the two coins are magically correlated, i.e. it so happens that they always yield the same result. Such a random map is clearly not implementable by local mechanisms: we definitely need some amount of (prior) communication between Alice and Bob in order to make it happen. Yet it can be, as in *Magic-Coins* (see Figure 2), perfectly non-signalling. While the setup cannot be used to communicate 'information' between distant places, it can be used to create spontaneous 'correlations' between them. We must forbid this from happening. In this respect, assuming only (non-uniform) continuity is the wrong direction to take.

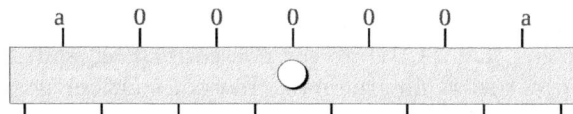

Fig. 2. *MagicCoins.* Inputs are ignored. Output random variables X_1 and X_n are both uniformly distributed and maximally correlated. As such *MagicCoins* is a localized random map, but can be modified into a translation-invariant random map over entire configurations using the method described for *Parity.* *MagicCoins* is non-signalling, but it must not be considered a valid PCA: it is not non-correlating.

3 Avoiding Spontaneous Correlations

From the previous discussion we are obliged to conclude that the formalization of a robust notion of the causality of a dynamics is indeed a non-trivial matter in a probabilistic setting. From the *MagicCoins* example, we draw the conclusion that such a notion must forbid the creation of spontaneous correlations between distant places. The following definition clarifies what is meant by (non-)correlation between subsystems.

Definition 2 (Independence, tensor, trace). *Let I and J be disjoint. Stating that X_I and X_J are independent is equivalent to stating that for any uv in $\Sigma^{I \cup J}$, $\rho_{I \cup J}(uv) = \rho_I(u)\rho_J(v)$, with u (resp. v) the restriction of uv to I (resp. J). In other words, when ρ is seen as a vector in $\mathbb{R}^{\Sigma^{I \cup J}} \simeq \mathbb{R}^{\Sigma^I} \otimes \mathbb{R}^{\Sigma^J}$, we have $\rho_{I \cup J} = \rho_I \otimes \rho_J$. Whether or not X_I and X_J are independent, we can always recover ρ_I as the marginal of $\rho_{I \cup J}$ by summation over every possible v, an operation which we denote Tr_J and call the trace-out/marginal-out operation. Namely we have that $\rho_I = Tr_J(\rho_{I \cup J})$, with $Tr_J(\rho_{I \cup J})(u) = \sum_v \rho_{I \cup J}(uv)$.*

A way to forbid spontaneous correlations is to require that, after one-time step of the global evolution G applied upon any initially fully determined configuration $\tilde{c}$, and for any two distant regions $I = -\infty \ldots x - 1$ and $J = x + 1 \ldots \infty$, the output $\rho = G\tilde{c}$ be such that $\rho_{I \cup J} = \rho_I \otimes \rho_J$. In other words remote regions remain independent. This formulation is somewhat cumbersome, because it seeks to capture statically a property that is essentially dynamical. The following definition clarifies what it means for a random map to be localized upon a subsystem.

Definition 3 (Extension, localization, tensor). *Let I and J be disjoint; let $S : \Sigma^I \to \Sigma^I$ and $T : \Sigma^J \to \Sigma^J$ be independent random maps. Then $m_{S \times T} = m_S \otimes m_T$, where $\otimes$ is the Kronecker/tensor product. We say that a random map from $\Sigma^{I \cup J}$ into itself is localized upon I if m_G is such a tensor product.*

We can then forbid spontaneous correlations directly in terms of dynamics:

Definition 4 (non-correlation). *A random map G over configurations is non-correlating if and only if for any output cell x, there exist random maps A, B acting over input cells $-\infty \ldots, x - 1$ and $x + 1, \ldots + \infty$ respectively, such that $m_{Tr_x \circ G} = m_{A \otimes B}$.*

For example, *Parity* (see Figure 1) is non-correlating by construction. Now, is it reasonable to say that PCAs are the non-correlating, shift-invariant random maps? Amazingly, this is not the case. Indeed, consider a small variation of *Parity*, which we call *GenNLBox* and define in Figure 3. Such a random map is clearly not implementable by local mechanisms: we definitely need some amount of communication between Alice and Bob in order to make it happen. It suffices to notice that *GenNLBox* is in fact a generalization of the NL box, which we recover for $n = 2$, see Figure 3*(a)*. But then, the NL box owes its name precisely to the fact that it is not implementable by local mechanisms. Formal proofs of this assertion can be found in the literature [2,12] and rely on the fact that the NL box (maximally) violates the Bell inequalities [3,14]. Yet *GenNLBox* (see Figure 3 *(a)*), was perfectly non-correlating. Hence, whilst the set-up cannot be used to communicate 'information' between distant places (it is non-signalling), nor to create spontaneous 'correlations' between distant places (it is non-correlating), we still must forbid it from happening!

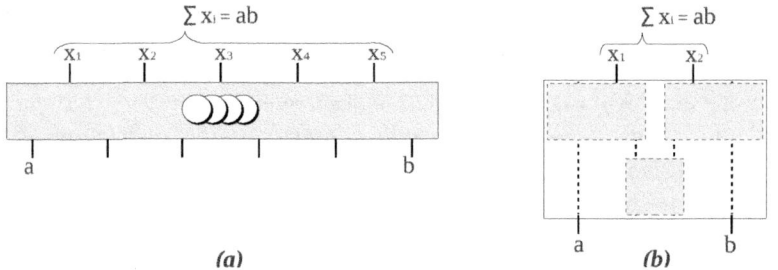

Fig. 3. *(a) GenNLBox.* Input cells $1 \ldots n - 1$ are ignored. Output random variables $(X_i)_{i=1\ldots n}$ have the property that for strict subset I of $1 \ldots n$, X_I is uniformly distributed, yet the global parity is always ab. As such *GenNLBox* is a localized random map, but can be modified into a translation-invariant random map over entire configurations using the method described for *Parity*. *(b)* NL box. The output random variables X_1 and X_2 are uniformly distributed, but their parity is always equal ab. No circuit of the displayed dotted form can meet these specifications [3,14].

4   Common Cause Principle

From the *GenNLBox* example of the previous Section, we are obliged to conclude that a robust notion of the causality of a dynamics in a probabilistic setting cannot be phrased just in terms of a non-signalling or a non-correlation property. Yet, this example has a virtue: it points towards similar questions raised in theoretical physics. Hence, this suggests looking at how those issues were addressed so far in theoretical physics.

Indeed, the NL box is generally regarded as 'unphysical' because it does not comply with Bell's [3] 'local causality', meaning that there is no prior shared resource that one may distribute between Alice and Bob (the outputs of the middle box of Figure 3 *(b)*), that will then allow them to perform the required task separately.

Bell's 'local causality' [3], 'Screening-off' [6], 'Stochastic Einstein locality' [8,4], are all similar conditions, which stem from improvements upon Reichenbach's 'Principle of Common Cause'[10,13], as was nicely reviewed in [9]. The common cause principle can be summarized as follows: "Two events can be correlated at a certain time if and only if, conditional upon the events of their mutual past, they become independent". In the context of this paper, this gives:

Definition 5 (Screening-off). *A random map over configurations G obeys the screening-off condition if and only if for any input cell i with values in Σ, we have that there exists random maps $(A_x, B_x)_{x \in \Sigma}$ acting over input cells $-\infty \ldots, i-1$ and $i+1, \ldots + \infty$ respectively, such that for any configuration c:*

$$Tr_i \rho_{G(c)} = \rho_{A_{c_i}(c)} \otimes \rho_{B_{c_i}(c)}.$$

Here input i is said to screen-off G.

This screening-off condition is physically motivated and does not suffer the problems that the non-signalling and non-correlation conditions had. Unfortunately however, it suffers most of the problems of the original, Standard-PCA definition: it is again incomplete and non-composable. For example, *Parity* does not obey the screening-off condition. Yet, *Parity* is a natural PCA, clearly implementable by local mechanisms as was shown in Figure 1. But the prior shared resource which is necessary in order to separate Alice from Bob is not present in the input cells, rather it is generated on-the-fly within one time-step of G, c.f. the circle-marked boxes. In other words, the reason why the screening-off condition rejects *Parity*, is because the condition is too stringent in demanding that screening-off events be made explicit in the inputs x. A more relaxed condition would be to require that x may be completed so as to then screen-off G.

Definition 6 (Screening-off-completable). *A random map over configurations G is screening-off-completable, or simply V-causal if and only if for any input cell i, we have that there exist independent random maps (with input/output ranges marked as sub/superscript indices) $A_{-\infty \ldots i-1}^{-\infty \ldots i-1, l'}, L_{l', l}^{i}, C_{i}^{l, r}, R_{r, r'}^{i+1}, A_{i+1 \ldots \infty}^{r', i+2 \ldots \infty}$ such that:*

$$m_G = (m_L \otimes m_R)(m_A \otimes m_C \otimes m_B).$$

Here C is said to screen-off G at i. See Figure 4 (a).

Again, is it reasonable to say that PCAs are the screening-off-completable, shift-invariant random maps? Again, this is not the case. Indeed, consider a small variation of *Parity*, which we call *VBox* and define in Figure 5. Such a random map is not implementable by local mechanisms: we need some amount of communication between Alice and Bob in order to make it happen. Yet *VBox* is perfectly V-causal, as shown in Figure 5. Hence, whilst the set-up is screening-off-completable, we still must forbid it, our condition is again too weak. A natural

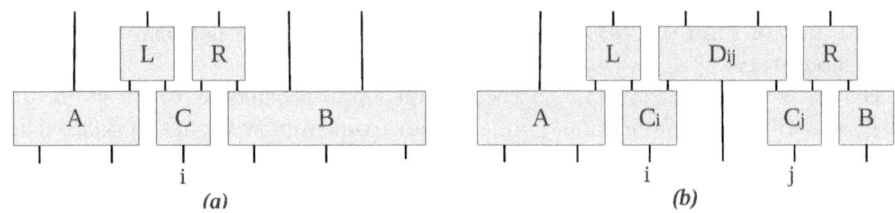

Fig. 4. *(a)* G is V-causal if it can be put in this form, for each i. *(b)* A strengthened condition: VV-causality.

family of conditions to consider is VV-causality (as defined in Figure 4 *(b)*), V^3-causality etc. This route seems a dead end; we believe that the *VBox* is the $k = 1$ instance of the following more general result:

Conjecture 1. *For all k, there exists a V^k-Box which is V^k-causal but not V^{k+1}-causal.*

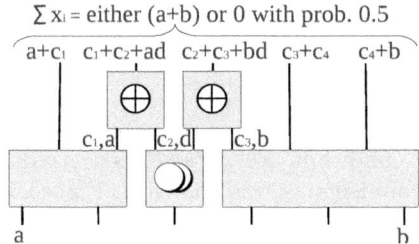

Fig. 5. V-box. Input cells $1 \ldots n - 1$ are ignored. Output random variables $(X_i)_{i=1\ldots n}$ have the property that for strict subset I of $1 \ldots n$, X_I is uniformly distributed, yet their global parity is, with equal probability, ab or 0. As such *VBox* is a localized random map, but can be modified into a translation-invariant random map over entire configurations using the method described for *Parity*. *VBox* is non-signalling, non-correlating and V-causal, but it must not be considered a valid PCA: it is not VV-causal. The V-box can be seen as a chain of two *GenNLBoxes*, chaining more of them yields a $V^k Box$.

5   Concluding Definition

Our best definition. We have examined several, well-motivated causality principles (non-signalling, non-correlation, V-causality) and shown, through a series of surprising counter-examples, that random maps with these properties are not necessarily implementable by local mechanisms. In the limit when k goes to infinity, V^k-causality turns into the following definition (assuming shift-invariance):

Definition 7 (Probabilistic Cellular Automata). *A random map over configurations G is a PCA if and only if*

$$m_G = \left(\bigotimes m_D \right)\left(\bigotimes m_C \right)$$

where the ith random map C has input i and outputs l_i, r_i, and the ith random map D has inputs r_{i-1}, l_i and output i, see Figure 6.

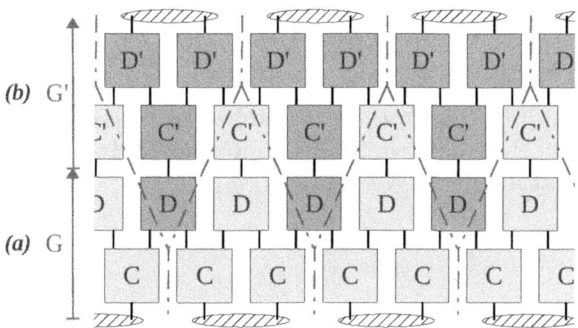

Fig. 6. *(a)* An operational definition of PCAs. Time flows upwards. Wires stand for cells. The first two layers of boxes describe one application of a PCA $m_G = (\bigotimes m_D)(\bigotimes m_C)$. *(b)* Its compositionality. The next two layers describe one application of a PCA $m_{G'} = (\bigotimes m_{D'})(\bigotimes m_{C'})$. The resulting composition is a PCA $m_{G'}m_G = \big(\bigotimes(m_{D'} \otimes m_D)m_{C'}m_D\big)\big(\bigotimes m_{C'}m_D(m_C \otimes m_C)\big)$ over the hatched supercells of size 2.

This definition has several advantages over Standard-PCAs. First, it is complete, because it captures exactly what is meant (up to grouping) by a shift-invariant random map implementable by local mechanisms. Second, it is composable, as shown in Figure 6. Third, it is less ad-hoc. Indeed, we have mentioned in the introduction that much mathematics is done by having axiomatic definitions and operational description to coincide (usually by means of structure versus characterization theorems). It could be said that the same effect has been achieved in this paper, but in a different manner. Indeed, in this paper we have considered the natural candidate axiomatic definitions of PCAs, whose limit is the natural operational description of PCAs, and we have discarded each of them — by means of counter-examples. In this sense, we have pushed this family of candidate axiomatic definitions to its limit, i.e. as far as to coincide with the operational description.

Still, we cannot really pretend to have reached an axiomatization, nor a characterization of PCAs: Definition 7 can be presented as just the square of two Standard-PCAs; or even more simply as a Standard-PCA with its two phases reversed: first a model of noise gets applied (a stochastic matrix is applied homogeneously on each individual cell), and second, a classical CA is applied. If anything, this paper has shown that it is rather unlikely that an axiomatization can be achieved. The authors are not in the habit of publishing negative results. However, the question of an characterization of PCAs à la Hedlund has been a long-standing issue: we suspect that many researchers have attempted to obtain such a result in vain. At some point, it becomes just as important to discard the possibility of a Theorem as to establish a new one. Moreover, advances on the issue of causality principles [3,4,8,10,13], have mainly arisen from the discovery of counter-examples (See [9] and the more recent [5]), to which this paper adds a number. An impressive amount of literature [2,12] focusses on the NL box counter-example (as regards its comparison with quantum information processing, but also its own information processing power) and raise the question of n-party extensions: the *GenNLBox* , the V-box, and the V^k-box could prove useful in this respect.

Acknowledgements. The authors would like to acknowledge enlightening discussions with Joe Henson, Jean Mairesse, Jacques Mazoyer, Nicolas Schabanel, Rafaël Sorkin and Reinhard Werner.

References

1. Arrighi, P., Nesme, V., Werner, R.: Unitarity plus causality implies localizability. In: QIP 2010 (long talk), ArXiv preprint: arXiv:0711.3975 (2010)
2. Beckman, D., Gottesman, D., Nielsen, M.A., Preskill, J.: Causal and localizable quantum operations. Phys. Rev. A 64, 052309 (2001)
3. Bell, J.S.: On the Einstein Podolsky Rosen paradox. Physics 1, 195 (1964)
4. Butterfield, J.: Stochastic Einstein locality revisited. The British Journal for the Philosophy of Science 58(4), 805 (2007)
5. Coecke, B., Lal, R.: Time-asymmetry and causal structure. Arxiv preprint arXiv:1010.4572 (2010)
6. Craig, D., Dowker, F., Henson, J., Major, S., Rideout, D., Sorkin, R.D.: A Bell inequality analog in quantum measure theory. Journal of Physics A: Mathematical and Theoretical 40, 501 (2007)
7. Hedlund, G.A.: Endomorphisms and automorphisms of the shift dynamical system. Math. Systems Theory 3, 320–375 (1969)
8. Hellman, G.: Stochastic Einstein-locality and the Bell theorems. Synthese 53(3), 461–503 (1982)
9. Henson, J.: Comparing causality principles. Studies in History and Philosophy of Science Part B: Studies in History and Philosophy of Modern Physics 36(3), 519–543 (2005)
10. Hofer-Szabo, G., Redei, M., Szabo, L.E.: On Reichenbach's common cause principle and Reichenbach's notion of common cause. The British Journal for the Philosophy of Science 50(3), 377 (1999)
11. Kari, J.: On the circuit depth of structurally reversible cellular automata. Fundamenta Informaticae 38(1-2), 93–107 (1999)
12. Popescu, S., Rohrlich, D.: Quantum nonlocality as an axiom. Foundations of Physics 24(3), 379–385 (1994)
13. Rédei, M., Summers, S.J.: Local primitive causality and the common cause principle in quantum field theory. Foundations of Physics 32(3), 335–355 (2002)
14. Werner, R.F., Wolf, M.M.: All-multipartite Bell-correlation inequalities for two dichotomic observables per site. Physical Review A 64(3), 32112 (2001)

Three Theorems on n-REA Degrees: Proof-Readers and Verifiers

Mingzhong Cai[*]

Department of Mathematics, Cornell University,
Ithaca, NY 14853, United States of America
yiyang@math.cornell.edu

Abstract. We first show that an n-**REA** degree is array recursive if and only if it is r.e. traceable. This gives an alternate proof that an n-**REA** degree has a strong minimal cover if and only if it is array recursive. Then we prove that an n-**REA** degree is strongly jump traceable if and only if it is strongly superlow. These two results both generalize corresponding equivalence theorems for the r.e. degrees. In these proofs, we provide an interesting technique to handle n-**REA** degrees, which also gives a new proof of an old result that every **FPF** n-**REA** degree is complete.

1   Introduction

Our story begins with array recursive degrees. The term *array recursive* degrees actually comes from its complement, the *array nonrecursive* (**ANR**) degrees, which were defined in [4] to generalize $\overline{\mathbf{GL_2}}$ degrees. In particular, **ANR** degrees share a lot of nice properties with $\overline{\mathbf{GL_2}}$ degrees. Examples of their common properties include the 1-generic bounding property, the cupping property and relative recursive enumerability (see [4] and [3]). Recall that the modulus function $m_K(n)$ of K (the halting problem) is defined as the least stage by which the initial segment of length n settles down in the standard recursive enumeration of K. A function g *dominates* f if $g(x) \geq f(x)$ for cofinitely many x.

Definition 1. *A degree* **a** *is array recursive if every function recursive in* **a** *is dominated by* m_K. *A degree is array nonrecursive* (**ANR**) *if it is not array recursive.*

In degree theory, a degree **a** is a *strong minimal cover* of **b** if $\mathcal{D}(< \mathbf{a}) = \mathcal{D}(\leq \mathbf{b})$, i.e., every degree strictly below **a** is below **b**. It is a very difficult and long-standing question to characterize all the degrees which have strong minimal covers. The notion of **ANR** degrees comes into play in the following theorem:

Theorem 2 ([4]). *No* **ANR** *degree has a strong minimal cover.*

In particular, this lead to the first major progress on the question of strong minimal covers: In [5], Ishmukhametov gave a characterization of the r.e. degrees

---

[*] The author was partially supported by NSF Grants DMS-0554855 and DMS-0852811.

with strong minimal covers by introducing the notion of r.e. traceability. Recall that W_e stands for the e-th r.e. set in any uniformly recursive indexing of r.e. sets, and $|X|$ denotes the number of elements in X.

Definition 3. *A degree* **a** *is* r.e. traceable *if there is a recursive function* $f(n)$ *such that for every function* $g(n) \leq_T$ **a**, *there is a recursive function* $h(n)$ *such that for every* n, $g(n) \in W_{h(n)}$ *and* $|W_{h(n)}| \leq f(n)$.

We call $W_{h(n)}$ an (r.e.) *trace* for $g(n)$ (*admitting* f). It was shown in [5] that:

Theorem 4. *Every* r.e. traceable degree has a strong minimal cover.

Theorem 5. *An* r.e. degree is r.e. traceable if and only if it is array recursive.

Together with Theorem 2 one can easily get:

Corollary 6. *An* r.e. degree has a strong minimal cover if and only if it is array recursive.

A natural generalization of r.e. degrees is the notion of iterates of the r.e. relation: A degree is 1-**REA** if it is r.e.; A degree is $(n + 1)$-**REA** if it is r.e. in and strictly above an n-**REA** degree. In [2], we generalized Corollary 6 to the n-**REA** degrees.

Theorem 7. *An* n-**REA** degree has a strong minimal cover if and only if it is array recursive.

Interestingly, the proof of Theorem 7 in [2] does not follow from the corresponding generalization of Theorem 5. In fact, it was not known whether Theorem 5 has a generalization to the n-**REA** degrees. Greenberg asked the author whether it is true that an n-**REA** degree is array recursive if and only if it is r.e. traceable. Here we give a positive answer.

Theorem 8. *An* n-**REA** degree is r.e. traceable if and only if it is array recursive.

In addition, it is not difficult to see that r.e. traceable degrees are downward closed in the Turing degrees, so we can actually get a little bit more from this result (compared to Theorem 7):

Corollary 9. *If a degree is below an array recursive* n-**REA** *one, then it has a strong minimal cover.*

Our second theorem is related to some recent research on two lowness notions, namely strongly jump traceable degrees and strongly superlow degrees. They were introduced aiming to give combinatorial characterizations of the *K-trivials*, an important notion in randomness. For details of the motivations and other examples, see [8].

An *order function* is an unbounded nondecreasing recursive function. A degree **a** is *strongly jump traceable* if for every order function $r(n)$ there is a trace $W_{h(n)}$ admitting $r(n)$ such that $\varphi_n^A(n) \in W_{h(n)}$ whenever $\varphi_n^A(n)$ converges (it is easy to see that this definition does not depend on the choice of the set A in the degree **a**). A degree **a** is *strongly superlow* if for every order function $r(n)$ there is a recursive limit approximation $\lambda(n, s)$ of A' (i.e., $\lim_s \lambda(n, s) = A'(n)$) such that the number of changes through the n-th column $\langle \lambda(n, s) \rangle_{s \in \omega}$ is bounded by $r(n)$. In other words, **a** is superlow with arbitrary order function as the recursive bound on the number of changes.

It is known that every strongly superlow degree is strongly jump traceable, and the other direction holds for the r.e. degrees (see [8, Section 8.4]). In [9], this is generalized to the n-r.e. degrees, which form a proper subclass of the n-**REA** degrees. In Section 4, we continue to generalize this result to the n-**REA** degrees, and interestingly the proof has a flavor similar to that of Theorem 8.

Theorem 10. *An n-**REA** degree is strongly jump traceable if and only if it is strongly superlow.*

The third theorem is a generalization of Arslanov's Completeness Criterion that **FPF** r.e. degrees are complete ([1]). This was already generalized to the n-**REA** degrees in [6], i.e., every **FPF** n-**REA** degree is above $\mathbf{0}'$.

Recall that a function f is *fixed-point-free* if $\varphi_e \neq \varphi_{f(e)}$ for any index e, and a degree is *fixed-point-free* (**FPF**) if it computes a fixed-point-free function. It is a classical result that a degree is **FPF** if and only if it computes a diagonally nonrecursive (DNR) function: f is DNR if $f(e) \neq \varphi_e(e)$ for any e with $\varphi_e(e) \downarrow$.

We will show the following theorem, which is a variation of the generalized completeness criterion in [6]:

Theorem 11. *Suppose **a** is an n-**REA** degree, then the following are equivalent:*

1. *a computes a function f which is DNR;*
2. *a computes a function f which dominates $\varphi_e(e)$, i.e., $f(e) \geq \varphi_e(e)$ holds for any e with $\varphi_e(e) \downarrow$.*

Note that (2) implies (and so is equivalent to) **a** being above $\mathbf{0}'$: For every $\varphi_e(e)$, one can effectively find an e' such that $\varphi_{e'}(e')$ converges to the number of steps in the computation of $\varphi_e(e)$ if it converges, and diverges if $\varphi_e(e)$ diverges. Then using $f(e')$ one can effectively tell whether $\varphi_e(e)$ converges or not. Moreover, it is trivial that (2) implies (1), and so we only need to show that (1) implies (2). We use Lewis' characterization of non-**FPF** degrees in terms of a weaker notion of traceability (see [7]):

Theorem 12. *For every degree **a**, the following are equivalent:*

1. *a is not **FPF**;*
2. *there is a recursive function $f(n)$ such that for every function $g(n)$ recursive in **a**, there is a recursive function $h(n)$ such that $|W_{h(n)}| \leq f(n)$ for every n, and $g(n) \in W_{h(n)}$ for infinitely many n.*

For simplicity we say that **a** is *weakly r.e. traceable* if (2) holds (and similarly $W_{h(n)}$ is a *weak trace* for $g(n)$). Now we only need to show the following:

Theorem 13. *Given an n-**REA** degree **a**, if for every function f recursive in **a** there are infinitely many e such that $f(e) < \varphi_e(e)$, then **a** is weakly r.e. traceable.*

In the setting of Theorem 11, this shows that the negation of (2) implies the negation of (1). We will give a proof of the above theorem in Section 5.

A final remark is that, the n-**REA** degrees seem to be strongly related to different notions of traceability. We hope that research along these lines may suggest other connections between the n-**REA** degrees and various combinatorial properties.

2   Basic Conventions and Notions

We use W_e^σ to denote the recursive enumeration with the index e and the oracle σ in $|\sigma|$ many steps. Without loss of generality, we can assume that no number x can be enumerated before step $x + 1$, and we often regard W_e^σ as a string of length $|\sigma|$.

It is worth noting here that we use $\succeq$ (analogously $\succ$, $\preceq$ and $\prec$) for the extension relation of strings, so the notation $W_e^\tau \succeq W_e^\sigma$ means that W_e^τ extends W_e^σ as a string, not as a set.

For A an m-**REA** set, we put $\emptyset = B_0 <_T B_1 <_T B_2 <_T \cdots <_T B_m = A$ where each B_{i+1} is r.e. in B_i. We say that B_i is at the i-th level of the enumeration. Similarly any binary string σ which is assumed to be an initial segment of B_i is said to be at *the i-th level*, or simply an *i-th level string*. In this paper, we always use the subscript of a string to denote its level. Given an m-**REA** set A with the sets B_i's as above, an i-th level string τ_i is *true* if it is an initial segment of B_i, and a string is *wrong* if it is not true.

When we say "let x ($\sigma \prec B_i$) be large (long) enough to have property P", we actually mean to find the least x (shortest σ) which has property P. Another y (τ) is *large enough* (*long enough*) with property P if it is larger than or equal to x (extends σ) in this setting.

For convenience, we use 0^t to denote the binary string of t zeros, and use W_e^t to denote $W_e^{0^t}$.

3   Proof of the First Theorem: Proof-Readers

For A an m-**REA** set, we first give detailed proofs for $m = 2$ and $m = 3$ to supply some intuition. Based on this intuition, we then sketch a full proof for any m. The following lemma will be very useful and we present it explicitly here for convenience. The proof is almost obvious by the Limit Lemma.

Lemma 14. *If* **a** *is array recursive, then for every function* $l(n) \leq_T$ **a***, there is a recursive function* $\lambda(n, s)$ *such that* $\lim_s \lambda(n, s)$ *exists and is greater than* $l(n)$ *for each* n*; in addition, the number of changes in each column of approximation* $\langle \lambda(n, s) \rangle_{s \in \omega}$ *is bounded by* n*.*

3.1 $m = 2$

Since r.e. traceable degrees have strong minimal covers and hence are array recursive, we only need to show that every array recursive 2-**REA** degree **a** is r.e. traceable, i.e., there is a recursive function $f(n)$ such that for every function $g(n) \leq_T \mathbf{a}$, there is an r.e. trace $W_{h(n)}$ for $g(n)$ admitting $f(n)$.

We pick $A \in \mathbf{a}$ a 2-REA set, i.e., $A = B_2 >_T B_1 >_T B_0 = \emptyset$ with $B_1 = W_i^\emptyset = W_i$ and $B_2 = W_j^{B_1} = W_j^{W_i}$.

Now given a function $g(n) = \varphi_e^A(n)$, we first define a function $l_0(n) \leq_T A$ as follows: Let $\sigma_2 \prec B_2$ be long enough to compute the value $g(n)$, i.e., $\varphi_e^{\sigma_2}(n) \downarrow = g(n)$. Then we let $\sigma_1 \prec B_1$ be long enough to enumerate all elements in σ_2, i.e., $W_j^{\sigma_1} \succcurlyeq \sigma_2$ (note that it is not necessary and usually not the case that $W_j^{\sigma_1} \prec B_2$). Finally let $l_0(n)$ be large enough to enumerate all elements in σ_1, i.e., $W_i^{l_0(n)} \succcurlyeq \sigma_1$.

Now by Lemma 14 we have a recursive function $\lambda_0(n, s)$ whose limit is greater than $l_0(n)$. We first try to use values from this function to compute $g(n)$: let x be the first value greater than or equal to $l_0(n)$ in the column $\langle \lambda_0(n, s) \rangle_{s \in \omega}$. It is easy to see that W_i^x extends σ_1, but it might contain some wrong information after the σ_1 part. So if we simply use W_i^x in the next level enumeration and find $W_j^{W_i^x}$, we might get something which is wrong, since the wrong "tail" of W_i^x could enumerate something which is not in B_2. If we use this wrong string in the computation, we might get a wrong answer for $g(n)$.

The solution is to define another function $l_1(n) \leq_T A$ to "correct" or to "proof-read" the wrong information in W_i^x. Now we fix functions l_0 and λ_0 as above. For each n, let x be the first value greater than or equal to $l_0(n)$ in the corresponding column of λ_0. Now W_i^x might contain some error, so we let t be the first number such that $W_i^x(t) \neq B_1(t)$ (if such t does not exist, then let t be the first number enumerated into B_1 which is greater than $|W_i^x|$). The only possible situation is that $W_i^x(t) = 0$ and $B_1(t) = 1$. We then let $l_1(n)$ be large enough to enumerate t into B_1, i.e., $t \in W_i^{l_1(n)}$. This function l_1 is recursive in A, and so by Lemma 14 there is a recursive function $\lambda_1(n, s)$ whose limit is greater than $l_1(n)$.

We say that W_i^x *requires proof-reading* and we call $W_i^{l_1(n)}$ (or any later enumeration which sees $t \in B_1$) a *proof-reader* for W_i^x.

With these two limit functions λ_0, λ_1 in hand, one can give a uniform enumeration $W_{h(n)}$ as follows: Fix n; for each pair (p, q) such that $0 \leq p, q \leq n$, we go through the columns $\langle \lambda_0(n, s) \rangle_{s \in \omega}$ and $\langle \lambda_1(n, s) \rangle_{s \in \omega}$ respectively for the p-th change and the q-th change. If we cannot find either one, the computation simply diverges. Now let x be the value of $\lambda_0(n, s)$ after the p-th change and y be the value of $\lambda_1(n, s)$ after the q-th change. Take the longest common initial segment of W_i^x and W_i^y, say τ_1, then enumerate $\tau_2 = W_j^{\tau_1}$ and compute $\varphi_e^{\tau_2}(n)$. If this computation does not converge, do nothing. If it converges, then enumerate the value into $W_{h(n)}$. It is easy to see that if p is the number of changes in the column when it is the first time that the value of $\lambda_0(n, s)$ is greater than or equal to $l_0(n)$ and q is similarly the number of changes in the column when it is the first time that the value $\lambda_1(n, s)$ is greater than or equal to $l_1(n)$, then this computation must give us the correct answer $g(n)$. So $g(n) \in W_{h(n)}$ and $|W_{h(n)}| \leq (n + 1)^2$, i.e., $W_{h(n)}$ is a trace for $g(n)$ admitting $(n + 1)^2$.

3.2 $m = 3$

Now we have $A = B_3 >_T B_2 >_T B_1 >_T B_0 = \emptyset$, $B_1 = W_i^{B_0}$, $B_2 = W_j^{B_1}$ and $B_3 = W_k^{B_2}$.

Given a function $g(n) = \varphi_e^A(n)$, similarly we first define a function $l_{00}(n) \leq_T A$ as the number large enough to enumerate $\sigma_1 \prec B_1$, which is long enough to enumerate $\sigma_2 \prec B_2$, which is long enough to enumerate $\sigma_3 \prec B_3$, which is long enough to compute $g(n)$.

Now we have, by Lemma 14, a limit function $\lambda_{00}(n, s)$. Using the same idea as in the previous case, we get a function $l_{01}(n)$ and a limit approximation $\lambda_{01}(n, s)$ such that with λ_{00} and λ_{01} one can get a true initial segment τ_1 of B_1 which is longer than σ_1.

For a similar reason, now if we use τ_1 in the enumeration, it might produce something with a wrong tail, and so we cannot use $W_j^{\tau_1}$ in the next step enumeration. Then we need another (second-level) proof-reading process. The problem is that now we are at the second level of the enumeration and a proof-reader of $W_j^{\tau_1}$ requires some true initial segment of B_1, which is not recursively given.

We fix such λ_{00} and λ_{01}, and we define a new function $l_{10}(n)$: for each n, use the first values in the approximations $\lambda_{00}(n, s)$ and $\lambda_{01}(n, s)$ which are greater than or equal to $l_{00}(n)$ and $l_{01}(n)$ respectively, and find τ_1 as above, then let $l_{10}(n)$ be large enough to enumerate some ξ_1 which is long enough to enumerate a proof-reader for $W_j^{\tau_1}$. With a limit function $\lambda_{10}(n, s)$, a similar recursive enumeration process gives us a different η_1 instead of ξ_1 and this η_1 itself needs proof-reading, i.e., it is long enough but may contain some wrong information which we cannot use in the next level enumeration. So we have another function l_{11} (and the corresponding limit function λ_{11}) which is large enough to enumerate a string at the first level to proof-read η_1 and give a true initial segment of B_1, which is long enough to enumerate some initial segment of B_2 which proof-reads $W_j^{\tau_1}$. After this second level proof-reading process, we have a true initial segment of B_2 and this initial segment is long enough to enumerate some string at the third level to compute $g(n)$.

So finally we have four functions recursive in A and four corresponding recursive approximations. It is easy to find a trace $W_{h(n)}$ for $g(n)$ in the same way and the recursive bound for the number of elements in $W_{h(n)}$ is $(n + 1)^4$.

3.3 General Case

With the above ideas in mind, we sketch the proof.

Now we have $A = B_m >_T B_{m-1} >_T \cdots >_T B_0 = \emptyset$ and each $B_{i+1} = W_{e_i}^{B_i}$. For simplicity of notions, we write W_i instead of W_{e_i}. We can show the following claim by induction on $i \in [1, m - 1]$:

Claim. To get a true and long enough τ_i in the procedure of computing $g(n)$, one needs 2^i functions recursive in A in its proof-reading process.

The base case is §3.1. For the inductive step, to get a true and long enough τ_i, we need two strings at the i-th level: a recursively-generated one ($W_{i-1}^{\tau_{i-1}}$) which is long enough but may contain errors in its tails, and a proof-reader $(W_{i-1}^{\xi_{i-1}})$ which corrects the first mistake in the first string. These two $i-1$-th level strings τ_{i-1} and ξ_{i-1} must be true and long enough. So by induction hypothesis, we need 2^{i-1} functions to generate each one. Therefore we need 2^i functions for a true and long enough τ_i.

In the end, note that to correctly compute $g(n)$, we only need a long enough string at the m-th level, so we need a true and long enough initial segment at the $m-1$-th level. Finally we have 2^{m-1} functions recursive in A and each has a corresponding recursive limit approximation as in Lemma 14. The construction of $W_{h(n)}$ is analogous to these in §3.1 and §3.2, and the recursive bound for $|W_{h(n)}|$ is $f(n) = (n+1)^{2^{m-1}}$.

4   Proof of the Second Theorem: Verifiers

We will follow the same strategy: we first present a detailed proof for $m = 2$ and sketch a proof for $m = 3$, then the general case will be clear by following the same pattern. We again do not try to write out a detailed full proof, because it is neither necessary to write out nor easy to read.

We will need the following lemma, which is quite easy to prove.

Lemma 15. *A degree a is strongly jump traceable if and only if for every order function $r(n)$ and every partial function $\varphi_e^A(n)$ there is a trace $W_{h(n)}$ admitting $r(n)$ such that if $\varphi_e^A(n) \downarrow$ then $\varphi_e^A(n) \in W_{h(n)}$.*

4.1 $m = 2$

We only need to show that strongly jump traceability implies strongly superlowness. Following the same notion, let $B_1 = W_i$ and $A = B_2 = W_j^{W_i}$. To show that A is strongly superlow, we need to give a limit computation of A' and guarantee that the number of changes in the approximation can be bounded by any given order function $r(n)$.

We first define a partial function $\varphi_{e_2}^A(n)$ as follows: given n, we try to compute $\varphi_n^A(n)$: If it diverges then $\varphi_{e_2}^A(n)$ diverges; if it converges then let $\varphi_{e_2}^A(n)$ be the initial segment $A \upharpoonright u$ where u is the use of the computation. Applying Lemma 15, we can get a trace $W_{h_2(n)}$ for $\varphi_{e_2}^A(n)$ admitting some recursive $r_2(n)$ which we will specify later. For each σ enumerated into $W_{h_2(n)}$, we can use it as an oracle and try to compute $\varphi_n^\sigma(n)$. If it converges then we might guess that $\varphi_n^A(n)$ converges. However, it is possible that some wrong σ enumerated into $W_{h_2(n)}$ makes $\varphi_n^{(\cdot)}(n)$ converge and so we might have a wrong guess on whether $\varphi_n^A(n)$ converges.

Now the solution is to use a *verifier* of such σ at a lower level of the enumeration. Given any σ at the second level, let $B_1 \upharpoonright u$ be the initial segment of B_1 which is long enough to enumerate a true initial segment of $A = B_2$ of length $|\sigma|$, and we call such $B_1 \upharpoonright u$ a *verifier* of σ. We define another partial function $\varphi_{e_1}^A(n, k)$ as follows: Enumerate $W_{h_2(n)}$ and wait for the k-th element to appear.

If such an element does not appear then $\varphi_{e_1}^A(n,k)$ diverges; if the k-th element is σ, then let $\varphi_{e_1}^A(n,k)$ be the verifier of such σ.

Then by Lemma 15 again we have a trace $W_{h_1(n,k)}$ for $\varphi_{e_1}^A(n,k)$ admitting some recursive $r_1(n,k)$ which we will also specify later. The intuition is that, for each σ enumerated into $W_{h_2(n)}$ as the k-th element, we need to enumerate a verifier of it in $W_{h_1(n,k)}$ for us to believe that σ is true.

To give a limit computation of A', we shall have, at each stage s, a guess as to whether $n \in A'$, and guarantee that our guess is eventually correct. At stage s, we can enumerate $W_{h_2(n)}$ up to s steps and let $\sigma^0, \sigma^1, \cdots, \sigma^i$ be all the elements that have appeared in $W_{h_2(n)}$, in the enumeration order. For each σ^j, we can enumerate $W_{h_1(n,j)}$ up to s steps. For each τ enumerated, we say that it is *verified* if $\tau \preccurlyeq W_i^s$, i.e., it is the initial segment of our current guess as to $W_i = B_1$.

We say that σ^j is *verified* (at stage s) if for the longest verified τ in $W_{h_1(n,j)}^s$, we have $\sigma^j \preccurlyeq W_j^\tau$, i.e., σ^j looks like a correct initial segment of A at this stage. We also call $W_{h_1(n,j)}$ the *verifier set* for σ^j.

So at this stage s, we guess that n is in A' if there is such a verified σ^j with $\varphi_n^{\sigma^j}(n) \downarrow$. Now we need to show two facts: 1, this is a limit computation of A'; 2, we can arrange r_1 and r_2 to make the number of changes in each column be bounded by any given order function $r(n)$.

For the first claim, if $\varphi_n^A(n)$ converges, then eventually a true initial segment σ of A which is long enough to make $\varphi_n(n)$ converge will be enumerated and it will eventually be verified (though not necessarily by its verifier), since eventually only true initial segments of B_1 are verified at the first level and the longest such is long enough to verify that σ is a true initial segment of A by our construction. In the other direction, if $\varphi_n^A(n)$ diverges, then for any σ enumerated into $W_{h_2(n)}$, eventually it cannot be verified, since the longest true initial segment τ in its verifier set $W_{h_1(n,j)}$ is long enough to see that σ is wrong.

Given a σ in $W_{h_2(n)}$ as the j-th element, we know that $|W_{h_1(n,j)}| \le r_1(n,j)$ and so we switch our mind at most $r_1(n,j)$ times for this σ. Then the total number of changes in the approximation of $A'(n)$ is bounded by $r_1(n,0) + r_1(n,1) + \ldots + r_1(n, r_2(n)) \le r_2(n) r_1(n, r_2(n))$ (the pairing function is monotone).

Now in order to make $r_2(n) r_1(n, r_2(n)) \le r(n)$, we can first find an order function r_2 such that $r_2(n) \le \sqrt{r(n)}$, and we pick an order function r_1 such that $r_1(n, r_2(n)) \le \sqrt{r(n)}$. The number of changes in the limit approximation above is then bounded by $r(n)$.

4.2 $m = 3$

To simplify our notations (and to save some letters for other uses) we have $A = B_3 = W_3^{B_2}$, $B_2 = W_2^{B_1}$ and $B_1 = W_1$.

We first define a partial function $\varphi_{e_3}^A(n)$ which outputs an initial segment of A which is long enough to compute $\varphi_n^A(n)$. Then we have a trace $W_{h_3(n)}$ for this partial function. Similarly we define $\varphi_{e_2}^A(n,k)$: for the k-th element σ_3 enumerated into $W_{h_3(n)}$, we output its verifier τ_2 at the second level, and we also get a trace $W_{h_2(n,k)}$ for it.

Now for any such verifier, we again need to verify it by a string at the first level. We define another $\varphi_{e_1}^A(n,k,l)$: for the k-th element σ_3 enumerated into $W_{h_3(n)}$ and for the l-th element τ_2 enumerated into $W_{h_2(n,k)}$, we output a verifier of τ_2 at the first level of the enumeration. Then we can get a trace $W_{h_1(n,k,l)}$ for it admitting $r_1(n,k,l)$.

Our limit computation is as follows: At each stage s, we enumerate every trace up to s steps. We say that a first level string σ_1 is *verified* (at stage s) if $\sigma_1 \preccurlyeq W_1^s$. An i-th level string σ_i is *verified* if $\sigma_i \preccurlyeq W_i^{\sigma_{i-1}}$ where σ_{i-1} is one of the longest verified strings in the verifier set of σ_i. We guess that $\varphi_n^A(n)$ converges if there is a verified third level string which converges at $\varphi_n^{(\cdot)}(n)$.

Similarly one can show that this is a limit computation for A': we prove by induction that eventually true initial segments and only true initial segments at each level can be verified: at the first level this is obvious; for level i, we eventually have, at level $i-1$, only true initial segments that are verified by induction hypothesis, and then according to our construction, the longest such true σ_{i-1} in each verifier set extends the verifier of the corresponding i-th level string σ_{i-1} is verifying, therefore only true initial segments at the i-th level can be verified in a long run.

In addition, it is easy to see that the number of changes in the column n of this limit computation is bounded by $r_3(n) \times r_2(n,r_3(n)) \times r_1(n,r_3(n),r_2(n,r_3(n)))$. In order to make it bounded by $r(n)$ we can easily make each term in the product bounded by the cubic root of $r(n)$.

4.3 General Case

With the discussion above, the construction and the verification in the general case are clear. In short, we will define partial functions $\varphi_{e_i}^A$ with their traces level by level. In the limit approximation, we guess at stage s that $\varphi_n^A(n)$ converges if there is a verified m-th level string which makes $\varphi_n^{(\cdot)}(n)$ converge. The number of changes is bounded by a product of m terms and one can bound each by the m-th root of $r(n)$. The details of the proof are omitted.

5 Proof of the Third Theorem: Proof-Readers again

In this section we prove Theorem 13. With the idea of the previous two theorems, we only sketch the proof for $m = 2$ and leave the other parts to the reader.

Again we let $B_1 = W_i$ and $A = B_2 = W_j^{W_i}$. Given $g(n) = \varphi_e^A(n)$, we first define $l_0(n)$ exactly the same way as in the proof of the first theorem in Section 3, i.e., it is large enough to enumerate some first level string which is long enough to enumerate a second level initial segment of B_2 which computes the value $g(n) = \varphi_n^A(n)$. By the property given, we know that $\varphi_n(n) > l_0(n)$ infinitely often, and so at these n's where $\varphi_n(n) > l_0(n)$ we may expect to use $\varphi_n(n)$ to compute $g(n)$ correctly. However, we have a similar problem that $\varphi_n(n)$ may enumerate some string with a "wrong tail", and so we need to define a proof-reader for it.

The trick here is that $\varphi_n(n)$ is not total, so we need to modify our strategy. Recursively in A, we define a sequence x_i as follows: let x_0 be the first number n that $\varphi_n(n)$ converges and is greater than $l_0(n)$; given x_i, let x_{i+1} be the first number $n > x_i$ that $\varphi_n(n)$ converges and is greater than $l_0(n)$. It is easy to see that this sequence is infinite, strictly increasing and recursive in A. Then we define a function $l_1(n)$ recursively in A such that $l_1(n)$ is large enough to enumerate a proof-reader for $\varphi_{x_n}(x_n)$. Then similarly $\varphi_n(n) > l_1(n)$ infinitely often.

We can now find a weak trace $W_{h(n)}$ for $g(n)$ as follows: Given n, we try to compute $\varphi_n(n)$ and in addition, every $\varphi_m(m)$ for $m \leq n$ simultaneously. If $\varphi_n(n) \downarrow = x$ and if any $\varphi_m(m)$ converges to y, then we use W_i^x with a proof-reader W_i^y (i.e., find their longest common initial segment) to enumerate a second level string to compute $\varphi_e^{(\cdot)}(n)$, and finally enumerate the value into $W_{h(n)}$ if it converges. It is easy to see that $|W_{h(n)}| \leq n$.

Then we need to show that infinitely often $g(n) \in W_{h(n)}$: for these n's with $\varphi_n(n) > l_1(n)$, it is easy to see that $n \leq x_n$ and $g(x_n)$ is in $W_{h(x_n)}$ by a correct proof-reading process.

In general, for an m-**REA** degree A, we need a similar proof-reading process as in Section 3, and our construction gives a recursive bound $f(n) = n^{2^{m-1}-1}$ for the weak r.e. traces.

References

1. Arslanov, M.: On some generalizations of a fixed point theorem. Izv. Vyssh. Uchebn. Zaved. Mat. (5), 9–16 (1981)
2. Cai, M.: Array nonrecursiveness and relative recursive enumerbility. To appear in the Journal of Symbolic Logic
3. Cai, M., Shore, R.A.: Domination, forcing, array nonrecursiveness and relative recursive enumerablity. To appear in the Journal of Symbolic Logic
4. Downey, R., Jockusch, C., Stob, M.: Array nonrecursive degrees and genericity. London Mathematical Society Lecture Notes Series, vol. 224, pp. 93–105. University Press (1996)
5. Ishmukhametov, S.: Weak recursive degrees and a problem of Spector. In: Arslanov, Lempp (eds.) Recursion Theory and Complexity, de Gruyter, pp. 81–89 (1999)
6. Jockusch, C.G., Lerman, M., Soare, R.I., Solovay, R.M.: Recursively Enumerable Sets Modulo Iterated Jumps and Extensions of Arslanov's Completeness Criterion. The Journal of Symbolic Logic 54(4), 1288–1323 (1989)
7. Lewis, A.E.M.: Π_1^0 classes, strong minimal covers and hyperimmune-free degrees. Bulletin of the London Mathematical Society 39(6), 892–910 (2007)
8. Nies, A.: Computability and Randomness. Clarendon Press, Oxford (2008)
9. Ng, K.M.: Computability, Traceability and Beyond, Ph.D. Thesis

A Computational Approach to an Alternative Working Environment for the Constructible Universe

Merlin Carl

Institute for Logic, Language and Computation, Universiteit van Amsterdam,
Postbus 94242, 1090 GE Amsterdam, The Netherlands
Mathematisches Institut, Rheinische Friedrich-Wilhelms-Universität Bonn,
Endenicher Allee 60, 53115 Bonn, Germany

Abstract. We exhibit aspects of fine structure theory related to infinitary computations and describe a new approach to fine structure based on it, the F-hierarchy, and its refinement, the hyperings. As a sample application, we prove a variant of a theorem of Magidor.

1 Infinitary Computations and Fine Structure

Infinitary computations attempt to generalize classical computability notions from the integers to the ordinals. Several machine models for infinitary computations have been proposed. Possibly the most well-known among them are the Infinite Time Turing Machines by Hamkins and Lewis studied in [5]. Another model are Ordinal Register Machines (ORM) introduced in [9] and further elaborated in [14]. These are basically register machines where each register can store an arbitrary ordinal at each time. At successor steps, the machine works like an ordinary register machine, while at limit times, the liminf of the former states is taken for each register separately.

Say a set x of ordinals is ORM-computable if there is an ORM-program P and a finite set of ordinals p such that P computes the characteristic function χ_x of x with initial input p. The computational power of these machines is then given by the following theorem:

Theorem 1. *A set x is ORM-computable iff $x \in \mathbf{L}$, where $\mathbf{L}$ is Gödel's constructible universe.*

Proof. See, e.g., [9].

This appears to be a very stable phenomenon: All approaches basically lead to $\mathbf{L}$, which suggests that $\mathbf{L}$ play the role that the 'intuitively computable' objects possess in classical recursion theory by the classical Church-Turing-Thesis.

In fact, since the beginning of research on infinite computations, there has been a strong relation to the constructible universe $\mathbf{L}$ and the tools developed for its analysis, particularly Jensen's fine structure theory [6]. In this original

B. Löwe et al. (Eds.): CiE 2011, LNCS 6735, pp. 21–30, 2011.

approach, bounded truth predicates, so called master codes, are used to reduce the complexity of relevant definitions. The methods and proofs of fine structure theory remain involved and challenging. Therefore, several attempts at its simplification have been made since its first introduction in [6], the most well-known of which are Silver machines, Σ^* theory and hyperfine structure theory [4]. An analysis of $\mathbf{L}$ using ORMs can be found in [14].

Here, we present another finestructural hierarchy based on the view of fine structural operators as generalized recursive operators. The main idea is to single out important 'macros' necessary for such arguments and build them directly into the hierarchy. This should lead to a convenient 'working environment' for fine structural arguments based on its analysis in terms of infinitary computations.

As an example, let us, for a first-order structure $\mathbf{M} = (M, \boldsymbol{R}, \boldsymbol{F})$ with relations $\boldsymbol{R}$ and functions $\boldsymbol{F}$ well-ordered by a binary relation $<_\mathbf{M}$, consider the canonical Skolem function $h_\mathbf{M} : \Gamma \times M^{<\aleph_0} \rightarrow M$, where Γ is the set of all first-order formulas in the language for $\mathbf{M}$. $h_\mathbf{M}(\phi, \boldsymbol{p})$ will be the $<_\mathbf{M}$-minimal $x \in M$ such that $\mathbf{M} \models \phi(x, \boldsymbol{p})$, if such an x exists, and otherwise be undefined. Thus, $h_\mathbf{M}$ can be seen as a μ-operator or a search procedure on $\mathbf{M}$. It is a shortcut for an algorithm that runs once through M in the order of $<_\mathbf{M}$, calling as a subroutine a truth predicate for $\mathbf{M}$ to test for each element whether or not $\phi(\cdot, \boldsymbol{p})$ holds and returning the first element where this test is successful.

Two such operators and two such relations occur frequently in fine-structural arguments, and on these we will base our approach: The need for the $\in$-relation is evident. The ternary comprehension operator $I(X, \phi, \boldsymbol{y})$ corresponds to a characteristic function and forms, for an appropriate triple consisting of a set X, a formula ϕ and a finite parameter sequence $\boldsymbol{y} \subset X$ the set $\{z \in X | \phi(z, \boldsymbol{y})\}$. Also, we will have a ternary well-order $<_{\alpha, x, y}$ for each $\alpha \in \text{On}$ on the objects under consideration. Finally, if $(X, \phi, \boldsymbol{y})$ is a triple as above, the Skolem function $S(X, \phi, \boldsymbol{y})$ returns the $<_X$-minimal $z \in X$ such that $\phi(z, \boldsymbol{y})$, if such z exists, otherwise $\varnothing$.

This approach lead to the F-hierarchy, which was first introduced by Koepke and van Eijmeren in [13], applied to a proof of the covering lemma in [7] and used for several fine-structural arguments in [1]. We will explain the hierarchy and use it for a proof of the approximation theorem, a structural theorem about $\mathbf{L}$ for the F-hierarchy analogous to a theorem of Magidor [10]. Due to lack of space, we keep our exposition brief: Proofs of technicalities are mostly left out and can be found in [13], [1] or [7].

2   The F-hierarchy

By our observations, the right language for our approach to $\mathbf{L}$ should be the first-order language $\mathcal{L}$ using countably many variable symbols $(v_n | n \in \omega)$, the logical symbols $=, \wedge, \vee, \neg, \exists, \forall$ and the brackets $(,)$, with ternary function symbols I and S for the interpretation and Skolem functions, a 2-ary relation symbol $\in$ for set membership and a ternary relation symbol $<$ for the well-order. Syntax and semantics are defined as usual in first-order logic. By $\mathcal{L}_0$, we mean the quantifier-free formulas of $\mathcal{L}$.

The F-hierarchy will now consist of structures of the form

$$\mathbf{F}_\alpha := (F_\alpha, \in, I|F_\alpha, S|F_\alpha, < |F_\alpha)$$

defined by ordinal recursion as follows:

Definition 1. *For $\alpha \leq \omega$, we let $F_\alpha := \mathbf{V}_\alpha$, $S|F_\alpha = I|F_\alpha = \varnothing$, $\in$ the usual set membership relation and $<$ any well-ordering of $\mathbf{V}_\omega$ compatible with $\in$ in the sense that $x \in y$ implies $x <_{F_\omega} y$.*

If $\alpha > \omega$ and $\mathbf{F}_\alpha$ has already been defined, we set, for a finite $\boldsymbol{x} \subset F_\alpha$ and ϕ an $\mathcal{L}_0$-formula: $I(F_\alpha, \phi, \boldsymbol{x}) := \{z \in F_\alpha | \mathbf{F}_\alpha \models \phi[z, \boldsymbol{x}]\}$. In this case, we call $(F_\alpha, \phi, \boldsymbol{x})$ a name for this set. Similarly, $S(F_\alpha, \phi, \boldsymbol{x})$ is the $<_{F_\alpha}$-minimal $z \in F_\alpha$ such that $\mathbf{F}_\alpha \models \phi[z, \boldsymbol{x}]$ if such z exists, otherwise 0. We then set $F_{\alpha+1} := \{I(F_\alpha, \phi, \boldsymbol{x}) | \boldsymbol{x} \subset F_\alpha$ is finite and ϕ is an $\mathcal{L}_0$-formula$\}$. For $x, y \in F_{\alpha+1}$, we let $x <_{F_\alpha} < y$ iff the lexically smallest name $(F_\beta, \psi, \boldsymbol{z}) =: N(x)$ for x is lexically smaller than that of y, where we use $\in$ on the first component, some fixed enumeration of formulas for the second and the lexical order obtained from $<_{F_\alpha}$ on the third. This fixes the structure $\mathbf{F}_{\alpha+1} = (F_{\alpha+1}, \in, I|F_{\alpha+1}, S|F_{\alpha+1}, < |F_{\alpha+1})$.

If λ is a limit ordinal and $\mathbf{F}_\gamma$ is defined for $\gamma < \lambda$, then $\mathbf{F}_\lambda$ is $\bigcup_{\gamma<\lambda} \mathbf{F}_\gamma$.

It is rather straightforward to check the following properties of this hierarchy:

Theorem 2. *For every pair of ordinals $\alpha < \gamma$, we have $F_\alpha \subseteq F_\gamma$ and $F_\alpha \in F_\gamma$. Furthermore, each F_γ is transitive. Finally, $F_\infty := \bigcup_{\alpha \in \mathrm{On}} F_\alpha = \mathbf{L}$.*

Proof. See, e.g., [7].

Let us call a map $E : \mathbf{F}_\alpha \to \mathbf{F}_\beta$ **fine** if $E|F_\omega = \mathrm{id}$ and it preserves the basic relations and operations of $\mathbf{F}_\alpha$, where $\alpha \leq \beta \leq \infty$. The F-hierarchy is associated with a canonical hull operator: If $X \subseteq \mathbf{L}$, then $F\{X\}$ denotes the closure of X under I and S, i.e., the intersection of all sets Y such that Y is closed under I and S and $X \subseteq Y$ (such sets are called constructibly closed, or c.c. for short). A very important property of constructible hierarchies is that sufficiently closed substructures of its levels are isomorphic to levels themselves. (For the $\mathbf{L}$-hierarchy, one, e.g., usually requires closure under Σ_1-Skolem functions.) This condensation property also holds for the F-hierarchy:

Theorem 3. *Let $X \subseteq \mathbf{L}$ be a set closed under I and S such that every element of X has also a name in X. Then there are a unique fine isomorphism σ and $\beta \in \mathrm{On}$ such that $\sigma : \mathbf{F}_\beta \simeq X$. The map σ is called the condensation map for X.*

2.1   Extension of Embeddings

For the next section, we will also need a technique for extending fine maps between F-levels to larger domains. The method is quite standard, so we merely sketch it. Details and proofs can, e.g., be found in [7]. (Note that our definition of a base is slightly different from the definition given there. However, both notions are equivalent.)

Definition 2. F_α *is called a base for* F_β, *if* $F\{\bar\alpha \cup \boldsymbol{p}\}$ *is isomorphic to some* $F_{\alpha(\bar\alpha,\boldsymbol{p})} \in F_\alpha$ *for all* $\bar\alpha < \alpha$, *and finite* $\boldsymbol{p} \subseteq F_\gamma$. *The corresponding collapsing map will be denoted by* $\sigma_{\bar\alpha,\boldsymbol{p}}$.

Now let $E : F_\alpha \to F_\beta$ be a fine map, where F_α is a base for F_γ. Then, over the index set $I := \{(\mu,\boldsymbol{p})|\mu < \alpha,$ and $\boldsymbol{p}$ is a finite subset of $F_\gamma\}$, we can define the structures $F_{\alpha(\mu,\boldsymbol{p})} \simeq F\{\mu \cup \boldsymbol{p}\}$, fine maps $\pi_{(\mu_1,\boldsymbol{p}_1),(\mu_2,\boldsymbol{p}_2)} : F_{\alpha(\mu_1,\boldsymbol{p}_1)} \to F_{\alpha(\mu_2,\boldsymbol{p}_2)}$ by $\pi_{(\mu_1,\boldsymbol{p}_1),(\mu_2,\boldsymbol{p}_2)}(x) := \sigma_{\mu_2,\boldsymbol{p}_2}^{-1} \circ \sigma_{\mu_1,\boldsymbol{p}_1}$ and form the directed system $\langle F_{\alpha(\mu,\boldsymbol{p})}, \pi_{(\mu_1,\boldsymbol{p}_1),(\mu_2,\boldsymbol{p}_2)}\rangle$. Its direct limit will be F_γ, let $\pi_{\mu,\boldsymbol{p}}$ be the limit map for the component with index $(\mu,\boldsymbol{p})$. By the base property of F_α, we can map it over component-wise to F_β and form the direct limit of this mapped directed system. This will be denoted by $\mathrm{Ext}(F_\gamma, E)$. It can be shown (see, e.g., [7]) that this limit, if well-founded, will be isomorphic to some F_δ, let $\hat\pi_{\mu,\boldsymbol{p}}$ be the corresponding limit maps. For $x \in F_\gamma$, pick $(\mu,\boldsymbol{p}) \in I$ such that $x \in F\{\mu \cup \boldsymbol{p}\}$ and set $\pi_E(x) := \hat\pi_{\mu,\boldsymbol{p}} \circ E \circ \sigma_{\mu,\boldsymbol{p}}^{-1}(x)$. This defines a fine map from F_γ to F_δ. In fact:

Lemma 1. *If* E, F_γ, F_δ *and* π_E *are as above, then there is* $\pi_E^+ \supset \pi_E$ *such that* $\pi_E^+ : F_{\gamma+1} \to F_{\delta+1}$ *is fine.*

3   A Sample Application: An Approximation Lemma for L

For this section, $\neg 0^\#$ denotes the statement that there is no automorphism of $\mathbf{L}$, i.e., no structure-preserving map $\sigma : \mathbf{L} \to \mathbf{L}$. Jensen's famous covering theorem shows how, in the absence of $0^\#$, uncountable sets X of ordinals can be approximated by elements of $\mathbf{L}$ from 'above'. It is a natural question whether something similar can be done from 'below', i.e., whether sets of ordinals can be 'exhausted' by elements of $\mathbf{L}$. This turns out to be the case: By a theorem of Magidor [10], if $\neg 0^\#$ and X is a set closed under primitive recursive set functions, there are $X_i \in L$ such that $\bigcup_{i\in\omega} X_i = X$. We prove a similar theorem in the context of the fine hierarchy, replacing p.r.-closure by the condition of constructible closure. Since we work in the fine hierarchy, we don't have to distinguish between definition complexities, which shortens and simplifies the proof. This sample application suggests that the 'working environment' has been chosen reasonably.

Definition 3. F_α *is an* ω*-base for* F_δ *if, for* $A \subset F_\delta$ *countable,* $\bar\alpha < \alpha$, *and* $\pi : F\{F_{\bar\alpha} \cup A\} \to_{\mathrm{coll}} F_\beta$, *then* $\beta < \alpha$.

Obviously, being an ω-base is a stronger property than being a base. This gives some information on the wellfoundedness of extensions.

Lemma 2. *Let* α *be a limit ordinal such that* $\mathrm{cf}(\alpha) > \omega$. *Suppose* $E : F_\alpha \to F_\beta$ *is fine and cofinal in* β *and that* F_α *is an* ω*-base for* F_δ. *Define the extension map* π_E *as described in the section on extension of embeddings. Then the direct limit of the mapped directed system is well-founded.*

Proof. We define a directed system S as follows: Its index set is $I = \{(\bar\alpha,p)|\bar\alpha < \alpha,p$ is a finite subset of $F_\delta\}$, it has associated structures $\pi_{\bar\alpha,p} : F\{F_{\bar\alpha} \cup p\} \to_{\mathrm{coll}}$

$F_{\zeta(\bar{\alpha},p)}$ (where $\zeta(\bar{\alpha},p) < \alpha$ because of the definition of ω-base), and maps $\pi_{(\alpha_1,p_1),(\alpha_2,p_2)} : F_{\zeta(\alpha_1,p_1)} \to F_{\zeta(\alpha_2,p_2)}$ defined as usual. The direct limit of S is F_δ. Since the relevant structures and maps are easily seen to be elements of F_α, the system S lifts up by E to a directed system S_E. Now, suppose the limit of S_E is ill-founded, so there are $i_1 < i_2 < ... \in I$ such that (set $\zeta(i_k) = \zeta(k)$ etc. for convenience) the limit of the directed sequence

$$F_{E(\zeta(1))} \to_{E(\pi_{1,2})} F_{E(\zeta(2))} \to_{E(\pi_{2,3})} \cdots$$

is ill-founded. Set $i_k = (\alpha_k, p_k)$, $\bar{\alpha} = \bigcup_{i\in\omega} \alpha_i$, $A = \bigcup_{i\in\omega} p_k$. So $A \subset F_\delta$ is countable and $\bar{\alpha} < \alpha$ since $\mathrm{cf}(\alpha) > \omega$.

Consider $h : F\{F_{\bar{\alpha}} \cup A\} \to_{\mathrm{coll}} F_\zeta$; $\zeta < \alpha$ by definition of ω-base. Define $\pi_k := \pi_{(k,\infty)} : F_{\zeta(k)} \to F_\zeta$ by $h \circ \pi_{i_k}^{-1}$. $\pi_{k,\infty} \in F_\alpha$ by the usual argument: It is the identity on α_k, maps the condensed finite parameter correctly into $h[A]$ and respects $\mathbf{S}_0$-terms. Hence it is definable in F_α, and—as α is a limit ordinal—an element of F_α. Hence all structures and maps mentioned are in F_α, and so, for each $k, \ell \in \omega$, $\pi_{k\ell} : F_{\zeta(k)} \to F_{\zeta(\ell)}$, $\pi_k : F_{\zeta(k)} \to F_\zeta$ and $\pi_\ell : F_{\zeta(\ell)} \to F_\zeta$ lift up to $E(\pi_{k\ell}) : E(F_{\zeta(k)}) \to E(F_{\zeta(\ell)})$, $E(\pi_k) : E(F_{\zeta(k)}) \to E(F_\zeta)$ and $E(\pi_\ell) : E(F_{\zeta(\ell)}) \to E(F_\zeta)$, respectively.

So the limit of the directed system can be embedded in $F_{E(\zeta)}$ and is thus not ill-founded.

From now on, we assume $\neg 0^\#$.

Definition 4. $X \subset F_\infty$ is *constructibly closed in the ordinals* (*"Ω-c.c."*) if $F\{X\} \cap F_\infty = X$.

Theorem 4. *Let X be constructibly closed in the ordinals. Then there are $\langle X_i | i \in \omega \rangle$ such that $X_i \in \mathbf{L}$ and $X = \bigcup_{i\in\omega} X_i$.*

The rest of the section is devoted to the proof of this theorem.

Suppose $X \subset \beta$ is as in the theorem statement, $\beta = \sup(X)$. Let $E : F_\alpha \to F\{X\}$ be the inverse of the collapsing map.

Theorem 5. $\mathrm{rng}(E) \cap \beta = X$.

Proof. Immediate by the fact that X is Ω-c.c.

The proof will be an induction on β. So suppose the theorem to be true for $\gamma < \beta$.

If β is a successor, say $\beta = \gamma + 1$, then $X \cap \gamma$ is obviously Ω-c.c. and thus a countable union of constructible sets by induction hypothesis, say $X \cap \gamma = \bigcup_{i\in\omega} Z_i$. Then $X = (X \cap \gamma) \cup \{\gamma\} = \{\gamma\} \cup \bigcup_{i\in\omega} Z_i$ witnesses our claim. For the same reason, we may without loss of generality assume that α is a limit ordinal.

If $\mathrm{cf}(\mathrm{otp}(X)) = \mathrm{cf}(\alpha) = \omega$, then let $\langle \alpha_i | i \in \omega \rangle$ be cofinal in α. Induction applies to each $X \cap E(\alpha_i)$, giving us $X_{ij} \in L$ for $i, j \in \omega$ such that $X = \bigcup_{i\in\omega} X \cap E(\alpha_i) = \bigcup_{i\in\omega} \bigcup_{j\in\omega} X_{ij}$. So assume $\mathrm{cf}(\alpha) > \omega$ from now on.

Theorem 6. *We can assume without loss of generality that $E|_{F_\alpha} \neq \mathrm{id}$.*

Proof. Otherwise, we have $X = \alpha \in L$ and the theorem statement is trivial.

Theorem 7. F_α *is not an ω-base for F_∞.*

Proof. Otherwise, by Lemma 1, we can lift E to a fine map $\pi_E : F_\infty \to F_\infty$ other than the identity, which contradicts the assumption $\neg 0^\#$.

So let ρ be minimal such that F_α is not an ω-base for F_ρ. Also, let $\bar{\alpha} < \alpha$ and let $A \subset F_\rho$ be a countable set such that $F\{F_{\bar{\alpha}} \cup A\} \to_{\text{coll}} F_\zeta$, $\zeta \geq \alpha$. Setting $\bar{A} = \text{coll}" A$, we get $F_\zeta = F\{F_{\bar{\alpha}} \cup \bar{A}\}$. Since ρ is minimal and $\zeta \leq \rho$, we actually have $\zeta = \rho$ and therefore may assume without loss of generality that $F_\rho = F\{F_{\bar{\alpha}} \cup A\}$.

If ρ is a limit ordinal, then $F_\rho = \bigcup_{i \in \omega} F\{F_{\bar{\alpha}} \cup A_i\}$, where the A_i are such that A_n is an initial segment of A_{n+1} for $n \in \omega$ and $\bigcup_{i \in \omega} A_i = A$ (this is possible since A is countable). So we actually have $\text{cf}(\rho) = \omega$.

Now we distinguish two cases:

Case 1: ρ is a successor, let $\rho = \gamma + 1$.

By minimal choice of ρ, F_α is an ω-basis for F_γ, so there is a fine embedding $\pi_E : F_\gamma \to F_\delta$ extending E. In particular $\pi_E"\alpha = X$, so $(\pi_E"F_\gamma) \cap \beta = X$. By theorem 4.23, π_E is fine up to F_γ, so there is a lifting $\pi_E^+ : F_\rho \to F_{\delta+1}$ extending π_E; in particular, $\pi_E^{+}"\alpha = X$, $(\pi_E^{+}"F_\rho) \cap \beta = X$. This implies: $X = \beta \cap \text{rng}(\pi_E^+) = \beta \cap \pi_E^{+}" F_\rho = \beta \cap \pi_E^{+}" F\{F_{\bar{\alpha}} \cup A\} = \beta \cap F\{\pi_E^{+}" F_{\bar{\alpha}} \cup \pi_E^{+}" A\} = \beta \cap F\{\pi_E^{+}" F_{\bar{\alpha}} \cup \pi_E^{+}" A\} = \beta \cap F\{E" F_{\bar{\alpha}} \cup \pi_E^{+}" A\}$.

Now $E" F_{\bar{\alpha}}$ is Ω-c.c. (since it is a fine image of an Ω-c.c. set) and $\bar{\alpha} < \alpha$, so $E(\bar{\alpha}) < \beta$ and we get from the induction hypothesis that $E" F_{\bar{\alpha}} = \bigcup_{i \in \omega} Y_i$, where $Y_i \in L$ for $i \in \omega$. We can assume without loss of generality that $i < j$ implies $Y_i \subset Y_j$. Thus: $X = \beta \cap F\{\bigcup_{i \in \omega} Y_i \cup \pi_E^{+}" A\} = \bigcup_{i \in \omega} \beta \cap F\{Y_i \cup \pi_E^{+}" A\} = \bigcup_{i \in \omega, q \in (\pi_E^{+}" A)^{<\omega}} F\{Y_i \cup q\} \cap \beta$, and $\beta \cap F\{Y_i \cup q\} \in L$, so we are done in this case.

Case 2: ρ is a limit ordinal.

We have already seen that $\text{cf}(\rho) = \omega$ in this case. Let $\langle \rho_i | i \in \omega \rangle$ be a sequence of ordinals cofinal in ρ, so $\rho_i < \rho$ for $i \in \omega$ and $\rho = \bigcup_{i \in \omega} \rho_i$. We consider separately the countably many directed systems $S_k, k \in \omega$: I_k, the index set of S_k is given by $I_k := \{(\tau, q) | \tau < \alpha, q$ is a finite subset of $F_{\rho_k}\}$, the structures and maps between them are defined as before.

Each system S_k has a direct limit isomorphic to the fine level F_{ρ_k}, remember $\bigcup_{i \in \omega} F_{\rho_k} = F_\rho$. Since ρ is minimal, F_α is an ω-base for each F_{ρ_k}, and by Lemma 1, the corresponding lifted direct limites are well-founded and hence isomorphic to fine levels. Name these levels $F_{\bar{\gamma}_k}$ and let $\pi_n^E : F_{\rho_k} \to F_{\bar{\gamma}_k}$ be the corresponding embeddings extending E, so $E \subseteq \pi_i^E$. We have $E" F_\alpha = X$, and hence for each $i \in \omega$, we have $\pi_i^{E}" F_\alpha = X$. Since these maps are order-preserving, this implies $\text{rng}(\pi_i^E) \cap F_\beta = X$. Observe that $F_\alpha \subseteq F_\rho = F\{F_{\bar{\alpha}} \cup A\} = \bigcup_{i \in \omega} F\{F_{\bar{\alpha}} \cup (A \cap F_{\rho_i})\}$. Let $\beta_n := \sup\{\pi_n^{E}" F\{F_{\bar{\alpha}} \cup (A \cap F_{\rho_n})\} \cap \beta\}$, so that $X = \bigcup_{n \in \omega} (\beta_n \cap \pi_n^{E}" F\{F_{\bar{\alpha}} \cup (A \cap F_{\rho_n})\})$. Let $X_n := (\beta_n \cap \pi_n^{E}" F\{F_{\bar{\alpha}} \cup (A \cap F_{\rho_n})\})$, so $X = \bigcup_{n \in \omega} X_n$ and $X_n = \beta_n \cap F\{\pi_n^{E}" F_{\bar{\alpha}} \cup \pi_n^{E}" (A \cap F_{\rho_n})\} = \beta_n \cap F\{E" F_{\bar{\alpha}} \cup \pi_n^{E}" (A \cap F_{\rho_n})\}$. Since

$\bar{\alpha} < \alpha$, $E" F_{\bar{\alpha}} = \bigcup_{i \in \omega} Y_i$, where $Y_i \in L$ by induction. Assume again without loss of generality that $i < j$ implies $Y_i \subset Y_j$. Then we can conclude:

$$
\begin{aligned}
X_n &= \beta_n \cap F\{\bigcup_{i \in \omega} Y_i \cup \pi_n^{E"}(A \cap (F_{\rho_n}))\} \\
&= \beta_n \cap \bigcup \{F\{Y_i \cup q\} | i \in \omega, q \in \pi_n^{E"}(A \cap F_{\beta_n})^{<\omega}\} \\
&= \bigcup \{\beta_n \cap F\{Y_i \cup q\} | i \in \omega, q \in \pi_n^{E"}(A \cap F_{\beta_n})^{<\omega}\},
\end{aligned}
$$

which is a countable union of sets in $\mathbf{L}$. So we have represented each X_n as a countable union of constructible sets, and hence $X = \bigcup_{n \in \omega} X_n$.

This concludes the proof of the theorem.

4   Hyperings

The F-hierarchy, though successful when applied to deep structural theorems like Jensen's covering lemma or the approximation theorem above, is still too coarse form some deeper theorems of constructible combinatorics: For proofs like that of the square-principle, it is crucial that 'essential' effects (like singularizations or collapses of local cardinals) only take place at limit stages of the hierarchy. This is because limit stages can be approximated conveniently from 'below'. For the F-hierarchy, this is not the case (see [1]). Therefore, it needs to be refined for such applications. This can be done by introducing restricted versions of the basic operators, thereby adapting ideas from hyperfine structure theory [4]. From the computational perspective, we add sequences of ORM-computable operators to interpolate between the stages of the former, coarse F-hierarchy.

The idea is here again that, for an analysis of $\mathbf{L}$ in terms of computability, only a rather limited number of macros and tools is actually used and that these tools can be directly build into the structures in question and the underlying language used. In this way, the intuition behind generalized concepts of computability can be used, while applications maintain their independence from the details of implementation. When the necessity for more tools arises, these can simply be amended to the framework. For a direct simulation of Silver machines on ORM's, see [14].

Definition 5. *A **location** is a countable sequence of the form $\langle F_\alpha, x_1, x_2, ... \rangle$, where each x_i is either a finite sequence of elements of F_α or Ω. If the ordinal in the first position is β, it is a β-**location**. If $s = \langle F_\alpha, x_1, x_2, ... \rangle$ is a location, the set $\mathrm{supp}(s) := \{i \in \omega | x_i \neq \mathbf{0}\}$ is called the **support** of s.*

*Let $s = \langle F_\alpha, x_1, ... \rangle$ be a location. The **structure corresponding to** s is defined thus: $\langle F_\alpha, I, S, I|x_1, N \circ I|x_1, S|x_1, N \circ S|x_1, I|x_2, N \circ I|x_2, S|x_2, N \circ S|x_2, ... \rangle$ For $X \subset F_\alpha$, the s-hull of X, written $F_s\{X\}$, is the closure of X under all functions belonging to s. $<_{\mathrm{loc}}$ denotes the lexical ordering (from the left) of locations.*

A location s_2 is a successor of a location s_1 if $(s_1)_0 = (s_2)_0$ and further $(s_2)_i = (s_1)_i$ or $(s_2)_i$ is the immediate lexical successor of $(s_1)_i$ for each $i \geq 1$. Here,

$(s)_i$ is the ith member of the sequence s. (Note that, in contrast to [4], these locations refer to F-levels. In this context, locations in the sense of Friedman and Koepke are a special case of the notion introduced here.)

Definition 6. *Suppose A is a class of locations, $<_H \subset A \times A$. Then $H := \langle A, <_H \rangle$ is a linear hypering of $\mathbf{L}$ if the following axioms are satisfied:*

1. *$<_H = <_{\mathrm{loc}}|H$ is a well-order,*
2. *for each location $\ell_\alpha^n(x)$, there is an α-location $s = \langle \alpha, y_1, y_2, ... \rangle$ in H such that $y_n = x$,*
3. *for each ordinal α, both $\ell_\alpha(\mathbf{0})$ and $\ell_\alpha(\Omega)$ are in H,*
4. *for $S \subset H$ a set of α-locations, $<_{\mathrm{loc}}$-$\sup\{S\} \in H$,*
5. *if $s_1, s_2 \in H$ are α-locations, H contains a (naturally unique) chain from s_1 to s_2, and*
6. *if $\alpha < \beta$, $\pi : F_\alpha \to F_\beta$ a fine map, then $s \in H_\alpha$ iff $\pi(s) \in H_\beta$.*

Linear hyperings satisfy some convenient extra properties resemblant of Silver machines and the hyperfine structure theory of Friedman and Koepke ([4]). For a linear hypering H, the hull operators $F_s\{X\}$ associated with each $s \in H$ are monotonic in the sense that $s_1 <_H s_2$ and $(s_1)_1 = F_\alpha$, we have $F_{s_1}\{X\} \subseteq F_{s_2}\{X \cup \{F_\alpha\}\}$ for each $X \subseteq \mathbf{L}$. They are compact in the sense that $x \in F_s\{X\}$ implies the existence of a finite $y \subseteq X$ with $x \in F_s\{y\}$, and have the condensation property: If X is a substructure of F_s for some $s \in H$, then its transitive collapse will be of the form $F_{\bar{s}}$ form some $s \geq_H \bar{s} \in H$. We call a linear hypering H slow if for the successor s^+ of $s \in H$ in $<_H$, we have $s = s^+$ almost everywhere (i.e., only finitely many scopes move). Then slow hyperings also have the following finiteness property: If $t = s^+$ in H, then there is finite $a \subset F_t$ such that $F_t\{X\} \subseteq F_s\{X \cup a\}$ for each $X \subseteq \mathbf{L}$.

This last property means that hull operators of a hypering grow very slowly in strength: In particular, singularizations may now only take place at limit locations, i.e., locations that are limits of the $<_H$-ordering.

An Example and Some Applications

A very canonical example of a hypering is the horizontal hypering H_2 from [1]: Here, the locations are of the form $s_\alpha^x := \langle F_\alpha, x, x, ... \rangle$ for a finite $x \subset F_\alpha$, while $<_{H_2}$ is just $<_{\mathrm{loc}}|H_2$. The disadvantage of this hypering is that it is not slow and hence does not satisfy the finiteness property, but only a countability property. This can easily be fixed by inserting the locations $s_{\alpha,n}^x := \langle F_\alpha, x^+, x^+, ..., x^+, x, x, ... \rangle$, where the first to nth element of s_α^x are replaced by x^+. This results in the so-called slowed-down horizontal hypering H_3.

The F-hierarchy, together with hyperings, seems strong enough to reproduce the whole of constructible combinatorics: For more basic statements like Cantor's generalized continuum hypothesis GCH or its strengthening, the combinatorical principle $\Diamond$ as well as structural theorems like covering or approximation, bare F-levels are sufficient. Arbitrary hyperings are sufficient for the construction of a coarse morass in $\mathbf{L}$, while for the construction of a full morass (see, e.g., [2]) or a square-sequence as introduced in [6], slow hyperings are both necessary and sufficient. Proofs for each of these can be found in [1].

5 Further Plans

The F-hierarchy and its refinements can be successfully applied to exhibit the constructible universe **L** with the result that proofs become sometimes considerably shorter and simpler. It will be subject to further work whether this holds as well for a variety of other principles of constructible combinatorics like higher-gap morasses or morass with square. However, the necessity of passing to hyperings for more sophisticated combinatorics is still somewhat unsatisfying and gives rise to the conjecture that the approach so far ignores operators that should be considered as basic and hence added to the structures as well. The F-hierarchy was build on the intuition that Skolem function can be equivalently treated as truth predicates, thus avoiding the use of master codes and other tools of classical fine structure. As this appears to be inadequate for some applications, the question arises which extra macros could and should be taken into account.

Another direction of future research is the generalization of the approach to relativized **L**-structures, particularly core models. Here, the hope is that the advantages of using the F-hierarchy for, e.g., a proof of the covering lemma allow simplifications in the context of these more involved models as well. Here, the main difficulty is to set up an appropriate notion of a fine ultrapower: Although some parts of the theory of, say, the Dodd-Jensen core model K^{DJ} like an iteration theory up to a Dodd-Jensen lemma can be reworked with out major difficulties for a rather ad hoc choice of an ultrapower concept, the preservation of finestructural parameters like standard parameters and even projecta becomes a surprisingly persistent obstacle. Here, a deeper analysis of the essential 'macros' will certainly be required.

References

1. Carl, M.: Alternative Finestructural and Computational Approaches to Constructibility. Submitted as PhD thesis at the university of Bonn (2010)
2. Devlin, K.: Constructibility. Perspectives in Mathematical Logic. Springer, Heidelberg (1984)
3. Donder, H.-D.: Coarse Morasses in **L**. In: Jensen, R.B., Prestel, A. (eds.) Set theory and model theory. Proceedings of a Symposium held in Bonn, June 1-3. Lecture Notes in Mathematics, vol. 872, pp. 37–54. Springer, Heidelberg (1979)
4. Friedman, S.D., Koepke, P.: An elementary approach to the fine structure of L. Bulletin of Symbolic Logic 4, 453–468
5. Hamkins, J.D., Lewis, A.: Infinite Time Turing Machines. Journal of Symbolic Logic 65(2), 567–604 (2000)
6. Jensen, R.B.: The fine structure of the constructible hierarchy. Annals of Mathematical Logic 4, 229–308 (1972)

7. Koepke, P.: A new finestructural hierarchy for the constructible niverse. CRM preprints, Barcelona 554 (2003)
8. Koepke, P.: Ordinal Computability and Fine Structure. Talk iven in the Oberseminar Mathematische Logik, Bonn (2005)
9. Koepke, P., Siders, R.: Register Computations on Ordinals. Archive for Mathematical Logic 47(6), 529–548 (2008)
10. Magidor, M.: Representing sets of ordinals as countable unions of sets in the core model. Transactions of the American Mathematical Society 317(1), 91–121 (1990)
11. Richardson, T.L.: Silver Machine approach to the constructible universe. Ph.D. thesis, University of California, Berkeley (1979)
12. Sacks, G.E.: Higher Recursion Theory. Perspectives in Mathematical Logic, vol. 2. Springer, Heidelberg (1990)
13. van Eijmeren, M., Koepke, P.: A refinement of Jensen's constructible hierarchy. In: Löwe, B., Piwinger, B., Räsch, T. (eds.) Classical and new paradigms of computation and their complexity hierarchies. Papers from the Conference Foundations of the Formal Sciences III (FotFS III) held at the Universität Wien, Vienna, September 21-24 (2001); Trends in Logic, vol. 23, Kluwer, 159-169 (2004)
14. Weckbecker, G.: Ordinal Register Machines and Constructibility. Dipl.-Math. Thesis. Bonn (2010)

Automatic Learners with Feedback Queries

John Case[1], Sanjay Jain[2,*], Yuh Shin Ong[2], Pavel Semukhin[3,**],
and Frank Stephan[2,4,***]

[1] Department of Computer and Information Sciences,
University of Delaware, Newark, DE 19716-2586, USA
case@cis.udel.edu
[2] Department of Computer Science, National University of Singapore,
Singapore 117417, Republic of Singapore
sanjay@comp.nus.edu.sg, yuhshin@gmail.com
[3] Department of Computer Science, University of Regina,
Canada
pavel@semukhin.name
[4] Department of Mathematics, National University of Singapore,
Singapore 119076, Republic of Singapore
fstephan@comp.nus.edu.sg

Abstract. Automatic classes are classes of languages for which a finite
automaton can decide whether a given element is in a set given by its
index. The present work studies the learnability of automatic families
by automatic learners which, in each round, output a hypothesis and
update a long term memory, depending on the input datum, via an automatic function, that is, via a function whose graph is recognised by a
finite automaton. Many variants of automatic learners are investigated:
where the long term memory is restricted to be the just prior hypothesis whenever this exists, cannot be of size larger than the size of the
longest example or has to consist of a constant number of examples seen
so far. Furthermore, learnability is also studied with respect to queries
which reveal information about past data or past computation history;
the number of queries per round is bounded by a constant. These models are generalisations of the model of feedback queries, given by Lange,
Wiehagen and Zeugmann.

1 Introduction

The present work carries on recent investigations of learnability properties in
connection with automatic structures and automatic families [8,14,15]. An advantage of an automatic family over general indexed families [1,18,20] is that
the first-order theory of automatic families, as well as of automatic structures
in general, is decidable [12,13,16]. Here in the first-order theory, the predicates

---

* Supported in part by NUS grant number C-252-000-087-001 and R252-000-420-112.
** Supported in part by NUS grant number R146-000-114-112.
*** Supported in part by NUS grant numbers R146-000-114-112 and R252-000-420-112.

(relations) and functions (mappings) allowed are automatic. Furthermore, relations and functions that are first-order defined from other automatic relations and functions are automatic again [12,13,16]. Also, automatic functions are linear time computable [8]. These nice properties of automatic structures make them not only a useful tool in learning theory but also in other areas such as model checking and Boolean algebras [6,16,22,23]. Common examples of automatic predicates from the prior literature are predicates to compare the length of strings, the lexicographic order (denoted by $\leq_{lex}$) and the length-lexicographic order (denoted by $\leq_{ll}$). Here x is *length-lexicographically less than* y iff either $|x| < |y|$ or $|x| = |y|$ and $x <_{lex} y$, where $|x|$ denotes the length of string x.

Furthermore, although the class of all regular languages is learnable using queries [4], this is not true for the case of inductive inference from positive data [1,11]. Hence, it is worth investigating more closely which classes of regular languages are learnable from positive data and which are not. For example, Angluin [3] considered learnability of the class of k-reversible languages. These studies were later extended [9]. In this context, it is useful to consider which automatic families are learnable and which are not. As noted by Jain, Luo and Stephan [14], even the class of 0-reversible languages is not automatic. However, some very nice sub-classes of pattern languages [2,8] are automatic families and learnable automatically, that is, by learners which are given using finite automata.

The underlying model of learnability we consider is inductive inference [1,11,21] and the main changes to the standard model of inductive inference are the following two: (1) the target class of languages for learning is an automatic family [12,13,15,16], that is, membership for the class to be learnt is recognised by a finite automaton in a uniform way; (2) the learner itself has to be automatic [14]. These learners will then be given by a function, where in each stage, the learner outputs a hypothesis and updates its long term memory based on a current input and its previous long term memory; this function has to be recognised by a finite automaton. Such learners satisfy much more realistic complexity bounds than learners which have access to the full history of all past data and computations. A further motivation for studying learners which are automatic is that in some situations (such as space exploration by robots), it may be more reasonable to have finite automata as a model rather than Turing machines. Another motivation for the work goes back to the programme of Khoussainov and Nerode [16] to find which results from computable model theory can be transferred to model theory based on finite automata.

The notion of learners with explicit bounds on the long term memory had already been studied previously in the setting of algorithmic learners [10,17]. Such memory restrictions were considered as too restrictive. This led to enrichment of the learners by allowing feedback queries and other instruments to access some, but not all information about the past [7,19,25]. The present work investigates these notions for the case of automatic learners learning automatic families.

Outline of the paper. Section 3 provides some examples to give some insight into the above definitions and notions. Section 4 provides the main results on learning with feedback queries. Theorem 7 shows that every automatic family satisfying Angluin's tell-tale condition has a feedback learner with only one query per round and with an additional long term memory permitted to be as large as the longest datum seen so far. Theorem 8 shows that there is a class which has a learner employing only one feedback query per round and without any long term memory but which does not have an automatic learner relying on long term memory only. Theorems 9 and 10 relate various memory types with feedback queries and investigate the hierarchies which result from counting the size of the bounded example memory and the number of feedback queries per round. Section 5 deals with learners using a marked memory space (see Section 2 for definition). Theorem 11 shows that the more general marked memory space of type 2 permits to learn any learnable class with a long term memory bounded by the size of the hypothesis. In contrast to this, Theorem 12 shows that such a result is not possible for a marked memory space of type 1. It is an open problem whether a learner using long term memory bounded by hypothesis size can be replaced by an iterative class-preserving learner [14]. Theorem 13 gives a partial answer that this is possible, if the iterative learner has additionally access to a marked memory space of type 1.

2 Learning with Feedback and Memory Limitations

The symbol $\mathbb{N}$ denotes the set of natural numbers, $\{0, 1, 2, \ldots\}$. Given two strings $x = x(0)x(1) \ldots x(n-1)$ and $y = y(0)y(1) \ldots y(m-1)$, over the alphabet Σ, we define the convolution [16], $conv(x, y)$, over the alphabet $(\Sigma \cup \{\diamond\})^2$ as follows (where $\diamond \notin \Sigma$). Let $p = \max\{n, m\}$, and $x' = x\diamond^{p-n}$, and $y' = y\diamond^{p-m}$. Then, $conv(x, y) = (x'(0), y'(0))(x'(1), y'(1)) \ldots (x'(p-1), y'(p-1))$. Similarly, one can define $conv$ on multiple arguments. A relation (predicate) R or a function f is called *automatic* if the set $\{conv(x_1, x_2, \ldots, x_n) : R(x_1, x_2, \ldots, x_n)\}$ and $\{conv(x_1, x_2, \ldots, x_m, y) : f(x_1, x_2, \ldots, x_m) = y\}$, respectively, are regular.

A family of languages, $\{L_\alpha : \alpha \in I\}$ is said to be *automatic* [16] iff (a) I (called the index domain) is regular, (b) there is a regular set D (called the domain) such that each $L_\alpha \subseteq D$ and (c) the set $\{conv(\alpha, x) : \alpha \in I \wedge x \in D \wedge x \in L_\alpha\}$ is regular. Here D and I are sets of strings over some finite alphabet.

Automatic structures are structures given by finitely many automatic relations and functions — where these structures can also be considered in a more general sense, when they are just isomorphic to a collection of finitely many automatic predicates and functions with corresponding regular domains.

Fix a domain $D \subseteq \Sigma^*$, where Σ is a finite alphabet. Let $\# \notin \Sigma$. A text is a mapping from $\mathbb{N}$ to $D \cup \{\#\}$. We let $T[n]$ denote $T(0)T(1) \ldots T(n-1)$. Content of a text T, $content(T)$, is $\{T(i) : i \in \mathbb{N}\} - \{\#\}$. Sequences are initial segments of texts. Content of a sequence $\sigma = T[n]$, denoted $content(\sigma)$, is $\{T(i) : i < n\} - \{\#\}$. Intuitively, $\#$ denotes pauses in the presentation of data. A text T is for a language L iff $content(T) = L$.

We will be considering learning of an automatic family $\mathcal{L} = \{L_\alpha : \alpha \in I\}$ by a learner using hypothesis space $\mathcal{H} = \{H_\beta : \beta \in J\}$, where $\mathcal{L}$ and $\mathcal{H}$ are automatic families, with I and J being regular sets and languages L_α, H_β being subsets of a regular set $D \subseteq \Sigma^*$, for Σ being a finite alphabet.

A learner is an algorithmic device mapping $\Gamma^* \times (J \cup \{?\}) \times (\Sigma^* \cup \{\#\})$ to $\Gamma^* \times (J \cup \{?\})$, where Γ is a finite alphabet. Intuitively, members of Γ^* represent the long term memory of the learner. Furthermore, ? represents that the learner repeats its previous hypothesis.

The basic model of learning [11] is given as follows. Fix an input text T for a target language L. The learner has initial memory $mem_0 \in \Gamma^*$ and initial hypothesis $\beta_0 \in J \cup \{?\}$. In stage n, the learner receives the input $T(n)$, updates its previous memory mem_n to mem_{n+1} and outputs a hypothesis β_{n+1}. For general learners as studied by Gold [11], there is no restriction on the learners except for the mapping $(mem_n, T(n)) \mapsto (mem_{n+1}, \beta_{n+1})$ being computable (here note that the learner does not know n, unless it stores it in its memory). The learner learns [11] a language L iff for all texts T for L, for β_n as defined above, there is an n such that (i) $H_{\beta_n} = L$, and (ii) for all $m \geq n$, $\beta_m \in \{\beta_n, ?\}$. The learner learns a class $\mathcal{L}$ of languages iff it learns each $L \in \mathcal{L}$.

For learning automatic families of languages a characterization based on Angluin's condition [1] determines when an automatic family is learnable [14].

Proposition 1 (Based on Angluin [1]). *An automatic family $\{L_\alpha : \alpha \in I\}$ is learnable by an algorithmic learner iff, for every $\alpha \in I$, there is a bound b_α such that, for all $\beta \in I$, the implication*

$$\{x \in L_\alpha : |x| \leq b_\alpha\} \subseteq L_\beta \subseteq L_\alpha \Rightarrow L_\beta = L_\alpha$$

holds. We call the set $\{x \in L_\alpha : |x| \leq b_\alpha\}$ a tell-tale set for L_α. This condition is called Angluin's tell-tale condition.

Therefore, the challenge is to study learnability by more restrictive learners. In the setting of automatic structures, it is natural that such learners are automatic [14], that is, for which the mapping $(mem_n, T(n)) \mapsto (mem_{n+1}, \beta_{n+1})$ is automatic. Hypothesis and updated memory of such learners can be computed in time *linear in their previous memory and current datum* [8]. The price paid is that the learner can no longer access the full past history of the data observed. In general, the requirement of a learner to be automatic is a real restriction [14].

Jain, Luo and Stephan [14] had considered various ways in which the size of the long term memory of the automatic learners is bounded in length. The length-restrictions considered are as follows. For the following, T is an arbitrary input text, mem_n and β_n denotes the long term memory and hypothesis of the learner just before getting input $T(n)$.

(a) the size of the hypothesis plus a constant; that is, for some constant c (independent of T), $|mem_n| \leq |\beta_n| + c$.

(b) the size of the longest datum observed so far plus a constant;, that is, for some constant c (independent of T), $|mem_n| \leq \max\{|T(i)| : i < n\} + c$.

(c) just constant size; that is, $|mem_n| \leq c$, for some constant c.

For the ease of notation, the "plus a constant" is omitted in the notations below. Note that the learner is not constrained regarding which alphabet it uses for its memory; therefore, it might, for example, store the convolution of up to k

number of examples in the case that the size of the longest datum seen so far is the memory bound (here k is some a priori fixed constant).

As such memory limited learners are quite restrictive, it is natural to consider mechanisms which provide some access to past data, besides what can be remembered in the memory by automatic learners. If one considers *fat texts*, see [21], where every data item is repeated infinitely often in the text, then Jain, Luo and Stephan [14] showed that automatic learners are able to learn all automatic families satisfying Angluin's tell-tale condition. Hence, the present work looks at criteria which are more powerful than just limited memory structure but less powerful than fat texts. These methods are based on active strategies of the learner, such as making feedback queries about whether some data item has already been seen in the past. While the mechanisms presented here are well-studied in the case of algorithmic learners, the combination of such mechanisms with automatic learners is novel. Also novel is the generalised model of memory space, which subsumes feedback learning and related criteria.

For the following definitions, $\mathcal{L} = \{L_\beta : \beta \in I\}$ refers to the language class being learnt. L_α refers to the target language, and a text T for L_α is the input given to the learner. $\mathcal{H} = \{H_\beta : \beta \in J\}$ refers to the family used by the learner as hypothesis space. Furthermore, mem_n and β_n denote the long term memory and hypothesis of the learner just before receiving input $T(n)$. We sometimes consider memory of the learner as a set. For this, $conv(x_1, x_2, \ldots, x_r)$ represents the set $\{x_1, x_2, \ldots, x_r\}$ (thus there are several representations of any set of size ≥ 2). This allows for automatic updating and testing of elements in a set. For ease of notation, we will often refer directly to the sets in these cases, rather than the representation.

Definition 2 (a) An automatic learner is called *iterative* iff $mem_n = \beta_n$.

(b) An automatic learner is a *learner with k-bounded example memory* iff $mem_n \subseteq content(T[n])$, number of elements in mem_n is at most k, and $mem_{n+1} \subseteq mem_n \cup \{T(n)\}$ (note that the memory of the learner here is interpreted as a set).

If k is not specified, we call the learner a bounded example memory learner.

For the following definition, we additionally provide an automatic learner with a different kind of memory (called marked memory space), which is a set of strings over a finite alphabet Δ. This marked memory will only grow (set inclusion wise) as the learner gets more data.

Definition 3 An *automatic learner* using *a marked memory space* is a learner, with an associated marked memory space, an automatic query function Q (asking k questions per round, for some constant k), and a marked memory space updater F defined as follows on input text T.

Initial memory of the learner is mem_0, initial hypothesis of the learner is β_0 and initial marked memory space is $Z_0 = \emptyset$. Furthermore, mem_n, Z_n and β_n denote the long term memory, the marked memory and hypothesis of the learner just before receiving input $T(n)$.

(a) Q is an automatic mapping from (Γ^*, Σ^*) to a subset of Δ^* of size k.

(b) Suppose, $Q(mem_n, T(n)) = S_n = \{y_1, y_2, \ldots, y_k\}$. Let $b_i = 1$ iff $y_i \in Z_n$. Then, the mapping, $(mem_n, T(n), b_1, b_2, \ldots, b_k) \mapsto (mem_{n+1}, \beta_{n+1})$ is automatic. Furthermore, $Z_{n+1} = Z_n \cup X_{n+1}$, where

(i) for Type 1 memory space, there is an automatic function F such that $F(mem_n, T(n), b_1, b_2, \ldots, b_k) = X_{n+1}$, and

(ii) for Type 2 memory space, there is an automatic function F such that for all $w \in \Delta^*$, $F(mem_n, T(n), b_1, b_2, \ldots, b_k, w) = 1$ iff $w \in X_{n+1}$.

Note that in case of Type 1 memory space, X_n is necessarily finite (with cardinality bounded by some fixed constant), whereas for Type 2 memory, X_n may be infinite.

Definition 4 (a) An *automatic feedback learner* is a special case of an automatic learner using a marked memory space, for which, in Definition 3, the domain D is used for memory instead of Δ^*, and $X_{n+1} = \{T(n)\}$.

When k is fixed as in Definition 3 above, we call the corresponding feedback learner, a k-feedback learner or a learner which uses k feedback queries.

(b) An *automatic learner using hypothesis queries* is a special case of a learner using a marked memory space, for which, in Definition 3, the index set J of the hypothesis space is used for memory instead of Δ^* and $X_{n+1} = \{\beta_{n+1}\}$. This allows a learner to check whether it had earlier issued a particular hypothesis.

Note that both feedback learners and learners using hypothesis queries can be implemented using Type 1 marked memory space. Many of these learning notions had been defined earlier without requiring that the learners are automatic. The general notion of learning which is underlying the notion of an automatic learner is due to Gold [11] and is called explanatory learning. The variant with an explicit long term memory as used here was introduced by Freivalds, Kinber and Smith [10]. The special case of iterative learning is quite popular and predates the definition of general memory limitations, it was introduced by Wiehagen [25] and later by Wexler and Culicover [24]. Bounded example memory was considered by Osherson, Stob and Weinstein [21]; Lange and Zeugmann [19] extended this study. Wiehagen [25] and Lange and Zeugmann [19] introduced and studied feedback learning; Case, Jain, Lange and Zeugmann [7] quantified the amount of feedback queries per round.

For many of these types of automatic learners for automatic families, one can choose the hypothesis space $\mathcal{H}$ to be equal to $\mathcal{L}$; this may sometimes cause a restriction, e.g., when the amount of the memory allowed to the learner depends on the size of the hypothesis or when the long term memory of the learner has to be the most recent hypothesis, as in the case of iterative learning. The main reason for hypothesis space not to be critical in many cases is that one can automatically convert the indices from one automatic family to another for the languages which are common to both automatic families. A result in the present work which depends on the choice of the hypothesis space is Theorem 13. In the case that the hypothesis space does not matter, often, for the ease of notation, the languages are given in place of the indices as conjectures of the learner. This stands in a contrast to the corresponding results for indexed families of recursive languages [19,20].

3   Some Illustrative Examples

We now provide some examples to give insight into the learning criteria and their properties. They show that learnability by automatic learners cannot be characterised from the inclusion structure of a family alone, as the inclusion structure in the class of co-singleton sets is independent of the alphabet size.

Example 5. *The family of all co-singleton sets $\{0,1\}^* - \{x\}$, with $x \in \{0,1\}^*$, is automatic. It does not have an automatic learner, as such a learner cannot memorise all the data observed [14]. However, it can be learnt by an automatic feedback learner (using one query per round), which converges to a hypothesis for $\{0,1\}^* - \{x\}$, for the length-lexicographically least member x of $\{0,1\}^*$ for which the feedback query answer remains negative forever.*

In contrast, the family of all sets $\{0\}^* - \{x\}$, with $x \in \{0\}^*$, has an automatic learner using memory bounded by the size of the longest datum seen so far. This is so because, automatic families defined over unary alphabet can be learnt by an automatic learner whenever they satisfy Angluin's tell-tale condition [14].

The next example deals with intervals of the lexicographic ordering; one could formulate similar results also with other automatic linear orderings. For the case of the lexicographic order, there is a difference between closed and open intervals.

Example 6. *The family of the closed intervals $L_{\mathrm{conv}(x,y)} = \{z \in \Sigma^* : x \leq_{lex} z \leq_{lex} y\}$ is automatic and can also be learnt by an automatic learner with 2-bounded example memory. Furthermore, it has an automatic iterative learner. Both learners memorise, either explicitly or implicitly by padding into the hypothesis, the lexicographically least and greatest data seen so far.*

The family of the open intervals $L_{\mathrm{conv}(x,y)} = \{z \in \Sigma^* : x <_{lex} z <_{lex} y\}$ is also automatic. However, it cannot be learnt as it violates Angluin's tell-tale condition. The open interval $L_{\mathrm{conv}(000,1)} = \{0^n : n > 3\}$ is the ascending union of the open intervals $L_{\mathrm{conv}(0^3,0^m)} = \{0^4, 0^5, \ldots, 0^{m-1}\}$; already Gold [11] observed that classes of this form cannot be learnt from positive data.

Further examples can be found in [14,15].

4   Learning with Feedback Queries

The following result shows that feedback queries together with a quite liberal long term memory permit to reach the full learning power, that is, every family satisfying Angluin's tell-tale condition can be learnt this way.

Theorem 7. *If automatic family $\mathcal{L}$ satisfies Angluin's tell-tale condition, then $\mathcal{L}$ can be learnt by an automatic learner using one-feedback query per round and a long term memory bounded by the longest word seen so far plus a constant.*

Thus, the learners considered in the above result are as general as recursive learners (see Proposition 1). Therefore, the next results compare various more restrictive models of learning with feedback and limitations on the long term memory. First, it is shown that there are cases where feedback queries are more important than any form of long term memory.

Theorem 8. *There is a class $\mathcal{L}$ satisfying the following two statements:*

(a) *An automatic learner without any long term memory, but using one-feedback query per round, can learn $\mathcal{L}$;*
(b) *No automatic learner learns $\mathcal{L}$.*

This result is witnessed by the class consisting of the set $L_\varepsilon = \{0, 1\}^+$ and all sets $L_x = \{y \in \{0, 1\}^* : |y| \leq |x|, y \neq x\} \cup \{y2^n : |y| = |x|\}$ with $x \in \{0, 1\}^+$.

Theorem 9. *There is a class $\mathcal{L}$ with the following properties:*

(a) $\mathcal{L}$ *can be learnt by an automatic learner with 1-bounded example memory;*
(b) $\mathcal{L}$ *can be learnt by an automatic learner using two feedback queries per round, along with a long term memory bounded by the size of the hypothesis;*
(c) $\mathcal{L}$ *cannot be learnt by an automatic iterative learner using feedback queries.*

This theorem is witnessed by the class $\mathcal{L}$ consisting of $L_0 = \{0\}^+$ and $L_{\text{conv}(x,y)} = \{x, y\} \cup \{0^{n+1} : x(n) = 1\}$, where $x \in \{0, 1\}^* \cdot \{1\}$ and $y \in \{0\}^{|x|} \cdot \{0\}^*$.

Theorem 10. *Let $k \geq 1$. Let $\mathcal{L} = \{F : \exists n [\emptyset \subsetneq F \subseteq \{0^m : (k+1)n \leq m < (k+1)(n+1)\}]\}$. Then the following statements hold:*

(a) *Some automatic learner can learn $\mathcal{L}$ using k-bounded example memory;*
(b) *Some automatic learner without any long term memory can learn $\mathcal{L}$ using k feedback queries;*
(c) $\mathcal{L}$ *cannot be learnt by any automatic learner using only $k-1$ bounded example memory (where the learner does not have any memory besides the examples memorised);*
(d) *An automatic learner without any long term memory cannot learn $\mathcal{L}$ using only $k - 1$ feedback queries.*

Parts (a) and (b) of the above theorem can be generalised to show that the class $\mathcal{L}$ can be learnt by an automatic learner which uses, for given $r \in \{0, 1, \ldots, k\}$, a bounded example memory of size r and $k - r$ feedback queries.

5   Learning Using a Marked Memory Space

An automatic learner using a marked memory space of type 1 can simulate a learner using feedback queries, as it could mark every datum observed in the memory space and then query the memory in place of doing a feedback query. Hence it follows from Theorem 7 that every class satisfying Anguin's tell-tale condition can be learnt by a learner with marked memory space of type 1 and a long term memory bounded by the size of the largest example seen so far plus a constant. Hence, the explorations in this section target at more restricted limitations of the long term memory combined with the usage of a marked memory space.

Theorem 11. *Every automatic family satisfying Angluin's tell-tale condition has an automatic learner, with long term memory bounded by hypothesis size plus a constant, using in addition a marked memory space of type 2.*

One might ask whether it is necessary to have a marked memory space of type 2 in the above result. The next result shows that in some cases this is indeed needed and a marked memory space of type 1 is not enough.

Theorem 12. *Let $\mathcal{L}$ be the class consisting of $\{0\}^+$ and all finite sets F with $\{\varepsilon\} \subseteq F \subseteq \{0\}^*$. $\mathcal{L}$ has an automatic learner. However, if an automatic learner is permitted to use only long term memory bounded by the size of its current hypothesis plus a marked memory space of type 1, then $\mathcal{L}$ is not learnable.*

It is open whether an automatic learner, with the memory bounded by the hypothesis size, can be made iterative [14]. Theorem 13 deals with the counterpart of this problem when a marked memory space of type 1 is also permitted.

Theorem 13. *If a class $\mathcal{L}$ has an automatic learner with its long term memory bounded by hypothesis size and not using any marked memory space, then $\mathcal{L}$ also has an automatic iterative learner using hypothesis queries and a new underlying automatic family as hypothesis space.*

References

1. Angluin, D.: Inductive inference of formal languages from positive data. Information and Control 45, 117–135 (1980)
2. Angluin, D.: Finding patterns common to a set of strings. Journal of Computer and System Sciences 21, 46–62 (1980)
3. Angluin, D.: Inference of reversible languages. Journal of the ACM 29, 741–765 (1982)
4. Angluin, D.: Learning regular sets from queries and counterexamples. Information and Computation 75, 87–106 (1987)
5. Blum, L., Blum, M.: Toward a mathematical theory of inductive inference. Information and Control 28, 125–155 (1975)
6. Blumensath, A., Grädel, E.: Automatic structures. In: 15th Annual IEEE Symposium on Logic in Computer Science (LICS), pp. 51–62. IEEE Computer Society, Los Alamitos (2000)
7. Case, J., Jain, S., Lange, S., Zeugmann, T.: Incremental concept learning for bounded data mining. Information and Computation 152, 74–110 (1999)
8. Case, J., Jain, S., Le, T.D., Ong, Y.S., Semukhin, P., Stephan, F.: Automatic learning of subclasses of pattern languages. In: Dediu, A.-H. (ed.) LATA 2011. LNCS, vol. 6638, pp. 192–203. Springer, Heidelberg (2011)

9. Fernau, H.: Identification of function distinguishable languages. Theoretical Computer Science 290, 1679–1711 (2003)
10. Freivalds, R., Kinber, E., Smith, C.H.: On the impact of forgetting on learning machines. Journal of the Association of Computing Machinery 42, 1146–1168 (1995)
11. Mark Gold, E.: Language identification in the limit. Information and Control 10, 447–474 (1967)
12. Hodgson, B.R.: Théories décidables par automate fini. Ph.D. thesis, University of Montréal (1976)
13. Hodgson, B.R.: Décidabilité par automate fini. Annales des sciences mathématiques du Québec 7(1), 39–57 (1983)
14. Jain, S., Luo, Q., Stephan, F.: Learnability of automatic classes. In: Dediu, A.-H., Fernau, H., Martín-Vide, C. (eds.) LATA 2010. LNCS, vol. 6031, pp. 321–332. Springer, Heidelberg (2010)
15. Jain, S., Ong, Y.S., Pu, S., Stephan, F.: On automatic families. TRB1/10, School of Computing, National University of Singapore (2010)
16. Khoussainov, B., Nerode, A.: Automatic presentations of structures. In: Leivant, D. (ed.) LCC 1994. LNCS, vol. 960, pp. 367–392. Springer, Heidelberg (1995)
17. Kinber, E., Stephan, F.: Language learning from texts: mind changes, limited memory and monotonicity. Information and Computation 123, 224–241 (1995)
18. Lange, S., Zeugmann, T.: Language learning in dependence on the space of hypotheses. In: Proceedings of the Sixth Annual Conference on Computational Learning Theory (COLT), pp. 127–136. ACM Press, New York (1993)
19. Lange, S., Zeugmann, T.: Incremental learning from positive data. Journal of Computer and System Sciences 53, 88–103 (1996)
20. Lange, S., Zeugmann, T., Zilles, S.: Learning indexed families of recursive languages from positive data: a survey. Theoretical Computer Science 397, 194–232 (2008)
21. Osherson, D., Stob, M., Weinstein, S.: Systems That Learn, An Introduction to Learning Theory for Cognitive and Computer Scientists. The MIT Press, Cambridge (1986)
22. Rubin, S.: Automatic Structures. Ph.D. Thesis, The University of Auckland (2004)
23. Rubin, S.: Automata presenting structures: a survey of the finite string case. The Bulletin of Symbolic Logic 14, 169–209 (2008)
24. Wexler, K., Culicover, P.W.: Formal Principles of Language Acquisition. The MIT Press, Cambridge (1980)
25. Wiehagen, R.: Limes-Erkennung rekursiver Funktionen durch spezielle Strategien. Elektronische Informationsverarbeitung und Kybernetik (EIK) 12, 93–99 (1976)

Splicing Systems: Accepting Versus Generating

Juan Castellanos[1], Victor Mitrana[2,*], and Eugenio Santos[2]

[1] Department of Artificial Intelligence, Faculty of Informatics,
Polytechnic University of Madrid,
28660 Boadilla del Monte, Madrid, Spain
jcastellanos@fi.upm.es
[2] Department of Organization and Structure of Information,
University School of Informatics, Polytechnic University of Madrid,
Crta. de Valencia km 7, 28031 Madrid, Spain
victor.mitrana@upm.es, esantos@eui.upm.es

Abstract. In this paper we propose a condition for rejecting the input word by an accepting splicing system which is defined by a finite set of forbidding words. More precisely, the input word is accepted as soon as a permitting word is obtained provided that no forbidding word has been obtained so far, otherwise it is rejected. Note that in the new variant of accepting splicing system the input word can be rejected if either no permitting word is ever generated (like in [10]) or a forbidding word has been generated and no permitting word had been generated before. We investigate the computational power of the new variants of accepting splicing systems. We show that the new condition strictly increases the computational power of accepting splicing systems. Rather surprisingly, accepting splicing systems considered here can accept non-regular languages, a situation that has never occurred in the case of (extended) finite splicing systems without additional restrictions.

1 Introduction

One of the basic mechanism by which genetic material is merged is the recombination of DNA sequences under the effect of enzymatic activities. This process has been formalized as a word rewriting operation as follows: the restriction enzymes have been approximated by a finite set of rules defining the restriction sites and the DNA sequences, on which the enzymes act, have been approximated by a finite set of words usually called axioms. This is actually the main idea of the *splicing* operation viewed as a language theoretical approach of the recombinant behavior of DNA under the influence of restriction enzymes and ligases considered by T. Head in [7]. Roughly speaking, the splicing operation is applied to two DNA sequences (represented by words) which are cut at specific sites (represented by splicing rules), and the first subword of one sequence is pasted to the second segment of the other and vice versa. A new formal device to generate languages based on the iteration of splicing operation has been

---

* Work partially supported by the Alexander von Humboldt Foundation.

B. Löwe et al. (Eds.): CiE 2011, LNCS 6735, pp. 41–50, 2011.

considered. Known as *splicing system*, this computation model has been vividly investigated in the last two decades. In spite of the vast literature devoted the topic, the real computational power of finite splicing systems is still partially unknown as the characterization of languages generated by these systems is an open problem. The problem is completely solved for extended splicing systems (a terminal alphabet is used for squeezing out the result), i.e. extended splicing systems are computationally equivalent to finite automata. Another large part of the research in this area has been focused on defining different types of splicing systems and investigating their computational power from a language generating point of view. Many variants of splicing systems have been defined and investigated; we mention here just a few of them: distributed splicing systems [3], splicing systems with multisets [5], splicing systems with permitting and forbidding contexts [6], programmed and evolving splicing systems [14]. Under certain circumstances, splicing systems are computationally complete and universal (see [15] for an overview). This result suggests the possibility to consider splicing systems as theoretical models of programmable universal DNA computers based on the splicing operation.

Several other works like [1], and the references therein, address two fundamental questions concerning splicing systems: *recognition*, which asks for an algorithm able to decide whether or not a given regular language is a splicing language, and *synthesis*, which asks for an effective procedure to construct a splicing system able to generate a given splicing language.

In [10] a novel look on splicing systems is proposed, namely splicing systems are viewed as language accepting devices and not generating ones. More precisely, a usual splicing system is used for accepting/rejecting an input word in accordance with some predefined accepting conditions. The new computational model was called *accepting splicing system*. It is rather strange that though the theory of splicing systems is mature and well developed, an accepting model based on the splicing operation has not considered so far with two exceptions:

– Work [9], where two well-known NP-complete problems were solved with a variant of accepting splicing systems with regular sets of splicing rules. This variant with finite sets of splicing rules was further investigated in [8].

– Work [2], where a splicing recognizer that computes by observing and contains a part exhibiting some similarity to the accepting splicing system defined in [10]. Two ways of iterating the splicing operation and two variants of accepting splicing system are investigated in [10]. Altogether, one obtains four models which are compared with each other as well as with the generating splicing systems from the computational power point of view.

This work is a continuation of [10]. While the accepting splicing systems considered in [10] reject the input word only if no word (considered as a permitting word) from a given finite set is obtained during the splicing process, in this paper we propose a similar condition for rejecting the input word. This condition is also defined by a finite set of words considered as forbidding words. More precisely, the input word is accepted as soon as a permitting word is obtained provided that no forbidding word has been obtained so far, otherwise it is rejected. Note that in the new variant of accepting splicing system the input word can be rejected if either no permitting word is ever generated (like in [10]) or a forbidding word has been generated and no permitting word had been generated before. The main goal of this paper is to investigate the computational power of the new variants of splicing systems. Clearly, the new variants are at least as powerful as the variants considered in [10]. We actually show that the new condition strictly increases the computational power of accepting splicing systems. Rather surprisingly, accepting splicing systems considered here can accept non-regular languages, a situation that has never occurred in the case of (extended) finite splicing systems without additional restrictions.

2   Basic Definitions and Notation

We start by summarizing the notions used throughout the paper. For all undefined notions the reader may consult [17]. An *alphabet* is a finite and nonempty set of symbols. Any finite sequence of symbols from an alphabet V is called *word* over V. The set of all words over V is denoted by V^*, the empty word is denoted by ε, and the length of the word x is denoted by $|x|$. If $w = xyz$ with x, y, z being non-empty words, then x is a prefix of w, z is a suffix of w, and y is a subword for w. Moreover, we write $x^{-1}w = yz$ and $wz^{-1} = xy$. By convention, if x is not a prefix (suffix) of y, then $x^{-1}y = y$ ($yx^{-1} = y$). For two sets of words A and B, we write $A^{-1}B = \{x^{-1}y \mid x \in A, y \in B\}$ and $AB^{-1} = \{xy^{-1} \mid x \in A, y \in B\}$. For a word x we denote by $\mathrm{Pref}_k(x)$, $\mathrm{Suff}_k(x)$ and $\mathrm{Inf}_k(x)$, the prefix, suffix and the set of subwords of x, respectively.

Note that we ignore the empty word when we define a language and the empty set when we define a class of languages.

A splicing rule over V is 4-tuple $[(u_1, u_2); (u_3, u_4)]$, with $u_1, u_2, u_3, u_4 \in V^*$. For a splicing rule $r = [(u_1, u_2); (u_3, u_4)]$ and a pair of words $x, y \in V^*$, we write

$$\sigma_r(x, y) = \{y_1 u_3 u_2 x_2 \mid x = x_1 u_1 u_2 x_2, y = y_1 u_3 u_4 y_2\}$$
$$\cup \{x_1 u_1 u_4 y_2 \mid x = x_1 u_1 u_2 x_2, y = y_1 u_3 u_4 y_2\}$$

for some $x_1, x_2, y_1, y_2 \in V^*$. This definition is extended to a set of splicing rules R and a language L by

$$\sigma_R(L) = \bigcup_{r \in R} \bigcup_{w_1, w_2 \in L} \sigma_r(w_1, w_2).$$

Without risk of confusion, we also denote for two languages L_1, L_2

$$\sigma_R(L_1, L_2) = \bigcup_{x_1 \in L_1} \bigcup_{x_2 \in L_2} \sigma_R(x_1, x_2), \text{ where } \sigma_R(x_1, x_2) = \bigcup_{r \in R} (\sigma_r(x_1, x_2).$$

A *generating splicing system* (*GenSS* for short) is a construct

$$H = (V, A, R),$$

where V is an alphabet, $A \subseteq V^*$ is the initial language, and R is a set of splicing rules over V. For a splicing system $H = (V, A, R)$ we set

$$
\begin{aligned}
&\sigma_R^0(A) = A, \\
&\sigma_R^{i+1}(A) = \sigma_R^i(A) \cup \sigma_R(\sigma_R^i(A)), i \geq 0, \\
&\sigma_R^*(A) = \bigcup_{i \geq 0} \sigma_R^i(A).
\end{aligned}
\qquad (*)
$$

When the set of splicing rules is clear, we omit the subscript. Then, the language generated by H is defined as $L(H) = \sigma_R^*(A)$. Adding a terminal alphabet T we get an *extended generating splicing system* $H = (V, T, A, R)$, $T \subseteq V$, which generates the language $L(H) = T^* \cap \sigma_R^*(A)$. As all systems considered in this paper are extended systems, we shall omit the word "extended". Given a generating splicing system H as above, we say that a word $w \in L(H)$ is a *proper word* of $L(H)$, if it is generated in at least one splicing step. Clearly each word in $L(H) \setminus A$ is proper. The class of languages generated by $GenSS$ is denoted by $\mathcal{L}(GenSS)$.

An important result in splicing theory is the so-called *Regularity Preserving Lemma* proved first in [4], as a consequence of a more general result, and then in [16] by a direct argument. It states that $GenSS$ with a finite set of rules and a finite initial language, i.e. A and R are both finite sets, generate exactly the class of regular languages [13]. When one allows the set of splicing rules (written as words like in [12]) to be described by regular expressions, we obtain computationally complete systems [12].

For a $GenSS$ $H = (V, T, A, R)$ we also introduce the following non-uniform variant of iterated splicing, where the splicing is only done with axioms. More precisely, in the non-uniform case splicing at any step occurs between a generated word in the previous step and an axiom, differently from the general case where splicing at any step occurs between any two words generated in the previous steps. We set

$$
\begin{aligned}
&\tau_R^0(A) = A, \\
&\tau_R^{i+1}(A) = \sigma_R(\tau_R^i(A), A), i \geq 0, \\
&\tau_R^*(A) = \bigcup_{i \geq 0} \tau_R^i(A).
\end{aligned}
\qquad (\diamond)
$$

The language generated by H in the non-uniform way is defined as $L_n(H) = \tau_R^*(A) \cap T^*$. The class of languages generated by $GenSS$ in the non-uniform way is denoted by $\mathcal{L}_n(GenSS)$.

Theorem 1. [13,10] *Both $\mathcal{L}(GenSS)$ and $\mathcal{L}_n(GenSS)$ equal the class of regular languages.*

We now introduce the definitions and terminology for accepting splicing systems. An *accepting splicing system* (*AccSS* for short) is a 6-tuple

$$\Gamma = (V, T, A, R, P, F),$$

where V is an alphabet, $H_\Gamma = (V, T, A, R)$ is a splicing system, while P and F are finite sets of words over V. The elements of P are called *permitting words* while those of F are called *forbidding words*.

Let $\Gamma = (V, T, A, R, P, F)$ be an *AccSS* and a word $w \in V^*$; we define the following iterated splicing that is slightly different from (*):

$$\sigma_R^0(A, w) = \{w\},$$
$$\sigma_R^{i+1}(A, w) = \sigma_R^i(A, w) \cup \sigma_R(\sigma_R^i(A, w) \cup A), i \geq 0,$$
$$\sigma_R^*(A, w) = \bigcup_{i \geq 0} \sigma_R^i(A, w).$$

Although this operation and that defined by (*) are denoted in the same way, there is no risk of confusion as that defined by (*) is an one-argument function while that defined here has two arguments. We say that the word $w \in T^*$ is accepted by Γ if there exists $k \geq 0$ such that

$$(i) \quad \sigma_R^k(A, w) \cap P \neq \emptyset,$$
$$(ii) \quad \sigma_R^k(A, w) \cap F = \emptyset.$$

The following short discussion is in order. The reason for this definition of $\sigma_R^*(A, w)$ is two fold: on the one hand, we maintain a certain uniformity in the definitions of the two ways of acceptance by *AccSS* (see below) and on the other hand, we forbid axioms to be considered as permitting or forbidding words unless they are obtained as proper words. This restriction avoids a "funny" situation in which an *AccSS* accepts either every word whenever an axiom is a final word, or no word whenever an axiom is a forbidding word.

Remark 1. *The following sequence of inclusions is immediate:*
$$(\sigma_R^*(A \cup \{w\}) \setminus A) \subseteq \sigma_R^*(A, w) \subseteq \sigma_R^*(A \cup \{w\}).$$
On the other hand, the next equality will be useful in the sequel.
$$\sigma_R^*(A, w) = \sigma_R^*(A \cup \{w\}) \setminus \{x \in A \mid x \neq w,$$
$$x \text{ is not a proper word of } \sigma_R^*(A \cup \{w\})\}.$$
The language accepted by an *AccSS* Γ is denoted by $L(\Gamma)$.

Remark 2. *Note that every AccSS $\Gamma = (V, T, A, R, P, F)$ with $F = \emptyset$ is actually an (extended) AccSS considered in [10]. This remark suggests to consider for an AccSS $\Gamma = (V, T, A, R, P, F)$, the language $L^\emptyset(\Gamma) = L(\Gamma')$, where $\Gamma' = (V, T, A, R, P, \emptyset)$.*

The class of languages accepted by *AccSS* and *AccSS* without forbidding words is denoted by $\mathcal{L}(AccSS)$ and $\mathcal{L}^\emptyset(AccSS)$, respectively.

For an accepting splicing system $\Gamma = (V, T, A, R, P, F)$ we also introduce the following non-uniform way of accepting words similar to the non-uniform way of generating a language by a *GenSS*. The computation of such a system is nondeterministic; moreover the working mode of such a system involves words originating from the input word and a finite amount of information given by the set of axioms.

For an *AccSS* $= (V, T, A, R, P, F)$ and a word $w \in V^*$ we define the following non-uniform variant of iterated splicing, where the splicing is only done with axioms, similarly to ($\diamond$):

$$\tau_R^0(A, w) = \{w\},$$
$$\tau_R^{i+1}(A, w) = \tau_R^i(A, w) \cup \sigma_R(\tau_R^i(A, w), A), i \geq 0,$$
$$\tau_R^*(A, w) = \bigcup_{i \geq 0} \tau_R^i(A, w).$$

The language accepted by Γ in the non-uniform way is defined by:

$$L_n(\Gamma) = \{w \in T^* \mid \exists k \geq 0(\tau_R^k(A, w) \cap P \neq \emptyset) \ \& \ (\tau_R^k(A, w) \cap PF = \emptyset)\}.$$

The class of languages accepted by $AccSS$ and $AccSS$ without forbidding words in the non-uniform way is denoted by $\mathcal{L}_n(AccSS)$ and $\mathcal{L}_n^\emptyset(AccSS)$, respectively.

3   Computational Power

The inclusions $\mathcal{L}_n^\emptyset(AccSS) \subseteq \mathcal{L}_n(AccSS)$ and $\mathcal{L}^\emptyset(AccSS) \subseteq \mathcal{L}(AccSS)$ are immediate from definitions. Furthermore, by Theorem 2 in [10] $\mathcal{L}_n^\emptyset(AccSS) \subset \mathcal{L}^\emptyset(AccSS)$ holds. The proof of this theorem can be easily completed to a proof for the inclusion $\mathcal{L}_n(AccSS) \subseteq \mathcal{L}(AccSS)$. Based on these observations we now state:

Proposition 1. $\mathcal{L}_n^\emptyset(AccSS) \subset \mathcal{L}_n(AccSS)$ and $\mathcal{L}^\emptyset(AccSS) \subset \mathcal{L}(AccSS)$.

Proof. It suffices to provide a language in $\mathcal{L}_n(AccSS) \setminus \mathcal{L}^\emptyset(AccSS)$. This language, say L, is defined by the regular expression a^+b^+. It is clear that a word is in L if and only if the following conditions are satisfied:
  (i) it does not contain the subword ba;
  (ii) it contains the subword ab.
We now construct an $AccSS$ that accepts L in the non-uniform way. Let

$$\Gamma = (\{a, b, \#, \$\}, \{a, b\}, \{\#\#, \$\$\}, R, \{\$ab\$\}, \{\#ba\#\}),$$

where $R = \{[(ba, \varepsilon); (\#, \#)], [(\varepsilon, ba\#); (\#, \#)], [(ab, \varepsilon); (\$, \$)], [(\varepsilon, ab\$); (\$, \$)]\}$.

By this construction, it is easy to note that if the input word contains the subword ba, then the forbidding word $\#ba\#$ is obtained in the second splicing step. As no permitting word can be produced in the first splicing step, actually no matter the input word, we infer that all input words as above are rejected by Γ. Consequently, a necessary (but not sufficient) condition for an input word to be accepted by Γ is to not contain the subword ba. On the other hand, a similar reasoning lead to the conclusion that the permitting word $\$ab\$$ is also obtained in the second splicing step provided that the input word does contain the subword ab. By these considerations, after the second step, Γ either accepts its input, provided it contains ab, but not ba, and rejects otherwise.

On the other hand, following [10], for every $AccSS$ $\Gamma = (V, T, A, R, P, \emptyset)$, there exists an integer $k > 0$ such that if $w \in L(\Gamma)$, with $|w| \geq k$, then $wyw \in L(\Gamma)$ for any $y \in T^*$. In conclusion, $\{a^n b^m \mid n, m \geq 1\} \notin \mathcal{L}^\emptyset(AccSS)$ which concludes the proof. $\square$

It is worth mentioning here the very simple and efficient way to define the language in the proof of the previous proposition by an accepting splicing system (after two splicing steps only) in comparison with a generating splicing system. Actually, the class $\mathcal{L}_n(AccSS)$ contains "almost" all regular languages. More precisely,

Proposition 2. *For every regular language $L \subseteq V^*$ and $\pounds \notin V$, the language $\pounds L \in \mathcal{L}_n(AccSS)$.*

Proof. Let $A = (Q, V\delta, q_0, Q_f)$ be a deterministic finite automaton accepting the language L. We construct the following $AccSS$:

$$\Gamma = (V \cup Q \cup \{\pounds, \$, \#\}, V \cup \{\pounds\}, Q\{\#\} \cup \{\#\#, \$\$\}, R, Q_f, \{\#\pounds\#\}),$$

where

$$R = \{[(X\pounds, \varepsilon); (\#, \#)] \mid X \in V \cup \{\pounds\}\} \cup \{[(\varepsilon, \pounds\#); (\#, \#)], [(\pounds, \varepsilon); (\$, \$)]\} \cup$$
$$\{[(\$, a); (q_0, \#)] \mid a \in V\} \cup \{[(qa, \varepsilon); (\delta(q, a), \#)] \mid a \in V\}.$$

As in the proof of the previous result, after the first two consecutive steps, the forbidding word $\#\pounds\#$ is obtained provided that the input word is of the form $x\pounds y$ with $|x| > 0$. Therefore, all input words of this form are rejected by Γ.

Let us now analyze the computation of Γ on an input word of the form $\pounds y$, $y \in V^*$. In the first two splicing steps, one obtains consecutively $\$y$ and then $q_0 y$. Note that the other by-product words are $\pounds\$$ and $\$\#$ that cannot be further spliced. From now on, a word qz, $q \in Q$, $z \in V^*$, is computed at some step if and only if $y = xz$ and $\delta(q_0, x) = q$. In conclusion, an input word $\pounds y$ is accepted by Γ if and only if $y \in L(A)$.

The proof is complete as soon as we note that every input word $y \in V^*$ is "inert" with respect to Γ, in the sense that no splicing can be done.   □

We now prove a result which is rather unexpected as this situation has never occurred so far in the case of finite splicing systems, namely finite splicing systems able to define non-regular languages.

Proposition 3. *The non-regular language $\{a^n b^n \mid n \geq 1\}$ lies in $\mathcal{L}(AccSS)$.*

Proof. Let $\Gamma = (V, \{a, b\}, A, R, P, F)$ be the $AccSS$ defined by

$$V = \{a, b, \#, \$, \text{¢}, \pounds, \yen\},$$
$$A = \{\#\#, \$\$, \pounds\pounds, \text{¢¢}, \text{¢}\yen, \yen\text{¢}\},$$
$$P = \{a\text{¢¢}b\}, \text{ and } F = \{\$ba\#, a\yen\yen, \yen\yen b\},$$

and R contains the following rules which are accompanied by their role:
    (i) $\{[(ba, \varepsilon); (\#, \#)], [(\varepsilon, ba\#); (\$, \$)]$. In two consecutive splicing steps the forbidding word $\$ba\#$ is generated, provided that the input word contains the subword ba.

(ii) $\{[(a,b);(\pounds,\pounds)],[(a,\pounds);(\text{\cent},\text{\cent})],[(\pounds,b);(\text{\cent},\text{\cent})]\}$. If the input word contains the subword ab, it is split into two parts $xa\pounds$ and $\pounds by$, with $x,y \in \{a,b\}^*$ in the first splicing step. In the next splicing step, the symbol $\pounds$ is replaced by $\text{\cent}$ in both parts mentioned above. Note that in the first two splicing steps, no permitting word can be obtained. Therefore, every input word containing the subword ba is rejected by Γ. In conclusion, we analyze the computation of Γ on an input word of the form $a^n b^m$ after getting the two words $a^n\text{\cent}$ and $\text{\cent}b^m$.

(iii) $\{[(\varepsilon,a\text{\cent});(\text{\cent},\text{\cent})],[(\text{\cent}b,\varepsilon);(\text{\cent},\text{\cent})]\}$. The number of occurrences of a and b in the two words mentioned above is decreased simultaneously. Note that the end and beginning marker, respectively, remains unchanged, namely $\text{\cent}$.

(iv) $\{[(\varepsilon,a\text{\cent});(\text{\cent},\yen)],[(\text{\cent}b,\varepsilon);(\yen,\text{\cent})]\}$. This marker can be changed to $\yen$.

(v) $\{[(a\text{\cent},\varepsilon);(\varepsilon,\text{\cent}b)],[(\yen,\varepsilon),(\varepsilon,\yen)]\}$. With these rules, Γ comes to taking a decision. If $n = m$, then both words $a\text{\cent}$ and $\text{\cent}b$ have been eventually generated and the permitting word $a\text{\cent}\text{\cent}b$ is finally obtained. Let us analyze the splicing step when the forbidding word $a\yen\yen$ is obtained. This means that in the previous splicing step, both word $a\yen$ and $\yen$ were obtained for the first time in the computation of Γ on the input word $a^n b^m$. Consequently, $n > m$ holds. It is worth noting that $a\text{\cent}$ and $\text{\cent}b$ are also available for splicing, so that the permitting word $a\text{\cent}\text{\cent}b$ and the forbidding word $a\yen\yen$ are obtained in the same splicing step. The case when the forbidding word $\yen\yen b$ is treated analogously.

In conclusion, the permitting word $a\text{\cent}\text{\cent}b$ is obtained before any forbidding word is obtained if and only if the input word is of the form $a^n b^m$ with $n = m$. $\square$

As it can be easily proved that the regular language $\{a^{2n} \mid n \geq 1\}$ does not belong to $\mathcal{L}(AccSS)$, we have:

Corollary 1. *The class of regular languages is incomparable with $\mathcal{L}(AccSS)$.*

We now consider an important subclass of regular languages that can be accepted by accepting splicing For a given $k > 0$, and an alphabet V, we consider a triple $S_k = (A,B,C)$, where A, B and C are sets of words over V of length k. A language L over V is called k-*locally testable in the strict sense* (k-LTSS for short) if there exists a triple $S_k = (A,B,C)$ over V as above such that for any $w \in V^*$ with $|w| \geq k$, $w \in L$ iff $[\text{Pref}_k(w) \in A, \text{Suff}_k(w) \in B, \text{Inf}_k(w) \subseteq C]$ ([11]). When L is specified by $S_k = (A,B,C)$, we write $L = L(S_k)$. A language L is called locally testable in the strict sense (LTSS) iff L is k-LTSS for some $k > 0$. Clearly, every k-LTSS language is regular. A k-LTSS language L over V is *prefix-disjoint* if there exists a triple $S_k = (A,B,C)$ such that $L = L(S_k)$ and $(V^{-1}L) \cap (C \cup B) = \emptyset$. A *suffix-disjoint* k-LTSS language is defined analogously.

Proposition 4. *Every prefix-disjoint or suffix-disjoint k-LTSS language belongs to $\mathcal{L}_n(AccSS)$ for any $k \geq 1$.*

Proof. We assume that $L = L(A,B,C)$ is a prefix-disjoint k-LTSS language over the alphabet V and $S_k = (A,B,C)$ satisfies the prefix-disjoint condition. We construct the $AccSS$ $\Gamma = (U,V,I,R,P,F)$, where

- $U = V \cup \{\langle x \rangle \mid x \in A \cup B \cup C\} \cup \{\text{¢}, \$\}$,
- $R = \{[(x, \varepsilon); (\langle x \rangle, \text{¢})] \mid x \in A\} \cup \{[(\varepsilon, ax\text{¢}); (\text{¢}, \text{¢})] \mid a \in V, x \in A\} \cup$
  $\{[(\langle x \rangle a, \varepsilon); (\langle y \rangle, \$)] \mid a \in V, y \in C \cup B, xa = by \text{ for some } b \in V\}$,
- $A = \{\langle x \rangle \text{¢} \mid x \in A\} \cup \{\langle x \rangle \$ \mid x \in C \cup B\}$,
- $P = \{\langle x \rangle \mid x \in B\}$ and $F = \{\text{¢}ax\text{¢} \mid a \in V, x \in A\}$.

The working mode of Γ can be easily understood as soon as one makes an analogy with the construction in the proof of Proposition 2, where the words in A play the role of the marker £, and the symbols $\langle x \rangle$ play the role of the states. □

4   Final Remarks

The results proposed here are intended to improve the picture concerning the computational power of the accepting models based on the splicing operation as a counterpart of the well investigated generating splicing systems. However, there is still room for improving the overall picture. For instance, the precise relationship between the class of regular languages and each of the classes $\mathcal{L}_n^{\emptyset}(AccSS)$, $\mathcal{L}^{\emptyset}(AccSS)$ and $\mathcal{L}_n(AccSS)$ is

Another area of interest concerns the decidability properties of accepting splicing systems. The next result is just a beginning.

Theorem 2. *The membership problem is decidable for $\mathcal{L}(AccSS)$.*

Proof. Algorithm 1 solves the membership problem for an arbitrary $AccSS$ $\Gamma = (V, T, A, R, P, F)$: Note that the condition in line 1 is algorithmically testable as

---

Algorithm 1 Membership algorithm. **Input:** $w \in T^*$, $|w| = n$, $w \notin P$

1: **if** $w \notin L^{\emptyset}(\Gamma)$ **then**
2:     **return false; halt;**
3: **else**
4:     **for all** $k \geq 1$ **do**
5:         $Q := \sigma_R^k(A, w)$;
6:         **if** $(Q \cap F \neq \emptyset)$ **then**
7:             **return false; halt;**
8:         **else**
9:             **if** $(Q \cap P \neq \emptyset)$ **then**
10:                 **return true; halt;**
11:             **end if**
12:         **end if**
13:     **end for**
14: **end if**

---

the membership problem for $\mathcal{L}^{\emptyset}(AccSS)$ is decidable (see [10]). Moreover, if the condition from line 1 is satisfied, then the algorithm eventually halts within the cycle **for**.   □

By [10], the emptiness and finiteness problems are decidable for $\mathcal{L}_n^{\emptyset}(AccSS)$. The status of these problems as well as of other decision problems for the accepting splicing systems considered here is still open.

Another investigation of interest in our view is to consider the accepting splicing systems introduced here as problem solvers like in [9]. To this aim, the property of an accepting splicing systems to make a decision after a finite number of splicing steps appears to be important. In other words, the rejection of the input word is always a consequence of reaching a forbidding word. None of the constructions proposed here has this property. Can each accepting splicing system be equivalently transformed into an accepting splicing system having this property?

References

1. Bonizzoni, P., Mauri, G.: Regular splicing languages and subclasses. Theoret. Comput. Sci. 340, 349–363 (2005)
2. Cavaliere, M., Jonoska, N., Leupold, P.: DNA splicing: computing by observing. Natural Computing 8, 157–170 (2009)
3. Csuhaj-Varjú, E., Kari, L., Păun, G.: Test tube distributed systems based on splicing. Computers and AI 15, 211–232 (1996)
4. Culik II, K., Harju, T.: Splicing semigroups of dominoes and DNA. Discrete Appl. Math. 31, 261–277 (1991)
5. Denninghoff, K.L., Gatterdam, R.W.: On the undecidability of splicing systems. Intern. J. Computer Math. 27, 133–145 (1989)
6. Freund, R., Kari, L., Păun, G.: DNA computing based on splicing. The existence of universal computers. Theory of Computing Syst. 32, 69–112 (1999)
7. Head, T.: Formal language theory and DNA: an analysis of the generative capacity of specific recombinant behaviours. Bull. Math. Biology 49, 737–759 (1987)
8. Loos, R., Malcher, A., Wotschke, D.: Descriptional complexity of splicing systems. Intern. J. Found. Comp. Sci. 19, 813–826 (2008)
9. Loos, R., Martín-Vide, C., Mitrana, V.: Solving SAT and HPP with accepting splicing systems. In: Runarsson, T.P., Beyer, H.-G., Burke, E.K., Merelo-Guervós, J.J., Whitley, L.D., Yao, X. (eds.) PPSN 2006. LNCS, vol. 4193, pp. 771–777. Springer, Heidelberg (2006)
10. Mitrana, V., Petre, I., Rogojin, V.: Accepting splicing systems. Theor. Comput. Sci. 411, 2414–2422 (2010)
11. McNaughton, R., Papert, S.: Counter-free automata. MIT Press, Cambridge (1971)
12. Păun, G.: Regular extended H systems are computationally universal. J. Automata, Languages, Combinatorics 1, 27–36 (1996)
13. Păun, G., Rozenberg, G., Salomaa, A.: Computing by splicing. Theoret. Comput. Sci. 168, 321–336 (1996)
14. Păun, G., Rozenberg, G., Salomaa, A.: Computing by splicing. Programmed and evolving splicing systems. In: IEEE Intern. Conf. on Evolutionary Computing, Indianapolis, pp. 273–277 (1997)
15. Păun, G., Rozenberg, G., Salomaa, A.: DNA Computing - New Computing Paradigms. Springer, Berlin (1998)
16. Pixton, D.: Regularity of splicing languages. Discrete Appl. Math. 69, 101–124 (1996)
17. Rozenberg, G., Salomaa, A. (eds.): Handbook of Formal Languages, vol. I-III. Springer, Berlin (1997)

Effective Categoricity of Injection Structures

Douglas Cenzer[1], Valentina Harizanov[2], and Jeffrey B. Remmel[3,*]

[1] Department of Mathematics, University of Florida, P.O. Box 118105, Gainesville
FL 32611, United States of America
cenzer@math.ufl.edu

[2] Department of Mathematics, George Washington University, Washington DC
20052, United States of America
harizanv@gwu.edu

[3] Department of Mathematics, University of California at San Diego, La Jolla, CA
92093, United States of America
jremmel@ucsd.edu

Abstract. We study computability theoretic properties of computable injection structures and the complexity of isomorphisms between these structures.

1 Introduction

Computable model theory deals with the algorithmic properties of effective mathematical structures and the relationships between such structures. Perhaps the most basic kind of relationship between two structures is that of isomorphism. It is natural to study the isomorphism problem in the context of computable mathematics by investigating the following question.

Given two effective structures which are isomorphic, what is the least complex isomorphism between them?

Let $\mathbb{N} = \{0, 1, 2, \ldots\}$ denote the natural numbers and $\mathbb{Z} = \{0, \pm 1, \pm 2, \ldots\}$ denote the integers. We let ω denote the order type of $\mathbb{N}$ under the usual ordering and Z denote the order type of $\mathbb{Z}$ under the usual ordering. In what follows, we restrict our attention to countable structures for computable languages. Hence, if a structure is infinite, we can assume that its universe is the set of natural numbers, $\mathbb{N}$. We recall some basic definitions. If $\mathcal{A}$ is a structure with universe A for a language $\mathcal{L}$, then $\mathcal{L}^A$ is the language obtained by expanding $\mathcal{L}$ by constants for all elements of A. The *atomic diagram* of $\mathcal{A}$ is the set of all atomic sentences and negations of atomic sentences from $\mathcal{L}^A$ true in $\mathcal{A}$. The *elementary diagram* of $\mathcal{A}$ is the set of all first-order elementary sentences of $\mathcal{L}^A$ true in $\mathcal{A}$. A structure $\mathcal{A}$ is *computable* if its atomic diagram is computable and a structure $\mathcal{A}$ is *decidable* if its elementary diagram is computable. We call two structures *computably isomorphic* if there is a computable function that is an isomorphism between them. A computable structure $\mathcal{A}$ is *relatively computably isomorphic* to

* Cenzer was partially supported by the NSF grant DMS-652372, Harizanov by the NSF grant DMS-0904101, and Remmel by the NSF grant DMS-0654060.

B. Löwe et al. (Eds.): CiE 2011, LNCS 6735, pp. 51–60, 2011.

a possibly noncomputable structure $\mathcal{B}$ if there is an isomorphism between them that is computable in the atomic diagram of $\mathcal{B}$. A computable structure $\mathcal{A}$ is *computably categorical* if every computable structure that is isomorphic to $\mathcal{A}$ is computably isomorphic to $\mathcal{A}$. A computable structure $\mathcal{A}$ is *relatively computably categorical* if every structure that is isomorphic to $\mathcal{A}$ is relatively computably isomorphic to $\mathcal{A}$. Similar definitions arise for other naturally definable classes of structures and their isomorphisms. For example, for any $n \in \omega$, a structure is Δ_n^0 if its atomic diagram is Δ_n^0, two Δ_n^0 structures are Δ_n^0 *isomorphic* if there is a Δ_n^0 isomorphism between them, and a computable structure $\mathcal{A}$ is Δ_n^0 *categorical* if every computable structure that is isomorphic to $\mathcal{A}$ is Δ_n^0 isomorphic to $\mathcal{A}$.

Among the simplest nontrivial structures are equivalence structures, i.e., structures of the form $\mathcal{A} = (\omega, E)$ where E is an equivalence relation. The complexity of isomorphisms between computable equivalence structures was recently studied by Calvert, Cenzer, Harizanov, and Morozov [2] where they characterized the computably categorical and also the relatively Δ_2^0 categorical equivalence structures. Cenzer, LaForte, and Remmel [4] extended this work by investigating equivalence structures in the Ershov hierarchy. More recently, Cenzer, Harizanov and Remmel [3] studied Σ_1^0 and Π_1^0 equivalence structures.

For any equivalence structure $\mathcal{A}$, we let $\mathrm{Fin}(\mathcal{A})$ denote the set of elements of $\mathcal{A}$ that lie in finite equivalence classes. For equivalence structures, it is natural to consider the different sizes of the equivalence classes of the elements in $\mathrm{Fin}(\mathcal{A})$ since such sizes code information into the equivalence relation. The *character* of an equivalence structure $\mathcal{A}$ is the set

$$\chi(\mathcal{A}) = \{(k,n) : n, k > 0 \text{ and } \mathcal{A} \text{ has at least } n \text{ equivalence classes of size } k\}.$$

This set provides a kind of skeleton for $\mathrm{Fin}(\mathcal{A})$. Any set $K \subseteq (\mathbb{N} - \{0\}) \times (\mathbb{N} - \{0\})$ such that for all $n > 0$ and k, $(k, n + 1) \in K$ implies $(k, n) \in K$, is called a *character*. We say a character K is *bounded* if there is some finite k_0 such that for all $(k, n) \in K$, $k < k_0$. Khisamiev [6] introduced the concepts of s-functions and s_1-functions in his work on Abelian p-groups with computable copies. In the book [1] by Ash and Knight, there is a discussion of equivalence structures in the context of Khisamiev's results.

Definition 1. *Let $f : \omega^2 \to \omega$. The function f is an s-function if the following hold:*

1. *for every $i, s \in \omega$, $f(i, s) \le f(i, s + 1)$;*
2. *for every $i \in \omega$, the limit $m_i = \lim_s f(i, s)$ exists.*

We say that f is an s_1-function if, in addition:

3. *for every $i \in \omega$, $m_i < m_{i+1}$.*

Calvert, Cenzer, Harizanov and Morozov [2] gave conditions under which a given character K can be the character of a computable equivalence structure. In particular, they observed that if K is a bounded character and $\alpha \le \omega$, then there is a computable equivalence structure with character K and exactly α infinite equivalence classes. To prove the existence of computable equivalence structures for unbounded characters K, they needed additional information given by s- and s_1-functions. They showed that if K is a Σ_2^0 character, $r < \omega$, and either (a) there is an s-function f such that

$$(k, n) \in K \Leftrightarrow \mathrm{card}(\{i : k = \lim_{s \to \infty} f(i, s)\}) \geq n \text{ or}$$

(b) there is an s_1-function f such that for every $i \in \omega$, $(\lim_s f(i, s), 1) \in K$, then there is a computable equivalence structure with character K and exactly r infinite equivalence classes.

In this paper, we study *injection structures*. Here an injection is just a one-to-one function and an injection structure $\mathcal{A} = (A, f)$ consists of a set A and an injection $f : A \to A$. $\mathcal{A}$ is a *permutation structure* if f is a permutation of A. Given $a \in A$, the orbit $\mathcal{O}_f(a)$ of a under f is

$$\mathcal{O}_f(a) = \{b \in A : (\exists n \in \mathbb{N})(f^n(a) = b \ \vee \ f^n(b) = a)\}.$$

The order $|a|_f$ of a under f is $\mathrm{card}(\mathcal{O}_f(a))$. Clearly the isomorphism type of a permutation structure $\mathcal{A}$ is determined by the number of orbits of size k for $k = 1, 2, \ldots, \omega$. By analogy with characters of equivalence structures, we define the *character* $\chi(\mathcal{A})$ of an injection structure $\mathcal{A} = (A, f)$ by

$$\chi(\mathcal{A}) = \{(k, n) : \mathcal{A} \text{ has at least } n \text{ orbits of size } k\}.$$

Injection structures (A, f) may have two types of infinite orbits, Z-orbits which are isomorphic to $(\mathbb{Z}, S)$ in which every element is in the range of f and ω-orbits which are isomorphic to (ω, S) and have the form $\mathcal{O}_f(a) = \{f^n(a) : n \in \mathbb{N}\}$ for some $a \notin \mathrm{Rng}(f)$. Thus injection structures are characterized by the number of orbits of size k for each finite k and by the number of orbits of types Z and ω.

We will examine the complexity of the set of elements with orbits of a given type in an injection structure $\mathcal{A} = (A, f)$, in particular, we shall study the complexity of $\mathrm{Fin}(\mathcal{A}) = \{a : \mathcal{O}_f(a) \text{ is finite}\}$.

It is clear from the definitions above that any computable injection structure (A, f) will induce a Σ_1^0 equivalence structure (A, E) in which the equivalence classes are simply the orbits of (A, f).

The outline of this paper is as follows. In Section 2, we investigate algorithmic properties of computable injection structures and their characters, characterize computably categorical injection structures, and show that they are all relatively computably categorical. We prove that a computable injection structure $\mathcal{A}$ is computably categorical if and only if it has finitely many infinite orbits.

In Section 3, we characterize Δ_2^0 categorical injection structures as those with either finitely many orbits of type ω, or with finitely many orbits of type Z. We show that they coincide with the relatively Δ_2^0 categorical structures. Finally, we prove that every computable injection structure is relatively Δ_3^0 categorical.

In the final section, we consider briefly two additional problems. The *spectrum question* is to determine the possible sets (or degrees of sets) that can be the set $\mathrm{Fin}(\mathcal{A})$ for some computable injection structure of a given isomorphism type.

For example, we show that for any c. e. degree **b**, there is a computable injection structure $\mathcal{A}$ such that $\mathrm{Fin}(\mathcal{A})$ has degree **b**. The other problem is to determine the complexity of the theory $\mathrm{Th}(\mathcal{A})$ of a computable injection structure $\mathcal{A}$, as well as the complexity of its elementary diagram $\mathrm{FTh}(\mathcal{A})$. It is easy to see that the character $\chi(\mathcal{A})$ is computable from the theory $\mathrm{Th}(\mathcal{A})$ and presumably the reverse is also true.

The notions and notations of computability theory are standard and as in Soare [7].

2   Computably Categorical Structures

In this section, we show that the characters of computable injection structures are exactly the c. e. characters. We characterize the computably categorical injections structures and show that they coincide with the relatively computably categorical structures.

Lemma 1. *For any computable injection structure* $\mathcal{A} = (\mathbb{N}, f)$,

(a) $\{(k, a) : a \in \mathrm{Rng}(f^k)\}$ *is a* Σ_1^0 *set,*
(b) $\{(a, k) : \mathrm{card}(\mathcal{O}_f(a)) \geq k\}$ *is a* Σ_1^0 *set,*
(c) $\{a : \mathcal{O}_f(a) \text{ is infinite}\}$ *is a* Π_1^0 *set,*
(d) $\{a : \mathcal{O}_f(a) \text{ has type } Z\}$ *is a* Π_2^0 *set,*
(e) $\{a : \mathcal{O}_f(a) \text{ has type } \omega\}$ *is a* Σ_2^0 *set, and*
(f) $\chi(\mathcal{A})$ *is a* Σ_1^0 *set.*

We say that a subset K of $(\mathbb{N} - \{0\}) \times (\mathbb{N} - \{0\})$ is a *character* if there is some injection structure with character K. This is equivalent to saying that $K \subseteq (\mathbb{N} - \{0\}) \times (\mathbb{N} - \{0\})$ and that for all $n > 0$ and all k, $(k, n + 1) \in K \implies (k, n) \in K$. The following is not difficult to prove.

Proposition 1. *For any* Σ_1^0 *character* K, *there is a computable permutation structure* (A, f) *with character* K *and with any specified countable number of orbits of types* ω *and* Z. *Furthermore,* $\{a : \mathcal{O}_f(a) \text{ is finite}\}$ *is computable.*

Proposition 1 shows that injection structures are simpler than equivalence structures in an important way. The characters are simpler, i.e., they are Σ_1^0 rather than Σ_2^0, and there is no distinction between characters which have or do not have s_1-functions.

Theorem 2. *If* $\mathcal{A} = (\mathbb{N}, f)$ *is a computable injection structure with finitely many infinite orbits, then* $\mathcal{A}$ *is relatively computably categorical.*

Proof. Assume $\mathcal{A}$ has m orbits of type ω and n orbits of type Z where $m, n \in \mathbb{N}$. Let $a_1, \ldots, a_m$ be elements of the m orbits of type ω, each not in the range of f. Let $b_1, \ldots, b_n$ be representatives of the n orbits of type Z. A Scott formula for a finite sequence $c_0, \ldots, c_r$ of elements is a conjunction of Δ_1^0 formulas of the following kinds. First, for each $t \leq r$, we have either

1. $f^k(c_t) = c_t$ for some minimal k, or
2. $f^k(a_i) = c_t$ for some unique i and k, or
3. $f^k(b_i) = c_t$ for some unique i and k, or
4. $f^k(c_t) = b_i$ for some unique i and k.

Nothing more needs to be said about the elements c_t which fall into cases (2), (3) or (4). For two elements c_s and c_t which fall into case (1) with the same value of k, we need to add either

5. $f^j(c_s) = c_t$ for some unique $j < k$, or
6. $(\forall j < k) f^j(c_s) \neq c_t$.

If two finite sequences satisfy the same Scott formula as defined above, then it is clear that there exists an automorphism of $\mathcal{A}$ mapping one sequence to the other while preserving the infinite orbits.

It was proved in [5] that any computable injection structure with finitely many infinite orbits is computably categorical. Our next result will show that these are the only computably categorical injection structures.

Theorem 3. *If the computable injection structure $\mathcal{A}$ has infinitely many infinite orbits, then it is not computably categorical.*

Proof. First we consider the cases where either $\mathcal{A}$ consists of an infinite number of orbits of type ω and no orbits of type Z, or $\mathcal{A}$ consists of an infinite number of orbits of type Z and no orbits of type ω.

Suppose first that $\mathcal{A} = (\mathbb{N} - \{0\}, f)$ consists of infinitely many orbits of type ω where $f((2i+1)2^n) = (2i+1)2^{n+1}$ for each $i \geq 0$ and $n \geq 0$. Thus the range of f is a computable set. Now we build a structure $\mathcal{B} = (\mathbb{N} - \{0\}, g)$ isomorphic to $\mathcal{A}$ such that the range of g is not computable, so that $\mathcal{A}$ cannot be computably isomorphic to $\mathcal{B}$.

Let C be a noncomputable c.e. set and let $C = \bigcup_s C_s$ where $\{C_s : s \in \omega\}$ is a computable sequence of finite sets such that each $C_s \subseteq \{0, 1, \ldots, s - 1\}$ and $\operatorname{card}(C_{s+1} - C_s) \leq 1$ for all s. The injection g is defined in stages g_s so that $Dom(g_s) = \{(2i+1)2^n : i, n \leq s\}$. Initially, we have $g_0(1) = 2$. Assume that at stage s, the function g_s will have orbits beginning with $2i + 1$ for each $i \notin C_s$. At stage $s + 1$, we first extend g_s as follows. For each orbit $\mathcal{O}$ of g_s, let m be the unique element of $\mathcal{O}$ not in the domain of g_s and let $I_{\mathcal{O}} = \{i : 2i + 1 \in \mathcal{O}\} = \{i_0 < i_1 < \cdots < i_t\}$. Then extend $\mathcal{O}$ by appending $\{(2i+1)2^{s+2} : i \in I_{\mathcal{O}}\}$ on the right, that is, by letting $g_{s+1}(m) = (2i_0 + 1)2^{s+2}$ and letting $g_{s+1}((2i_r + 1)2^{s+2}) = (2i_{r+1} + 1)2^{s+2}$ for each $r < t$. Next, we add a new orbit by letting $g_{s+1}(2s+3)2^n = (2s+3)2^{n+1}$ for all $n \leq s+1$. Finally, we look to see if some number $i \in C_{s+1} - C_s$. If so, then we append the orbit of $2i + 1$ to the end of the orbit of $2s + 3$ by letting $g_{s+1}((2s + 3)2^{s+2}) = 2i + 1$. Let $g = \bigcup_s g_s$. It is clear that $(\mathbb{N} - \{0\}, g)$ is a computable injection structure isomorphic to $\mathcal{A}$ and that $\mathbb{N} - \operatorname{Rng}(g) = \{2i+1 : i \notin C\}$. Thus $\mathcal{B} = (\mathbb{N} - \{0\}, g)$ is not computably isomorphic to $\mathcal{A}$.

Next suppose that $\mathcal{A} = (\mathbb{N} - \{0\}, f)$ consists of infinitely many orbits of type Z where every orbit of type Z is computable. We shall build a structure $\mathcal{B} = (\mathbb{N} - \{0\}, g)$ in which the orbit of 1 is not computable. The construction is similar to that given above with several modifications. First, to make the orbits have type Z, we extend the orbits at stage $s+1$ to the right when s is even and to the left when s is odd. Second, when i appears in C_{s+1}, we append the orbit of $2i+3$ to the orbit of 1. In this way, we have $i \in C$ if and only if $2i+3 \in \mathcal{O}_g(1)$, so that $\mathcal{O}_g(1)$ is not computable.

Now let $\mathcal{A} = (\mathbb{N} - \{0\}, f)$ be a computable injection structure with infinitely many infinite orbits. Suppose first that $\mathcal{A}$ has infinitely many orbits of type ω and that $\mathrm{Rng}(f)$ is computable. Let $\mathcal{A}_0$ be the restriction of $\mathcal{A}$ to the orbits of type ω and let $\mathcal{B}_0$ be a computable structure isomorphic to $\mathcal{A}_0$ but not computably isomorphic to $\mathcal{A}_0$ as above. By Proposition 1, there is a computable injection structure $\mathcal{C}$ with $\chi(\mathcal{C}) = \chi(\mathcal{A})$ and which has the same number of orbits of type Z as $\mathcal{A}$. Then $\mathcal{B} = \mathcal{B}_0 \oplus \mathcal{C}$ is a computable injection structure which is isomorphic to $\mathcal{A}$ but is not computably isomorphic to $\mathcal{A}$ since any isomorphism would have to map $\mathcal{A}_0$ to $\mathcal{B}_0$. The argument when $\mathcal{A}$ has infinitely many orbits of type Z is similar.

The following corollary is immediate.

Corollary 1. *For any computable injection structure $\mathcal{A}$,*

1. *$\mathcal{A}$ is computably categorical if and only if $\mathcal{A}$ is relatively computably categorical.*
2. *$\mathcal{A}$ is computably categorical if and only if $\mathcal{A}$ has finitely many infinite orbits.*

We also have the following corollary to the proof of Theorem 3.

Corollary 2. *Let $\mathbf{d}$ be a c. e. degree.*

1. *If $\mathcal{A}$ is a computable injection structure which has infinitely many orbits of type ω, then there is a computable injection structure $\mathcal{B} = (B, g)$ isomorphic to $\mathcal{A}$ in which $\mathrm{Rng}(g)$ is a c. e. set of degree $\mathbf{d}$.*
2. *If $\mathcal{A}$ is a computable injection structure which has infinitely many infinite orbits of type Z, then there is a computable injection structure $\mathcal{B} = (B, g)$ isomorphic to $\mathcal{A}$ in which $\mathcal{O}_g(1)$ is of type Z and is a c. e. set of degree $\mathbf{d}$.*

3 Δ_2^0 Categorical Structures

Theorem 4. *Suppose that the computable injection structure $\mathcal{A}$ either does not have infinitely many orbits of type ω or does not have infinitely many orbits of type Z. Then $\mathcal{A}$ is relatively Δ_2^0 categorical.*

Proof. Recall from Lemma 1 that $\{a : \mathcal{O}(a) \text{ is infinite}\}$ is a Π_1^0 set. Under the assumption of our theorem, $\{a : \mathcal{O}(a) \text{ has type } \omega\}$ and $\{a : \mathcal{O}(a) \text{ has type } Z\}$ will be Δ_2^0 sets. Thus, given isomorphic structures $\mathcal{A} = (A, f)$ and $\mathcal{B} = (B, g)$, we may use an oracle for $\mathbf{0}'$ to partition A and B into three sets each: the orbits of finite type, the orbits of type ω, and the orbits of type Z.

First suppose that $\mathcal{A}$ consists of infinitely many orbits of type ω and only finitely many orbits of type Z. We shall construct an isomorphism $h : A \to B$ which is computable in $\mathbf{0}'$. First let $c_1 < \cdots < c_t$ be representatives of the orbits of type Z in $\mathcal{A}$ and $d_1 < \cdots < d_t$ be the representatives of the orbits of type Z in $\mathcal{B}$. Then define $h(c_i) = d_i$ for $i = 1, \ldots, t$ and extend h in the obvious way to map the orbits of $c_1, \ldots, c_t$ to the orbits of $d_1, \ldots, d_t$, respectively. This map will be computable in $\mathbf{0}'$ since the orbit of each c_i and d_i is computable in $\mathbf{0}'$.

Next let A_{fin} denote the set of elements in $\mathcal{A}$ which have finite orbits and let B_{fin} be the set of elements in $\mathcal{B}$ which have finite orbits. A simple back and forth argument will allow us to show that we can construct an isomorphism from (A_{fin}, f) onto (B_{fin}, g) that is computable in $\mathbf{0}'$. That is, at any given stage s, assume that we have defined an isomorphism h_s on a finite set of orbits of (A_{fin}, f) onto a finite set of orbits of (B_{fin}, g). Then let $a \in A_{\mathrm{fin}}$ be the least element not in the domain of h_s. We can compute its orbit $\{a, f(a), f^2(a), \ldots, f^{n-1}(a)\}$ in (A_{fin}, f). Then search through the elements of (B_{fin}, g) until we find a b not in the range of h_s such that b has an orbit of size n and define $h_{s+1}(f^i(a)) = g^i(b)$ for $i = 0, 1 \ldots, n - 1$. Next let d be the least element of B_{fin} which is not in the range of h_s and not in the orbit of b. We can compute its orbit $\{d, g(d), g^2(d), \ldots, g^{m-1}(d)\}$ in (B_{fin}, g). Then we search for a $c \in A_{\mathrm{fin}}$ such that c is not in the domain of h_s or in the orbit of a, and has an orbit of size m. Then we set $h_{s+1}(f^i(c)) = g^i(d)$ for $i = 0, 1 \ldots, m - 1$. Since A_{fin} and B_{fin} are c. e. sets, it follows that h restricted to A_{fin} is computable in $\mathbf{0}'$.

Let A_ω and B_ω denote the set elements which are in orbits of type ω in $\mathcal{A}$ and $\mathcal{B}$, respectively. It is easy to see that $A - A_\omega$ and $B - B_\omega$ are c. e. sets. Since $\mathrm{Rng}(f)$ and $\mathrm{Rng}(g)$ are a c. e. sets, we may use an oracle for $\mathbf{0}'$ to compute a list $a_0, a_1, \ldots$ of $A_\omega - \mathrm{Rng}(f)$ and similarly a list $b_0, b_1, \ldots$ of $B_\omega - \mathrm{Rng}(g)$. Then we extend the isomorphism h by mapping A_ω to B_ω as follows. Given $a \in A_\omega$, compute the unique i and n such that $a = f^n(a_i)$, and let $h(a) = g^n(b_i)$.

Next suppose that $\mathcal{A}$ consists of infinitely many orbits of type Z and only finitely many orbits of type ω. Then again we shall construct an isomorphism $h : A \to B$ which is computable in $\mathbf{0}'$. First let $c_1 < \cdots < c_t$ be starts of the orbits of type ω in $\mathcal{A}$ and $d_1 < \cdots < d_t$ be starts of the orbits of type ω in $\mathcal{B}$. Then define $h(c_i) = d_i$ for $i = 1, \ldots, t$ and extend h in the obvious way to map the orbits of $c_1, \ldots, c_t$ to the orbits of $d_1, \ldots, d_t$, respectively. This map will be computable in $\mathbf{0}'$ since $A - \mathrm{Rng}(f)$ and $B - \mathrm{Rng}(g)$ are computable in $\mathbf{0}'$. Then we can use the back and forth argument to define an isomorphism h from A_{fin} onto B_{fin} computably in $\mathbf{0}'$.

Let A_Z and B_Z be the set of elements that lie in orbits of type Z in $\mathcal{A}$ and $\mathcal{B}$, respectively. In this case, it is easy to see that $A - A_Z$ and $B - B_Z$ are c. e. sets. Since the orbit of any element in $\mathcal{A}$ and $\mathcal{B}$ is computable in $\mathbf{0}'$, it follows that we can use an oracle for $\mathbf{0}'$ to compute a list $a_0, a_1, \ldots$ of representatives for the orbits. That is, observe that $\{(x, y) : \mathcal{O}(x) = \mathcal{O}(y)\}$ is a c. e. set. Let a_0 be the least element in A_Z and, for each i, let a_{i+1} be the least $a \in A_Z$ which is not in the same orbit as any of $a_0, \ldots, a_i$. Similarly we may compute a list $b_0, b_1, \ldots$ of representatives for the orbits of B_Z. Then an isomorphism h may be defined on A_Z as follows. Given $a \in A_Z$, compute the unique i and n such that either $a = f^n(a_i)$ or $a_i = f^n(a)$. In the first instance, let $h(a) = g^n(b_i)$ and in the second instance, let $h(a) = b$ for the unique b such that $g^n(b) = b_i$.

In each case, if $\mathcal{B}$ is not computable, we can nevertheless use an oracle for $\mathcal{B}'$ to compute the isomorphisms described above. Hence $\mathcal{A}$ is, in fact, *relatively* Δ_2^0 categorical.

Theorem 5. *Suppose that the computable injection structure $\mathcal{A}$ has infinitely many orbits of type ω and infinitely many orbits of type Z. Then $\mathcal{A}$ is not Δ_2^0 categorical.*

Proof. It clearly suffices to only consider the case when when $\mathcal{A}$ has no finite orbits. Let $\mathcal{A} = (A, f)$ be an injection structure in which the union of the orbits of type ω is a computable subset of A. We will construct a computable injection structure $\mathcal{B} = (\mathbb{N} - \{0\}, g)$ which is isomorphic to $\mathcal{A}$ in which the union of orbits of type ω is a Σ_2^0 set which is not Δ_2^0, so that $\mathcal{A}$ and $\mathcal{B}$ cannot be Δ_2^0 isomorphic.

It is well-known [7] that $Fin = \{e : W_e \text{ is finite}\}$ is a Σ_2^0 complete set. Thus for any Σ_2^0 set C, there is a function $F : \mathbb{N} \times \mathbb{N} \rightarrow \{0, 1\}$ such that, for all i, $i \in C \iff \{s : F(i, s) = 1\}$ is finite. Let C be an arbitrary Σ_2^0 set. We will define g such that $\mathcal{O}_g(2i + 1)$ has type ω if and only if $i \in C$. The orbits of $\mathcal{B} = (B, g)$ will be exactly $\{\mathcal{O}_g(2i + 1) : i \in \mathbb{N}\}$.

Initially we have $g_0(1) = 2$. After stage s, the function g_s will have orbits $\mathcal{O}_{g_s}(2i + 1) = \{(2i + 1)2^n : n \le 2s + 2\}$ for $i \le s$. At stage $s + 1$, we first extend each orbit of g_s as follows. Fix $i \le s$, let a be the unique element of $\mathcal{O}(2i+1)$ not in the range of g_s and let b be the unique element of $\mathcal{O}(2i + 1)$ not in the domain of g_s. In either case, we let $g_{s+1}(b) = (2i + 1)2^{2s+3}$. If $F(i, s + 1) = 0$, then $g_{s+1}((2i + 1)2^{2s+3}) = (2i + 1)2^{2s+4}$, and if $F(i, s + 1) = 1$, then let $g_{s+1}((2i + 1)2^{2s+4}) = a$. Next, we add a new orbit by letting $g((2s + 3)2^n) = (2s + 3)2^{n+1}$ for all $n \le 2s + 3$. Let $g = \bigcup_s g_s$. If $i \in C$, then $\{s : F(i, s) = 1\}$ is finite, say of cardinality m. Then by the construction, there is no element b such that $g^{m+1}(b) = 2i + 1$ and hence $\mathcal{O}(2i + 1)$ has type ω. If $i \notin C$, then $\{s : F(i, s) = 1\}$ is infinite. Thus for any m, there is an element b such that $g^{m+1}(b) = 2i + 1$ and hence $\mathcal{O}(2i + 1)$ has type Z. It is clear that $(\mathbb{N} - \{0\}, g)$ is a computable injection structure where all orbits are infinite and where $\mathcal{O}_g(2i + 1)$ of type ω if and only if $i \in C$. If C is not a Δ_2^0 set, then $\mathcal{B}$ cannot be Δ_2^0 isomorphic to $\mathcal{A}$. $\qed$

The following corollary is immediate.

Corollary 3. *For any computable injection structure $\mathcal{A}$,*

1. *$\mathcal{A}$ is Δ_2^0 categorical if and only if $\mathcal{A}$ is relatively Δ_2^0 categorical.*
2. *$\mathcal{A}$ is Δ_2^0 categorical if and only if $\mathcal{A}$ has either finitely many orbits of type ω or finitely many orbits of type Z.*

We also have the following corollary to the proof of Theorem 5.

Corollary 4. *For any Σ_2^0 set C, there exists a computable injection structure $\mathcal{A} = (A, f)$ in which the set of elements with orbits of type ω is a Σ_2^0 set with Turing degree equal to $\deg(C)$.*

Theorem 6. *Any computable injection structure $\mathcal{A}$ is relatively Δ_3^0 categorical.*

4   Conclusions and Future Research

In this paper, we have shown that the characters of computable injection structures are exactly the c.e. characters. We have completely determined the Δ_n^0 and relative Δ_n^0 categoricity of the computable injection structures, That is, we have characterized the computably categorical injection structures and shown that they are identical with the relatively computably categorical structures. We have characterized the Δ_2^0 categorical injection structures and shown that they are identical with the relatively Δ_2^0 categorical structures. We have shown that every computable injection structure is relatively Δ_3^0 categorical.

Corollaries 2 and 4 are results about the *spectra* of natural relations on computable injection structures. For a computable injection structure $\mathcal{A} = (A, f)$ and any cardinal $k \leq \omega$, we may consider the possible Turing degrees of $\{a : \operatorname{card}(\mathcal{O}_f(a)) = k\}$ as well as of $\{a : \mathcal{O}_f(a) \text{ has type } \omega\}$ and $\{a : \mathcal{O}_f(a) \text{ has type } Z\}$. For example, we know that for any computable injection structure $\mathcal{A}$, $\operatorname{Fin}(\mathcal{A})$ is a c.e. set. Thus, a natural question is to ask whether for any computable injection structure $\mathcal{A}$ and any c.e. Turing degree $\mathbf{c}$, there exists a computable injection structure $\mathcal{B}$ which is isomorphic to $\mathcal{A}$ such that $\operatorname{Fin}(\mathcal{B})$ has degree $\mathbf{c}$. Clearly, this is not possible for any computable injection structure. For example, if $\mathcal{A}$ has only finitely many infinite orbits, then $A - \operatorname{Fin}(A)$ is c.e. so that $\operatorname{Fin}(A)$ must be computable. Similarly, if $\mathbf{c} \neq \mathbf{0}$, then $\operatorname{Fin}(A)$ must be infinite if we are to have a computable injection structure $\mathcal{B}$ which is isomorphic to A such that $\operatorname{Fin}(\mathcal{B})$ of degree $\mathbf{c}$. However, we can prove the following theorem.

Theorem 7. *Let $\mathbf{c}$ be a c.e. degree. Let $\mathcal{A} = (A, f)$ be a computable injection structure such that $\operatorname{Fin}(\mathcal{A})$ is infinite, A has infinitely many orbits of size k for every $k \in \omega$, and A has infinitely many infinite orbits. Then there is a computable injection structure $\mathcal{B} = (B, g)$ such that $\mathcal{B}$ is isomorphic to $\mathcal{A}$ and $\operatorname{Fin}(\mathcal{B})$ is of degree $\mathbf{c}$.*

A more refined question is what sets can be realized as $\operatorname{Fin}(\mathcal{A})$ for a computable injection structure. An easy observation here is that $\operatorname{Fin}(\mathcal{A})$ can never be a simple c.e. set. That is, if $\operatorname{Fin}(\mathcal{A})$ is not computable, then $\operatorname{Fin}(\mathcal{A})$ must be infinite and also there must be some infinite orbits. But then each infinite orbit of $\mathcal{A}$ is a c.e. set in the complement of $\operatorname{Fin}(\mathcal{A})$. The question remains then whether any non-simple c.e. set can be $\operatorname{Fin}(\mathcal{A})$ for some computable injection structure $\mathcal{A}$.

Other questions include whether Theorem 7 can be extended to structures with specific characters.

Question 1. For which c.e. sets C, can we have $\operatorname{Rng}(g) = C$ in a computable injection structure with infinitely many orbits of type ω?

Question 2. For which c. e. sets C is there a computable injection structure $\mathcal{A}$ such that C is an orbit of $\mathcal{A}$?

Another area of future research is the decidability of injection structures and their theories.

Recall that for any structure $\mathcal{A}$, $\mathrm{Th}(\mathcal{A})$ denotes the first-order theory of $\mathcal{A}$, and $\mathrm{FTh}(\mathcal{A})$ denotes the elementary diagram of $\mathcal{A}$. In the case of equivalence structures Cenzer, Harizanov, and Remmel [3] showed that the character of an equivalence structure together with the number of infinite classes effectively determines its theory. Similarly, they showed that the character together with the function mapping any element to the size of its equivalence class effectively determines its elementary diagram.

The following result is not difficult.

Proposition 2. *For any injection structure $\mathcal{A}$, the character $\chi(\mathcal{A})$ is computable from the theory $\mathrm{Th}(\mathcal{A})$. Hence if $\mathrm{Th}(\mathcal{A})$ is decidable, then $\chi(\mathcal{A})$ is computable.*

It should be the case, as it was for equivalence structures, that $\mathrm{Th}(\mathcal{A})$ and $\chi(\mathcal{A})$ have the same Turing degree. We also conjecture that any computably categorical injection structure is decidable.

References

1. Ash, C.J., Knight, J.: Computable Structures and the Hyperarithmetical Hierarchy. In: Studies in Logic and the Foundations of Mathematics, vol. 144, North-Holland Publishing Co., Amsterdam (2000)
2. Calvert, W., Cenzer, D., Harizanov, V., Morozov, A.: Effective categoricity of equivalence structures. Annals of Pure and Applied Logic 141, 61–78 (2006)
3. Cenzer, D., Harizanov, V., Remmel, J.B.: Σ_1^0 and Π_1^0 equivalence structures. Annals of Pure and Applied Logic 162, 490–503 (2011)
4. Cenzer, D., LaForte, G., Remmel, J.B.: Equivalence structures and isomorphisms in the difference hierarchy. Journal of Symbolic Logic 74, 535–556 (2009)
5. Cenzer, D., Remmel, J.B.: Feasibly categorical abelian groups. In: Clote, P., Remmel, J.B. (eds.) Feasible Math II, Proceedings 1992 Cornell Workshop, Birkhauser, pp. 91–153 (1995)
6. Khisamiev, N.G.: Constructive Abelian groups, In: Ershov, Y.L., Goncharov, S.S., Nerode, A., Remmel, J.B. (eds.) Handbook of Recursive Mathematics, vol. 2, pp. 1177–1231. North-Holland, Amsterdam (1998)
7. Soare, R.I.: Recursively Enumerable Sets and Degrees. Springer, Berlin (1987)

Consistency and Optimality

Yijia Chen[1], Jörg Flum[2], and Moritz Müller[3]

[1] Department of Computer Science and Engineering, Shanghai Jiao Tong University,
Dongchuan Road, No. 800, 200240 Shanghai, China
yijia.chen@cs.sjtu.edu.cn
[2] Abteilung für mathematische Logik, Albert-Ludwigs-Universität Freiburg,
Eckerstraße 1, 79104 Freiburg, Germany
joerg.flum@math.uni-freiburg.de
[3] Centre de Recerca Matemàtica, Campus de Bellaterra,
Edifici C 08193, Bellaterra, Spain
mmueller@crm.cat

Abstract. Assume that the problem Q_0 is not solvable in polynomial time. For theories T containing a sufficiently rich part of true arithmetic we characterize $T \cup \{\mathrm{Con}_T\}$ as the minimal extension of T proving for some algorithm that it decides Q_0 as fast as any algorithm $\mathbb{B}$ with the property that T proves that $\mathbb{B}$ decides Q_0. Here, Con_T claims the consistency of T. Moreover, we characterize problems with an optimal algorithm in terms of arithmetical theories.

1 Introduction

By Gödel's Second Incompleteness Theorem a consistent, computably enumerable and sufficiently strong theory T cannot prove its own consistency Con_T. In other words, $T \cup \{\mathrm{Con}_T\}$ is a proper extension of T.

In Bounded Arithmetic one studies the complexity of proofs in terms of the computational complexity of the concepts involved in the proofs (see e.g. [1, Introduction]). Stronger theories allow reasoning with more complicated concepts. For example, a computational problem may be solvable by an algorithm whose proof of correctness needs tools not available in the given theory; moreover, stronger theories may know of faster algorithms solving the problem. When discussing these issues with the authors, Sy-David Friedman asked whether $T \cup \{\mathrm{Con}_T\}$ can be characterized in this context as a minimal extension of T. We could prove the following result (all terms will be defined in the paper).

Theorem 1. *Let Q_0 be a decidable problem not in* PTIME. *Then there is a finite true arithmetical theory T_0 and a computable function F assigning to every computably enumerable theory T with $T \supseteq T_0$ an algorithm $F(T)$ such that (a) and (b) hold:*

(a) T_0 proves that $F(T)$ is as fast as any algorithm T-provably deciding Q_0.

(b) For every theory T^* with $T^* \supseteq T$ the following are equivalent:

    (i) T^* proves Con_T.

B. Löwe et al. (Eds.): CiE 2011, LNCS 6735, pp. 61–70, 2011.
© Springer-Verlag Berlin Heidelberg 2011

(ii) The algorithm $F(T)$ T^*-provably decides Q_0.

(iii) There is an algorithm such that T^* proves that it decides Q_0 and that it is as fast as any algorithm T-provably deciding Q_0.

Hence, by merely knowing the extension T of T_0 we are able to compute the algorithm $F(T)$, which is, provably in T_0, as fast as any algorithm T-provably deciding Q_0; however, in order to prove that $F(T)$ decides Q_0 we need the full strength of $T \cup \{\mathrm{Con}_T\}$. In this sense, $T \cup \{\mathrm{Con}_T\}$ is a minimal extension of T.

The content of the different sections is the following. In Section 3, by a standard diagonalization technique we derive a result showing for every computably enumerable set D of algorithms the existence of an algorithm that on every input behaves as some algorithm in D and that is as fast as every algorithm in D (see Lemma 2). In Theorem 7 of Section 4 we characterize problems with an optimal algorithm in terms of arithmetical theories. Finally Section 5 contains a proof of Theorem 1.

Many papers in computational complexity, older and recent ones, address the question whether hard problems have *optimal* or *almost optimal* algorithms. Although Levin [5] observed that there exists an optimal algorithm that finds a satisfying assignment for every satisfiable propositional formula, it is not known whether the class of satisfiable propositional formulas or the class of tautologies have an almost optimal algorithm.

Krajíček and Pudlák [4] showed for the latter class that an almost optimal algorithm exists if and only if "there exists a finitely axiomatized fragment T of the true arithmetic such that, for every finitely axiomatized consistent theory S, there exists a deterministic Turing machine $\mathbb{M}$ and a polynomial p such that for any given n, in time $\leq p(n)$ the machine $\mathbb{M}$ constructs a proof in T of $\mathrm{Con}_S(\underline{n})$." Here $\mathrm{Con}_S(\underline{n})$ claims that no contradiction can be derived from S by proofs of lengths at most n.

Hartmanis [2] and Hutter [3] considered 'provable' algorithms, where 'provable' refers to a computably enumerable, more or less specified true theory T. Hartmanis compares the class of problems decidable within a given time bound with the class of problems T-provably decidable within this time bound and he studies time hierarchy theorems in this context. Hutter constructs an algorithm "which is the fastest and the shortest" deciding a given problem. As Hutter says, Peter van Emde Boas pointed out to him that it is not provable that his algorithm decides the given problem and that his proof is a "meta-proof which cannot be formalized within the considered proof system" and he adds that "a formal proof of its correctness would prove the consistency of the proof system, which is impossible by Gödel's Second Incompleteness Theorem."

2   Some Preliminaries

First we fix some notations and introduce some basic concepts. We consider problems as subsets of Σ^*, the set of strings over the alphabet $\Sigma = \{0, 1\}$. For an algorithm $\mathbb{A}$ and a string $x \in \Sigma^*$ we let $t_{\mathbb{A}}(x)$ denote the running time of

$\mathbb{A}$ on x. In case $\mathbb{A}$ does not halt on x, we set $t_{\mathbb{A}}(x) := \infty$. If $t_{\mathbb{A}}(x)$ is finite, we denote by $\mathbb{A}(x)$ the output of $\mathbb{A}$ on x.

If $\mathbb{A}$ and $\mathbb{B}$ are algorithms, then $\mathbb{A}$ *is as fast as* $\mathbb{B}$ if there is a polynomial p such that for every $x \in \Sigma^*$

$$t_{\mathbb{A}}(x) \leq p\big(t_{\mathbb{B}}(x) + |x|\big). \tag{1}$$

Note that here we do not require that $\mathbb{A}$ and $\mathbb{B}$ decide the same $Q \subseteq \Sigma^*$.

An algorithm deciding Q is *optimal* if it is as fast as every other algorithm deciding Q, that is, if it has no superpolynomial speedup infinitely often. An algorithm $\mathbb{A}$ deciding Q is *almost optimal* if (1) holds for every other algorithm deciding Q and every $x \in Q$ (hence nothing is required of the relationship between $t_{\mathbb{A}}(x)$ and $t_{\mathbb{B}}(x)$ for $x \notin Q$).

We do not distinguish algorithms from their codes by strings and we do not distinguish strings from their codes by natural numbers. However, we do not fix a computation model (Turing machines, random access machines,...) for algorithms. We state the results in such a way that they hold for every standard computation model.

3   Diagonalizing over Algorithms

In computability theory diagonalization techniques are used in various contexts. We will make use of the following result.

Lemma 2 (Diagonalization Lemma). *Let D be a computably enumerable and nonempty set of algorithms. Then there is an algorithm $\mathbb{A}$ such that (a) and (b) hold.*

(a) *The algorithm $\mathbb{A}$ halts precisely on those inputs on which at least one algorithm in D halts, and in that case it outputs the same as some algorithm in D; more formally, for all $x \in \Sigma^*$*

  – $t_{\mathbb{A}}(x) < \infty \iff t_{\mathbb{D}}(x) < \infty$ *for some $\mathbb{D} \in D$;*

  – *if $t_{\mathbb{A}}(x) < \infty$, then there is $\mathbb{D} \in D$ with $\mathbb{A}(x) = \mathbb{D}(x)$.*

(b) *There is a $d \in \mathbb{N}$[1] such that for all $\mathbb{D} \in D$ there is a $c_{\mathbb{D}}$ such that for all $x \in \Sigma^*$*

$$t_{\mathbb{A}}(x) \leq c_{\mathbb{D}} \cdot \big(t_{\mathbb{D}}(x) + |x|\big)^d.$$

Moreover, there is a computable function that maps any algorithm $\mathbb{E}$ enumerating the set D of algorithms to an algorithm $\mathbb{A}$ satisfying (a) and (b).

In particular, if all algorithms in D decide $Q \subseteq \Sigma^*$, then $\mathbb{A}$ is an algorithm deciding Q as fast as every $\mathbb{D} \in D$.

Proof. Let the algorithm $\mathbb{E}$ enumerate the set D of algorithms, that is, $\mathbb{E}$, once having been started, eventually prints out exactly the algorithms in D. For each $i \in \mathbb{N}$ we denote by $\mathbb{E}_i$ the last algorithm printed out by $\mathbb{E}$ in i steps; in particular, $\mathbb{E}_i$ is undefined if $\mathbb{E}$ hasn't printed any algorithm in i steps.

Algorithm $\mathbb{A}$ is defined as follows.

---

[1] As the proof shows the constant $d \in \mathbb{N}$ does not even depend on D but it depends on the concrete machine model one uses.

$\mathbb{A}(x)$ // $x \in \Sigma^*$

   1. $\ell \leftarrow 0$
   2. **for** $i = 0$ **to** ℓ
   3.   **if** $\mathbb{E}_i$ is defined **then** simulate the $(\ell - i)$th step
   4.   of $\mathbb{E}_i$ on x
   5.   **if** the simulation halts **then** halt and output
   6.   accordingly
   7. $\ell \leftarrow \ell + 1$
   8. goto 2.

Of course (the code of) $\mathbb{A}$ can be computed from (the code of) $\mathbb{E}$. It is easy to see that $\mathbb{A}$ satisfies (a). Furthermore, there are constants $c_0, d_0 \in \mathbb{N}$ such that for all $x \in \Sigma^*$ and every $\ell \in \mathbb{N}$, lines 2–6 take time at most

$$c_0 \cdot (\ell + |x|)^{d_0}. \tag{2}$$

To verify (b), let $\mathbb{D} \in D$ and $i_{\mathbb{D}}$ be the minimum $i \in \mathbb{N}$ with $\mathbb{E}_i = \mathbb{D}$. Fix an input $x \in \Sigma^*$. For

$$\ell = i_{\mathbb{D}} + t_{\mathbb{E}_{i_{\mathbb{D}}}}(x) \text{ and } i = i_{\mathbb{D}}$$

the simulation in line 3 halts if it didn't halt before. Therefore

$$t_{\mathbb{A}}(x) \le O\left(\sum_{\ell=0}^{i_{\mathbb{D}}+t_{\mathbb{D}}(x)} (\ell + |x|)^{d_0} \right) \qquad \text{(by (2))}$$

$$\le O\left((i_{\mathbb{D}} + t_{\mathbb{D}}(x) + |x|)^{d_0+1} \right) \le c_{\mathbb{D}} \cdot \left(t_{\mathbb{D}}(x) + |x| \right)^{d_0+1}$$

for an appropriate constant $c_{\mathbb{D}} \in \mathbb{N}$ only depending on $\mathbb{D}$. $\qquad\qquad\square$

The preceding proof uses the idea underlying standard proofs of a result due to Levin [5]. Even more, Levin's result is also a consequence of Lemma 2:

Example 3 (Levin [5]). Let $F : \Sigma^* \to \Sigma^*$ be computable. An *inverter of F* is an algorithm $\mathbb{I}$ that given y in the image of F halts with some output $\mathbb{I}(y)$ such that $F(\mathbb{I}(y)) = y$. On inputs not in the image of F, the algorithm $\mathbb{I}$ may do whatever it wants.

Let $\mathbb{F}$ be an algorithm computing F. For an arbitrary algorithm $\mathbb{B}$ define $\mathbb{B}^*$ as follows. On input y the algorithm $\mathbb{B}^*$ simulates $\mathbb{B}$ on y; if the simulation halts, then by simulating $\mathbb{F}$ it computes $F(\mathbb{B}(y))$; if $F(\mathbb{B}(y)) = y$, then it outputs $\mathbb{B}(y)$, otherwise it does not stop. Thus if $\mathbb{B}^*$ halts on $y \in \Sigma^*$, then it outputs a preimage of y and

$$t_{\mathbb{B}^*}(y) \le O\left(t_{\mathbb{B}}(y) + t_{\mathbb{F}}(\mathbb{B}(y)) + |y| \right). \tag{3}$$

Furthermore, if $\mathbb{B}$ is an inverter of F, then so is $\mathbb{B}^*$.

Let $D := \{\mathbb{B}^* \mid \mathbb{B} \text{ is an algorithm}\}$. Denote by $\mathbb{I}_{\text{opt}}$ an algorithm having for this D the properties of the algorithm $\mathbb{A}$ in Lemma 2. By the previous remarks it is easy to see that $\mathbb{I}_{\mathrm{opt}}$ is an inverter of F. Moreover, by Lemma 2 (b) and (3), we see that for any other inverter $\mathbb{B}$ of F there exists a constant $c_{\mathbb{B}}$ such that for all y in the image of F

$$t_{\mathbb{I}_{\mathrm{opt}}}(y) \leq c_{\mathbb{B}} \cdot \big(t_{\mathbb{B}}(y) + t_{\mathbb{F}}(\mathbb{B}(y)) + |y|\big)^d.$$

In this sense $\mathbb{I}_{\mathrm{opt}}$ is an optimal inverter of F.

4   Algorithms and Arithmetical Theories

To talk about algorithms and strings we use *arithmetical formulas*, that is, first-order formulas in the language $L_{\mathrm{PA}} := \{+, \cdot, 0, 1, <\}$ of Peano Arithmetic Arithmetical sentences are *true* (*false*) if they hold (do not hold) in the standard L_{PA}-model. For a natural number n let $\dot{n}$ denote the natural L_{PA}-term without variables denoting n (in the standard model).

Recall that an arithmetical formula is Δ_0 if all quantifiers are bounded and it is Σ_1 if it has the form $\exists x_1 \ldots \exists x_m \psi$ where ψ is Δ_0.

We shall use a Δ_0-formula

$$\mathrm{Run}(u, x, y, z)$$

that defines (in the standard model) the set of tuples (u, x, y, z) such that u is an algorithm that on input x outputs y by the (code of a complete finite) run z; recall that we do not distinguish algorithms from their codes by strings and strings from their codes by natural numbers.

For the rest of this paper we fix a $Q_0 \subseteq \Sigma^*$ and an algorithm $\mathbb{A}_0$ deciding Q_0.

The formula

$$\mathrm{Dec}_{Q_0}(u) := \forall x \exists y \exists z \, \mathrm{Run}(u, x, y, z) \, \wedge$$
$$\forall x \forall y \forall y' \forall z \forall z' \big((\mathrm{Run}(\dot{\mathbb{A}}_0, x, y, z) \wedge \mathrm{Run}(u, x, y', z')) \to y = y'\big)$$

defines the set of algorithms deciding Q_0.

Let L_{all} with $L_{\mathrm{PA}} \subset L_{\mathrm{all}}$ be a language containing countably many function and relation symbols of every arity ≥ 1 and countably many constants. A *theory* is a set T of first-order L_{all}-sentences.

Definition 4. Let T be a theory.
(a) An algorithm $\mathbb{A}$ *T-provably decides* Q_0 if T proves $\mathrm{Dec}_{Q_0}(\dot{\mathbb{A}})$.
(b) T is *sound for Q_0-decision* means that for every algorithm $\mathbb{A}$

if $\mathbb{A}$ T-provably decides Q_0, then $\mathbb{A}$ decides Q_0.

(c) T is *complete for Q_0-decision* means that for every algorithm $\mathbb{A}$

if $\mathbb{A}$ decides Q_0, then $\mathbb{A}$ T-provably decides Q_0.

For a computably enumerable sound theory T that proves $\text{Dec}_{Q_0}(\dot{\mathbb{A}}_0)$ the set

$$D(T) := \{\mathbb{D} \mid \mathbb{D} \ T\text{-provably decides } Q_0\} \tag{4}$$

is a computably enumerable set of algorithms deciding Q_0. Thus, by Lemma 2 for $D = D(T)$ we get an algorithm $\mathbb{A}$ deciding Q_0 as fast as every algorithm in $D(T)$. If in addition T is complete for Q_0-decision, then $D(T)$ would be the set of all algorithms deciding Q_0 and thus $\mathbb{A}$ would be an optimal algorithm for Q_0. So, the problem Q_0 would have an optimal algorithm if we can find a computably enumerable theory that is both sound and complete for Q_0-decision. Unfortunately, there is no such theory as shown by the following proposition. We relax these properties in Definition 6 and show in Theorem 7 that the new ones are appropriate to characterize problems with optimal algorithms.

Proposition 5. *There is no computably enumerable theory that is sound and complete for Q_0-decision.*

Proof. We assume that there is a computably enumerable theory T that is sound and complete for Q_0-decision and derive a contradiction by showing that then the halting problem for Turing machines would be decidable.

For every Turing machine $\mathbb{M}$ we consider two algorithms. On every input $x \in \Sigma^*$ the first algorithm $\mathbb{B}_0(\mathbb{M})$ first checks whether x codes a run of $\mathbb{M}$ accepting the empty input tape and then it simulates $\mathbb{A}_0$ on x (recall $\mathbb{A}_0$ is the fixed algorithm deciding Q_0). If x codes an accepting run, then $\mathbb{B}_0(\mathbb{M})$ reverses the answer $\mathbb{A}_0(x)$ of $\mathbb{A}_0$ on x, otherwise it outputs exactly $\mathbb{A}_0(x)$. Clearly $\mathbb{B}_0(\mathbb{M})$ decides Q_0 if and only if $\mathbb{M}$ does not halt on the empty input tape.

The second algorithm $\mathbb{B}_1(\mathbb{M})$, on every input $x \in \Sigma^*$ first checks exhaustively whether $\mathbb{M}$ halts on the empty input tape; if eventually it finds an accepting run, then it simulates $\mathbb{A}_0$ on x and outputs accordingly. It is easy to verify that $\mathbb{B}_1(\mathbb{M})$ decides Q_0 if and only if $\mathbb{M}$ halts on the empty input tape.

As T is sound for Q_0-decision, it proves at most one of $\text{Dec}_{Q_0}(\mathbb{B}_0(\dot{\mathbb{M}}))$ and $\text{Dec}_{Q_0}(\mathbb{B}_1(\dot{\mathbb{M}}))$, and as it is complete for Q_0-decision it proves at least one of these sentences. Hence, given $\mathbb{M}$, by enumerating the T-provable sentences we can decide whether $\mathbb{M}$ halts on the empty input tape. □

Definition 6. A theory T is *almost complete for Q_0-decision* if for every algorithm $\mathbb{A}$ deciding Q_0 there is an algorithm T-provably deciding Q_0 that is as fast as $\mathbb{A}$.

Theorem 7. *The following are equivalent for $Q_0 \subseteq \Sigma^*$:*
(i) Q_0 has an optimal algorithm;
(ii) There is a computably enumerable and arithmetical theory T that is sound and almost complete for Q_0-decision.

Proof. (i) $\Rightarrow$ (ii): We set $T := \{\text{Dec}_{Q_0}(\dot{\mathbb{A}})\}$ where $\mathbb{A}$ is an optimal algorithm for Q_0. Then T is a computably enumerable true arithmetical theory. Truth implies soundness and almost completeness follows from the optimality of $\mathbb{A}$.

(ii) $\Rightarrow$ (i): Let T be as in (ii). Then the set $D(T)$ defined by (4) is a computably enumerable set of algorithms deciding Q_0 (by soundness). By Lemma 2 for $D = D(T)$ we get an algorithm $\mathbb{A}$ deciding Q_0 as fast as every algorithm in $D(T)$ and hence by almost completeness as fast as any algorithm deciding Q_0. Thus, $\mathbb{A}$ is an optimal algorithm for Q_0. $\square$

A result related to the implication (ii) $\Rightarrow$ (i) is shown by Sadowski in [7]. He shows assuming that there does not exist an almost optimal algorithm for the set TAUT of all propositional tautologies, that for every theory T there exists a subset of TAUT in PTIME which is not T-provably in PTIME (cf. [7, Definition 7.5]).

5   Proof of Theorem 1

Recall that $Q_0 \subseteq \Sigma^*$ and that $\mathbb{A}_0$ is an algorithm deciding Q_0. A theory T is Σ_1-complete if every true arithmetical Σ_1-sentence is provable in T. The following result is a consequence of Lemma 2.

Lemma 8. *Assume $Q_0 \notin$ PTIME. Let T be a computably enumerable Σ_1-complete theory such that T proves $\mathrm{Dec}_{Q_0}(\mathbb{\dot{A}}_0)$. Then there is an algorithm $\mathbb{A}$ such that:*
(a) The algorithm $\mathbb{A}$ is total (i.e., $t_{\mathbb{A}}(x) < \infty$ for all $x \in \Sigma^*$) and as fast as every algorithm T-provably deciding Q_0;
(b) T is consistent if and only if $\mathbb{A}$ decides Q_0.
Moreover, there is a computable function diag *that maps any algorithm $\mathbb{E}$ enumerating some Σ_1-complete theory T proving $\mathrm{Dec}_{Q_0}(\mathbb{\dot{A}}_0)$ to an algorithm $\mathbb{A}$ with (a) and (b).*

Proof. For an algorithm $\mathbb{B}$ let $\mathbb{B}\|\mathbb{A}_0$ be the algorithm that on input $x \in \Sigma^*$ runs $\mathbb{B}$ and $\mathbb{A}_0$ on x in parallel and returns the first answer obtained. Then

$$t_{\mathbb{B}\|\mathbb{A}_0} \leq O\Big(\min\{t_{\mathbb{B}}, t_{\mathbb{A}_0}\}\Big). \tag{5}$$

Claim 1. If T is consistent and proves $\mathrm{Dec}_{Q_0}(\mathbb{\dot{B}})$, then $\mathbb{B}\|\mathbb{A}_0$ decides Q_0.

Proof of Claim 1: By contradiction, assume that T is consistent, proves $\mathrm{Dec}_{Q_0}(\mathbb{\dot{B}})$ and $\mathbb{B}\|\mathbb{A}_0$ does not decide Q. Then $\mathbb{B}\|\mathbb{A}_0$ and $\mathbb{A}_0$ differ on some input $x \in \Sigma^*$. Thus $t_{\mathbb{B}}(x) \leq t_{\mathbb{A}_0}(x)$ and in particular $\mathbb{B}$ halts on x. Therefore, the following Σ_1-sentence φ is true

$$\varphi := \exists x \exists y \exists y' \exists z \exists z' \big(\mathrm{Run}(\mathbb{\dot{A}}_0, x, y, z) \wedge \mathrm{Run}(\mathbb{\dot{B}}, x, y', z') \wedge \neg y = y'\big).$$

By Σ_1-completeness T proves φ. However, φ logically implies $\neg \mathrm{Dec}_{Q_0}(\mathbb{\dot{B}})$ and thus T is inconsistent, a contradiction. $\dashv$

The set

$$D_1(T) := \Big\{\mathbb{B}\|\mathbb{A}_0 \mid T \text{ proves } \mathrm{Dec}_{Q_0}(\mathbb{\dot{B}})\Big\}$$

is nonempty as $\mathbb{A}_0 \| \mathbb{A}_0 \in D_1(T)$ by assumption. Let $\mathbb{A}$ be the algorithm obtained for $D = D_1(T)$ by Lemma 2. Then the statement (a) of our lemma holds by (5) and Lemma 2 (b).

For consistent T, by Claim 1 the set $D_1(T)$ only contains algorithms deciding Q_0, thus $\mathbb{A}$ decides Q_0 by Lemma 2.

If T is inconsistent, let $\mathbb{B}_{\mathrm{bad}}$ be an algorithm that accepts every input in the first step. Then $\mathbb{B}_{\mathrm{bad}} \| \mathbb{A}_0 \in D_1(T)$ by inconsistency of T. Thus, by Lemma 2 (b), the algorithm runs in polynomial time and thus does not decide Q_0.

As from an algorithm enumerating T we effectively get an algorithm enumerating $D_1(T)$, by Lemma 2 it should be clear that a computable function diag as claimed exists.   □

Remark 9. As the preceding proof shows we only need the assumption $Q_0 \notin$ PTIME in the proof of the implication from right to left in (b).

Proof of Theorem 1: Let Q_0 be a decidable problem not in PTIME. Among others, the finite true arithmetical theory T_0 claimed to exist in Theorem 1 will contain a formalization of Lemma 8.

We choose a Σ_1-formula $\mathrm{Prov}(x, y)$ defining (in the standard model) the set of pairs (m, n) such that algorithm m enumerates a theory[2] that proves the sentence n.

We let

$$\mathrm{Con}(x) := \neg \mathrm{Prov}\big(x, \ulcorner \neg 0 \dot{=} 0 \urcorner\big)$$

(here $\ulcorner \varphi \urcorner$ denotes the Gödel number of φ). If $\mathbb{E}$ enumerates a theory T, we write Con_T for $\mathrm{Con}(\dot{\mathbb{E}})$.[3]

Let $f : \mathbb{N} \to \mathbb{N}$ be the function given by

$$f(m) := \ulcorner \mathrm{Dec}_{Q_0}(\dot{m}) \urcorner.$$

Both, this function f and the function diag, from Lemma 8 are computable and hence, Σ_1-definable in the standard model. For better readability we shall use f and diag like function symbols in arithmetical formulas.

Further, let the arithmetical formula As-fast-as(x, y) define the pairs (n, m) such that algorithm n is as fast as algorithm m and let $\mathrm{Ptime}(x)$ define the set of polynomial time algorithms. Finally, we set

$$\mathrm{Afap}(x, y) := \forall z \big(\mathrm{Prov}(x, f(z)) \to \mathrm{As\text{-}fast\text{-}as}(y, z) \big).$$

Then for an algorithm $\mathbb{E}$ enumerating a theory T the statement "the algorithm $F(T)$ is as fast as any algorithm T-provably deciding Q_0," that is, the statement (a) in Theorem 1 is formalized by the sentence

$$\mathrm{Afap}\big(\dot{\mathbb{E}}, F(\dot{T})\big). \tag{6}$$

---

[2] We may assume that every enumeration algorithm enumerates a theory by deleting those printed strings that are not sentences.

[3] The notation is ambiguous, as the definition depends on the choice of $\mathbb{E}$, however not the arguments to follow.

Recall that Robinson introduced a finite, Σ_1-complete arithmetical theory R. Let e-Rob(x) be a Σ_1-formula expressing that the algorithm x enumerates a theory extending $R \cup \{\text{Dec}_{Q_0}(\dot{\mathbb{A}}_0)\}$.

We now define the theory T_0. It extends R by the following sentences (s1)–(s5):

(s1) $\forall x \big(\text{e-Rob}(x) \rightarrow \text{Afap}(x, \text{diag}(x))\big)$,
   (a formalization of Lemma 8 (a))

(s2) $\forall x \big((\text{Con}(x) \wedge \text{e-Rob}(x)) \rightarrow \text{Dec}_{Q_0}(\text{diag}(x))\big)$,
   (a formalization of part of Lemma 8 (b))

(s3) $\forall x (\text{Ptime}(x) \rightarrow \neg\text{Dec}_{Q_0}(x))$,
   (Q_0 is not in PTIME)

(s4) $\forall x \big(\neg\text{Con}(x) \rightarrow \forall y(\text{Sent}(y) \rightarrow \text{Prov}(x, y))\big)$
   (every inconsistent theory proves every sentence; here Sent(y) is a Δ_0-formula defining the first-order L_{all}-sentences)

(s5) $\forall x \forall y \big((\text{As-fast-as}(x, y) \wedge \text{Ptime}(y)) \rightarrow \text{Ptime}(x)\big)$
   (if algorithm x is as fast as the polynomial algorithm y, then it is polynomial too).

Let T be a computably enumerable extension of T_0 and let $\mathbb{E}$ be an algorithm enumerating T. We claim that for the algorithm

$$F(T) := \text{diag}(\mathbb{E})$$

(see Lemma 8) the statements (a) and (b) of Theorem 1 hold.

The arithmetical sentence $F(T) = \text{diag}(\dot{\mathbb{E}})$ is Σ_1 and true, so T_0 proves it by Σ_1-completeness (as $T_0 \supseteq R$). By the same reason, T_0 proves e-Rob$(\dot{\mathbb{E}})$. As T_0 contains (s1), T_0 proves Afap$(\dot{\mathbb{E}}, F(T))$; that is, T_0 proves that $F(T)$ is as fast as any algorithm T-provably deciding Q_0. Thus (a) in Theorem 1 holds.

We turn to (b). Let T^* be a theory with $T^* \supseteq T$.

(i) $\Rightarrow$ (ii): So, we assume that T^* proves Con_T. We already know that T_0, and hence T^*, proves e-Rob$(\dot{\mathbb{E}})$. As T^* contains (s2), for $x = \dot{\mathbb{E}}$ we see that T^* proves $\text{Dec}_{Q_0}(\text{diag}(\dot{\mathbb{E}}))$ and thus $\text{Dec}_{Q_0}(F(T))$; that is, $F(T)$ T^*-provably decides Q_0.

(ii) $\Rightarrow$ (iii): Immediate by part (a) of the theorem.

(iii) $\Rightarrow$ (i): Let $\mathbb{A}$ be an algorithm such that T^* proves $\text{Dec}_{Q_0}(\dot{\mathbb{A}})$ and Afap$(\dot{\mathbb{E}}, \dot{\mathbb{A}})$; the latter means that T^* proves

$$\forall z(\text{Prov}(\dot{\mathbb{E}}, f(z)) \rightarrow \text{As-fast-as}(\dot{\mathbb{A}}, z)). \tag{7}$$

Let $\mathbb{B}$ be an algorithm such that T^* proves

$$\text{Ptime}(\dot{\mathbb{B}}). \tag{8}$$

Then T^* proves the following implications:

$$\neg\text{Con}_T \rightarrow \text{Prov}(\dot{\mathbb{E}}, f(\dot{\mathbb{B}})) \qquad \text{(by (s4) and as Sent}(f(\dot{\mathbb{B}})) \text{ is } \Sigma_1)$$

$$\neg\text{Con}_T \rightarrow \text{As-fast-as}(\dot{\mathbb{A}}, \dot{\mathbb{B}}) \qquad\qquad\qquad\qquad \text{(by (7))}$$

$$\neg\text{Con}_T \rightarrow \text{Ptime}(\dot{\mathbb{A}}) \qquad\qquad\qquad\qquad \text{(by (8) and (s5))}$$

$$\neg\text{Con}_T \rightarrow \neg\text{Dec}_{Q_0}(\dot{\mathbb{A}}) \qquad\qquad\qquad\qquad \text{(by (s3))}.$$

As T^* proves $\mathrm{Dec}_{Q_0}(\mathbb{A})$, we see that T^* proves Con_T.   □

We close with an application to Zermelo-Fraenkel set theory ZFC. Here we add the usual ZFC-definitions of the symbols of L_{PA} as new axioms.

Corollary 10. *Assume* ZFC *is consistent. Then there exist a problem Q and an algorithm $\mathbb{A}$ satisfying (a) and (b).*
(a) There is no algorithm deciding Q and being as fast as every other algorithm deciding Q.
(b) The algorithm $\mathbb{A}$ decides Q and is as fast as any algorithm that ZFC-*provably decides Q.*

Proof. Messner [6] proved that there is a problem Q, even decidable in exponential time, that does not have an almost optimal algorithm. In particular, then Q satisfies (a) and $Q \notin$ PTIME. We choose $\mathbb{A}$ according to Lemma 8 for $Q_0 := Q$ and $T := $ ZFC; then (b) holds.   □

Acknowledgments

The authors thank the John Templeton Foundation for its support under Grant #13152, *The Myriad Aspects of Infinity*. Yijia Chen is affiliated with BASICS and MOE-MS Key Laboratory for Intelligent Computing and Intelligent Systems which is supported by National Nature Science Foundation of China (61033002).

References

1. Cook, S.A., Nguyen, P.: Logical Foundations of Proof Complexity. Cambridge University Press, Cambridge (2010)
2. Hartmanis, J.: Relations between diagonalization, proof systems, and complexity gaps. Theoretical Computer Science 8, 239–253 (1979)
3. Hutter, M.: The fastest and shortest algorithm for all well-defined problems. International Journal of Foundations of Computer Science 13, 431–443 (2002)
4. Krajicèk, J., Pudlák, P.: Propositional proof systems, the consistency of first order theories and the complexity of computations. The Journal of Symbolic Logic 54, 1063–1079 (1989)
5. Levin, L.: Universal search problems (in Russian). Problemy Peredachi Informatsii 9, 115–116 (1973)
6. Messner, J.: On optimal algorithms and optimal proof systems. In: Meinel, C., Tison, S. (eds.) STACS 1999. LNCS, vol. 1563, pp. 361–372. Springer, Heidelberg (1999)
7. Sadowski, Z.: On an optimal propositional proof system and the structure of easy subsets. Theoretical Computer Science 288, 181–193 (2002)

Cupping and Diamond Embeddings: A Unifying Approach

Chengling Fang[1], Jiang Liu[2,*] and Guohua Wu[1,**]

[1] Division of Mathematical Sciences, School of Physical and Mathematical Sciences,
Nanyang Technological University, 21 Nanyang Link, Singapore 637371
`fang0032@e.ntu.edu.sg`, `guohua@ntu.edu.sg`
[2] State Key Laboratory of Computer Science, Institute of Software, Chinese
Academy of Sciences, 4# South Fourth Street, Zhong Guan Cun, Beijing 100190,
Peoples' Republic of China
`liuj@ios.ac.cn`

Abstract. In this paper, we prove that for any nonzero cappable degree $\mathbf{c}$, there is a d.c.e. degree $\mathbf{d}$ and a c.e. degree $\mathbf{b} < \mathbf{d}$ such that $\mathbf{c}$ cups $\mathbf{d}$ to $\mathbf{0}'$, caps $\mathbf{b}$ to $\mathbf{0}$ and for any c.e. degree $\mathbf{w}$, either $\mathbf{w} \leq \mathbf{b}$ or $\mathbf{w} \vee \mathbf{d} = \mathbf{0}'$. This result has several well-known theorems as direct corollaries, including Arslanov's cupping theorem, Downey's diamond theorem, Downey-Li-Wu's complementation theorem, and Li-Yi's cupping theorem, etc.

1 Introduction

A d.c.e. degree $\mathbf{d}$ is said to have almost universal cupping property, if it cups all c.e. degrees not below $\mathbf{d}$ to $\mathbf{0}'$. This notion was first proposed by Liu and Wu in their paper [8], where the existence of such degrees is proved. In this paper, we prove that any nonzero cappable degree can have an almost universal cupping degree as a complement.

Theorem 1. *Let $\mathbf{c}$ be a nonzero c.e. cappable degree. Then there is an almost universal cupping degree $\mathbf{d}$ and a c.e. degree $\mathbf{b} < \mathbf{d}$ such that $\mathbf{c}$ cups $\mathbf{d}$ to $\mathbf{0}'$, caps $\mathbf{b}$ to $\mathbf{0}$ and for any c.e. degree $\mathbf{w}$ below $\mathbf{d}$, $\mathbf{w} \leq \mathbf{b}$.*

Theorem 1 implies Downey-Li-Wu's complementation theorem ([5]), which states that any cappable c.e. degree is complemented in the d.c.e. degrees, and also Arslanov's cupping theorem [2] and Downey's diamond theorem [4].

We now see how Theorem 1 implies Li-Yi's cupping theorem [7], which states that there are incomplete d.c.e. degrees $\mathbf{d}_1$ and $\mathbf{d}_2$ such that any nonzero c.e. degree cups one of them to $\mathbf{0}'$. Fix a nonzero c.e. cappable degree $\mathbf{c}$, apply Theorem 1 to obtain corresponding degrees $\mathbf{b}$ and $\mathbf{d}$, and then apply Theorem 1 once again, to $\mathbf{b}$ now, to obtain degrees $\mathbf{e}$ and $\mathbf{f}$, where $\mathbf{e} < \mathbf{f}$, $\mathbf{f}$ is an almost universal cupping degree, $\mathbf{e}$ isolates $\mathbf{f}$, and $\mathbf{b}$ and $\mathbf{e}$ (hence $\mathbf{d}$ and $\mathbf{f}$) form a

* The second author is partially supported by grants NSFC-90718041, NSFC-60736017, NSFC-60970031 and NSFC 91018012 from China.
** The third author is partially supported by NTU grant RG37/09, M52110101.

B. Löwe et al. (Eds.): CiE 2011, LNCS 6735, pp. 71–80, 2011.
© Springer-Verlag Berlin Heidelberg 2011

minimal pair. Then for any nonzero c.e. degree $\mathbf{w}$, it is easy to see that $\mathbf{w}$ cups one of $\mathbf{d}$ and $\mathbf{f}$ to $\mathbf{0}'$, as if $\mathbf{w}$ is below $\mathbf{b}$ (or below $\mathbf{e}$), then $\mathbf{w}$ cups $\mathbf{f}$ ($\mathbf{d}$ respectively) to $\mathbf{0}'$, and if $\mathbf{w}$ is below neither $\mathbf{b}$ nor $\mathbf{e}$, then $\mathbf{w}$ cups both $\mathbf{f}$ and $\mathbf{d}$ to $\mathbf{0}'$. We comment here that the latter property could not be obtained from Li-Yi's cupping theorem.

As the incomplete maximal d.c.e. degree constructed by Cooper, Harrington, Lachlan, Lempp and Soare in [3] has the almost universal cupping property, we ask whether every nonzero capable c.e. degree can have such maximal d.c.e. degrees as complements.

Our terminology and notation are standard and follow Soare [9].

2   Requirements

Fix a c.e. set C in the nonzero cappable c.e. degree $\mathbf{c}$ and fix an enumeration $\{C_s\}_{s\in\omega}$ of C. To prove Theorem 1, we will construct a c.e. set B, a d.c.e set D and partial computable (p.c.) functionals $\{\Gamma_e\}_{e\in\omega}$, $\{\Delta_e\}_{e\in\omega}$, $\{\Theta_e\}_{e\in\omega}$ to satisfy the following requirements:

$\quad \mathcal{G}_e\colon K = \Gamma_e(B \oplus D \oplus W_e)$ or $W_e = \Theta_e(B)$,
$\quad \mathcal{N}_e\colon \Phi_e(C) = \Phi_e(B) = f$ total $\;\rightarrow\; f$ is computable,
$\quad \mathcal{R}_e\colon W_e = \Lambda_e(B \oplus D) \;\rightarrow\; W_e = \Delta_e(B)$,

where $\{(W_e, \Lambda_e)\}$ is an effective list of pair (W, Λ) such that W is a c.e. set and Λ is a p.c. functional, $\{\Phi_e\}_{e\in\omega}$ is an effective list of p.c. functionals.

Let $\mathbf{b}$ be the degree of B and $\mathbf{d}$ be the degree of $B \oplus D$. It is easy to see that all the $\mathcal{G}$ and $\mathcal{R}$-requirements guarantee that $\mathbf{d}$ is an almost universal cupping degree. All the $\mathcal{N}$-requirements guarantee that $\mathbf{c} \cap \mathbf{b} = \mathbf{0}$, hence $\mathbf{c} \cap \mathbf{d} = \mathbf{0}$, otherwise, if $\mathbf{c} \cap \mathbf{d} > \mathbf{0}$, then there is a c.e. degree $\mathbf{a}$ such that $\mathbf{c} \cap \mathbf{d} > \mathbf{a} > \mathbf{0}$, by the $\mathcal{R}$-requirements, $\mathbf{a} \leq \mathbf{b}$, then $\mathbf{c} \cap \mathbf{b} \geq \mathbf{a} > \mathbf{0}$, which contradicts to that $\mathbf{c} \cap \mathbf{b} = \mathbf{0}$.

3   Strategies

In this section, we describe all the basic strategies and the interactions between them. All the strategies are handled on a priority tree.

3.1   A $\mathcal{G}$-Strategy

Let η be a $\mathcal{G}_e$-strategy. *For convenience, we use Γ_η to denote $\Gamma_{e(\eta)}$, use $\gamma_\eta(x)$ to denote $\gamma_{e(\eta)}(B \oplus D \oplus W_e; x)$ and use W_η to denote $W_{e(\eta)}$.* The task of a basic $\mathcal{G}_e$-strategy η is to construct a p.c. functional Γ_η such that $W_e \oplus B \oplus D$ can compute K correctly via Γ_η. η works as follows:

1. At η-stage s, if Γ_η is empty, then we define $\Gamma_\eta(B \oplus D \oplus W_\eta; x)[s] \downarrow= K_s(x)$ for $x < s$. Otherwise, check whether there is some $x < s$ such that $\Gamma_\eta(B \oplus D \oplus W_\eta; x)[s-1] \downarrow\neq K_s(x)$. If so, then we choose the least such x, put $\gamma_\eta(x)[s-1]$ into D and undefine all $\Gamma_\eta(B \oplus D \oplus W_\eta; z)$ for $z \geq x$ with $\Gamma_\eta(B \oplus D \oplus W_\eta; z)[s-1] \downarrow$. If not, go to step 2.

2. Find the least $x < s$ such that $\Gamma_\eta(B \oplus D \oplus W_\eta; x)[s-1] \uparrow$. Then, we define $\Gamma_\eta(B \oplus D \oplus W_\eta; x)[s]$ with an appropriate use $\gamma_\eta(x)[s]$.

In the construction, the interactions between the $\mathcal{G}$-strategies and the $\mathcal{R}$-strategies will make the definition of $\Gamma_\eta(B \oplus D \oplus W_\eta)$ even more subtle, as $\mathcal{R}$-strategies may enumerate (or extract) numbers into (from) D, and request the γ_η-uses big or the same as before.

3.2 An $\mathcal{N}$-Strategy

In [1], Ambos-Spies et al. showed a c.e. degree **c** is cappable iff it is not promptly simple. Here, a set A is called *promptly simple* if there is a computable function p such that for all $i \in \omega$

$$|W_i| = \infty \;\Rightarrow\; \exists x \exists s [x \in W_{i, \text{ at } s} \text{ and } A_s \lceil x \neq A_{p(s)} \lceil x],$$

where $W_{i, \text{ at } s}$ is the set of natural numbers entering W_i at stage s. According to this, the basic idea of an $\mathcal{N}_e$-strategy, α say, is to build a partial computable function p_α threatening C to be promptly simple. We try to satisfy the following subrequirements:

$$\mathcal{S}_{\alpha,i}: |W_i| = \infty \;\Rightarrow\; \exists x \exists s [x \in W_{i, \text{ at } s} \text{ and } C_s \lceil x \neq C_{p_\alpha(s)} \lceil x]$$

A substrategy $\mathcal{S}_{\alpha,i}$ constructs a computable function f_α^i simultaneously such that if $\mathcal{S}_{\alpha,i}$ is not satisfied and $\Phi_\alpha(B) = \Phi_\alpha(C)$ is total, then f_α^i will be defined as a total function and computes $\Phi_\alpha(B) = \Phi_\alpha(C)$ correctly.

We say a $\mathcal{S}_{\alpha,i}$-requirement is satisfied at stage s if p_α is constructed such that there are x and $t < s$ satisfying $x \in W_{i, \text{ at } t}$ and $C_t \lceil x \neq C_{p_\alpha(t)} \lceil x$. As C is cappable, there must exist some i (least) such that $\mathcal{S}_{\alpha,i}$-requirement cannot be satisfied.

An $\mathcal{N}$-strategy is a standard gap-cogap argument, which can be found in Ambos-Spies et al.'s paper [1], or Soare's book [9], or Downey et al.'s paper [5].

In the following, we describe how α works. For convenience, we use Φ_α to denote $\Phi_{e(\alpha)}$. As usual, the length agreement functions are defined as follows:

$$l(\alpha, s) = \max\{x \mid \forall y < x[\Phi_\alpha(C; y)[s] \downarrow = \Phi_\alpha(B; y)[s] \downarrow]\} \text{ and}$$
$$m(\alpha, s) = \max\{0; l(\alpha, t) \mid t < s \text{ is an } \alpha\text{-stage}\}.$$

Here, s is an α-expansionary stage if $s = 0$ or s is an α-stage such that $l(\alpha, s) > m(\alpha, s)$, where a stage s is an α-stage if α is visited at stage s.

Let s be an α-expansionary stage. If there are some i and x such that

(O1) $\mathcal{S}_{\alpha,i}$ is not satisfied;

(O2) x enters W_i at stage s;

(O3) $\exists y [y < \ell(\alpha, s) \; \& \; f_\alpha^i(y)[s] \uparrow \; \& \; x > \max\{\varphi_\alpha(C_s; y')[s] : y' \leq y\}]$,

then let i be the least one and *open a gap* for $\mathcal{S}_{\alpha,i}$ as follows:

(1) For those $y' \leq y$, where y is as in (O3), if $f_\alpha^i(y')[s] \uparrow$, then define $f_\alpha^i(y') = \Phi_e(C; y')[s]$;

(2) Set the restraint $r(\alpha, i, s) = 0$, and initialize all substrategies $\mathcal{S}_{e,j}$ with $j > i$.

Let v be the next α-expansionary stage and *close the gap* as follows:

(C1) Define $p_\alpha(s) = v$.
(C2) Set $r(\alpha, i, v) = v$.

During a gap, if C has a change below x, then by $p_\alpha(s) = v$, $C_s \lceil x \neq C_{p_\alpha(s)} \lceil x$, $\mathcal{S}_{\alpha,i}$ is satisfied. In this case, we say that the gap is *closed successfully*. If there is no such C-change, then the gap is *closed unsuccessfully*.

Suppose that $\Phi_\alpha(B) = \Phi_\alpha(C) = f$ is total. Since C has a cappable degree, there is a least i such that $\mathcal{S}_{e,i}$ cannot be satisfied. Then $\mathcal{S}_{\alpha,i}$ opens infinitely many gaps, and each one is closed unsuccessfully. Let

$$s_0 < v_0 < s_1 < v_1 < \cdots < s_n < v_n < \cdots$$

be the stages at which $\mathcal{S}_{\alpha,i}$ opens and closes gaps alternatively. We prove below that f_α^i computes f correctly.

Fix y and let $f_\alpha^i(y)$ be defined at stage s_n. Then we open an $\mathcal{S}_{\alpha,i}$-gap at stage s_n. That is,

(1) s_n is an α-expansionary stage,
(2) some x enters W_i at stage s_n,

and there is some y with

(3) $y < \ell(\alpha, s_n)$, $f_\alpha^i(y)[s_n] \uparrow$ and $x > \varphi_\alpha(C; y)[s_n]$,
(4) $f_\alpha^i(y)$ is defined as $\Phi_\alpha(C; y)[s_n]$ at the end of stage s_n.

Then at stage v_n, we close this gap by defining $p_\alpha(s_n) = v_n$, restraining numbers less than v_n from entering B till stage s_{n+1}. Since C has no change below $\varphi_\alpha(C_{s_n}; y)[s_n]$ inside this gap (otherwise, $\mathcal{S}_{\alpha,i}$ will be satisfied), $\Phi_\alpha(C; y)[s_n] = \Phi_\alpha(C; y)[v_n]$, and hence

$$f_\alpha^i(y) = \Phi_\alpha(B; y)[s_n] = \Phi_\alpha(C; y)[s_n] = \Phi_\alpha(C; y)[v_n] = \Phi_\alpha(B; y)[v_n].$$

Now numbers less than v_n are restrained from entering B between stages v_n and s_{n+1}, and as a result, the computation $\Phi_\alpha(B; y)[v_n]$ is preserved and hence

$$f_\alpha^i(y) = \Phi_\alpha(B; y)[v_n] = \Phi_\alpha(B; y)[s_{n+1}] = \Phi_\alpha(C; y)[s_{n+1}].$$

By induction, we have for all $m \geq n$,

$$f_\alpha^i(y) = \Phi_\alpha(B; y)[s_m] = \Phi_\alpha(C; y)[s_m] = \Phi_\alpha(C; y)[v_m] = \Phi_\alpha(B; y)[v_m].$$

Since both $\Phi_\alpha(B; y)$ and $\Phi_\alpha(C; y)$ converge, we have

$$f_\alpha^i(y) = \Phi_\alpha(B; y) = \Phi_\alpha(C; y) = f(y).$$

Note that if $\Phi_\alpha(C)$ is partial, then (O3) may be prevented from opening a gap infinitely many times. In this case, the $\mathcal{N}$-requirement is satisfied vacuously.

Say that $\mathcal{S}_{\alpha,i}$ *requires attention* at an α-expansionary stage s if one of the following holds:

(1) $\mathcal{S}_{\alpha,i}$ is inside a gap.

(2) $\mathcal{S}_{\alpha,i}$ is inside a cogap. There are two subcases:

> **(2A)** There is some $y \in \text{dom}(f_\alpha^i)$ such that $C_s \lceil \varphi_\alpha(C; y)[v] \neq C_v \lceil \varphi_\alpha(C; y)[v]$, where v is the last α-expansionary stage.
>
> **(2B)** $\mathcal{S}_{\alpha,i}$ is ready to open a gap.

In case (2), (2A) has priority higher than (2B). It may happen that (2A) prevents $\mathcal{S}_{\alpha,i}$ from opening a gap (2B) for almost all times. In this case, $\text{dom}(f_\alpha^i)$ is finite, and there is some $y \in \text{dom}(f_\alpha^i)$ with $\Phi_\alpha(C; y) \uparrow$. $\mathcal{S}_{\alpha,i}$ has two outcomes $g_i <_L d_i$, where g denotes the case that $\mathcal{S}_{\alpha,i}$ opens and closes infinitely many gaps during the construction (as a result, $f_{\alpha,i}$ is totally defined), and d denotes that there is some $y \in \text{dom}(f_\alpha^i)$ with $\Phi_\alpha(C; y) \uparrow$.

Say that $\mathcal{S}_{\alpha,i}$ *receives attention* at an α-expansionary stage s as follows if $\mathcal{S}_{\alpha,i}$ requires attention at this stage:

Case 1: (1) happens.

Then close the gap, define $p_\alpha(v) = s$, and initialize all nodes with lower priority. Stop stage s. If C has a change below x (x is defined in (O2) at stage v, where v is the stage at which the gap is opened), then we say that the gap is closed successfully, and declare that $\mathcal{S}_{\alpha,i}$ is satisfied. Otherwise. The gap is closed unsuccessfully.

Case 2: (2A) happens. Then $\mathcal{S}_{\alpha,i}$ has outcome d_i.

Case 3: (2B) happens. Then $\mathcal{S}_{\alpha,i}$ opens a gap, extends the definition of f_α^i according to (O3). $\mathcal{S}_{\alpha,i}$ has outcome g_i.

In the construction, we don't put $\mathcal{S}$ strategies on the priority tree. We just attach the outcomes of $\mathcal{S}$ to α. Thus, α has outcomes

$$g_0 <_L d_0 <_L g_1 <_L d_1 <_L \cdots <_L g_i <_L d_i <_L \cdots <_L d <_L f,$$

which are described as follows:

f. f denotes the case in which there are only finitely many α-expansionary stages *(and hence, α is satisfied trivially)*.

d. d denotes the case in which there are infinitely many α-expansionary stages, and α's substrategies can require attention only finitely many times *(if so,then $\Phi_\alpha(C)$ is not total because (O3) fails for almost all times and hence there is some z such that $\Phi_\alpha(C; z) \uparrow$, $\Phi_\alpha(C)$ is not total.)*

g_i. g_i denotes the case in which the substrategy $\mathcal{S}_{\alpha,i}$ opens (and closes) gaps infinitely often. As described above, $f_{\alpha,i}$ is totally defined.

d_i. d_i denotes the case in which $\mathcal{S}_{\alpha,i}$ can open gaps only finitely often and there is some $y \in \text{dom}(f_{\alpha,i})$ such that $\Phi_\alpha(C; y)$ diverges.

3.3 An $\mathcal{R}$-Strategy

Let β be an $\mathcal{R}_e$-strategy. We use Λ_β and W_β to denote $\Lambda_{e(\beta)}$ and $W_{e(\beta)}$ respectively. The basic idea of an $\mathcal{R}$-strategy is the *isolation strategy* considered by Wu in [10]. To approximate the equation $W_\beta = \Lambda_\beta(B \oplus D)$, define the length of agreement functions at a β-stage as follows:

$l(\beta, s) = \max\{x : (\forall y < x)[W_{\beta,s}(y) = \Lambda_\beta(B \oplus D; y)[s] \downarrow]\}$, and
$m(\beta, s) = \max\{0; l(\beta, t) \mid t < s \text{ is a } \beta\text{-stage}\}$.

A stage s is called a β-expansionary stage if $s = 0$ or s is a β-stage such that $l(\beta, s) > m(\beta, s)$. β defines a partial computable functional Δ_β (we intend to ensure that $\Delta_\beta(B)[s]$ and $W_\beta[s]$ agree on $l(\beta, s)$ at each β-expansionary stage) and once a disagreement between $\Delta_\beta(B)$ and W_β appears, at y say, β will force a disagreement between W_β and $\Lambda_\beta(B \oplus D)$ by recovering a computation $\Lambda_\beta(B \oplus D; y)$ to a previous one. β has three outcomes:

∞: There are infinitely many β-expansionary stages and $\Delta_\beta(B)$ and W_β agree at all β-expansionary stages. *In this case, $\Delta_\beta(B)$ will be defined as a total function, and it computes W_β correctly. This satisfies the $\mathcal{R}_e$ requirement.*

 f: There are only finitely many β-expansionary stages, and hence $W_\beta \neq \Lambda_\beta(B \oplus D)$. *In this case, the $\mathcal{R}_e$ requirement is satisfied vacuously. $\Delta_\beta(B)$ will be defined at only finitely many stages, and hence $\Delta_\beta(B)$ is partial.*

 d: A disagreement between $\Delta_\beta(B)$ and W_β appears and a disagreement between W_β and $\Lambda_\beta(B \oplus D)$ is created and preserved. *In this case, the $\mathcal{R}_e$ requirement is satisfied vacuously.*

These three outcomes have the priority $\infty <_L f <_L d$. As specified in [10], when numbers are removed from D, we also enumerate those stages, at which these numbers are enumerated into D, into B, for the consistency between distinct $\mathcal{R}$-strategies.

3.4 The Interaction between one $\mathcal{R}$-Strategy and one $\mathcal{G}$-Strategy

We now consider the interaction between an $\mathcal{R}$-strategy and a $\mathcal{G}$-strategy. Let β be a $\mathcal{R}_e$-strategy and η be a $\mathcal{G}_{e'}$-strategy with $\eta \subset \beta$. To code K into $B \oplus D \oplus W_\eta$, η may enumerate infinitely many numbers into D, and these enumerations may destroy the disagreements created by β infinitely many times, and β fails to satisfy the $\mathcal{R}_e$ requirement. To solve this problem, β uses the threshold strategy to lift the γ-uses to ensure the computations are clear of the enumeration of γ-uses. That is, when β is firstly visited, it chooses a threshold k_β, and in the remainder of the construction, whenever a number $m \leq k_\beta$ enters K, we reset β by cancelling all the parameters of β, except k_β.

Now β works by running cycles (infinitely), with cycle 0 started first, and each cycle n constructs a functional Δ_β^n correspondingly, and once it finds a difference between $\Delta_\beta(B)$ and W_β, cycle n initiates a new cycle $n + 1$. The cycle n works as follows:

1. Wait for a β-expansionary stage s_0^n.

2. For any $x < l(\beta, s_0^n)$ with $\Delta_\beta^n(B; x)$ undefined currently, define $\Delta_\beta^n(B; x)[s_0^n]$ $= W_{\beta, s_0^n}(x)$ with use $\delta_\beta^n(B; x)[s_0^n] = s_0^n$. Go to step 1, and simultaneously, wait for a β-expansionary stage $s_1^n > s_0^n$ such that $\Delta_\beta^n(B)$ and W_β differ at some y.

3. At stage s_1^n, put $\gamma_\eta(k_\beta)[s_1^n]$ into D. For any $z \leq \gamma_\eta(k_\beta)[s_1^n]$ with $\Theta_{\beta\eta}(B; z) \uparrow$ currently, define $\Theta_{\beta\eta}(B; z)[s_1^n] = W_{\eta, s_1^n}(z)$ with use $\theta_{\beta\eta}(z)[s_1^n] = s_1^n$. Wait for a β-expansionary stage $s_2^n > s_1^n$ such that W_η changes below $\gamma_\eta(k_\beta)[s_1^n]$ (*we want to lift the γ_η-uses by this change instead of putting $\gamma_\eta(k_\beta)[s_1^n]$ into D*). Start cycle $n + 1$.

4. At stage s_2^n, extract all numbers entering D after stage s_0^n out of D to make a disagreement between W_β and $\Lambda_\beta^{B \oplus D}$ at y, and put $s_0^n + 1$ into B to undefine all $\Delta_{\beta'}^i(B; y)$ which is (re)defined between stages s_0^n and s_2^n, where β' is an $\mathcal{R}$-strategy with $\beta' \subset \beta$.

If β runs cycle n to step 4 above, then all $\gamma_\eta(k)$ for $k \geq k_\beta$ will be defined to be greater than $\lambda_\beta(B \oplus D; y)[s_0^n]$.

For the full β-strategy, there are four cases:

g_η: β runs infinitely many different cycles, and each cycle n eventually reaches step (3) and waits at step (3) forever. In this case, $\Theta_{\beta\eta}(B)$ is totally defined and it computes W_η correctly. Hence $\mathcal{G}_{e'}$ is satisfied at β as W_η is computable in B. Note that in this case, the $\mathcal{R}_e$-requirement is not satisfied at β.

(n, ∞): Cycle n goes back from step (2) to step (1) infinitely often. Then there are infinitely many β-expansionary stages, and $\Delta_\beta^n(B)$ and W_β agree at all these stages. In this case, $\Delta_\beta^n(B)$ is total and it computes W_β correctly. In this case, $\mathcal{R}_e$-requirement is satisfied at β, and β enumerates $\gamma_\eta(k_\beta)$ into D only finitely often.

(n, f): There are only finitely many β-expansionary stages. Let s be the largest β-expansionary stage and n be the largest started cycle at this stage. Then after stage s, β will wait at cycle n forever. In this case, as there are only finitely many β-expansionary stages, W_β and $\Lambda_\beta(B \oplus D)$ are not equal. Hence the $\mathcal{R}_e$-requirement is satisfied vacuously, and β enumerates $\gamma_\eta(k_\beta)$ into D only finitely often.

d: There is some cycle n which can reach step 4, then a disagreement between $\Lambda_\beta(B \oplus D)$ and W_β is created and this disagreement will be preserved forever. Hence $\mathcal{R}_e$-requirement is satisfied at β. Again, β enumerates $\gamma_\eta(k_\beta)$ into D only finitely often.

The priority of these outcomes is as follows: $g_\eta <_L (0, \infty) <_L (0, f) <_L \cdots <_L (n, \infty) <_L (n, f) <_L \cdots <_L d$. Note that if β has outcome g_η, then the $\mathcal{R}_e$ is not satisfied at β, and hence a backup strategy β' say, for the $\mathcal{R}_e$-requirement ought to be set below this outcome g_β. As β' knows that $\gamma_\eta(k_\beta)$-use goes to infinity, β' only believes a computation $\Lambda_{\beta'}(B \oplus D; y)[s]$ if $\gamma_\eta(k_\beta)[s] > \lambda_{\beta'}(B \oplus D; y)[s]$. We call such a computation a β'-believable computation. As η's further enumeration into D can not affect the action of β', β' works in the usual way. So in the definition of the length agreement function, we use only believable computations.

3.5   Interactions between one $\mathcal{R}$-strategy and two $\mathcal{G}$-strategies

Let β be an $\mathcal{R}_e$-strategy, η_1, η_2 be $\mathcal{G}_{e_i}$-strategies with $\eta_1 \subset \eta_2 \subset \beta$. We consider the interactions between β and η_1, η_2 and this interaction can be easily generalised to the interaction between β and several $\mathcal{G}$-strategies, while the key ideas are the same.

Again, β is to make its computations clear of all the enumerations of γ_{η_i}-uses into D for $i = 1, 2$. The basic idea is to iterate the threshold strategy described in the previous section to deal with η_2 first, and then η_1. First, as before, β will choose a threshold k_β when it is firstly visited, and then start cycle 0. In general, cycle n works as follows.

1. Wait for a β-expansionary stage s_0^n.
2. For any $y < l(\beta, s_0^n)$ with $\Delta_\beta^n(B; y)$ not defined currently, define $\Delta_\beta^n(B; y)[s_0^n]$ $= W_{\beta, s_0^n}(y)$ with use $\delta_\beta^n(y)[s_0^n] = s_0^n$. Go to step 1, and simultaneously, wait for a β-expansionary stage $s_1^n > s_0^n$ such that $\Delta_\beta^n(B)$ and W_β differ at a number y.
3. At stage s_1^n, choose the least such y as y_n, and we call y_n a potential witness found by cycle n.
3.2. At stage s_1^n, put $\gamma_{\eta_2}(k_\beta)[s_1^n]$ into D. For any $z \le \gamma_{\eta_2}(k_\beta)[s_1^n]$ with $\Theta_{\beta\eta_2}(B; z)$ not defined currently, define $\Theta_{\beta\eta_2}(B; z)[s_1^n] = W_{\eta_2, s_1^n}(z)$ with use $\theta_{\beta\eta_2}(z)[s_1^n]$ $= s_1^n$. Start new cycle $n + 1$, and simultaneously, wait for a β-expansionary stage $s_2^n > s_1^n$ such that W_{η_2} changes below $\gamma_{\eta_2}(k_\beta)[s_1^n]$.
3.1. We say that cycle n pass the threshold k_β for η_2. At stage s_2^n, put $\gamma_{\eta_1}(k_\beta)[s_2^n]$ into D. Note that all the cycles started after stage s_1^n are cancelled. For any $z \le \gamma_{\eta_1}(k_\beta)[s_2^n]$ with $\Theta_{\beta\eta_1}(B; z)$ not defined currently, define $\Theta_{\beta\eta_1}(B; z)[s_2^n]$ $= W_{\eta_1, s_2^n}(z)$ with use $\theta_{\beta\eta_1}(z)[s_2^n] = s_2^n$. Start new cycle $n + 1$, and simultaneously, wait for a β-expansionary stage $s_3^n > s_2^n$ such that W_{η_1} changes below $\gamma_{\eta_1}(k_\beta)[s_2^n]$.
4 At stage s_3^n, cancel all cycles $n' > n$. We say that cycle n pass the threshold k_β for η_1. Extract all the numbers being enumerated into D after stage s_0^n out of D, and put s_0^n into B for the consistency between β and other strategies.

β-strategy now has the following outcomes:

g_2: g_2 denotes the outcome that all cycles reach 3.2, but only finitely many of them reach 3.1. That is, only finitely many cycles pass the threshold k_β for η_2. In this outcome, $\Theta_{\beta\eta_2}(B)$ is total and computes W_{η_2} correctly. The $\mathcal{G}_{e_2}$-requirement is satisfied at β. Note that in this case, β's actions have only finite impact on η_1.

g_1: g_1 denotes the outcome that infinitely many cycles reach 3.1, but none of them reaches 4. That is, no cycle passes the threshold k_β for η_1. In this case, $\Theta_{\beta\eta_1}(B)$ is total and computes W_{η_1} correctly. The $\mathcal{G}_{e_1}$-requirement is satisfied at β. Note that in this case, β's actions have infinite impact on η_2, as $\gamma_{\eta_2}(k_\beta)$ is driven to infinity, and η_2 is injured at β. Below outcome g_1, both $\mathcal{G}_{e_2}$ and $\mathcal{R}_e$ will have back-up strategies.

(n, ∞): Some cycle n goes back from step 2 to step 1 infinitely often. In this case, there is a stage large enough after which cycle n can never reach 3.2, and hence $\Delta_\beta^n(B)$ is total and computes W_β correctly. The $\mathcal{R}_e$-requirement is satisfied at β.

(n, f): After some stage, β waits forever at cycle n at step 1. Again, the $\mathcal{R}_e$-requirement is satisfied at β.

d: This outcome denotes the case that some cycle n reaches to step 4. That is, cycle n passes the threshold k_β for both η_2 and η_1. In this case, $\Lambda_\beta(B \oplus D)$ and W_β differ at y_n, and the computation $\Lambda_\beta(B \oplus D; y_n)$ is preserved forever. Thus $\mathcal{R}_e$-requirement is satisfied at β. In this case, β's actions have only finite impact on η_2 and η_1.

The priority of these outcomes is in what follows $g_1 <_L g_2 <_L (0, \infty) <_L (0, f) <_L \cdots <_L (n, \infty) <_L (n, f) <_L \cdots <_L d$.

Note that numbers are enumerated into D only when 3.2 or 3.1 are reached, at those stages when outcome g_2 or g_1 is true. So below these two outcomes, strategies know that the related γ-uses go to infinite, and hence it is okay for these strategies to use the believable computations. Numbers can be extracted from D only when step 4 is reached, where for the consistency between β and other strategies, a small number is enumerated into B. It is different from [8], where numbers are extracted from D when β takes outcome g_1. Correspondingly, we need to enumerate numbers into B at these stage. It is this subtle difference that allows us to combine the construction of almost universal cupping degrees with the gap-cogap argument to prove Theorem 1.

3.6   Interaction between $\mathcal{N}$-Strategies and $\mathcal{R}$-Strategies

We now consider the interactions between $\mathcal{N}$-strategies and $\mathcal{R}$-strategies. Surprisingly, these two kinds of strategies are consistent, and the interactions between them are quite simple. To see this, let α and β be an $\mathcal{N}$-strategy and an $\mathcal{R}$-strategy respectively.

A nontrivial case is $\alpha^\frown g_i \subseteq \beta$ for some i. In this case, at any β-stage, α opens gap g_i, and at these stages, β either puts numbers into D, or puts numbers into B (i.e. β performs the disagreement argument by extracting numbers from D and putting some number into B simultaneously). The former actions do not have any impact on α, and the latter actions, enumerating numbers into B, are allowed as the gap g_i is open at the current stage. As β acts only when it is visited, β's action does not injure α.

On the other hand, as we have seen in the $\mathcal{R}$-strategy, β acts only when it is visited, to recover a computation to a previous one and ensure that no $\mathcal{G}$-strategy can change it later. That is, when a corresponding W changes below $\gamma(k)$ at stage s_1 say (so $\gamma(k)$ is enumerated into D before stage s_1, at stage s_0 say), this change undefines related γ-uses, which allows us to redefine these uses as big numbers. This new definition of $\Gamma(B, D, W; x)$, $x \geq k$, depends on $\gamma(k)$ in D, and if later, if $\gamma(k)$ is removed from D, this newly defined $\Gamma(B, D, W; x)$ is again undefined, as W has a change below $\gamma(k)$, and $\gamma(k)$ is not in D now.

So once W changes below $\gamma(k)$, β can recover a computation at any later stage. This ensures that β acts only at β-stages, at which the gap g_i is open. So if β is on the true path of the construction, the gap g_i will be open infinitely many times, and hence β can act all these stages.

This completes the description of the basic strategies and possible interactions between them.

References

1. Ambos-Spies, K., Jockusch Jr., C.G., Shore, R.A., Soare, R.I.: An algebraic decomposition of the recursively enumerable degrees and the coincidence of several classes with the promptly simple degrees. Trans. Amer. Math. Soc. 128, 109–128, 537-569 (1984)
2. Arslanov, M.M.: Structural properties of the degrees below $0'$. Dokl. Akad, Nauk SSSR(N. S.) 283, 270–273 (1985)
3. Cooper, S.B., Harrington, L., Lachlan, A.H., Lempp, S., Soare, R.I.: The d.r.e. degrees are not dense. Ann. Pure Appl. Logic 55, 125–151 (1991)
4. Downey, R.: D.r.e. degrees and the nondiamond theorem. Bull. London Math. Soc. 21, 43–50 (1989)
5. Downey, R., Li, A., Wu, G.: Complementing cappable degrees in the difference hierachy. Annals of pure and applied logic 125, 101–118 (2004)
6. Li, A., Song, Y., Wu, G.: Universal cupping degrees. In: Cai, J.-Y., Cooper, S.B., Li, A. (eds.) TAMC 2006. LNCS, vol. 3959, pp. 721–730. Springer, Heidelberg (2006)
7. Li, A., Yi, X.: Cupping the recursively enumerable degrees by d.r.e. degrees. Proc. London Math. Soc. 79, 1–21 (1999)
8. Liu, J., Wu, G.: Degrees with almost universal cupping property. In: Ferreira, F., Löwe, B., Mayordomo, E., Mendes Gomes, L. (eds.) CiE 2010. LNCS, vol. 6158, pp. 266–275. Springer, Heidelberg (2010)
9. Soare, R.I.: Recursively Enumerable Sets and Degrees. Springer, Berlin (1987)
10. Wu, G.: Isolation and lattice embeddings. Journal of Symbolic Logic 67, 1055–1064 (2002)
11. Yates, C.E.M.: A minimal pair of recursively enumerable degrees. J. Symbolic Logic 31, 159–168 (1966)

On the Kolmogorov Complexity of Continuous Real Functions

Amin Farjudian

Division of Computer Science, University of Nottingham Ningbo, 199 Taikang East
Road, Ningbo, 315100, China
Amin.Farjudian@nottingham.edu.cn

Abstract. Kolmogorov complexity was originally defined for finitely-representable objects. Later, the definition was extended to real numbers based on the behaviour of the sequence of Kolmogorov complexities of finitely-representable objects—such as rational numbers—used to approximate them. The idea will be taken further here by extending the definition to *functions* over real numbers. Any real function can be represented as the limit of a sequence of finitely-representable enclosures, such as polynomials with rational coefficients. The asymptotic behaviour of the sequence of Kolmogorov complexities of the enclosures in such a sequence can be considered as a measure of practical suitability of the sequence as the candidate for representation of that real function. Based on that definition, we will prove that for any growth rate imaginable, there are real functions whose Kolmogorov complexities have higher growth rates. In fact, using the concept of *prevalence*, we will prove that 'almost every' real function has such a high-growth Kolmogorov complexity. Moreover, we will present an *asymptotic* bound on the Kolmogorov complexities of total single-valued computable real functions.

1 Introduction

In computer science, *Kolmogorov complexity* of an object provides a measure of the complexity of its *description*, hence the alternative name *descriptive complexity*. In fact, Solomonoff [20,21] was the first to develop the concept, for the purpose of studying the complexity of *finite* objects, such as finite sequences of binary digits [15].

There have been attempts at extending the concept to non-finite objects. Regarding *real numbers*, there are at least two main approaches in the literature: one based on the *real Turing machine* (RTM) of Blum, Shub, and Smale [2]; and another based on an effective setting, such as that of Böhm et al. [3] or the *Type-2 Theory of Effectivity* (TTE) of Weihrauch [23].

A real Turing machine is very similar to an ordinary Turing machine, except that each of its registers is capable of holding the exact value of a real number at any time, and the machine is capable of carrying out arithmetic operations on real numbers in unit time. Montaña and Pardo [16] and Ziegler and Koolen [24]

B. Löwe et al. (Eds.): CiE 2011, LNCS 6735, pp. 81–91, 2011.

study Kolmogorov complexity over sequences of real numbers based on the theory of real Turing machines. This approach is very elegant but too abstract to address the issue of *effective representation* of real numbers.

Type-2 theory of effectivity is another framework for studying computability over real numbers, within which the issue of effective representation is addressed. Hence, it provides a theoretical foundation for *exact real computation* [3,8,7,23]. In this framework, each ideal object (such as non-finitely-representable real numbers, functions, manifolds, etc) is represented as the limit of a sequence of *finitely-representable* approximations.

As an example, consider the case of real numbers. Each real number can be represented as the limit of a sequence of rational numbers. Another choice for a *basis* of representation of real numbers is the set of *arbitrary precision* floating point numbers [18,14,13,12]. In fact, it is in this setting that the Kolmogorov complexity of real numbers has been studied by Cai and Hartmanis [4] and Staiger [22].

In a *sound* and *complete* exact framework, the result of any computation must be provided to within any *accuracy* that is demanded. For instance, in one viable protocol one could represent accuracy using integers, and interpret an accuracy n as 'being within the radius of 2^{-n} of the exact result'. Thus, the concrete indicator of accuracy is arbitrary. Nonetheless, the expectation is that with 'higher' accuracy demanded, one needs more computational resources to satisfy the querying party.

The inspiration for the framework of this paper comes from the work of Cai and Hartmanis [4]. Every real number $x \in [0,1]$ can be approximated by a sequence $\langle x_i \rangle_{i \in \mathbb{N}}$ of arbitrary precision binary floating point numbers in $[0,1]$, such that:

1. each x_n has a binary representation of the form $r_{x_n} = 0.d_1^{x_n} d_2^{x_n} \ldots d_n^{x_n}$ where $\forall i \in \mathbb{N} : d_i^{x_n} \in \{0,1\}$. In other words, each x_n has at least one representation of maximum length $2 + n$. If we ignore the leading "0." part, then we can say that each x_n has a representation of length n.
2. $\forall n \in \mathbb{N} : |x - x_n| < 2^{-n}$.

Of course, for each $n \in \mathbb{N}$, there may be more clever ways of describing x_n other than just writing down its binary expansion. Therefore, for each real number $x \in [0,1]$ and each $n \in \mathbb{N}$, there exists a rational number $x_n \in [0,1]$ which has a *description* of length at most n and satisfies $|x - x_n| < 2^{-n}$. If we let $K(x_n)$ denote the length of the shortest possible description of x_n then $K(x_n) \leq n$. For the real number x, the Kolmogorov complexity $K_{\mathbb{R}}(x)$ can be defined as:

$$K_{\mathbb{R}}(x) := \frac{1}{2} \left(\liminf_{n \to \infty} \frac{K(x_n)}{n} + \limsup_{n \to \infty} \frac{K(x_n)}{n} \right)$$

It should be clear that $\forall x \in [0,1] : 0 \leq K_{\mathbb{R}}(x) \leq 1$. Cai and Hartmanis [4] prove that:

(i) For Lebesgue-almost every x in $[0,1]$, $K_{\mathbb{R}}(x) = 1$.
(ii) For every $t \in [0,1]$, the set $K_{\mathbb{R}}^{-1}(t)$ is uncountable and has Hausdorff dimension t.
(iii) The graph of $K_{\mathbb{R}}$ is a fractal.

We try to address the cost of effective representation of functions by extending the definition of Kolmogorov complexity to the set $C[0,1]$ of continuous real functions from $[0,1]$ to $\mathbb{R}$. *The main result of this paper states that no matter what rate of growth one considers, 'almost all' functions in the Banach space $C[0,1]$ have Kolmogorov complexities with higher growth rates.* As such, this result can be regarded as an extension of item (i) to the case of the function space $C[0,1]$.

Remark 1. Due to lack of space, most of the proofs are omitted from this printed version. However, an extended abstract including all the proofs is available online [9].

2 Kolmogorov Complexity of a Continuous Real Function

By *effective representation* of functions in $C[0,1]$ we mean the representation of each such element as the limit of a (*not necessarily computable*) sequence of finitely-representable objects. *Domain theory* [10,1,8,6] provides a suitable setting for this purpose. Of course, the actual structure that we introduce will not be a domain, but concepts such as approximation will be developed in accordance with domain theory. To start, we use *enclosures* to approximate functions:

Definition 1 ($[f,g]$: function enclosure)

1. For $f,g \in \mathbb{R}^{[0,1]}$ we define the function enclosure $[f,g]$ by

$$[f,g] := \left\{ h \in R^{[0,1]} \,\middle|\, \forall x \in [0,1] : f(x) \le h(x) \le g(x) \right\}$$

   (*Obviously if $\exists x_0 \in [0,1] : g(x) < f(x)$ then $[f,g]$ will be empty.*)
2. The function f (respectively g) is called the **lower boundary** (respectively **upper boundary**) of the enclosure $[f,g]$.
3. Enclosures H_1 and H_2 are said to be **consistent** if $H_1 \cap H_2 \neq \emptyset$. Otherwise, they are said to be **inconsistent**.

Furthermore, tighter enclosures provide better approximations of a function. This is expressed using the *width* of an enclosure:

Definition 2 (w: width operator). *Consider $f,g \in C[0,1]$. For the enclosure $[f,g]$ the width is defined as* $\mathrm{w}([f,g]) := \max\left\{ g(t) - f(t) \,\middle|\, t \in [0,1] \right\}$.

Note that $\mathrm{w}([f,g])$ is well defined as $[0,1]$ is compact and both f and g are assumed to be continuous.

Definition 3 ($\Gamma(h)$: graph of a function enclosure)

Let $h = [f, g]$ be an enclosure. By the graph of h we mean the set of points in $[0, 1] \times \mathbb{R}$ lying between the graphs of its lower and upper boundaries. Set theoretically, this is just the union of all functions in the enclosure, i. e. $\Gamma(h) := \cup h$.

The set of continuous-function enclosures under the reverse inclusion forms a poset:

Definition 4 ($\mathbb{FE}$). *We denote the set of non-empty continuous-function enclosures by $\mathbb{FE}$, i. e. $\mathbb{FE} := \{[f, g] \mid f, g \in C[0, 1], \forall t \in [0, 1] : f(t) \le g(t)\}$. We define an order $\sqsubseteq$ over this set as follows: $\forall h_1, h_2 \in \mathbb{FE} : h_1 \sqsubseteq h_2 \Leftrightarrow h_2 \subseteq h_1$. The pair $(\mathbb{FE}, \sqsubseteq)$ is a partial order which we simply denote by $\mathbb{FE}$.*

An element $h \in \mathbb{FE}$ is maximal if and only if $\mathrm{w}(h) = 0$, in which case $h = [f, f]$, for some $f \in C[0, 1]$. We treat maximal elements as singletons and for simplicity, refer to $[f, f]$ as f. Using this convention one may talk about chains of *function enclosures* that converge to *functions* in $C[0, 1]$. A sequence $\langle [f_i, g_i] \rangle_{i \in \mathbb{N}}$ of enclosures is called a *chain* if $\forall i \in \mathbb{N} : [f_i, g_i] \sqsubseteq [f_{i+1}, g_{i+1}]$. The enclosure $[f, g]$ is said to be the limit of such a chain if $f = \lim_{n \to \infty} f_n$ and $g = \lim_{n \to \infty} g_n$, where the limits are taken with respect to the supremum norm in $C[0, 1]$.

Note that the partial order $\mathbb{FE}$ is not closed under limit of chains, hence it is not *complete*. As an example, take the chain $\{[f_i, g_i] \mid i \in \mathbb{N}\}$ defined (for all $i \in \mathbb{N}$ and $x \in [0, 1]$) by $f_i(x) = \lambda x.0$ and

$$g_i(x) = \begin{cases} x/(i+1) & \text{if} \quad 0 \le x \le 1 - 1/(i+2) \\ (i+1)x - i & \text{if} \quad 1 - 1/(i+2) \le x \le 1 \end{cases}$$

It should be clear that $\forall i \in \mathbb{N} : [f_i, g_i] \sqsubseteq [f_{i+1}, g_{i+1}]$, but 'the limit' of $\langle g_i \rangle_{i \in \mathbb{N}}$ is the non-continuous function

$$x \mapsto \begin{cases} 0 & \text{if} \quad x \in [0, 1) \\ 1 & \text{if} \quad x = 1 \end{cases}$$

Nonetheless, if $\{h_i \mid i \in \mathbb{N}\}$ is a non-descending chain such that $\lim_{i \to \infty} \mathrm{w}(h_i) = 0$, then the chain has a limit in $\mathbb{FE}$.

Basic concepts in function enclosure arithmetic [19,5] will be used in this paper. The intuition behind this arithmetic is very simple as operators on functions mimic operators of ordinary interval arithmetic [17]. For instance, addition can be easily overloaded with function enclosures as in $[f_1, g_1] + [f_2, g_2] := [f_1 + f_2, g_1 + g_2]$. With other operations (such as $\times$, etc), a bit of tweaking is needed, similar to the case of ordinary interval arithmetic.

It is possible to be more practically minded and restrict oneself to a set of *finitely-representable* function enclosures, and still be able to approximate every function in $C[0, 1]$. For instance, one may consider the set of enclosures $[f, g]$ such that f and g are polynomials with rational coefficients.

Of course objects are finitely-representable only relative to some specific language $L \subseteq \Sigma^*$, for some alphabet Σ. To make the presentation easier, we assume that our alphabet is rich enough to include:

- 0 and 1
- symbols for arithmetic operators
- symbols for forming lists, pairs and lambda expressions

These can all be encoded by appropriate Turing machines, and as we study asymptotic behaviour, the inclusion or exclusion of these elements in Σ will not affect our results. Nonetheless, it is crucial that we keep the alphabet *finite*, as we will see from the proof of Proposition 2.

We say that an enclosure h is *approximated* by another enclosure g if g is finitely-representable and $g \sqsubseteq h$.[1] A countable set $\mathcal{B}$ of finitely-representable enclosures is called a *basis* for $\mathbb{FE}$ if each $h \in \mathbb{FE}$ is the limit of a chain $\langle h_i \rangle_{i \in \mathbb{N}}$ of elements in $\mathcal{B}$. Examples of bases are:

- set of polynomials with rational coefficients
- set of piecewise affine functions, where the end-points of each affine piece have rational coordinates

Now let $\mathcal{B}$ be a basis for $\mathbb{FE}$. We call $\rho : \mathbb{N} \to \mathcal{B}$ a *binary* representation of $f \in C[0,1]$ if $\forall n \in \mathbb{N} : \rho(n) \sqsubseteq [f, f] \wedge \mathrm{w}(\rho(n)) < 2^{-n}$.

Remark 2. All our results can be easily generalised to any base. Thus, to save space, we will refer to binary representations simply as representations.

Notation 1. *We reserve the notation $K(x)$ to denote the Kolmogorov complexity of any finitely-representable object x. This includes objects such as integer numbers, finite strings, arbitrary-precision floating-point numbers, finitely-representable functions enclosures, etc.*

Proposition 1. *For every function $f \in C[0,1]$ there exists a representation $\hat{f}$ of f of minimal Kolmogorov complexity, i.e. for any other representation of f such as $\rho : \mathbb{N} \to \mathcal{B}$:*

$$\forall n \in \mathbb{N} : K(\hat{f}(n)) \leq K(\rho(n))$$

Proof. Assume that $\mathcal{B}$ is enumerated as in $\langle h_i \rangle_{i \in \mathbb{N}}$. In order to define $\hat{f}$ over a certain $n \in \mathbb{N}$, first let X_n be those elements of $\mathcal{B}$ that approximate f and have width smaller than 2^{-n}, i.e. $X_n := \{h \in \mathcal{B} \mid h \sqsubseteq [f, f], \mathrm{w}(h) < 2^{-n}\}$ and consider the set $K(X_n) = \{K(h) \mid h \in X_n\}$ which is the image of X_n under K. This is a subset of $\mathbb{N}$, hence it has a least element n_0. All that remains is to *use the axiom of choice* and assign to $\hat{f}(n)$ one of the elements in $K^{-1}(n_0)$, or without the axiom of choice we can choose the element in $K^{-1}(n_0)$ with the smallest index in the enumeration $\langle h_i \rangle_{i \in \mathbb{N}}$ of $\mathcal{B}$. □

---

[1] Note that our definition of approximation is different from the *way-below* relation used in domain theory [1, Definition 2.2.1]. In particular, in our framework any finitely-representable enclosure g approximates itself.

Definition 5 (optimal representation). *The representation $\hat{f}$ as in Proposition 1 is called an* optimal *representation of the function f.*

Definition 6 ($K_C(f)$). *Let $\hat{f} : \mathbb{N} \to \mathcal{B}$ be an optimal representation of $f \in C[0,1]$. The* Kolmogorov complexity function *of f is defined as:*

$$K_C(f) : \mathbb{N} \to \mathbb{N}$$
$$n \mapsto K(\hat{f}(n))$$

We consider the asymptotic growth of $K_C(f)$ for functions in $C[0,1]$. This way we can study the set of functions f for which $K_C(f)$ is bounded in some way, e. g. by some polynomial, exponential, etc.

We first prove in Proposition 2 that there are functions in $C[0,1]$ with arbitrarily fast-growth Kolmogorov complexity. In the proof we seem to go to some length to manufacture 'one' such function. But, then—as is usually the case in mathematics—in Section 5 we will show that in fact 'almost all' functions in $C[0,1]$ have this property.

Proposition 2. *For any given $\theta : \mathbb{N} \to \mathbb{N}$, there exists a function f in $C[0,1]$ whose Kolmogorov complexity is above θ over infinitely many points. In other words:*

$$\forall m \in \mathbb{N} : \exists n \geq m : K_C(f)(n) \geq \theta(n)$$

Proof. See [9]. □

For instance, in Proposition 2, by taking $\theta(n)$ to be:

1. 2^n, one can show that there exists a real function $f \in C[0,1]$ whose Kolmogorov complexity $K_C(f)$ is not dominated by any polynomial.
2. $n!$, one can show that there exists a real function $f \in C[0,1]$ whose Kolmogorov complexity $K_C(f)$ is not dominated by any exponential function.

3 Invariant Ideals

Consider the poset $(\mathbb{N}^\mathbb{N}, \preceq)$ in which $\preceq$ is the pointwise ordering on functions: $f \preceq g \Leftrightarrow \forall n \in \mathbb{N} : f(n) \leq g(n)$, and define the operators $\vee, \wedge : \mathbb{N}^\mathbb{N} \times \mathbb{N}^\mathbb{N} \to \mathbb{N}^\mathbb{N}$ by $(f \vee g)(n) = \max(f(n), g(n))$ and $(f \wedge g)(n) = \min(f(n), g(n))$. This way one obtains a lattice, in which certain ideals are of interest to our discussion. For instance, the ideal of all functions smaller than some polynomial or exponential function. Remember that the principal ideal $\downarrow f$ is defined as $\downarrow f := \{g \in \mathbb{N}^\mathbb{N} \mid g \preceq f\}$.

Definition 7 (invariant ideal). *We call a set $\mathcal{U} \subseteq \mathbb{N}^\mathbb{N}$ a translation invariant proper ideal—*invariant ideal, *to be brief—of $\mathbb{N}^\mathbb{N}$ if*

1. *$\mathcal{U}$ contains the identity function: $\lambda n.n \in \mathcal{U}$.*
2. *$\mathcal{U}$ is a lower set, i. e. $\forall f \in \mathcal{U}, g \in \mathbb{N}^\mathbb{N} : g \preceq f \Rightarrow g \in \mathcal{U}$*
3. *$\mathcal{U}$ is closed under addition*
4. *$\mathcal{U}$ is closed under translation, i. e. $\forall f \in \mathcal{U}, k \in \mathbb{N} : (\lambda n.f(n+k)) \in \mathcal{U}$.*

5. *There exists a countable set of functions B such that $\mathcal{U} = \cup\{\downarrow f \mid f \in B\}$. Such a set B will be referred to as a* basis *for $\mathcal{U}$.*

Proposition 3. *For any invariant ideal $\mathcal{U}$, the following are true:*

(i) $\mathcal{U} \neq \emptyset \wedge \mathcal{U} \neq \mathbb{N}^{\mathbb{N}}$.

(ii) $\mathcal{U}$ includes every constant function $\lambda n.p$ and affine function $\lambda n.pn + q$, for every $p, q \in \mathbb{N}$.

Proof

(i) $\mathcal{U} \neq \emptyset$ because it contains identity. Now consider a countable basis $B = \{f_0, f_1, \ldots\}$ for $\mathcal{U}$ and define the function $h : \mathbb{N} \to \mathbb{N}$ by $h(n) = f_n(n) + 1$. Then $h \notin \mathcal{U}$.

(ii) Let p and q be natural numbers. The invariant ideal $\mathcal{U}$ contains the identity and is closed under translation, therefore $\lambda n.n + p \in \mathcal{U}$. Moreover, $\mathcal{U}$ is a lower set, thus $\lambda n.p \in \mathcal{U}$.

As $\mathcal{U}$ is closed under addition, adding identity p times to itself would give $\lambda n.pn \in \mathcal{U}$, and by adding the constant function $\lambda n.q$, we obtain $\lambda n.pn + q \in \mathcal{U}$. $\qquad\square$

From now on we reserve the symbol $\mathcal{U}$ to denote an invariant ideal. Examples of bases for invariant ideals are the set of polynomials with integer coefficients, or the set of exponential functions with integer parameters.

For each invariant ideal, the set $K_C^{-1}(\mathcal{U})$ consists of those functions in $C[0,1]$ whose Kolmogorov complexity function $K_C(f)$ is a member of $\mathcal{U}$.

Proposition 4. *For any invariant ideal $\mathcal{U}$ the following are true:*

(a) the set $K_C^{-1}(\mathcal{U})$ is closed under arithmetic operations. (Note that for division $h = f/g$, we require that g be nowhere zero, i. e. we do not allow division by zero.)

(b) the set $K_C^{-1}(\mathcal{U})$ is an F_σ set, i. e. it is the union of a countable family of closed sets.

(c) the set $K_C^{-1}(\mathcal{U})$ is Borel.

(d) let $\mathcal{U}^c$ denote the complement of $\mathcal{U}$, then:

$$\forall f \in K_C^{-1}(\mathcal{U}^c), \ g \in K_C^{-1}(\mathcal{U}), \ \tau \in \mathbb{R} \setminus \{0\} : (\tau f + g) \in K_C^{-1}(\mathcal{U}^c)$$

Proof. See [9]. $\qquad\square$

4 Prevalence and Shyness

Consider a topological real vector space V and let Φ be a predicate defined over V with support $S \subseteq V$, i.e. $\forall s \in V : (\Phi(s) \Leftrightarrow s \in S)$. In the case $\dim(V) = k < \infty$, one may use the k-dimensional Lebesgue measure Λ_k in order to express statements such as 'Φ holds almost everywhere in V', which would mean $\Lambda_k(V \setminus S) = 0$.

Now let V be an infinite-dimensional, separable[2] Banach space. Over such a space, we do not have any measure with properties similar to that of Lebesgue measure over Euclidean spaces. To be more precise: "any translation-invariant measure μ over V which is not identically zero has the property that all open sets have infinite measure" [11, page 2].

A translation-invariant alternative for 'almost every' in such infinite-dimensional spaces is 'prevalence', as introduced by Hunt et al. [11]. In other words, when V is infinite-dimensional, the statement 'S is a prevalent subset of V' gives us the same quality of information as the statement 'Φ holds almost everywhere in V' would give, were V finite-dimensional.

Here we use the (slightly less general variant of the) definition of prevalence which is based on the concept of a *probe*:

Definition 8 (probe). *Let $S \subseteq C[0,1]$. A finite-dimensional subspace P of $C[0,1]$ is said to be a probe for S if for all $f \in C[0,1]$, Lebesgue-almost-every point in the hyperplane $f + P$ belongs to S.*

For our purposes, prevalence means having a Borel subset with a probe:[3]

Definition 9 (prevalent/shy sets). *Let $S \subseteq C[0,1]$:*

1. *if S is a Borel set, then S is said to be prevalent if it has a probe.*
2. *if S is not a Borel set, then S is said to be prevalent provided it has a Borel subset $S' \subseteq S$ that has a probe.*
3. *S is said to be shy if and only if its complement in $C[0,1]$ is prevalent.*

5   Main Theorem

Consider an invariant ideal $\mathcal{U}$, with $B = \{\phi_i \mid i \in \mathbb{N}\}$ as its basis. Define the function $\theta : \mathbb{N} \to \mathbb{N}$ as:

$$\theta(n) := \begin{cases} 1 + \phi_0(0) & \text{if } n = 0 \\ 1 + \max(\{\theta(j) \mid 0 \le j \le n-1\} \cup \{\phi_k(n) \mid 0 \le k \le n\}) & \text{if } n \ge 1 \end{cases} \tag{1}$$

This way we get:

$$\forall m, n \in \mathbb{N} : n \ge m \Rightarrow \phi_m(n) < \theta(n) \tag{2}$$

Now consider the function f obtained by applying Proposition 2 over this θ. The Kolmogorov complexity of f dominates θ, and by implication—using (2)—is strictly greater than each ϕ_m, over infinitely many points. This means that f belongs to $K_C^{-1}(\mathcal{U}^c)$, and therefore:

---

[2] A topological space is called *separable* if it contains a countable dense subset.
[3] Definition 9 is indeed less general than the one given in [11], according to which a Borel set may be prevalent without having a probe.

Lemma 1. $K_C^{-1}(\mathcal{U}^c) \neq \emptyset$.

In fact, we will prove that for any invariant ideal $\mathcal{U}$, 'almost every' function in $C[0,1]$ belongs to $K_C^{-1}(\mathcal{U}^c)$, where $\mathcal{U}^c$ is the set complement of $\mathcal{U}$ in $\mathbb{N}^{\mathbb{N}}$:

Theorem 1. *For any invariant ideal $\mathcal{U}$, the set $K_C^{-1}(\mathcal{U}^c)$ is a prevalent subset of $C[0,1]$.*

Proof. See [9]. □

6 An Asymptotic Bound on the Kolmogorov Complexities of Computable Functions

Let us denote the set of all computable functions in $C[0,1]$ by $\tilde{C}[0,1]$ and assume that this set is enumerated as a sequence $\langle f_i \rangle_{i \in \mathbb{N}}$, with their respective Kolmogorov complexities enumerated as $\langle \phi_i \rangle_{i \in \mathbb{N}}$. We define θ as in (1) and (using Proposition 2) obtain a function f whose Kolmogorov complexity strictly dominates the Kolmogorov complexity of every computable f_i, over infinitely many points. In fact, we can go further: let $\mathcal{U}_\theta$ be the smallest invariant ideal that includes θ. It can be proved in the usual way that such an invariant ideal does exist. The set $K_C^{-1}(\mathcal{U}_\theta)$ is a shy subset of $C[0,1]$ which includes $\tilde{C}[0,1]$.

Just by using the closure properties as demanded by the definition of an invariant ideal, one can show that for *every* $f \in K_C^{-1}(\mathcal{U}_\theta{}^c)$ and *every* $f_i \in \tilde{C}[0,1]$, there exists an infinite set $J \subseteq \mathbb{N}$ such that $\forall j \in J : \phi_i(j) < K_C(f)(j)$:

Theorem 2. *The Kolmogorov complexity function of every $f \in K_C^{-1}(\mathcal{U}_\theta{}^c)$ is an asymptotic upper bound for the Kolmogorov complexity function of any computable function g in $\tilde{C}[0,1]$, i.e. there exists an infinite set $J \subseteq \mathbb{N}$ such that $\forall j \in J : K_C(g)(j) \leq K_C(f)(j)$.*

7 Summary and Discussion

We have defined a notion of Kolmogorov complexity for functions in $C[0,1]$ by drawing inspiration from the work of Cai and Hartmanis [4]. Essentially, we take into account the representation of an infinite object—such as a real number or a real function—as the limit of a sequence of finite approximations, and then base the definition of its Kolmogorov complexity on the growth rate of the descriptive complexities of those finite approximations.

Although the material is about a subject almost exclusive to computer science, the space $C[0,1]$ over which we studied the concept of Kolmogorov complexity includes both computable and non-computable functions. Therefore, it is not possible to interpret this result in a purely Turing-computable framework. Yet, it touches upon the issue of representation, and one could argue that ease of representation for an object is a desirable feature. In this respect, Theorem 1 can be interpreted as stating that 'almost all' real functions 'are hard to represent effectively'.

Moreover, in light of this result, even if all of $C[0,1]$ is supplied to us by an oracle, computations involving almost all real functions are infeasible, as long as we are restricted by finitely-representable approximations.

8   Future Work

There is no computable operator which takes an $f \in C[0,1]$ as input and returns $K_C(f)$, although upper bounds can be provided. Therefore, even though in the computable sense $K_C(f)$ is inaccessible, it will be interesting to see whether the nature of $K_C(f)$ (or any of its upper bounds) reveals anything about the analytic/algebraic properties of f.

Acknowledgements. I am very grateful to Jan Duracz for his proof-reading and helpful suggestions.

References

1. Abramsky, S., Jung, A.: Domain theory. In: Abramsky, S., Gabbay, D.M., Maibaum, T.S.E. (eds.) Handbook of Logic in Computer Science, vol. 3, pp. 1–168. Clarendon Press, Oxford, Oxford (1994)
2. Blum, L., Shub, M., Smale, S.: On a theory of computation and complexity over the real numbers; NP completeness, recursive functions and universal machines. Bulletin of the American Mathematical Society (new series) 21(1), 1–46 (1989)
3. Böhm, H.J., Cartwright, R., Riggle, M., O'Donnell, M.J.: Exact real arithmetic: A case study in higher order programming. In: Proceedings of the 1986 ACM Conference on LISP and Functional Programming, pp. 162–173. ACM, New York (1986); held at MIT, Cambridge, MA
4. Cai, J., Hartmanis, J.: On Hausdorff and topological dimensions of the Kolmogorov complexity of the real line. Journal of Computer and System Sciences 49(3), 605–619 (1994)
5. Duracz, J.A., Končný, M.: Polynomial function enclosures and floating point software verification. In: CFV 2008, Sydney, Australia (August 2008)
6. Edalat, A., Lieutier, A.: Domain theory and differential calculus (functions of one variable). In: Proceedings of 17th Annual IEEE Symposium on Logic in Computer Science (LICS 2002), Copenhagen, Denmark, pp. 277–286 (2002)
7. Edalat, A., Sünderhauf, P.: A domain theoretic approach to computability on the real line. Theoretical Computer Science 210, 73–98 (1999)
8. Escardó, M.H.: PCF extended with real numbers: a domain theoretic approach to higher order exact real number computation. Ph.D. thesis, Imperial College (1997)
9. Farjudian, A.: On the Kolmogorov complexity of continuous real functions (2011), an extended abstract available at, http://www.cs.nott.ac.uk/~avf/AuxFiles/2011-Farjudian-Kolmogorov-Real-Fun.pdf
10. Gierz, G., Hofmann, K.H., Keimel, K., Lawson, J.D., Mislove, M.W., Scott, D.S.: Continuous Lattices and Domains. In: Encycloedia of Mathematics and its Applications, vol. 93, Cambridge University Press, Cambridge (2003)
11. Hunt, B.R., Sauer, T., Yorke, J.A.: Prevalence: A translation-invariant "almost every" on infinite-dimensional spaces. Bulletin of the American Mathematical Society 27(2), 217–238 (1992)
12. Končný, M., Farjudian, A.: Compositional semantics of dataflow networks with query-driven communication of exact values. Journal of Universal Computer Science 16(18), 2629–2656 (2010)
13. Končný, M., Farjudian, A.: Semantics of query-driven communication of exact values. Journal of Universal Computer Science 16(18), 2597–2628 (2010)

14. Lambov, B.: Reallib: An efficient implementation of exact real arithmetic. Mathematical Structures in Computer Science 17(1), 81–98 (2007)
15. Li, M., Vitányi, P.: An Introduction to Kolmogorov Complexity and Its Applications. Springer, Heidelberg (1997)
16. Montaña, J.L., Pardo, L.M.: On Kolmogorov complexity in the real Turing machine setting. Information Processing Letters 67, 81–86 (1998)
17. Moore, R.E.: Interval Analysis. Prentice-Hall, Englewood Cliffs (1966)
18. Müller, N.T.: The iRRAM: Exact arithmetic in C++. In: Blank, J., Brattka, V., Hertling, P. (eds.) CCA 2000. LNCS, vol. 2064, pp. 222–252. Springer, Heidelberg (2001)
19. Neumaier, A.: Taylor forms - use and limits. Reliable Computing 9(1), 43–79 (2003)
20. Solomonoff, R.: A formal theory of inductive inference. Information and Control 7(1), 1–22 (1964)
21. Solomonoff, R.: A formal theory of inductive inference. Information and Control 7(2), 224–254 (1964)
22. Staiger, L.: The Kolmogorov complexity of real numbers. Theoretical Computer Science 284(2), 455–466 (2002)
23. Weihrauch, K.: Computable Analysis, An Introduction. Springer, Heidelberg (2000)
24. Ziegler, M., Koolen, W.M.: Kolmogorov complexity theory over the reals. Electronic Notes in Theoretical Computer Science 221, 153–169 (2008); Proceedings of the Fifth International Conference on Computability and Complexity in Analysis (CCA 2008)

Defining Languages by Forbidding-Enforcing Systems

Daniela Genova

Department of Mathematics and Statistics, University of North Florida,
Jacksonville, FL 32224, USA
d.genova@unf.edu

Abstract. Motivated by biomolecular computing, forbidding-enforcing
systems (fe-systems) were first used to define classes of languages (fe-
families) based on boundary conditions. This paper presents a new model
of fe-systems in which fe-systems define single languages (fe-languages)
based on forbidden and enforced subwords. The paper characterizes well-
known languages by fe-systems, investigates the relationship between fe-
families and fe-languages, and describes how an fe-system can generate
the solution to the k-colorability problem and model splicing.

Keywords: fe-languages, fe-families, formal languages, k-colorability problem, biomolecular computing, splicing.

1 Introduction

The rapid growth of biomolecular computing is interconnected with the quest
for new ways to define computation. Many computational models were defined
within the field of natural computing, e.g., self-assembly, splicing, membrane
systems (see [9,11,12,14,16]). Most of these models are rooted in classical formal
language theory. When computation is carried out by biomolecules, the nondeterministic behavior of molecules in a biochemical reaction inspires new nondeterministic ways of defining languages. Motivated by such non-determinism,
and by abstracting molecules to strings and sets of molecules to languages,
the authors of [2,4,15] introduced forbidding-enforcing systems (fe-systems) as
language-defining systems, where "everything that is not forbidden is allowed",
contrasting the determinism of grammars and automata where "everything that
is not allowed is forbidden". They showed that fe-systems can generate solutions to computational problems such as SAT and Hamiltonian Path Problem,
represent duplex DNA molecules and model splicing by an enzyme.

Forbidding-enforcing systems have been defined in the framework of membrane systems (see [1]), where the authors show that the additional restrictions
imposed by membranes cause fe-systems to define yet other new classes of languages, and have been also defined to model self-assembly of graphs (see [5]).

A topological investigation of fe-families can be found in [8]. Such a study
of formal languages comes to interest only with the introduction of fe-systems,

B. Löwe et al. (Eds.): CiE 2011, LNCS 6735, pp. 92–101, 2011.
© Springer-Verlag Berlin Heidelberg 2011

since, as it is shown in [8], none of the Chomsky families of languages correspond to an open or a closed set in the defined metric space.

This paper uses the concept of forbidding and enforcing to present a new way of defining languages, where one forbidding-enforcing system defines one language instead of a family of languages. Such a motivation comes from laboratory setting, where after a "wet" computation is performed the solution will be a specific language of words (DNA molecules) over the DNA alphabet.

The paper is organized as follows. Section 2 defines fe-languages and their basic properties are discussed in Section 3. The relationship between fe-languages and fe-families is investigated in Section 4. Section 5 provides characterizations of well-known languages by fe-systems. Section 6 discusses forbidding through enforcing. Generating solutions to the k-colorability problem is discussed in Section 7 and Section 8 discusses modeling splicing by fe-systems.

2   Forbidding-Enforcing Systems

A finite set of symbols (*alphabet*) is denoted by A and the free monoid consisting of all words over A is denoted by A^*. A subset of A^* is called a *language*. The *length* of a word $w \in A^*$ is denoted by $|w|$ and A^m is the set of all words of length m, whereas $A^{\leqslant m}$ is the set of all words of length at most m. The empty word, denoted by λ has length 0. The language A^+ consists of all words over A with positive length.

The word $y \in A^*$ is a *factor* (*subword*) of $x \in A^*$, if there exist $s, t \in A^*$, such that $x = syt$. The set of subwords of a word x is denoted by $sub\,(x)$ and the set of subwords of a language L by $sub\,(L)$, where $sub\,(L) = \cup_{x \in L}\, sub\,(x)$.

This paper uses the definitions and notation for fe-families as in [4] and for maximal languages as introduced in [8]. For more details about properties of fe-systems defining fe-families of languages, the reader is referred to [4,2,3,15,6,8]. Where not explicitly stated, assume that an alphabet A is given.

2.1   Forbidding Systems, f-languages

This paper introduces an fe-systems model, in which one forbidding-enforcing system defines a single language as opposed to a family of languages. For the single language model, only the definition of a forbidding set (Definition 1) remains the same as in [4].

Definition 1. A *forbidding set* $\mathcal{F}$ is a family of finite nonempty subsets of A^+; each element of a forbidding set is called a *forbidder*.

Definition 2. A word w is *consistent with a forbidder* F, denoted by $w\,con\,F$, if $F \not\subseteq sub\,(w)$. A word w is *consistent with a forbidding set* $\mathcal{F}$ denoted by $w\,con\,\mathcal{F}$, if $w\,con\,F$ for all $F \in \mathcal{F}$. If w is not consistent with $\mathcal{F}$, the notation is $w\,ncon\,\mathcal{F}$. The language $L(\mathcal{F}) = \{w \mid w\,con\,\mathcal{F}\}$ is said to be *defined* by the forbidding set $\mathcal{F}$. A language L is a *forbidding language* or *f-language*, if there is a forbidding set $\mathcal{F}$ such that $L = L(\mathcal{F})$.

Example 1. Let $\mathcal{F} = \{\{ab, ba\}, \{aa, bb\}\}$. Then $L(\mathcal{F}) = \{a^n, b^n, ab^n, a^nb, ba^n, b^na \mid n \geq 0\}$.

Example 2. Let $A = \{a, b\}$ and $\mathcal{F} = \{\{b\}\}$. Then $L(\mathcal{F}) = a^*$.

Example 3. Let $A = \{a, b\}$ and $\mathcal{F} = \{\{bb\}\}$. Then $L(\mathcal{F})$ contains words where any two b's are separated by at least one a. Note that $a^* \subseteq L(\mathcal{F})$.

The first part of the following remark simply says that if nothing is forbidden, then everything is allowed.

Remark 1. 1. $L(\mathcal{F}) = A^*$ if and only if $\mathcal{F}$ is empty.
 2. The empty word λ is in $L(\mathcal{F})$ for every $\mathcal{F}$.

2.2   Enforcing Systems, e-languages

Definition 3. An *enforcing set* $\mathcal{E}$ is a family of ordered pairs called *enforcers* (x, Y), such that $x \in A^*$ and $Y = \{y_1, \ldots, y_n\}$ where $y_i \in A^+$ for $i = 1, \ldots, n$, $x \in sub(y_i)$ and $x \neq y_i$ for every $y_i \in Y$. A word w *satisfies an enforcer* (x, Y) $(w \, sat \, (X, Y))$, if $w = uxv$ for some $u, v \in A^*$ implies that there exists $y_i \in Y$ and $u_1, u_2, v_1, v_2 \in A^*$ such that $y_i = u_2xv_2$ and $w = u_1u_2xv_2v_1$. A word w *satisfies an enforcing set* $\mathcal{E}$ $(w \, sat \, \mathcal{E})$, if w satisfies every enforcer in that set. If w does not satisfy $\mathcal{E}$, the notation is $w \, nsat \, \mathcal{E}$. For an enforcing set $\mathcal{E}$ the language of all words that satisfy it is denoted by $L(\mathcal{E})$. A language L is called an *e-language* if there exists an enforcing set $\mathcal{E}$ such that $L = L(\mathcal{E})$.

In the case that $x \notin sub(w)$, w is said to satisfy the enforcer trivially. We call enforcers in which $x = \lambda$ *brute*. In this case, a word from Y has to be a subword of w in order for w to satisfy the enforcer.

Remark 2. If $y \in Y$ then $y \, sat \, (x, Y)$.

Remark 3. $L(\mathcal{E}) = A^*$ if and only if $\mathcal{E} = \emptyset$.

An enforcer (x, Y) is called *strict* if $|Y| = 1$. The following example shows how a brute strict enforcer and an infinite set of strict enforcers may not be satisfied by any finite word.

Example 4. Let $\mathcal{E} = \{(\lambda, \{a\})\} \cup \{(a^i, \{a^{i+1}\}) \mid i \geq 1\}$. Then, $L(\mathcal{E}) = \emptyset$.

2.3   Forbidding-Enforcing Systems, fe-languages

Preserving the idea of a forbidding-enforcing system from [4], an analogous definition for a forbidding-enforcing language (fe-language) is presented.

Definition 4. A *forbidding-enforcing system* is an ordered pair $(\mathcal{F}, \mathcal{E})$, such that $\mathcal{F}$ is a forbidding set and $\mathcal{E}$ is an enforcing set. The language $L(\mathcal{F}, \mathcal{E})$ defined by this system consists of all words that are consistent with $\mathcal{F}$ and satisfy $\mathcal{E}$, i.e., $L(\mathcal{F}, \mathcal{E}) = L(\mathcal{F}) \cap L(\mathcal{E})$. A language L is called an *fe-language*, if there exists an fe-system $(\mathcal{F}, \mathcal{E})$, such that $L = L(\mathcal{F}, \mathcal{E})$.

Example 5. 1. Let $\mathcal{F} = \{\{ba\}\}$ and $\mathcal{E}_1 = \{(\lambda, \{a\})\} \cup \{(a^i, \{a^{i+1}, a^i b^i\}) \mid i \geq 1\}$. Then, $L_1 = L(\mathcal{F}, \mathcal{E}_1) = \{a^n b^m \mid n \leq m \text{ and } n, m \geq 1\}$.

2. Let $\mathcal{F} = \{\{ba\}\}$ and $\mathcal{E}_2 = \{(\lambda, \{b\})\} \cup \{(b^i, \{b^{i+1}, a^i b^i\}) \mid i \geq 1\}$. Then, $L_2 = L(\mathcal{F}, \mathcal{E}_2) = \{a^n b^m \mid n \geq m \text{ and } n, m \geq 1\}$.

3 Basic Properties of Forbidding-Enforcing Systems

The proposition below includes some immediate properties of forbidding and enforcing sets and the languages that they define. These properties are reminiscent of the ones proved for fe-families of languages [4,15].

Proposition 1. *Let $\mathcal{F}$ and $\mathcal{F}'$ be forbidding sets, $\mathcal{E}$ and $\mathcal{E}'$ be enforcing sets, and u and w be words.*

1. *If $u \in sub(w)$ and $w \, con \, \mathcal{F}$, then $u \, con \, \mathcal{F}$.*
2. *If $\mathcal{F}' \subseteq \mathcal{F}$, then $L(\mathcal{F}) \subseteq L(\mathcal{F}')$.*
3. *If $\mathcal{E}' \subseteq \mathcal{E}$, then $L(\mathcal{E}) \subseteq L(\mathcal{E}')$.*
4. *If $\mathcal{F}' \subseteq \mathcal{F}$ and $\mathcal{E}' \subseteq \mathcal{E}$, then $L(\mathcal{F}, \mathcal{E}) \subseteq L(\mathcal{F}', \mathcal{E}')$.*
5. *$L(\mathcal{F} \cup \mathcal{F}') = L(\mathcal{F}) \cap L(\mathcal{F}')$.*
6. *$L(\mathcal{E} \cup \mathcal{E}') = L(\mathcal{E}) \cap L(\mathcal{E}')$.*
7. *$L(\mathcal{F} \cup \mathcal{F}', \mathcal{E} \cup \mathcal{E}') = L(\mathcal{F}, \mathcal{E}) \cap L(\mathcal{F}', \mathcal{E}')$.*

Example 6. Consider the fe-systems from example 5. It follows from Property 7 above that $L = L_1 \cap L_2 = \{a^n b^n \mid n \geq 1\} = L(\mathcal{F}, \mathcal{E}_1 \cup \mathcal{E}_2)$.

4 Relationship between f-families and f-languages

Given an alphabet A consider a generating tree T_{A^*}. The root of the tree is λ. If u is a node in the tree, then ua is a child of u for all $a \in A$. Clearly, this tree contains all words in A^*. For more details on properties of T_{A^*} see [6]. It is obvious that for every language $L \subseteq A^*$, if the vertices that are not in L are removed from T_{A^*}, then the vertices of the resulting graph are precisely the words in L, but the resulting graph is not necessarily a tree. The resulting graphs for factorial languages, i.e. those for which $L = sub(L)$ however, are trees, as observed in the next section.

Consider again the forbidding set $\mathcal{F} = \{\{aa, bb\}, \{ab, ba\}\}$. It was discussed in [4,2,15,6,8], where it was used to define a family of languages. In [8] it was observed that $\mathcal{L}(\mathcal{F})$ has four maximal languages. Note that $L(\mathcal{F})$ from Example 1, is precisely the union of these four languages. The next theorem states that this is always the case. Figure 1 depicts the generating tree for $L(\mathcal{F})$.

Theorem 1. *Let $\mathcal{F}$ be a forbidding set. Let $\mathcal{M}(\mathcal{F})$ be the set of maximal languages for $\mathcal{F}$ and $L(\mathcal{F})$ the $\mathcal{F}$-language. Then, $L(\mathcal{F}) = \cup_{L \in \mathcal{M}(\mathcal{F})} L$.*

Proof. Let $\mathcal{F}$ be given. Consider $w \in L(\mathcal{F})$. Since $F \not\subseteq sub(w)$ for every $F \in \mathcal{F}$, w is in some language $K \in \mathcal{L}(\mathcal{F})$. Hence, w is in some maximal language L in $\mathcal{M}(\mathcal{F})$. It follows that $L(\mathcal{F}) \subseteq \cup_{L \in \mathcal{M}(\mathcal{F})} L$. Conversely, let $w \in \cup_{L \in \mathcal{M}(\mathcal{F})} L$. Then, $w \in L$ for some $L \in \mathcal{M}(\mathcal{F})$. It follows that for every $F \in \mathcal{F}$, $F \not\subseteq sub(w)$. Thus, $w \in L(\mathcal{F})$. □

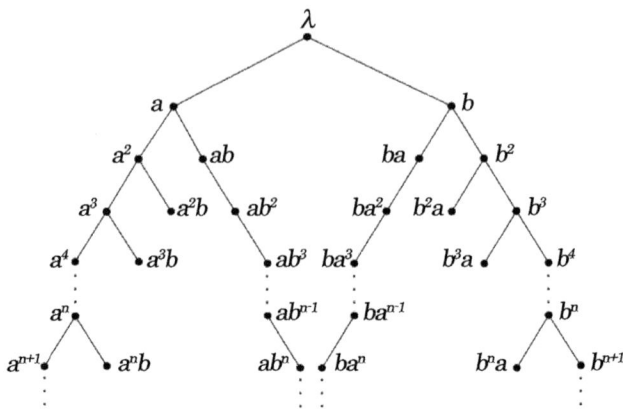

Fig. 1. The tree representing $L(\mathcal{F})$ from Example 1

5   Characterizing Formal Languages by Forbidding-Enforcing Systems

This section presents some characterizations of formal languages by fe-systems. It begins with the straightforward characterization of local languages.

Proposition 2. *L is local if and only if there is* $\mathcal{F} = \{\{v_1\}, \ldots, \{v_n\}\}$, *such that* $L = L(\mathcal{F})$.

Proof. A language L is local if and only if there exists a finite set of words $H = \{v_1, \ldots, v_n\}$ such that $L = A^* \backslash A^* H A^*$, which holds if and only if no word from L has a subword from H, i.e., if and only if $L = L(\mathcal{F})$, where $\mathcal{F} = \{\{v_1\}, \ldots, \{v_n\}\}$. □

The following proposition states that not all languages are f-languages.

Proposition 3. *There exists a language that is not a forbidding language.*

Proof. Let $L = \{a, ba\}$. Observe that the word b is forbidden if and only if $\{b\}$ is a forbidder, in which case ba is forbidden. Thus, L is not an f-language. □

This leads to the conclusion that if a factor (subword) u of a word x from a language L is such that $u \notin L$, then forbidding u leads to forbidding x which means that L cannot be an f-language. Thus, a finite language is not necessarily an f-language, but it is an fe-language, as shown later in this section.

The following lemma states that only factorial languages can be f-languages.

Lemma 1. *Let* $\mathcal{F}$ *be a forbidding set. Then,* $L(\mathcal{F})$ *is factorial.*

Proof. Let $\mathcal{F}$ be given and $w \in L(\mathcal{F})$. Let $u \in sub(w)$. If $F \subseteq sub(u)$ for some $F \in \mathcal{F}$, then $F \subseteq sub(w)$, which contradicts our assumption. Hence $u \in L(\mathcal{F})$ and $L(\mathcal{F})$ is factorial. □

The converse question, whether every factorial language is an f-language, is investigated next. In the case of f-families, it was shown in [8] that even though every maximal language in an f-family is factorial, not every family of languages for which the maximal languages are factorial is an f-family.

Remark 4. Every factorial language L can be represented by a directed tree T_L where the labels of the vertices are exactly the words in L.

Given a factorial language L, consider the tree T_L obtained from T_{A^*} as follows. For each $w \notin L$ let u be the longest prefix of w, such that $u \in L$ and let $w = uav$ where $a \in A$ and $u, v \in A^*$. Then, the tree T_L does not contain vertex ua along with all paths that begin at ua, i.e., the entire branch rooted at ua is removed from T_{A^*}. It is easy to see that the labels of the vertices of T_L are precisely the words in L.

Lemma 2. *Let L be factorial. Then, L is an f-language.*

Proof. Let L be a factorial language. From the tree T_L construct $\mathcal{F} = \{\{ua\} \mid u \in V(T_L), ua \notin V(T_L), a \in A\}$. Then, $w \in L(\mathcal{F})$ if and only if $x \notin sub(w)$ for all $\{x\} \in \mathcal{F}$ if and only if $w \in V(T_L)$, i.e., $w \in L$. Consequently, $L = L(\mathcal{F})$. □

The next theorem follows from Lemmas 1 and 2 above.

Theorem 2. *A language is factorial if and only if it is an f-language.*

Some well-known languages can be characterized by enforcing systems only. Example 4 shows that an enforcing set that generates only infinite words, defines the empty language. Hence, the empty language is an e-language.

Proposition 4. *There exists a non-semilinear language that is an e-language.*

Proof. Let $A = \{a\}$ and $L = \{a^{2^n} \mid n \geq 0\}$. Then, the enforcing set $\mathcal{E} = \{(\lambda, \{a, aa\})\} \cup \{(a^{2^i+1}, \{a^{2^{i+1}}\}) \mid i \geq 1\}$ defines L, i.e., $L = L(\mathcal{E})$. □

Proposition 5. *There exists a non-regular linear language that is an fe-language.*

Proof. Let $\mathcal{F} = \{\{ba\}\}$, $\mathcal{E} = \{(\lambda, \{a\}), (\lambda, \{b\})\} \cup \{(a^i, \{a^{i+1}, a^i b^i\}) \mid i \geq 1\} \cup \{(b^i, \{b^{i+1}, a^i b^i\}) \mid i \geq 1\}$. Then, as noted in Example 6, $L = \{a^n b^n \mid n \geq 1\} = L(\mathcal{F}, \mathcal{E})$. □

Proposition 6. *There exists a non-linear context-free language that is an fe-language.*

Proof. Let $L_1 = \{a^n b^n \mid n \geq 1\}$ and consider $L = L_1 L_1$. Then, $L = L(\mathcal{F}, \mathcal{E})$ where $\mathcal{F} = \{\{ba^i b^i a\} \mid i \geq 1\}$ and $\mathcal{E} = \{(\lambda, \{a, b\})\} \cup \{(a^i, \{a^{i+1}, ba^i b^i, a^i b^i a\}) \mid i \geq 1\} \cup \{(b^i, \{b^{i+1}, ba^i b^i, a^i b^i a\}) \mid i \geq 1\} \cup \{(b^j a^i b^i, \{b^{j+1} a^i b^i, a^j b^j a^i b^i\}) \mid j \geq 1, i \geq 1\} \cup \{(a^i b^i a^j, \{a^i b^i a^{j+1}, a^i b^i a^j b^j\}) \mid j \geq 1, i \geq 1\}$. □

Proposition 7. *There exists a non-context-free language that is an fe-language.*

Proof. Let $\mathcal{F} = \{\{ba\}, \{ca\}, \{ac\}, \{cb\}\}$ and $\mathcal{E} = \{(\lambda, \{a, b, c\})\} \cup \{(a^i, \{a^{i+1}, a^i b^i c^i\}), (b^i, \{b^{i+1}, a^i b^i c^i\}), (c^i, \{c^{i+1}, a^i b^i c^i\}) \mid i \geq 1\}$. Then, $L(\mathcal{F}, \mathcal{E}) = \{a^n b^n c^n \mid n \geq 1\}$.   □

Example 2 and Propositions 5, 6 and 7 prove the following statement. Assume that FIN, REG, LIN, CF, and CS denote the classes of finite, regular, linear, context-free, and context-sensitive languages respectively.

Theorem 3. *For every* $X \in \{REG - FIN, LIN - REG, CF - LIN, CS - CF\}$ *there exists* $L \in X$ *such that* L *is an fe-language.*

We conclude this section with an fe-characterization of finite languages.

Proposition 8. *Every finite language is an fe-language.*

Proof. Let $L = \{x_1, x_2, \ldots, x_n\}$ be a finite language with $m = max\{|w| \mid w \in L\}$. Construct $\mathcal{F} = \{\{w\} \mid w \in A^{m+1}\}$ and $\mathcal{E} = \{(w, \{u \mid u \in L$ and $w \in sub(u)\} \cup \{v_w \mid v_w \in A^{m+1}$ and $w \in sub(v_w)\}) \mid w \notin L$ and $w \in A^{\leq m}\}$. We show that $L = L(\mathcal{F}, \mathcal{E})$. Assume that $w \in L$. Obviously, $w \, con \, \mathcal{F}$. If there is a (x, Y) such that $x \in sub(w)$, then $w \in Y$. Hence, $w \, sat \, \mathcal{E}$. Consequently, $L \subseteq L(\mathcal{F}, \mathcal{E})$. Conversely, if $w \notin L$, then either $|w| \geq m + 1$ which implies that $w \, ncon \, \mathcal{F}$ or $|w| \leq m$ which implies that there exists an enforcer $(w, Y) \in \mathcal{E}$ and therefore, $w \, nsat \, \mathcal{E}$. In either case, $w \notin L(\mathcal{F}, \mathcal{E})$. Hence, $L(\mathcal{F}, \mathcal{E}) \subseteq L$. Consequently, $L = L(\mathcal{F}, \mathcal{E})$.   □

Using the proof of the above proposition, we now construct an fe-system that defines the non-forbidding language $L = \{a, ba\}$ from the proof of Proposition 3. The fe-system $(\mathcal{F}, \mathcal{E})$ where $\mathcal{F} = \{\{aaa\}, \{aab\}, \ldots, \{bbb\}\}$ and $\mathcal{E} = \{(\lambda, \{a, ba, aaa\}), (b, \{ba, bbb\}), (aa, \{aaa\}), (ab, \{aba\}), (bb, \{bba\})\}$ is such that $L = L(\mathcal{F}, \mathcal{E})$.

6   Forbidding through Enforcing

The concept of minimal connect introduced below is used to replace forbidders by enforcers.

Definition 5. Let F be a finite set of words. A word y is called a *connect* of F if $F \subseteq sub(y)$. A word x is called a *minimal connect* of F if x is a connect of F and for every connect y of F it holds that $y \in sub(x)$ implies $y = x$. The set of minimal connects of F is denoted by $C_{min}(F)$.

Theorem 4. *For every forbidding set* $\mathcal{F}$ *there exists an enforcing set* $\mathcal{E}$, *such that* $L(\mathcal{F}) = L(\mathcal{E})$.

Proof. Let $\mathcal{F}$ be a forbidding set and $a \in A$ be some symbol. Given $F \in \mathcal{F}$, consider $C_{min}(F)$ and for every $u \in C_{min}(F)$ define $\mathcal{E}_{F_u} = \{(u, \{ua\}), (ua, \{uaa\}), \ldots, (ua^n, \{ua^{n+1}\}), \ldots\}$. Let $\mathcal{E}_F = \cup_{u \in C_{min}(F)} \mathcal{E}_{F_u}$ and $\mathcal{E} = \cup_{F \in \mathcal{F}} \mathcal{E}_F$. If $w \, nsat \, \mathcal{E}$, then there is an enforcer $(ua^i, \{ua^{i+1}\})$ which is not satisfied, which implies that

$ua^i \in sub(w)$ and one of its copies in w is not enclosed in ua^{i+1}. Hence, $u \in sub(w)$ and since $u \in C_{min}(F)$ for some $F \in \mathcal{F}$, $w\ ncon\ \mathcal{F}$. Conversely, if $w\ ncon\ \mathcal{F}$ there exists an $F \in \mathcal{F}$ and $u \in C_{min}(F)$ such that $u \in sub(w)$. Since w is a finite word, it cannot satisfy the enforcing set $\mathcal{E}_{F_u}$. Hence $w\ nsat\ \mathcal{E}$.   □

Remark 5. Note that the above result does not necessarily make forbidding sets obsolete. Even though every forbidding set may be replaced by an enforcing set, replacing a finite forbidding set with an infinite number of enforcers may be undesirable.

7 k-colorability Problem

Forbidding-enforcing systems can generate solutions to computational problems. In [7] a more general, categorical description of the model was presented and a solution to the k-colorability problem defined by an fe-system in the category of sets was presented to illustrate information processing capabilities of fe-systems. This section shows how a solution to this well-known NP-complete problem can be described by fe-systems defining languages.

The k-colorability problem asks whether given a graph and a finite set of k colors it is possible to assign one color to each vertex in such a way that adjacent vertices have distinct colors. Such assignment of colors is called a k-coloring.

Let $G = (V, E)$ be a graph with n vertices, i.e. $V = \{v_1, v_2, \ldots, v_n\}$ and C be a set of k colors, i.e. $C = \{c_1, c_2, \ldots, c_k\}$. A k-coloring can be viewed as a word over the alphabet $A = V \cup C$ with specific properties.

fe-system construction. A combination of brute enforcing and forbidding is used to design an fe-system that corresponds to assigning exactly one color to a vertex. The enforcing set $\mathcal{E} = \{(\lambda, \{vc_1, vc_2, \ldots, vc_k\}) \mid v \in V\}$ ensures that every vertex is assigned at least one color. The forbidding set $\mathcal{F} = \{\{vc, vc'\} \mid v \in V$ and $c, c' \in C$ with $c \neq c'\}$ allows only these vertices that are assigned at most one color. Thus, every word in $L(\mathcal{F}, \mathcal{E})$ contains every vertex colored in exactly one color. Also, no two adjacent vertices should be colored the same. This is obtained from the forbidding set $\mathcal{F}' = \{\{uc, vc\} \mid \{u, v\} \in E$ and $c \in C\}$.

Following the above notation and construction we have the next result.

Theorem 5. *G is k-colorable if and only if $L(\mathcal{F} \cup \mathcal{F}', \mathcal{E}) \neq \emptyset$. Furthermore, any word $w \in L(\mathcal{F} \cup \mathcal{F}', \mathcal{E})$ represents exactly one k-coloring of G.*

Proof. Assume that G is k-colorable. Then, there exists a k-coloring c. Construct the word $w = v_1 c_{i_1} v_2 c_{i_2} \ldots v_n c_{i_n}$ where $c_{i_j} \in C$ and is the color c_i from the coloring c which corresponds to vertex v_j. Then, by construction of the fe-system $w \in L(\mathcal{F} \cup \mathcal{F}', \mathcal{E})$. Conversely, if the fe-language is not empty, there is $w \in L(\mathcal{F} \cup \mathcal{F}', \mathcal{E})$ which is a word over A. Since $w\ sat\ \mathcal{E}$ there is at least one $vc_j \in sub(w)$ for every $v \in V$ and some $c_j \in C$. Furthermore, no adjacent vertex to v can be colored by the same color because $w\ con\ \mathcal{F}'$ and no subword of w is of the form

vc_i where $j \neq i$ since $w \, con \, \mathcal{F}$. Observe that vc_j may have many copies in w, but w represents exactly one k-coloring of G. $\qquad\qquad\qquad\qquad\qquad\qquad\qquad\qquad\qquad\qquad\qquad$ □

8   Modeling Splicing

In [2,4] the authors show how fe-systems can define fe-families that describe cutting by an enzyme, recombination, and can model splicing. For details about the computational model of splicing systems, the reader is referred to [9,10,13,14]. This section shows how fe-systems defining fe-languages can be used to model these operations.

Cutting by an enzyme and recombination. Let $xuvy$ be a string modeling a double stranded DNA molecule where uv models the restriction site for some restriction enzyme, which upon cutting between u and v produces either two sticky overhangs or two blunt ends. Then, the language $L(\mathcal{F})$, where $\mathcal{F} = \{\{uv\}\}$ contains all molecules resulting after the cutting by this enzyme, as well as all other molecules that do not have the uv restriction site. Consequently, starting from $L(\mathcal{F})$, as a next step, one can model recombination by the enforcing set $\mathcal{E} = \{(u, \{uv\} \cup \{ua \mid a \in A\}), (v, \{uv\} \cup \{av \mid a \in A\})\}$. This contrasts the fe-families model presented in [2,4] where both overhangs and ligated strands are present in the fe-family. Another difference is that the above $\mathcal{E}$ is finite, whereas the enforcing set for the fe-families is infinite.

Splicing. Consider the words $x_1u_1v_1y_1$ and $x_2u_2v_2y_2$ and the splicing rule $r = (u_1, v_1; u_2, v_2)$, according to which the two words can be spliced to obtain the word $x_1u_1v_2y_2$, and by symmetry, the word $x_2u_2v_1y_1$. Observe that the enforcing set $\mathcal{E}_r = \{(u_1, \{u_1v_1, u_1v_2\} \cup \{u_1a \mid a \in A\}), (u_2, \{u_2v_1, u_2v_2\} \cup \{u_2a \mid a \in A\}), (v_1, \{u_1v_1, u_2v_1\} \cup \{av_1 \mid a \in A\}), (v_2, \{u_1v_2, u_2v_2\} \cup \{av_2 \mid a \in A\})\}$ models splicing for the splicing rule r and words over the alphabet A. This system is finite as opposed to the fe-family model which uses infinitely many enforcers to model one splicing rule.

9   Concluding Remarks

Similar to grammars and automata, forbidding-enforcing systems are structures that define languages. Unlike grammars and automata where a language is defined by generating or accepting every word symbol by symbol, forbidding-enforcing systems define a language by imposing restrictions on its subwords. Use of forbidding-enforcing systems, in which one fe-system defines a single language as opposed to a family of languages presents a new way of defining languages.

It was shown that the Chomsky classes of languages have representatives that can be defined by fe-systems. Characterizations of finite languages, local languages, and factorial languages by fe-systems were presented. It will be interesting to investigate which other classes of languages can be defined by fe-systems. Further applications of fe-systems to biomolecular computing should be investigated both theoretically and experimentally.

Acknowledgement

The author thanks the anonymous referee whose comments and suggestions improved a previous version of this paper. This work has been partially supported by a UNF Faculty Development Scholarship Grant.

References

1. Cavaliere, M., Jonoska, N.: Forbidding and enforcing in membrane computing. Natural Computing 2, 215–228 (2003)
2. Ehrenfeucht, A., Hoogeboom, H.J., Rozenberg, G., van Vugt, N.: Forbidding and enforcing. In: Winfree, E., Gifford, D.K. (eds.) DNA Based Computers V. AMS DIMACS, Providence, RI, vol. 54, pp. 195–206 (2001)
3. Ehrenfeucht, A., Hoogeboom, H.J., Rozenberg, G., van Vugt, N.: Sequences of languages in forbidding-enforcing families. Soft Computing 5(2), 121–125 (2001)
4. Ehrenfeucht, A., Rozenberg, G.: Forbidding-enforcing systems. Theoretical Computer Science 292, 611–638 (2003)
5. Franco, G., Jonoska, N.: Forbidding and Enforcing Conditions in DNA Self-assembly of Graphs. Nanotechnology: Science and Computation, Natural Computing Series, Part I 105–118 (2006)
6. Genova, D.: Forbidding and Enforcing of Formal Languages, Graphs and Partially Ordered Sets, PhD Thesis, University of South Florida (2007)
7. Genova, D., Jonoska, N.: Defining structures through forbidding and enforcing constraints. Physica B: Condensed Matter 394(2), 306–310 (2007)
8. Genova, D., Jonoska, N.: Topological Properties of Forbidding-Enforcing Systems. Journal of Automata, Languages and Combinatorics 11(4), 375–397 (2006)
9. Head, T.: Formal language theory and DNA: an analysis of the generative capacity of specific recombinant behaviors. Bull. Math. Biology 49(6), 737–759 (1987)
10. Head, T., Păun, G., Pixton, D.: Language theory and molecular genetics: generative mechanisms suggested by DNA recombination. In: Rozenberg, G., Salomaa, A. (eds.) Handbook of Formal Languages, vol. 2, pp. 295–360. Springer, Berlin (1996)
11. Jonoska, N., Sa-Ardyen, P., Seeman, N.C.: Computation by self-assembly of DNA graphs. Journal of Genetic Programming And Evolvable Machines 4(2), 123–138 (2003)
12. Păun, G.: Membrane Computing. An Introduction. Springer, Berlin (2002)
13. Păun, G., Rozenberg, G., Salomaa, A.: Computing by splicing. Theoretical Computer Science 168, 321–336 (1996)
14. Păun, G., Rozenberg, G., Salomaa, A.: DNA Computing, new computing paradigms. Springer, Heidelberg (1998)
15. van Vugt, N.: Models of Molecular Computing, PhD thesis, Leiden University (2002)
16. Winfree, E., Yang, X., Seeman, N.C.: Universal Computation via Self-assembly of DNA: Some Theory and Experiments. In: Landweber, L., Baum, E. (eds.) DNA computers II. AMS DIMACS series, vol. 44, pp. 191–198 (1998)

Axiomatizing Resource Bounds for Measure

Xiaoyang Gu[1], Jack H. Lutz[2], Satyadev Nandakumar[3], and James S. Royer[4],[*]

[1] LinkedIn Corporation, Mountain View CA 94043, United States of America
xgu@linkedin.com
[2] Department of Computer Science, Iowa State University, Ames, IA 50011,
United States of America
lutz@cs.iastate.edu
[3] Department of Computer Science and Engineering, Indian Institute of Technology
Kanpur, Kanpur, UP 280 016, India
satyadev@cse.iitk.ac.in
[4] Department of Electrical Engineering and Computer Science, Syracuse University,
Syracuse NY 13244, United States of America
royers@ecs.syr.edu

Abstract. Resource-bounded measure is a generalization of classical Lebesgue measure that is useful in computational complexity. The central parameter of resource-bounded measure is the *resource bound* Δ, which is a class of functions. Most applications of resource-bounded measure use only the "measure-zero/measure-one fragment" of the theory. For this fragment, Δ can be taken to be a class of type-one functions. However, in the full theory of resource-bounded measurability and measure, the resource bound Δ also contains type-two functionals. To date, both the full theory and its zero-one fragment have been developed in terms of a list of example resource bounds. This paper replaces this list-of-examples approach with a careful investigation of the conditions that suffice for a class Δ to be a resource bound.

1 Introduction

Resource-bounded measure is a generalization of classical Lebesgue measure theory that allows us to quantify the "sizes" (measures) of interesting subsets of various complexity classes. This quantitative capability has been useful in computational complexity because it has intersected informatively with reducibilities, completeness, randomization, circuit-size, and many other central ideas of complexity theory. Resource-bounded measure has given us a generalization of the probabilistic method that works inside complexity classes (leading, for example, to improved lower bounds on Boolean circuit size [14] and the densities

---

[*] A version of this paper was presented in the Tenth Intenational Workshop on Logic in Computational Complexity, 2009, which did not have published proceedings. The first author's research was supported in part by National Science Foundation Grants 0344187, 0652569 and 0728806. The second author's research was supported in part by Spanish Government MEC Project TIN 2005-08832-C03-02. The fourth author's research was supported in part by National Science Foundation Grant CCR-0098198.

B. Löwe et al. (Eds.): CiE 2011, LNCS 6735, pp. 102–111, 2011.

of complete problems [16]) and new complexity-theoretic hypotheses (e.g., the hypothesis that NP is a non-measure 0 subset of exponential time) with many plausible consequences, i.e., significant explanatory power. The somewhat outdated survey papers [4,2,2,5,17,22] and more recent papers in the bibliography [10] give a more detailed account of the scope of resource-bounded measure and its applications.

The central parameter in resource-bounded measure is the *resource bound*, which is a class Δ of functions. When Δ is unrestricted, resource-bounded measure coincides with classical Lebesgue measure on the Cantor space **C** of all decision problems. On the other hand, when Δ only contains functions satisfying a suitable complexity constraint, resource-bounded measure consists of the following two theories.

1. A theory of Δ-*measure*. This is a "Δ-constructive" measure theory on **C**.
2. A theory of *measure in* a complexity class $R(\Delta)$. This is a theory that Δ-measure imposes on the "result class" $R(\Delta)$.

Typically, one proves a result on measure in $R(\Delta)$ by proving a corresponding result on Δ-measure. This, together with the fact that the Δ-measure result implies a corresponding Δ'-measure result for every resource bound $\Delta' \supseteq \Delta$, provides resource-bounded measure a substantial underlying unity.

Of the hundred or so papers that have been written about resource-bounded measure since 1992, none gives a definition of the term "resource bound". This approach to resource bounds has been healthy for the initial development of a theory intended as a tool, but, as Socrates taught us in *Euthyphro* [23], a list of examples leaves us far short of understanding a concept. More pragmatically, as the list grows, it becomes ever more burdensome to verify that a theorem about a general resource bound Δ actually holds for all examples in the list. This paper shows that there is a simple and natural set of axioms with the following two properties. First, adequacy: Any class Δ satisfying the axioms can be used as a resource bound for measure. Second, generality: The most extensively used resource bounds satisfy the axioms. We thus propose to *define* a resource bound to be a class Δ satisfying the axioms.

What makes our task challenging is the fact that, in order to define resource-bounded measurability and measure [15] a resource bound Δ must contain not only functions on discrete domains like $\{0,1\}^*$ and $\mathbb{N}$, but also type-2 *functionals* that take functions as arguments. It has been a major undertaking to define what it means for such functionals to be feasible (computable in polynomial time) and to verify that the definition is robust [12,8]. The second author [15] has defined type-2 versions of the other time and space resource bounds that have been extensively used in resource-bounded measure (the quasi-polynomial time and space hierarchies). However, these definitions have not been proven to be robust.

Fortunately, it turns out that an *existing* set of axioms can be adapted to our purpose. Mehlhorn's *basic feasible functionals* [19] were originally defined as a function algebra, i.e., the *smallest* class containing some initial functions and enjoying these closure properties. The main contribution of the present paper is to demonstrate that, if we just discard the "smallest" proviso in Mehlhorn's scheme and define a resource bound to be *any* class of functionals containing the initial functions and having the closure properties of his definition, then we will indeed have a definition that is sufficient for the development of measurability and measure in [15]. We prove new function algebra characterizations of these quasi-polynomial time and space classes, thereby establishing that they are robust.

Two additional remarks on related work are in order here. First, there has been work on resource-bounded measure that is not captured by our axiomatization. The notable examples here are the measures in "small" complexity classes (e.g., the polynomial time class P) developed by Moser [20] (building on pioneering work of Mayordomo [18] and Allender and Strauss [1], [24]), the measures in probabilistic classes (e.g., the randomized exponential time class BPE) developed by Moser [21], and the measures in "large" complexity classes (e.g., the doubly developed by Harkins and Hitchcock [9]. To date, this work has all been confined to measure 0/measure 1 results. Future developments of general measurability and measure in these settings may necessitate – and guide – generalizations of the axiomatization presented here. It is open whether such an axiomatization exists.

The other line of related work that we mention is Dai's outer measure approach to measurability and measure in complexity classes [7]. This approach is simpler than that of [15] and the present paper in that it does not require type-two functionals. On the other hand, the approach of [7] only seems to yield theory 2 in the second paragraph of this introduction, so that all results are "local" to a particular complexity class. The unity provided by theory 1 above, i.e., a "global" Δ-measure on all of Cantor space, is a substantial advantage of our our present approach. Only future research will determine whether a single approach can achieve both the simplicity of [7] and the unity of [15].

2   Preliminaries

We use a binary alphabet $\{0, 1\}$ in this paper. A string is an element in $\{0, 1\}^*$. For every $w \in \{0, 1\}^*$, $|w|$ is the length of the string w and $w[i]$ denotes the ith bit of w. The *Cantor* space $\mathbf{C} = \{0, 1\}^\infty$ is the set of all infinite binary sequences. For an $S \in \mathbf{C}$, $S[i]$ is the ith bit of S, and $S[0..n-1]$ is the n-bit prefix of S. The standard enumeration of $\{0, 1\}^*$ is the enumeration of all strings in $\{0, 1\}^*$ in the lexicographic order such that the 0th element in the enumeration is λ. The binary encoding function is ntob : $\mathbb{N} \rightarrow \{0, 1\}^*$ such that for all $n \in \mathbb{N}$, ntob(n) is the nth string in the standard enumeration. The binary decoding function bton : $\{0, 1\}^* \rightarrow \mathbb{N}$ is the inverse of the binary encoding function. For example, bton$(\lambda) = 0$ and bton$(01) = 4$. A language is a subset of $\{0, 1\}^*$. The characteristic sequence of L is the $S \in \mathbf{C}$ such that $S[i] = 1 \iff$ ntob$(i) \in L$. The characteristic function of L is $\chi_L : \{0, 1\}^* \rightarrow \{0, 1\}$ such that $\chi_L(x) = 1 \iff x \in L$. When no ambiguity arises, we also use L for the characteristic sequence of L. We write $w \sqsubseteq A$ if string w is a prefix of a string/sequence A.

The binary notational successor functions are $s_0, s_1 : \{0, 1\}^* \rightarrow \{0, 1\}^*$ such that $s_0(u) = u0$ and $s_1(u) = u1$ for all $u \in \{0, 1\}^*$. The binary successor function is $s : \{0,1\}^* \to \{0,1\}^*$ such that for all $u \in \{0,1\}^*$, $s(u) = \text{ntob}(\text{bton}(u) + 1))$. The binary predecessor function is $\text{pred} : \{0,1\}^* \to \{0,1\}^*$ such that $\text{pred}(u) = \text{ntob}(\max\{\text{bton}(u) - 1, 0\})$ The smash function is $\# : \Sigma^{*2} \to \{0,1\}^*$ such that for all $u, v \in \{0,1\}^*$, $\#(u, v) = 1^{|u| \cdot |v|}$.

We also define the following hierarchy of functions. Let $g_0 = 2n$ and let $g_i(n) = 2^{g_{i-1}(\log n)}$ for all $i \in \mathbb{Z}^+$. Note that $g_1(n) = n^2$ and that $g_2(n) = n^{\log n}$. For $i \in \mathbb{N}$, let G_i be the class of functions that contains g_i and is closed under composition. We use G_i to represent different growth rates. G_1 represents polynomial growth rates $(O(n^c))$ and G_2 represents quasi-polynomial growth rates $(O(n^{\log^c n}))$. For each $i \in \mathbb{N}$, we call it quasi^i-polynomial for those growth rates bounded by a function in G_i.

3   Type–2 Functionals

In 1965, Cobham characterized type-1 polynomial-time computable functions using limited/bounded recursion on notation [6,26]. He proved that the class of polynomial-time computable functions is the smallest class of functions containing the constant 0 function, the binary notational successor functions, and the smash function that is closed under composition and limited recursion on notation. Mehlhorn extended the characterization of polynomial-time computability to type-2 functionals. We denote type-two functionals by $\mathcal{T}_2$.

Definition 1 (Mehlhorn [19]). *F is defined from G, H, K by limited recursion on notation if for all f, x, w, $F(f, x, \lambda) = G(f, x)$ and $F(f, x, wb) = H(f, x, wb, F(f, x, w))$ for bit b, obeying the constraint that $|F(f, x, w)| \leq |K(f, x, w)|$.*

In addition, we also use the following type-2 schemes from Kapron and Cook. [13].

Definition 2 (Kapron and Cook [13]). *F is defined from H, G_1, ..., G_l by functional composition if for all f, x, $F(f, x) = H(f, G_1(f, x), \ldots, G_l(f, x))$. F is defined from G by expansion if for all f, g, x, y, $F(f, g, x, y) = G(f, x)$.*

We start our investigation with Kapron and Cook's Basic Feasible Functionals, which are considered to be the type-2 analogue of polynomial-time computable functions.

Definition 3 (Kapron and Cook [13]). *Let X be a set of type-two functionals. The class of basic feasible functionals defined from X (BFF(X)) is the smallest class of functionals that contains X, all polynomial-time functions of type-one and the application functional $\mathbf{Ap}$, defined by $\mathbf{Ap}(f, x) = f(x)$, and is closed under functional composition, expansion, and limited recursion on notation. The basic feasible functionals are BFF($\varnothing$).*

Mehlhorn proved that the BFF's have the *Ritchie-Cobham property*, namely, $F \in$ BFF if and only if there exists an oracle Turing machine M and $G \in$ BFF such that for all input f and x, the running time of $M(f, x)$ is bounded by $|G(f, x)|$. Mehlhorn's result serves as partial evidence that the functional algebraic notion of BFF is robust. On top of this, Kapron and Cook defined a notion of type-2 polynomial-time computability based on oracle Turing machines that does not require the use of BFF time-bound like the one in Mehlhorn's result. The Basic Feasible Functionals capture the notion of an intuitively feasible class of type-two functionals. Basic Feasible Functionals and their probabilistic versions show up in cryptography, for instance, in many constructions of pseudo-random generators from one way functions. In [11] the authors remark that many cryptographic adversaries can be formalized as type-2 probabilistic feasible functionals or circuits.

First, we generalize Kapron and Cook's definition of second-order polynomials to the following.

Definition 4. *Let $i \in \mathbb{Z}^+$. First-order variables are elements of the set $\{n_1, n_2, \dots\}$. Second-order variables are elements of the set $\{L_1, L_2, \dots\}$. Second-order quasi^i-polynomials are defined inductively: any $c \in \mathbb{N}$ is a second-order quasi^i-polynomial; first-order variables are second-order quasi^i-polynomials; and if P, Q are second-order quasi^i-polynomials and L is a second-order variable, then $P + Q$, $P \cdot Q$, $L(P)$, and $g_i(P)$ are second-order quasi^i-polynomials.*

Second-order quasi^1-polynomials are the second-order polynomials defined by Kapron and Cook. Second-order quasi^2-polynomials are second-order quasi-polynomials. They also defined a notion of the length for type-1 functions.

Definition 5 (Kapron and Cook[13]). *For any $f : \{0,1\}^* \to \{0,1\}^*$, the length of f is the function $|f| : \mathbb{N} \to \mathbb{N}$ defined by $|f|(n) = \max_{|w| \le n} |f(w)|$.*

Note that $|f|$ is non-decreasing. With the above two definitions, Kapron and Cook defined the following notion of polynomial-time bounded oracle Turing machine computation.

Definition 6. *A type-two functional F is basic poly time if there is an oracle Turing machine M and a second-order polynomial P such that M computes F, and for all $\boldsymbol{f}$ and $\boldsymbol{x}$, the running time of $M(\boldsymbol{f}, \boldsymbol{x})$ is bounded by $P(|f_1|, \dots, |f_k|, |x_1|, \dots, |x_l|)$.*

Strongly confirming the robustness of the notion of BFFs, they proved the following.

Theorem 1 (Kapron and Cook [13]). *A functional F is BFF if and only if it is basic poly time.*

In this paper, we extend Mehlhorn's functional algebraic notion of feasible functionals to quasi-feasible functionals with the following definition.

Definition 7. *Let $X \subseteq \mathcal{T}_2$ and let $i \in \mathbb{Z}^+$. The class of basic i-feasible functionals defined from X (BFF$_i(X)$) is the smallest class of functionals containing X, all polynomial-time computable functions in $1^{g_i(|x|)}$, and $BFF(\varnothing)$, which is closed under functional composition, expansion, and limited recursion on notation. The basic i-feasible functionals are elements of the class BFF$_i(\varnothing)$.*

In the flavor of Kapron and Cook, we extend their oracle Turing machine based notion of feasible computation to the following.

Definition 8. *Let $i \in \mathbb{Z}^+$. A functional F is* basic quasi^i-polynomial time *if there is an oracle Turing machine M and a second-order quasi^i-polynomial P such that M computes F, and for all $\boldsymbol{f}$ and $\boldsymbol{x}$, the running time of $M(\boldsymbol{f}, \boldsymbol{x})$ is bounded by $P(|f_1|, \ldots, |f_k|, |x_1|, \ldots, |x_l|)$. Analogously, a functional F is* quasi^i-polynomial space *if an oracle machine M's running space is bounded by such a P.*

The following theorem is a corollary of Kapron and Cook's proof of theorem 1.

Theorem 2. *Let $i \in \mathbb{Z}^+$. A functional F is BFF_i if and only if it is basic quasi^i-polynomial time.*

In the machine model, the time bound is based on both the input length and on the length of query answers. This is why we need to have $g_i(P)$ in the definition of second-order quasi^i-polynomials. The condition in the definition that a single second-order quasi^i-polynomial has to work for all input $\boldsymbol{f}$ prohibits an oracle Turing machine from using extra running time when the input function $\boldsymbol{f}$ is pathologically long. An oracle Turing machine M that computes a quasi^i-polynomial time functional, on any x, can only utilize an amount of time that is quasi^i-polynomial in the length of $\boldsymbol{f}$ it can provide evidence for, which can be much less than length of $\boldsymbol{f}$ depending on the type-0 inputs. More formally, let Q_x be the set of all queries made by M with $\boldsymbol{f}$ and x as input. Let P be the time bound of M. Let $f_{Q_x}(y) = \boldsymbol{f}(y)$ if $y \in Q_x$ and 0 otherwise. Then the running time $T_M(\boldsymbol{f}, x) \leq P(|f_{Q_x}|, |x|)$ for all $\boldsymbol{f}$ and x. The key idea behind Kapron and Cook's proof is that it is possible to find the oracle query q_{max} made by $M(\boldsymbol{f}, x)$ that maximizes $|\boldsymbol{f}(q_{max})|$ in BFF. And the inability to compute the length of $\boldsymbol{f}$ (in unary) in BFF is what makes their proof very involved. We will see in the following that the situation with polynomial space-bounded computation is much simpler precisely for the reason that, as we will soon prove in Lemma 1, the length functional in unary for arbitrary $\boldsymbol{f}$ is actually computable in polynomial space. First, we develop the definitions of computation feasible in terms of space.

In 1972, D. B. Thompson characterized the class of type-1 polynomial-space computable functions as the smallest class that contains the constant 0 function, the binary successor function, the smash function, and is closed under (type-1) composition and (type-1) bounded recursion [25]. We extend type-1 bounded recursion as follows.

Definition 9. *F is defined from G, H, K by* bounded recursion *(BR) if for all $\boldsymbol{f}$, $\boldsymbol{x}$, n, $F(\boldsymbol{f}, \boldsymbol{x}, 0) = G(\boldsymbol{f}, \boldsymbol{x})$, and $F(\boldsymbol{f}, \boldsymbol{x}, n+1) = H(\boldsymbol{f}, \boldsymbol{x}, n, F(\boldsymbol{f}, \boldsymbol{x}, n))$ obeying the constraint that $F(\boldsymbol{f}, \boldsymbol{x}, n) \leq K(\boldsymbol{f}, \boldsymbol{x}, n)$.*

Definition 10. *Let $X \subseteq T_2$ and let $i \in \mathbb{Z}^+$. The* class of *basic i-feasible space functionals defined from X ($\mathrm{BFSF}_i(X)$) is the smallest class of functionals containing X, all polynomial-time computable functions in T_1, $1^{g_i(|x|)}$, and the application functional $\mathbf{Ap}$, defined by $\mathbf{Ap}(f, x) = f(x)$, and which is closed under functional composition, expansion, and bounded recursion. The* basic i-feasible space functionals *are $\mathrm{BFSF}_i(\varnothing)$.*

Lemma 1. $L : (f, x) \mapsto 1^{|f|(|x|)}$ *is basic i-feasible space for all* $i \geq 1$.

Theorem 3. *A functional* F *is basic i-feasible space if and only if it is quasi^i-polynomial space.*

4 Resource Bounds

In the initial development of a theory of resource-bounded measure [15], a list of examples of resource-bounds were given based on an oracle Turing machine model of type-2 computation that is not known to be robust. In this section, we axiomatize the definition of a resource bound by adapting the axioms of Mehlhorn's basic feasible functionals and verify that most extensively used resource bounds are indeed resource bounds under this definition. Let K^k be the canonical Σ_k^P-complete language [3] and let χ_k be its characteristic function.

Definition 11. *A resource bound is a class* Δ *of functionals of type no more than 2 that is closed under BFF.*

Theorem 4. *Let* $i \in \mathbb{Z}^+$.

1. $p_i = BFF_i$ *is a resource bound.*
2. *Let* $k \geq 2$. *Define* $\Delta_k^{P_i} = BFF_i(\{\chi_{k-1}\})$. *Then* $\Delta_k^{P_i}$ *is a resource bound.*
3. *Define* $p_i$space $= BFSF_i$. *Then* $BFF(p_i$space$) = p_i$space, *i.e.,* $p_i$space *is a resource bound.*

5 Adequacy for Measure

The general theory of resource-bounded measurability and measure developed in [15] consists of the basic definitions, reviewed below, and proofs that the resulting Δ-measure and measure in $R(\Delta)$ have the fundamental properties of a measure (e.g., additivity, measurability of measure-0 sets, etc.). The main shortcoming of the list-of-examples approach is evident in these proofs: Each time that a functional is asserted to be Δ-computable, it is *incumbent on the reader to check* that this holds for each of the infinitely many resource bounds Δ in the list.

Our main task in the present section is to re-prove these theorems in a more satisfactory manner. Our proofs here assume only that Δ is a resource bound, as defined in section 3, and they *explicitly* prove that the relevant functionals are Δ-computable, using only the axioms (closure properties) defining resource bounds.

To put the matter simply, the proofs in [15] are measure-theoretically rigorous, but their generality is tedious and limited. Our contribution here is to make these proofs and the scope of their validity explicit. For this reason, the proofs given in the present section focus on the Δ-computability of various type-two functionals, referring to [15] for the non-problematic, measure-theoretic parts of the proofs. We first review the definitions necessary for the development of a resource-bounded measure.

A *probability measure* on **C** is a function $\nu : \{0,1\}^* \to [0,1]$ such that $\nu(\lambda) = 1$ and, for all $w \in \{0,1\}^*$, $\nu(w) = \nu(w0) + \nu(w1)$. For strings $v, w \in \{0,1\}^*$, if $\nu(w) > 0$, we write $\nu(v|w)$ for the conditional probability of v given w. The *uniform probability measure* is μ such that $\mu(w) = 2^{-|w|}$ for all $w \in \{0,1\}^*$.

Let ν be a probability measure on **C**. A *ν-martingale* is a function $d : \{0,1\}^* \to [0,\infty)$ with the property that for all $w \in \{0,1\}^*$, $d(w)\nu(w) = d(w0)\nu(w0) + d(w1)\nu(w1)$. We use **1** for the *unit martingale* defined by $\mathbf{1}(w) = 1$ for all $w \in \{0,1\}^*$, which is a ν-martingale for every probability measure ν.

Definition 12. *Let ν be a probability measure. Let d be a ν-martingale. Let $A \subseteq \{0,1\}^*$. We say that d covers A if there is an $n \in \mathbb{N}$ such that $d(A[0..n-1]) \geq 1$. We say that d succeeds on A if $\limsup\limits_{n\to\infty} d(A[0..n-1]) = \infty$ We say that d succeeds strongly on A if $\liminf\limits_{n\to\infty} d(A[0..n-1]) = \infty$ The set covered by d (the unitary success set) is $S^1[d] = \{A \mid d \text{ covers } A\}$. The success set of d is $S^\infty[d] = \{A \mid d \text{ succeeds on } A\}$. The strong success set of d is $S^\infty_{\mathrm{str}}[d] = \{A \mid d \text{ succeeds strongly on } A\}$.*

We use real-valued functions (probability measures, martingales, etc.) on discrete domains of natural numbers $\mathbb{N}$ and strings $\{0,1\}^*$ extensively. Let D be a discrete domain. A *computation* of a function $f : D \to \mathbb{R}$ is a function $\hat{f} : \mathbb{N} \times D \to \mathbb{Q}$ such that, for all $r \in \mathbb{N}$ and $x \in D$, $|\hat{f}(r,x) - f(x)| \leq 2^{-r}$, where $\hat{f}(r,x) = \hat{f}(r,x)$. For such a function f, there is a unique computation $\hat{f}$ of f such that $\hat{f}(r,x) = a \cdot 2^{-r}$ for some integer a for all $r \in \mathbb{N}$ and $x \in D$. We call this particular $\hat{f}$ the *canonical computation* of f. Whenever a function f is involved as a parameter in the computation of a type-2 functional, the functional operates on the canonical $\hat{f}$ instead of the f itself.

Definition 13 (Lutz [15]). *Let Δ be a resource bound. A Δ-probability measure on **C** is a probability measure ν on **C** such that ν is Δ-computable and there is a Δ-computable function $l : \mathbb{N} \to \mathbb{N}$ such that, for all $w \in \{0,1\}^*$, $\nu(w) = 0$ or $\nu(w) \geq 2^{-l(|w|)}$. We say that ν is weakly positive, if ν has the latter property.*

Definition 14 (Lutz [14,15]). *A constructor is a function $\delta : \{0,1\}^* \to \{0,1\}^*$ such that $x \sqsubset_{\neq} \delta(x)$ for all $x \in \{0,1\}^*$. The result of δ is the unique language $R(\delta)$ such that $\delta^k(\lambda) \sqsubseteq R(\delta)$. If Δ is a resource bound, then the result class $R(\Delta)$ of Δ is the set of all languages $R(\delta)$ such that $\delta \in \Delta$.*

The martingale splitting operators defined by Lutz [15] are instrumental in developing the general theory of resource-bounded measurability and measure in complexity classes.

Definition 15 (Lutz [15]). *Let X^+ and X^- be disjoint subsets of **C**, then a ν-splitting operator for (X^+, X^-) is a functional $\Phi : \mathbb{N} \times \mathcal{D}_\nu \to \mathcal{D}_\nu \times \mathcal{D}_\nu$, such that $\Phi(r,d) = (\Phi^+_r(d), \Phi^-_r(d))$ has the following properties for all $r \in \mathbb{N}$ and $d \in \mathcal{D}_\nu$.*

1. $X^+ \cap S^1[d] \subseteq S^1[\Phi_r^+(d)]$,
2. $X^- \cap S^1[d] \subseteq S^1[\Phi_r^-(d)]$,
3. $\Phi_r^+(d)(\lambda) + \Phi_r^-(d)(\lambda) \leq d(\lambda) + 2^{-r}$.

If Δ is a resource bound, a Δ-ν-splitting operator for (X^+, X^-) is a ν-splitting operator for (X^+, X^-) that is Δ-computable. Let $X \subseteq \mathbf{C}$. A Δ-ν-measurement of X is a Δ-ν-splitting operator for (X^+, X^-). A ν-measurement of X in $R(\Delta)$ is a Δ-ν-splitting operator for $(R(\Delta) \cap X, R(\Delta) - X)$. If Φ is a ν-splitting operator, then we write $\Phi_\infty^+ = \inf_{r \in \mathbb{N}} \Phi_r^+(\mathbf{1})(\lambda)$, $\Phi_\infty^- = \inf_{r \in \mathbb{N}} \Phi_r^-(\mathbf{1})(\lambda)$.

We now can define the resource-bounded measurabilities.

Definition 16. *A set $X \subseteq \mathbf{C}$ is ν-measurable in $R(\Delta)$, and we write $X \in \mathcal{F}_{R(\Delta)}^\nu$, if there exists a ν-measurement Φ of X in $R(\Delta)$. In this case, the ν-measure of X in $R(\Delta)$ is the real number $\nu(X|R(\Delta)) = \Phi_\infty^+$. ($\nu(X|R(\Delta))$ does not depend on the choice of Φ [15].)*

Definition 17. *A set $X \subseteq \mathbf{C}$ is Δ-ν-measurable, and we write $X \in \mathcal{F}_\Delta^\nu$, if there exists a Δ-ν-measurement ϕ of X. In this case, the Δ-ν-measure of X is the real number $\nu_\Delta(X) = \Phi_\infty^+$. ($\nu_\Delta(X)$ does not depend on the choice of Φ [15].)*

We will defer the proofs and the complete development to the full version of this paper. Here, to shed some light on the flavor of the rest of the development in the full version of this paper, we establish as an example, the fundamental theorem that the cylinders are Δ-measurable for all resource bound Δ.[1] First, we state a lemma that is useful for the proof of Theorem 5.

Lemma 2. *Let type-2 functional $B : (\{0,1\}^* \to \mathbb{R}) \times (\mathbb{N} \to \mathbb{N}) \to (\{0,1\}^* \times \{0,1\}^* \to \mathbb{R})$ be such that for every weakly positive probability measure $\nu : \{0,1\}^* \to [0,1]$, $l : \mathbb{N} \to \mathbb{N}$, and $w, v \in \{0,1\}^*$, $B(\nu,l)(w,v)$ be defined as $\nu(w|v)$ if v is a prefix of w and $\nu(w) \geq 2^{-l(|w|)}$, as 1 if w is a prefix of v and $\nu(w) \geq 2^{-l(|w|)}$, and as 0 otherwise. Then B is BFF over all weakly positive probability measures ν and all $l : \mathbb{N} \to \mathbb{N}$.*

Theorem 5 (Lutz [15]). *Let Δ be a resource bound. If ν is a Δ-probability measure on $\mathbf{C}$, then for each $w \in \{0,1\}^*$, the cylinder $\mathbf{C}_w$ is Δ-ν-measurable, with $\nu_\Delta(\mathbf{C}_w) = \nu(w)$.*

References

1. Allender, E., Strauss, M.: Measure on small complexity classes with applications for BPP. In: Proceedings of the 35th FOCS, pp. 807–818 (1994)
2. Ambos-Spies, K., Mayordomo, E.: Resource-bounded measure and randomness. In: Sorbi, A. (ed.) Complexity, Logic and Recursion Theory. Lecture Notes in Pure and Applied Mathematics, pp. 1–47. Marcel Dekker, New York (1997)
3. Balcázar, J.L., Díaz, J., Gabarró, J.: Structural Complexity I, 2nd edn. Springer, Berlin (1995)

---

[1] An updated version of [15] will be posted in the Computing Research Repository.

4. Buhrman, H., Torenvliet, L.: On the structure of complete sets. In: Proceedings of the Ninth Annual Structure in Complexity Theory Conference, pp. 118–133 (1994)
5. Buhrman, H., Torenvliet, L.: Complete sets and structure in subrecursive classes. In: Proceedings of Logic Colloquium 1996. LNL, vol. 12, pp. 45–78 (1998)
6. Cobham, A.: The intrinsic computational difficulty of functions. In: Logic, Methodology and Philosophy of Science II. North Holland, Amsterdam (1965)
7. Dai, J.J.: An outer-measure approach for resource-bounded measure. Theory of Computing Systems 45(1), 64–73 (2009)
8. Danner, N., Royer, J.: Adventures in time and space. Logical Methods in Computer Science 3(1:9), 1–53 (2007)
9. Harkins, R.C., Hitchcock, J.M.: Dimension, halfspaces, and the density of hard sets. In: Lin, G. (ed.) COCOON 2007. LNCS, vol. 4598, pp. 129–139. Springer, Heidelberg (2007)
10. Hitchcock, J.M.: Resource-Bounded Measure Bibliography(current January, 2011), http://www.cs.uwyo.edu/~jhitchco/bib/rbm.html
11. Impagliazzo, R., Kapron, B.: Logics for reasoning about cryptographic constructions. J. Comput. Syst. Sci. 72(2), 286–320 (2006)
12. Irwin, R., Kapron, B., Royer, J.: On characterizations of basic feasible functionals, part i. Journal of Functional Programming 11(1), 117–153 (2001)
13. Kapron, B., Cook, S.: A new characterization of type 2 feasibility. SIAM Journal on Computing 25, 117–132 (1996)
14. Lutz, J.H.: Almost everywhere high nonuniform complexity. J. Comput. Syst. Sci. 44(2), 220–258 (1992)
15. Lutz, J.H.: Resource-bounded measure. In: Proceedings of the 13th IEEE Conference on Computational Complexity, pp. 236–248 (1998)
16. Lutz, J.H., Mayordomo, E.: Measure, stochasticity, and the density of hard languages. SIAM Journal on Computing 23(4), 762–779 (1994)
17. Lutz, J.H., Mayordomo, E.: Twelve problems in resource-bounded measure. Bulletin of EATCS 68, 64–80 (1999)
18. Mayordomo, E.: Measuring in PSPACE. In: Proceedings of the 7th International Meeting of Young Computer Scientists. Topics in Computer Science, vol. 6, pp. 93–100. Gordon and Breach (1994)
19. Mehlhorn, K.: Polynomial and abstract subrecursive classes. Journal of Computer and System Science 12, 147–178 (1976)
20. Moser, P.: Martingale families and dimension in P. Theoretical Computer Science 400(1-3), 46–61 (2008)
21. Moser, P.: Resource-bounded measure on probabilistic classes. Information Processing Letters 106(6), 241–245 (2008)
22. Pavan, A.: Comparison of reductions and completeness notions. SIGACT News 34(2), 27–41 (2003)
23. Plato: Dialogues, vol. 1. Yale University Press, New Haven and London (1989)
24. Strauss, M.: Measure on p: Strength of the notion. Information and Computation 136(1), 1–23 (1997)
25. Thompson, D.B.: Subrecursiveness: machine independent notions of computability in restricted time and storage. Mathematical Systems Theory 6, 3–15 (1972)
26. Weihrauch, K.: Teilklassen primitiv-rekursiver Wortfunktionen. Ph.D. thesis, Rheinische Friedrich-Wilhelms-Universität Bonn (1973)

Complexity Issues for Preorders on Finite Labeled Forests

Peter Hertling[1] and Victor Selivanov[2,*]

[1] Institut für Theoretische Informatik, Mathematik und Operations Research,
Universität der Bundeswehr München, 85577 Neubiberg, Germany
peter.hertling@unibw.de
[2] A.P. Ershov Institute of Informatics Systems, Lavrentyev av. 6, 630090
Novosibirsk, Russia
vseliv@iis.nsk.su

Abstract. We prove that three preorders on the finite k-labeled forests are polynomial time computable. Together with an earlier result of the first author, this implies polynomial-time computability for an important initial segment of the corresponding degrees of discontinuity of k-partitions on the Baire space. Furthermore, we show that on ω-labeled forests the first of these three preorders is polynomial time computable as well while the other two preorders are NP-complete.

1 Introduction

As is well-known, reducibilities serve as useful tools for understanding the complexity (or non-computability) of decision problems on discrete structures. In computable analysis, many problems of interest turn out to be non-computable, even discontinuous. Thus, there is a need for tools to measure their non-computability or discontinuity. Accordingly, also in this context of decision problems on continuous structures people employed some reducibility notions.

Weihrauch [23,24] (see also the thesis [22] supervised by Weihrauch) introduced a topological reducibility relation $\leq_2$ for functions f between products of the Cantor space and a discrete space. Independently, Hirsch [10] gave a similar definition for functions between arbitrary topological spaces. Weihrauch [23,24] also introduced a generalisation of this reducibility relation to sets of functions. A computability-theoretic version of this generalisation, transferred via representations to computational problems on arbitrary represented spaces, is now called Weihrauch reducibility; see [4].

These notions have turned out to be very useful for understanding the non-computability and discontinuity of important computation problems in computable analysis [9,3] and constructive mathematics [24,4]. The first author has also introduced slightly weaker topological reducibility relations $\leq_0$ and $\leq_1$ for

---

* Both authors were supported by DFG-RFBR (Grant 436 RUS 113/1002/01, 09-01-91334).

B. Löwe et al. (Eds.): CiE 2011, LNCS 6735, pp. 112–121, 2011.

functions between arbitrary topological spaces in [7,8,9]. The notions are non-trivial even for the case of discrete spaces $Y = k = \{0, \ldots, k-1\}$ with k points, $1 < k < \omega$. We call functions $f : X \to k$ k-*partitions of* X because they are in a natural bijective correspondence with the partitions $(A_0, \ldots, A_{k-1})$ of X where $A_i = f^{-1}(i)$. For $k = 2$ the relation $\leq_0$ coincides with the classical Wadge reducibility [11].

In [7] (without proofs) and [8,9] (with proofs) the first author gave a "combinatorial" characterization of the important initial segment of the degree structures under these reducibilities of k-partitions of the Baire space $\mathbb{B} = \omega^\omega$ formed by the k-partitions of $\mathbb{B}$ the components of which are finite Boolean combinations of open sets. Namely, he introduced preorders $\leq_0, \leq_1, \leq_2$ on the set $\mathcal{F}_k$ of finite k-labeled forests (precise definitions are given in the next section) such that, for each $i \leq 2$, the structure of the topological $\leq_i$-degrees of the specified initial segment is isomorphic to the quotient-poset of $(\mathcal{F}_k \setminus \{\emptyset\}; \leq_i)$. In fact, he showed that with any k-partition f in the initial segment just described one can associate a non-empty finite k-labeled forest $B(f)$ such that, for any k-partitions f and g on the Baire space one has $f \leq_i g$ if, and only if, $B(f) \leq_i B(g)$. Furthermore, the mapping $f \mapsto B(f)$ to the set of finite non-empty k-forests is onto. This result provides a natural naming system for the specified initial segment of topological $\leq_i$-degrees of k-partitions of the Baire space. The second author has extended this result for $\leq_0$ to a much larger initial segment of k-partitions [21]. Note that the Baire space is important because it is commonly used in computable analysis [25] to represent many other spaces of interest.

The structure $(\mathcal{F}_k; \leq_0)$ and its extension to the structure $(\mathcal{P}_k; \leq_0)$ of finite k-labeled posets are also important due to their close relationship to the Boolean hierarchy of k-partitions that extends the classical Boolean (or difference) hierarchy of sets [13,12,20] and to some other fields of discrete mathematics like clones of functions on k [19,18].

The mentioned results motivated the study of $(\mathcal{F}_k; \leq_0)$ and $(\mathcal{P}_k; \leq_0)$ in a series of publications. In [14] it was shown that for any $k \geq 3$ the first-order theory of the quotient-poset of $(\mathcal{F}_k; \leq_0)$ is undecidable. It is even computably isomorphic to first-order arithmetic [15]. In [17] the same was shown for $\leq_1$ and $\leq_2$. In [16] a complete definability theory for the quotient-poset of $(\mathcal{F}_k; \leq_0)$ was developed. According to [19], the quotient-poset of $(\mathcal{P}_k; \leq_0)$ is universal, i.e., it contains any countable poset as a substructure.

The naming systems $(\mathcal{F}_k; \leq_i)$ and $(\mathcal{P}_k; \leq_i)$ are obviously computable, and the natural next step in understanding their computational properties is to look at their complexity. In [18] it is shown that the relation $\leq_0$ on $\mathcal{P}_k$ is NP-complete. In this paper we answer some natural complexity questions about the relations $\leq_0, \leq_1, \leq_2$ on $\mathcal{F}_k$, and also on the set $\mathcal{F}_\omega$ of finite ω-labeled posets. Note that in [9] it is shown that the structure of $\leq_2$-degrees of functions $f : \mathbb{B} \to \omega$ such that all $f^{-1}(k)$ are finite Boolean combinations of open sets is isomorphic to the quotient-poset of $(\mathcal{F}_\omega; \leq_2)$.

Our main results are summarized in the following two theorems.

Theorem 1. *The relation $\leq_0$ is polynomial-time computable on $\mathcal{F}_k$ for $k < \omega$ and on $\mathcal{F}_\omega$.*

Theorem 2. *The relations $\leq_1$ and $\leq_2$ are polynomial-time computable on $\mathcal{F}_k$ for $k < \omega$, and they are NP-complete on $\mathcal{F}_\omega$.*

In Section 2 we recall some relevant definitions and facts. In Sections 3, 4 and 5 we prove the results for $\leq_0$, $\leq_1$ and $\leq_2$, respectively. In Section 6 we mention some related open questions.

2 Preliminaries

We use some standard notation and terminology on posets which may be found e.g. in [5]. Throughout this paper, k denotes an arbitrary integer, $k \geq 2$, which is identified with the set $\{0, \ldots, k-1\}$. A *poset* is a partially ordered set $(P, \leq)$, that is, a set P with a binary relation $\leq$ on it that is reflexive, transitive, and anti-symmmetric. As usual, by $x < y$ we mean $(x \leq y \wedge x \neq y)$. For an element $x \in P$ the *upper cone of* x is the set $\uparrow x := \{y \in P \mid x \leq y\}$, and the *lower cone of* x is the set $\downarrow x := \{y \in P \mid y \leq x\}$. A *chain* is a poset $(P, \leq)$ such that additionally $x \leq y$ or $y \leq x$, for all $x, y \in P$. By a *forest* we mean a finite poset in which every upper cone $\uparrow x$ (endowed with the order obtained by restricting $\leq$ to $\uparrow x$) is a chain. A *tree* is a forest that has a largest element, that is, an element x satisfying $y \leq x$ for all $y \in P$. When such an element exists then it is uniquely determined, and it is called the *root* of the tree. The cardinality of a forest F is denoted $|F|$.

A *k-labeled forest* (or just a k-forest) is a triple $(P; \leq, c)$ consisting of a forest $(P; \leq)$ and a labeling $c : P \to k$. We call a k-forest $(P; \leq, c)$ *repetition-free* iff $c(x) \neq c(y)$ whenever y is an immediate successor of x in P, i.e., whenever $x < y$ and there does not exist a $z \in P$ with $x < z$ and $z < y$. We often simplify the notation of a k-forest to P. By default, we denote the labeling in a given k-forest by c. We are mainly interested in the set $\mathcal{F}_k$ of finite k-labeled forests, but also in the set $\mathcal{T}_k$ of finite k-labeled trees and in the set $\mathcal{C}_k$ of finite k-labeled chains. The finite k-labeled chains are identified with the finite non-empty words over the alphabet k. Any such word u is a sequence $u(0) \cdots u(n-1)$ of letters where $n = |u|$ is the length of u. We will view this as a k-labeled chain of length n with $u(0)$ being the label of the root of the chain.

For any k-forests $F, G \in \mathcal{F}_k$ and $i < k$, let $p_i(F)$ denote a k-tree obtained from F by adding a new greatest element with the label i. Let $F \sqcup G$ be the disjoint union of F, G. Note that any tree is of the form $p_i(F)$, and that any proper forest, i.e., a forest which is neither empty nor a tree, is of the form $F \sqcup G$ where F and G are nonempty forests. The singleton k-tree with label i is identified with i.

Definition 3. *1. A 0-morphism (resp., 1-morphism, resp., 2-morphism) $\varphi : (P; \leq_P, c_P) \to (Q; \leq_Q, c_Q)$ between k-forests is a monotone function $\varphi : (P; \leq_P) \to (Q; \leq_Q)$ that satisfies $c_P = c_Q \circ \varphi$ (resp., $c_P = f \circ c_Q \circ \varphi$ for* some $f : k \to k$, resp., $\forall x, y \in P((x \leq_P y \wedge c_P(x) \neq c_P(y)) \to c_Q(\varphi(x)) \neq c_Q(\varphi(y))))$.

2. We write $P \leq_0 Q$ (resp. $P \leq_1 Q$, $P \leq_2 Q$) to denote that there exists a 0-morphism (resp. 1-morphism, 2-morphism) $\varphi : P \to Q$.

It is easy to see that any 0-morphism is a 1-morphism and any 1-morphism is a 2-morphism. Therefore, $\leq_0$ implies $\leq_1$ and $\leq_1$ implies $\leq_2$. The relations $\leq_0, \leq_1, \leq_2$ are reflexive and transitive on $\mathcal{F}_k$, i.e., they are preorders. In the same way we define the preorders $\leq_0, \leq_1, \leq_2$ on the set $\mathcal{F}_\omega$ of finite ω-labeled forests (in which any natural number may be used as a label).

Let ω^* be the set of finite sequences (strings) of natural numbers. The empty string is denoted by $\emptyset$, the concatenation of strings σ, τ by $\sigma\tau$, the length of σ by $|\sigma|$. By ω^+ we denote the set of finite non-empty strings in ω. By $\sigma \sqsubseteq \tau$ we denote that the string σ is an initial segment of the string τ. For any n, $1 < n < \omega$, let n^* be the set of finite strings of elements of $\{0, \ldots, n-1\}$, $n^* \subseteq \omega^*$. E.g., 2^* is the set of finite strings of 0's and 1's. In computer science people often consider the sets A^* and A^+ of finite (respectively, finite non-empty) words over a finite alphabet A. Mathematically, these sets are of course the same as n^* and n^+ respectively, where n is the cardinality of A.

Since we plan to deal with the complexity of the introduced preorders, we need to represent them by finite words over a finite alphabet. For k-labeled forests we use the alphabet $k \cup \{(,)\}$. A k-labeled forest consists of finitely many k-labeled trees, and we will represent it as a concatenation of these trees, in an arbitrary order. A k-labeled tree T consists of a root with the label $c(T)$ and a forest F consisting of the trees appended to the root. It will be represented by the string $(c(T)F)$. For example, the 4-labeled forest $(\{1, 2, 3, 4, 5\}, \leq)$, with the partial order relation $\leq$ given by the set of pairs $\{(2, 1), (3, 1), (4, 1), (4, 3)\}$ and with the labeling c defined by $c(i) := i \mod 4$ can be represented by the string $(1(2)(3(0)))(1)$. Note that any string representing a k-labeled forest with n elements has length exactly $3n$, for any $k < \omega$. In order to represent ω-labeled forests, we proceed in the same way, but represent the labels in binary form. Thus, we use the alphabet $\{(,), 0, 1\}$. Since we wish to compare two forests, and the Turing machine expects the input on one input tape, we will give representations of the two forests to the Turing machine, separated by some special symbol, e.g., $\#$.

3   The Complexity of $\leq_0$

The following theorem is our first main result. It implies Theorem 1.

Theorem 4. *There is a Turing machine which, given two k-labeled forests F and G, checks in time $O(m + n + m^2 n + mn^2)$ (where $m := |F|$ and $n := |G|$) whether $F \leq_0 G$ or not, for any $k < \omega$. The same is true for ω-labeled forests (where now m resp. n are the lengths of representations of F and G, respectively).*

Thus, the relation $\leq_0$ is computable in cubic time on $\mathcal{F}_k$, for any $k < \omega$, and on $\mathcal{F}_\omega$.

Proof. We consider the case of k-labeled forests, for some $k < \omega$. The proof in the case of ω-labeled forests is similar. The Turing machine is based on the following sequence of statements for two forests F and G, which is similar to an observation in [20]:

1. If $|F| = 0$ then $F \leq_0 G$ is true.
2. If $|F| \geq 1$ and $|G| = 0$ then $F \leq_0 G$ is false.
3. If $F = p_i(F_0)$ and $G = p_i(G_0)$ then $F \leq_0 G$ iff $F_0 \leq_0 G$.
4. If $F = p_i(F_0)$, $G = p_j(G_0)$, and $i \neq j$ then $F \leq_0 G$ iff $F \leq_0 G_0$.
5. If $G = G_0 \sqcup G_1$ then $F \leq_0 G$ iff $F \leq_0 G_0 \vee F \leq_0 G_1$.
6. If $F = F_0 \sqcup F_1$ then $F \leq_0 G$ iff $F_0 \leq_0 G \wedge F_1 \leq_0 G$.

It is quite straightforward to construct a recursive algorithm and, thus, a Turing machine, that distinguishes cases following the above sequence of statements in order to decide for two given forests F and G whether $F \leq_0 G$ or not. Let $t(m, n)$ denote the maximum time needed by this Turing machine, given a forest F with $|F| = m$ and a forest G with $|G| = n$. An analysis of this algorithm shows that there exists a constant $c > 0$ such that for all $m, n \geq 0$

$$t(0, n) \leq c \cdot (1 + n) \quad \text{and} \quad t(m, 0) \leq c \cdot (1 + m),$$

and for all $m, n \geq 1$

$$
\begin{aligned}
t(m, n) \leq\ & c \cdot (m + n) \\
& + \max\{ t(m - 1, n), \\
& \qquad\quad t(m, n - 1), \\
& \qquad\quad \max\{t(m, n_1) + t(m, n_2) \mid n_1 \geq 1, n_2 \geq 1, n_1 + n_2 = n\}, \\
& \qquad\quad \max\{t(m_1, n) + t(m_2, n) \mid m_1 \geq 1, m_2 \geq 1, m_1 + m_2 = m\}\}
\end{aligned}
$$

Using these recursive inequalities, one shows by induction for $m, n \geq 1$

$$t(m, n) \leq 2c \cdot (m^2 n + mn^2).$$

This proves the claim $t(m, n) \in O(m + n + m^2 n + mn^2)$. □

The previous theorem may be used to establish polynomial-time computability of some other natural relations and functions on $\mathcal{F}_k$. We give some examples related to minimal k-forests. Recall that a *minimal k-forest* is a finite k-forest not $\leq_0$-equivalent to a k-forest of lesser cardinality. As observed in [20], any finite k-forest is equivalent to a unique (up to isomorphism) minimal k-forest. The next characterization of the minimal k-forests from [20] is a kind of inductive definition (by induction on the cardinality) of the minimal k-forests.

Proposition 5. *1. The empty k-forest is minimal, and any singleton k-forest is minimal.*
2. A non-singleton k-tree $(T; \leq, c)$ is minimal iff the k-forest $(F; \leq_F, c|_F)$ obtained by deleting the root from T is minimal, and if $c(\mathrm{root}(T)) \neq c(y)$ for all immediate predecessors y of the root in T.

3. *A non-empty, proper (i.e., not $\leq_0$-equivalent to a k-tree) k-forest is minimal iff all its k-trees are minimal and pairwise incomparable under $\leq_0$.*

The first item in the next lemma was observed in [20], the other follows from the previous proposition.

Lemma 6.   *1. Any two $\leq_0$-equivalent k-trees have the same root label.*
  2. If F is a minimal k-forest and $x \in F$ then the lower cone $\downarrow x$ (with the induced partial order and labeling) is minimal.

If $F = F_0 \sqcup \cdots \sqcup F_n$ where the summands are k-trees and $i < k$, let $q_i(F)$ denote the k-forest $G_0 \sqcup \cdots \sqcup G_m$ where the sequence $(G_0, \ldots, G_m)$ is obtained from $(F_0, \ldots, F_n)$ by deleting all roots labeled by i (note that $q_i(F)$ may be empty). Relate to any k-forest $F = F_0 \sqcup \cdots \sqcup F_n$ where the summands are k-trees the k-forest $F^* = F_{i_0} \sqcup \cdots \sqcup F_{i_m}$ where i_0 is the smallest $i \leq n$ with $F_i \not\leq_0 F_{i+1} \sqcup \cdots \sqcup F_n$, $i_1 > i_0$ is the smallest number $i \leq n$ (if any) with $F_i \not\leq_0 F_{i+1} \sqcup \cdots \sqcup F_n \sqcup F_{i_0}$, $i_2 > i_1$ is the smallest number $i \leq n$ (if any) with $F_i \not\leq_0 F_{i+1} \sqcup \cdots \sqcup F_n \sqcup F_{i_0} \sqcup F_{i_1}$, and so on. The next lemma follows from the definitions and from Theorem 4.

Lemma 7.   *1. For any $F \in \mathcal{F}_k$ and $i < k$, $|q_i(F)| \leq |F|$ and $|F^*| \leq |F|$.*
  2. For any $F = F_0 \sqcup \cdots \sqcup F_n$, $F_{i_0}, \ldots, F_{i_m}$ are pairwise $\leq_0$-incomparable and $F^* \equiv_0 F$.
  3. The function $F \mapsto q_i(F)$ is computable in linear time.
  4. The function $F \mapsto F^*$ is computable in cubic time.

The first two statements follow directly from the definitions. Due to lack of space, we omit the proofs of the other two statements.

Theorem 8. *There is a function* $\min : \mathcal{F}_k \to \mathcal{F}_k$ *computable in time* $O(n^4)$ *such that, for any* $F \in \mathcal{F}_k$, $\min(F)$ *is minimal and* $\min(F) \equiv_0 F$.

Proof. The function min is given by the following recursive algorithm:

1. if $|F| = 0$ or $|F| = 1$ then set $\min(F) = F$;
2. if $F = p_i(G)$ is not a singleton, then set $\min(F) = p_i(q_i(\min(G))^*)$;
3. if $F = F_0 \sqcup \cdots \sqcup F_m$ where the summands are k-trees and $m > 0$ then set $\min(F) = (\min(F_0) \sqcup \cdots \sqcup \min(F_m))^*$.

The correctness of the algorithm follows from the above proposition and lemmas. Let $t(n)$ be the maximum time needed for computing $\min(F)$ for some k-labeled forest F with $n = |F|$. Using Lemma 7, one shows $t(n) \in O(n^4)$.   □

4   The Complexity of $\leq_1$

Lemma 9. *The relations $\leq_1$ and $\leq_2$ coincide on $\mathcal{C}_\omega$ and also, for each $k < \omega$, on $\mathcal{C}_k$.*

Proof. For the nontrivial direction, let $u \leq_2 v$ via $\varphi : u \to v$ where u, v are non-empty words over the alphabet ω, hence also over the alphabet k for some $k < \omega$. It suffices to find a function $f : k \to k$ such that $u(n) = f(v(\varphi(n)))$ for each $n < |u|$. Let $a_0, \ldots, a_m$ be an enumeration without repetitions of all the letters in u. For each $i \leq m$, let $A_i = \{n < |u| : u(n) = a_i\}$ and $B_i = \{v(\varphi(n)) : n \in A_i\}$. Since φ is a 2-morphism, $B_0, \ldots, B_m$ are pairwise disjoint subsets of k. Let f be a function of k that, for each $i \leq m$, sends all elements of B_i to a_i. Then f has the desired property.   □

Proposition 10. *The relations $\leq_1$ and $\leq_2$ on $\mathcal{C}_\omega$ are NP-hard.*

Proof. By Lemma 9, it suffices to find a polynomial time reduction of 3-SAT to $\leq_2$ on $\mathcal{C}_\omega$ (in this proof we assume familiarity of the reader with some common knowledge from complexity theory, see e.g. [1,2]). We have to relate in polynomial time to any 3-CNF $C = C(x_0, \ldots, x_n)$ words u, v over the alphabet ω such that C is satisfiable iff $u \leq_2 v$. In "mnemonic notation", u will in fact be a word over the alphabet $\{0, 1, T, F\}$, and v will be a word over the alphabet $\{0, 1, x_0, \bar{x}_0 \ldots, x_n, \bar{x}_n\}$). A straightforward coding turns u, v into words over the alphabet ω.

We may assume that $C = D_0 \wedge \cdots \wedge D_m$ where any D_j is a disjunction $x_{p_j}^{a_j} \vee x_{q_j}^{b_j} \vee x_{r_j}^{c_j}$ of exactly three literals, $p_j, q_j, r_j \leq n$, $0 \leq a_j, b_j, c_j \leq 1$, $x^1 = x$ and $x^0 = \bar{x}$, and no disjunction contains a letter with its negation. For each $j \leq m$, let $\tilde{D}_j$ denote the word $x_{p_j}^{a_j} x_{q_j}^{b_j} x_{r_j}^{c_j}$. Define the words u' and v' by

$$u' := (TF01)^{n+1}(T01)^{m+1}$$
$$v' := x_0 \bar{x}_0 x_0 01 \cdots x_n \bar{x}_n x_n 01 \tilde{D}_0 01 \cdots \tilde{D}_m 01.$$

Note that $|u'| = 4(n + 1) + 3(m + 1)$ and $|v'| = 5(n + 1) + 5(m + 1)$. Finally, let $u := (01)^M u'$ and $v := (01)^M v'$ where $M = |v'| + 1$. Obviously, the function $C \mapsto (u, v)$ is computable in linear time.

We call $x_i \bar{x}_i x_i 01$ the *i-th variable* factor of v ($0 \leq i \leq n$), and $\tilde{D}_j 01$ the *j-th disjunction* factor of v ($0 \leq j \leq m$). Note that the variable (resp. disjunction) factors of v are in the natural bijective correspondence with the factors $TF01$ (resp. $T01$) of u. For each $i \leq n$, let t_i (resp. l_i) be the position in u (resp. in v) of the letter T (resp. of the first entry of x_i) in the i-th factor $TF01$ (resp. of the i-th variable factor). For each $j \leq m$, let s_j be the position in u of the letter T in the j-th factor $T01$.

Let C be satisfiable, i.e., true for some assignment $\alpha : \{x_0, \ldots, x_n\} \to \{T, F\}$. Define a monotone function $\varphi : |u| \to |v|$ as follows: φ sends the position of any entry of 0 or 1 in u to the corresponding position of the same letter in v (note that the numbers of entries of these letters in u and v coincide); for any $i \leq n$, if $\alpha(x_i) = T$ then $\varphi(t_i) = l_i$ and $\varphi(t_i + 1) = l_i + 1$, otherwise $\varphi(t_i) = l_i + 1$ and $\varphi(t_i + 1) = l_i + 2$; for any $j \leq m$, φ sends s_j to the position of the first true literal in the j-th disjunction factor. Then φ is a 2-morphism from u to v.

Conversely, let $\varphi : u \to v$ be a 2-morphism; we have to find a satisfying assignment α for C. Since u, v are repetition-free, φ is injective. Since $M > |v'|$,

$\varphi(0, \ldots, 2M - 1)$ contains some positions for both 0 and 1, hence φ cannot send any position of T or F to a position of 0 or 1. Since any factor of v without $0, 1$ has length 3, the positions of T, F in the i-th factor $TF01$ can go only to positions of $x_i \bar{x}_i x_i$ in the corresponding i-variable factor (for each $i \leq n$), and, for each $j \leq m$, $\varphi(s_j)$ is a position of some literal in the j-th disjunctive factor. For any $i \leq n$, let $\alpha(x_i) = T$ iff $\varphi(t_i) = l_i$. Then α is a desired satisfying assignment.   □

Theorem 11. *For any $k < \omega$, the relation $\leq_1$ on $\mathcal{F}_k$ is computable in cubic time. The relation $\leq_1$ on $\mathcal{F}_\omega$ is NP-complete.*

Proof. Let $k < \omega$. For any $G = (Q; \leq, c) \in \mathcal{F}_k$ and $f : k \to k$, we define $G_f := (Q; \leq, f \circ c)$. From the definition of $\leq_1$ we observe that $F \leq_1 G$ iff there exists a function $f : k \to k$ such that $F \leq_0 G_f$. Let $\{f_1, \ldots, f_{k^k}\}$ be an enumeration without repetition of $\{f \mid f : k \to k\}$. Since $G \mapsto (G_{f_1}, \ldots, G_{f_{k^k}})$ is computable in linear time and $\leq_0$ is computable in cubic time, $\leq_1$ on $\mathcal{F}_k$ is computable in cubic time.

For $\mathcal{F}_\omega$, $\leq_1$ is NP-hard by the previous proposition. It remains to show that it is in NP. The corresponding nondeterministic algorithm is obvious: given ω-labeled forests $F = (P; \leq_P, c_P)$ and $G = (Q; \leq_Q, c_Q)$, guess a function $f : c_Q(Q) \to c_P(P)$ and a function $\varphi : P \to Q$ and check (in polynomial time) whether φ is a 0-morphism from $(P; \leq_P, c_P)$ to $(Q; \leq_Q, f \circ c_Q)$.   □

Note that by Theorem 11 the relation $\leq_1$ on $\mathcal{F}_\omega$ is fixed-parameter tractable (compare [6]) if the largest label of the input forests is taken as the parameter.

5   The Complexity of $\leq_2$

Theorem 12. *For any $k < \omega$, the relation $\leq_2$ on $\mathcal{F}_k$ is computable in time $O(n^{k+3})$. The relation $\leq_2$ on $\mathcal{F}_\omega$ is NP-complete.*

Proof (Sketch). Let $k < \omega$. Let us denote by $f :\subseteq k \to k$ an arbitrary function whose domain $\mathrm{dom}(f)$ and whose range may be any subsets of $k = \{0, \ldots, k-1\}$. For $F = (P; \leq_P, c_P), G = (Q; \leq_Q, c_Q) \in \mathcal{F}_k$, and such a function $f :\subseteq k \to k$, let $F \leq_2^f G$ mean that there is a 2-morphism $\varphi : F \to G$ such that for all $x \in P$, if $c_Q(\varphi(x)) \in \mathrm{dom}(f)$ then $c_P(x) = f(c_Q(\varphi(x)))$. Note that $F \leq_2 G$ is equivalent to $F \leq_2^{f_\emptyset} G$ where $f_\emptyset$ is the partial function from k to k with $\mathrm{dom}(f) = \emptyset$. Therefore it suffices to show that the relation $\leq_2^f$ is computable in time $O(n^{3+k-|\mathrm{dom}(f)|})$, for any function $f :\subseteq k \to k$. Similarly to the proof of Theorem 4, one can construct a natural recursive algorithm based on the following sequence of statements for k-forests F and G and a function $f :\subseteq k \to k$. This sequence of statements is based on the recursive definition of $\leq_2$ in [7,8,9]:

1. If $|F| = 0$ then $F \leq_2^f G$ is true.
2. If $|F| \geq 1$ and $|G| = 0$ then $F \leq_2^f G$ is false.
3. If $F = p_i(F_0)$, $G = p_j(G_0)$, $j \in \mathrm{dom}(f)$, and $i = f(j)$ then $F \leq_2^f G$ iff $F_0 \leq_2^f G$.

4. If $F = p_i(F_0)$, $G = p_j(G_0)$, $j \in \text{dom}(f)$, and $i \neq f(j)$ then $F \leq_2^f G$ iff $F \leq_2^f G_0$.
5. If $F = p_i(F_0)$, $G = p_j(G_0)$, and $j \notin \text{dom}(f)$ then $F \leq_2^f G$ iff ($F \leq_2^f G_0 \vee (F_0 \leq_2^g G$ where g is defined by $\text{dom}(g) := \text{dom}(f) \cup \{j\}$, and by $g(l) := f(l)$ for $l \in \text{dom}(f)$, and $g(j) := i)$).
6. If $G = G_0 \sqcup G_1$ then $F \leq_2^f G$ iff $F \leq_2^f G_0 \vee F \leq_2^f G_1$.
7. If $F = F_0 \sqcup F_1$ then $F \leq_2^f G$ iff $F_0 \leq_2^f G \wedge F_1 \leq_2^f G$.

Note that in comparison with the sequence of statements in the proof of Theorem 4, there is an additional case, namely the fifth case. If f is total, i.e., if $|\text{dom}(f)| = k$, then this case plays no role, and one arrives at the same bound for the time as for the relation $\leq_0$. But if f is not total, this case has to be considered. One shows by induction that for any function $f :\subseteq k \to k$, the algorithm works in time $O(m+n+m^2 n^{1+k-|\text{dom}(f)|}+mn^{2+k-|\text{dom}(f)|})$, for any input forests F, G with $m := |F|, n := |G|$. This is shown first for all total functions f, i.e., with $k - |\text{dom}(f)| = 0$, then for all functions f with $k - |\text{dom}(f)| = 1$, then for all functions with $k - |\text{dom}(f)| = 2$, and so on.

For $\mathcal{F}_\omega$, $\leq_2$ is NP-hard by Proposition 10, so it remains to show that it is in NP. The corresponding nondeterministic algorithm is obvious: given F, G, guess a function $\varphi : F \to G$ and check (in polynomial time) whether φ is a 2-morphism. □

6   Some Open Questions

Some interesting questions related to the topic of this paper remain open. We have shown that on $\mathcal{F}_k$ the preorders $\leq_0$ and $\leq_1$ are computable in cubic time and $\leq_2$ is computable in time $O(n^{k+3})$. It would be interesting to determine the exact complexity classes of these preorders. One may also ask for the complexity of some further natural functions related to the quotient posets $\mathbb{F}_k^i$ of the preorders $(\mathcal{F}_k; \leq_i)$, $i \leq 2$. This applies for example to the rank and spectrum functions defined as follows. The quotient-posets $\mathbb{F}_k^i$ are known [8] to be well posets of rank ω, i.e., they have neither infinite antichains nor infinite descending chains but they have infinite ascending chains. The *rank function* $rk_i : \mathcal{F}_k \to \omega$ measures the rank of the $\leq_i$-equivalence class $[F]_i$ in $\mathbb{F}_k^i$. By the *spectrum function* on $\mathbb{F}_k^i$ we mean the function $sp_i : \omega \to \omega$ where $sp_i(n)$ is the number of elements of rank n in $\mathbb{F}_k^i$.

References

1. Balcázar, J.L., Díaz, J., Gabarró, J.: Structural complexity. I. In: EATCS Monographs on Theoretical Computer Science, vol. 11, Springer, Berlin (1988)
2. Balcázar, J.L., Díaz, J., Gabarró, J.: Structural complexity II. In: EATCS Monographs on Theoretical Computer Science, vol. 22, Springer, Berlin (1990)
3. Brattka, V., Gherardi, G.: Effective choice and boundedness principles in computable analysis. Bulletin of Symbolic Logic 17(1), 73–117 (2011)

4. Brattka, V., Gherardi, G.: Weihrauch degrees, omniscience principles and weak computability. J. Symb. Log. 76(1), 143–176 (2011)
5. Davey, B., Priestley, H.: Introduction to Lattices and Order, 2nd edn. Cambridge University Press, Cambridge (2002)
6. Downey, R.G., Fellows, M.R.: Parameterized Complexity Theory. Monographs in Computer Science. Springer, New York (1999)
7. Hertling, P.: A topological complexity hierarchy of functions with finite range. Technical Report 223, Centre de recerca matematica, Institut d'estudis catalans, Barcelona, Workshop on Continuous Algorithms and Complexity (October 1993)
8. Hertling, P.: Topologische Komplexitätsgrade von Funktionen mit endlichem Bild. Informatik Berichte 152, FernUniversität Hagen, Hagen (December 1993)
9. Hertling, P.: Unstetigkeitsgrade von Funktionen in der effektiven Analysis. PhD thesis, Fachbereich Informatik, FernUniversität Hagen (1996)
10. Hirsch, M.D.: Applications of topology to lower bound estimates in computer science. In: Hirsch, M.W., et al. (eds.) From topology to computation: Proceedings of the Smalefest, pp. 395–418. Springer, New York (1993)
11. Kechris, A.S.: Classical descriptive set theory.. Graduate Texts in Mathematics, vol. 156. Springer, Berlin (1995)
12. Kosub, S.: NP-partitions over posets with an application to reducing the set of solutions of NP problems. Theory Comput. Syst. 38(1), 83–113 (2005)
13. Kosub, S., Wagner, K.W.: The Boolean Hierarchy of NP-Partitions. In: Reichel, H., Tison, S. (eds.) STACS 2000. LNCS, vol. 1770, pp. 157–168. Springer, Heidelberg (2000)
14. Kudinov, O.V., Selivanov, V.L.: Undecidability in the Homomorphic Quasiorder of Finite Labeled Forests. In: Beckmann, A., Berger, U., Löwe, B., Tucker, J.V. (eds.) CiE 2006. LNCS, vol. 3988, pp. 289–296. Springer, Heidelberg (2006)
15. Kudinov, O.V., Selivanov, V.L.: Undecidability in the homomorphic quasiorder of finite labelled forests. J. Log. Comput. 17(6), 1135–1151 (2007)
16. Kudinov, O.V., Selivanov, V.L.: A Gandy Theorem for Abstract Structures and Applications to First-Order Definability. In: Ambos-Spies, K., Löwe, B., Merkle, W. (eds.) CiE 2009. LNCS, vol. 5635, pp. 290–299. Springer, Heidelberg (2009)
17. Kudinov, O.V., Selivanov, V.L., Zhukov, A.V.: Undecidability in Weihrauch Degrees. In: Ferreira, F., Löwe, B., Mayordomo, E., Mendes Gomes, L. (eds.) CiE 2010. LNCS, vol. 6158, pp. 256–265. Springer, Heidelberg (2010)
18. Kwuida, L., Lehtonen, E.: On the homomorphism order of labeled posets. Order (2010)
19. Lehtonen, E.: Labeled posets are universal. Eur. J. Comb. 29(2), 493–506 (2008)
20. Selivanov, V.: Boolean hierarchies of partitions over a reducible base. Algebra Logika 43(1), 77–109 (2004); Translation in Algebra and Logic 43(1), 44-61
21. Selivanov, V.L.: Hierarchies of Δ_2^0-measurable k-partitions. Mathematical Logic Quarterly 53(4–5), 446–461 (2007)
22. von Stein, T.: Vergleich nicht konstruktiv lösbarer Probleme in der Analysis. Diplomarbeit, Fakultät für Informatik, FernUniversität Hagen (March 1989)
23. Weihrauch, K.: The degrees of discontinuity of some translators between representations of the real numbers. Informatik Berichte 129, FernUniversität Hagen, Hagen (July 1992)
24. Weihrauch, K.: The TTE-interpretation of three hierarchies of omniscience principles. Informatik Berichte 130, FernUniversität Hagen, Hagen (September 1992)
25. Weihrauch, K.: Computable Analysis. Springer, Berlin (2000)

Randomness and the Ergodic Decomposition

Mathieu Hoyrup[1]

LORIA, INRIA Nancy-Grand Est, 615 rue du jardin botanique, BP 239, 54506
Vandœuvre-lès-Nancy, France
mathieu.hoyrup@loria.fr

Abstract. The interaction between algorithmic randomness and ergodic
theory is a rich field of investigation. In this paper we study the particular case of the ergodic decomposition. We give several positive partial
answers, leaving the general problem open. We shortly illustrate how
the effectivity of the ergodic decomposition allows one to easily extend
results from the ergodic case to the non-ergodic one (namely Poincaré recurrence theorem). We also show that in some cases the ergodic measures
can be computed from the typical realizations of the process.

1 Introduction

The goal of the paper is to study the interaction between the theory of algorith-
mic randomness, started by Martin-Löf [9], and ergodic theory (i.e. restricting
to shift-invariant measures). The first results in this direction were obtained by
V'yugin [12], who proved that Birkhoff's ergodic theorem and a weak form of
Shannon-McMillan-Breiman theorem hold for each Martin-Löf random sequence.
Recently several improvements of the first result have been achieved [10,6,1].

A classical result from ergodic theory, called the *ergodic decomposition*, states
that given a stationary process, almost every realization is actually a typical
realization of an *ergodic* process. The full process can be decomposed as the
combination of a collection of ergodic processes. It is natural to ask the question
whether every Martin-Löf random sequence (with respect to the stationary measure) statistically induces an ergodic measure, and if the sequence is Martin-Löf
random with respect to it.

We give three orthogonal cases in which we can give a positive answer: (i) when
the decomposition of the measure is computable, (ii) when the decomposition of
the measure is supported on an effective compact class of ergodic measures, (iii)
when the decomposition of the measure is finite. Observe that the three cases
are mutually incomparable. We leave the general problem open.

As a side result, we give sufficient conditions to infer the statistics of the
system from the observation; formally we give a sufficient condition on an ergodic
measure to be computable relative to its random elements.

In Section 2 we give the necessary background on computability and randomness. In Section 3 we develop results about randomness and combinations
of measures that will be applied in the sequel, but are of independent interest
(outside ergodic theory). We start Section 4 with a reminder on the ergodic
decomposition and then we present our results relating it to randomness.

B. Löwe et al. (Eds.): CiE 2011, LNCS 6735, pp. 122–131, 2011.

2   Preliminaries

We assume familiarity with algorithmic randomness and computability theory.

All the results stated in this paper hold on effectively compact computable metric spaces X and for computable maps $T : X \rightarrow X$ (as defined in computable analysis, see [13]), but for the sake of simplicity we formulate them only on the Cantor space $X = \{0,1\}^{\mathbb{N}}$ and for the shift transformation $T : X \rightarrow X$ defined by $T(x_0 x_1 x_2 \ldots) = x_1 x_2 x_3 \ldots$. The Cantor space is endowed with the product topology, generated by the cylinders $[w]$, $w \in \{0,1\}^*$. Implicitly, measures are probability measures defined on the Borel σ-algebra, and ergodic measures are stationary (i.e., shift-invariant) ergodic Borel probability measures. The set $\mathcal{P}(X)$ of probability measures over X is endowed with the weak* topology, given by the notion of weak convergence: measures P_n converge to P if for every $w \in \{0,1\}^*$, $P_n[w] \rightarrow P[w]$.

A name for a real number r is an infinite binary sequence encoding, in some canonical effective way, a sequence of rational numbers q_n such that $|q_n - r| < 2^{-n}$ for all n. A name for a probability measure P is the interleaving, in some canonical effective way, of names for the numbers $P[w]$, $w \in \{0,1\}^*$. A computable probability measure is a measure admitting a computable name: in other words, the numbers $P[w]$ are uniformly computable.

Let X, Y be any spaces among $\{0,1\}^{\mathbb{N}}$, $\mathbb{R}$ and $\mathcal{P}(X)$. A function $f : X \rightarrow Y$ is computable if there is an oracle machine that, given a name of $x \in X$ as an oracle, outputs a name of $f(x)$ (the computation never halts). Computable functions are continuous. f is computable on a set $A \subseteq X$ if the same holds for all $x \in A$ (nothing is required to the machine when $x \notin A$). An object y is **computable relative to** an object x if the function $x \mapsto y$ is computable on $\{x\}$, i.e. if there is an oracle machine that on any name of x as oracle, produces a name of y.

An open subset U of the Cantor space is effective if there is a (partial) computable function $\varphi : \mathbb{N} \rightarrow \{0,1\}^*$ such that $U = \bigcup_{n \in \mathbb{N}} [\varphi(n)]$. An effective compact set is the complement of an effective open set. Let $K \subseteq X$ be an effective compact set and $f : K \rightarrow Y$ a function computable on K.

Fact 1 (Folklore). $f(K)$ is an effective compact set.

Fact 2 (Folklore). If f is moreover one-to-one then $f^{-1} : f(K) \rightarrow K$ is computable on $f(K)$.

The product of two computable metric spaces has a natural structure of computable metric space.

Fact 3 (Folklore). If $K \subseteq X$ is an effective compact set and $f : K \times Y \rightarrow \overline{\mathbb{R}}$ is lower semi-computable, then the function $g : Y \rightarrow \overline{\mathbb{R}}$ defined by $g(y) = \inf_{x \in K} f(x,y)$ is lower semi-computable.

If f, g are real-valued functions, $f \stackrel{*}{<} g$ means that there exists $c \geq 0$ such that $f \leq cg$. $f \stackrel{*}{=} g$ means that $f \stackrel{*}{<} g$ and $g \stackrel{*}{<} f$.

2.1   Effective Randomness

Martin-Löf [9] was the first one to define a sound individual notion of random infinite binary sequence. He developed his theory for any computable probability measure on the Cantor space. This theory was then extended to non-computable measures by Levin [8], and later by [3,7] on general spaces ([5] was an extension to topological spaces, but for computable measures).

We will use the most general theory: we will be interested in randomness on the Cantor space and on the space of Borel probability measures over the Cantor space, for arbitrary (i.e. not necessarily computable) probability measures. In particular, we will use the notion of uniform test of randomness, introduced by Levin [8] and further developed in [3,4,7].

On a computable metric space X endowed with a probability measure P, there is a set $\mathcal{R}_P$ of P-random elements satisfying $P(\mathcal{R}_P) = 1$, together with a canonical decomposition (coming from the universal P-test) $\mathcal{R}_P = \bigcup_n \mathcal{R}_P^n$ where $\mathcal{R}_P^n$ are uniformly effective compact sets relative to P, $\mathcal{R}_P^n \subseteq \mathcal{R}_P^{n+1}$ and $P(\mathcal{R}_P^n) > 1 - 2^{-n}$. The sets $X \setminus \mathcal{R}_P^n$ constitute a universal Martin-Löf test. A P-test is a function $t : X \to [0, +\infty]$ which is lower semi-computable relative to P, such that $\int t \, dP \leq 1$.

A function $f : X \to Y$ is P-***layerwise computable*** if there is an oracle machine that, given n as input and a name of $x \in \mathcal{R}_P$ as an oracle, outputs a name of $f(x)$ (the computation never halts). Nothing is required to the machine when x is not P-random. When f is P-layerwise computable, for every P-random x, $f(x)$ is computable relative to x in a way that is not fully uniform, but uniform on each set $\mathcal{R}_P^n$.

Such a machine can be thought of as a probabilistic algorithm, but here the randomness is not part of the algorithm but of the input. Formally, it is the same notion, but usually, "succeeding with high probability" means that if we run the program on a given input several times, independently, it will succeed most of the times; here, the algorithm is deterministic and it will succeed on most inputs.

Lemma 1. *Let P be a computable measure, $f : X \to Y$ a P-layerwise computable function and $Q = f_*P$ the push-forward of P under f.*

1. *Q is computable and $f : \mathcal{R}_P \to \mathcal{R}_Q$ is onto.*
2. *If $f : X \to Y$ is moreover one-to-one then $f : \mathcal{R}_P \to \mathcal{R}_Q$ is one-to-one and f^{-1} is Q-layerwise computable.*

Proof. We only prove that f^{-1} is Q-layerwise computable, the other statements are proved in [6]. There is $c \in \mathbb{N}$ such that $\mathcal{R}_Q^n \subseteq f(\mathcal{R}_P^{n+c})$ for all n. Let $n \in \mathbb{N}$. $f : \mathcal{R}_P^{n+c} \to Y$ is one-to-one and computable so by Fact 2, $f^{-1} : f(\mathcal{R}_P^{n+c}) \to X$ is computable. As $\mathcal{R}_Q^n \subseteq f(\mathcal{R}_P^{n+c})$, $f^{-1} : \mathcal{R}_Q^n \to X$ is computable. This is uniform in n.

3   Randomness and Continuous Combination of Measures

The material developed here will be used to investigate the algorithmic content of the ergodic decomposition.

Given a *countable* class of probability measures P_i and real numbers $\alpha_i \in [0,1]$ such that $\sum_i \alpha_i = 1$, the convex combination $P = \sum_i \alpha_i P_i$ is again a probability measure. This can be generalized to *continuous* classes of measures, as we briefly recall now.

Let m be a probability measure over $\mathcal{P}(X)$. The set function P defined by $P(A) = \int Q(A) \, dm(Q)$ for measurable sets A is a probability measure over X, called the barycenter of m. It satisfies $\int f \, dP = \int (\int f \, dQ) \, dm(Q)$ for $f \in L^1(X, P)$. When m is computable, so is P. We can think of P as the measure describing the following process: first pick some measure Q at random according to m; then run the process with distribution Q.

Probabilistically, picking a sequence according to P or decomposing into these two steps are equivalent. We are interested in whether the algorithmic theory of randomness fits well with this intuition: are the P-random sequences the same as the sequences that are Q-random for some m-random Q?

Remark 1. Let $f : X \to [0, +\infty]$ be a lower semi-computable function. Let $F : \mathcal{P}(X) \to [0, +\infty]$ be defined by $F(Q) = \int f \, dQ$. F is lower semi-computable and $\int F \, dm = \int f \, dP$. As a result, F is a m-test if and only if f is a P-test.

Theorem 1. *Let $m \in \mathcal{P}(\mathcal{P}(X))$ be computable, and P be the barycenter of m. For $x \in X$, the following are equivalent:*

1. *x is P-random,*
2. *x is Q-random for some m-random Q.*

In other words,

$$\mathcal{R}_P = \bigcup_{Q \in \mathcal{R}_m} \mathcal{R}_Q.$$

Proof

Let $f(x) = \inf_Q t_m(Q).t_Q(x)$. f is lower semi-computable by Fact 3 ($\mathcal{P}(X)$ is effectively compact). As $\int f \, dP = \int (\int f \, dQ) \, dm(Q) \leq \int t_m(Q)(\int t_Q \, dQ) \, dm(Q) \leq 1$, f is a P-test, so if x is P-random then it is Q-random for some m-random measure Q.

Conversely, let $T_P(Q) = \int t_P \, dQ$ where t_P is a universal P-test. By Remark 1, T_P is an m-test so if Q is m-random then $T_P(Q) < \infty$, so t_P is a (multiple of a) Q-test. As a result, $\mathcal{R}_Q \subseteq \mathcal{R}_P$.

4   Randomness and Ergodic Decomposition

4.1   Background from Ergodic Theory

A sequence $x \in \{0,1\}^{\mathbb{N}}$ is **generic** if for each $w \in \{0,1\}^*$, the frequency of occurrences of w in x converges. If x is generic, we denote by Q_x the set function which maps each cylinder $[w]$ to the limit frequency of occurrences of w in x. Q_x extends to a probability measure over the Cantor space, which we also denote by Q_x. If x is generic then Q_x is stationary, i.e. $Q_x(A) = Q_x(T^{-1}(A))$ for every measurable set A. Birkhoff's ergodic theorem states that given a stationary measure P, P-almost every x is generic. A stationary measure P is ergodic if the only invariant sets have measure 0 or 1. Formally, if $T^{-1}(A) = A$ then $P(A) = 0$ or 1, for every measurable set A. If P is stationary ergodic then $Q_x = P$ for P-almost every x.

The ergodic decomposition theorem states that given a stationary probability measure P, the measure Q_x is ergodic for P-almost every x. There are mainly two proofs of this fact. One of them uses Choquet theorem from convex analysis (see [11]): the set of stationary probability measures is a convex compact set whose extreme points are exactly the ergodic measures. Then any point in that set, i.e. any invariant measure, can be expressed as a barycenter over the ergodic measures. More precisely, for any invariant measure P there is a unique probability measure m_P over $\mathcal{P}(X)$ which gives full weight to the ergodic measures, and such that $P(A) = \int Q(A) \, dm_P(Q)$ for every Borel set A. We will call m_P the **Choquet measure** associated to P.

4.2   Randomness and Ergodic Theorems

An algorithmic version of Birkhoff's ergodic theorem was eventually proved by V'yugin [12]: given a shift-invariant probability measure P, every P-random sequence is generic, and if P is moreover ergodic then $Q_x = P$ for every P-random sequence x (it was proved for computable measures, but it still works for non-computable measures). The proof was not immediate to obtain from the classical proof of Birkhoff's theorem, which is in a sense not constructive. In this paper we are interested in an algorithmic version of the ergodic decomposition theorem, which again cannot be proved directly.

More precisely, given a stationary measure P, we are interested in the following questions:

- if x is P-random, is Q_x ergodic?
- if x is P-random, is x also Q_x-random?
- if x is P-random, is Q_x an m_P-random measure?
- does any converse implication hold?

We give positive partial answers to these questions, leaving the general problem open. We will use the following lemmas (the first one was proved in [12]).

Lemma 2. *Let P be an ergodic stationary probability measure. For every $x \in \mathcal{R}_P$, $Q_x = P$.*

Lemma 3. *Let P be a stationary probability measure and m_P the associated Choquet measure. Every m_P-random measure is ergodic and stationary.*

Proof. It is known that the set of ergodic stationary measure is a G_δ-set. It is moreover effective, i.e. it is an intersection of effective open sets. As it has m_P-measure one, it contains $\mathcal{R}_{m_P}$.

4.3   First Answer: Effective Decomposition

A stationary probability measure P is always computable relative to its associated Choquet measure m_P. The converse does not always hold (see Section 4.4 for a counter-example).

Definition 1. *A computable stationary probability measure P is **effectively decomposable** if its Choquet measure is computable.*

As an application of Theorem 1, we directly get a result when P is effectively decomposable (i.e. when $m := m_P$ is computable).

Corollary 1. *Let P be a computable stationary probability measure that is effectively decomposable. For $x \in X$, the following are equivalent:*

1. *x is P-random,*
2. *x is Q-random for some m-random Q.*

In other words, the following are equivalent:

1. *x is P-random,*
2. *x is generic, Q_x-random and Q_x is m-random.*

We also have the following characterization.

Theorem 2. *Let P be a computable stationary probability measure. The following are equivalent.*

1. *P is effectively decomposable,*
2. *the function $X \to \mathcal{P}(X), x \mapsto Q_x$ is P-layerwise computable.*

Proof. $1 \Rightarrow 2$. In any probability space (Y, μ) with random elements $\mathcal{R}_\mu = \bigcup_n \mathcal{R}_\mu^n$, we define $d_\mu(y) = \min\{n : y \in \mathcal{R}_\mu^n\}$ ($d_\mu(y) = +\infty$ if y is not μ-random). $d : \mathcal{P}(Y) \times Y \to [0, +\infty]$ which maps (μ, y) to $d_\mu(y)$ is lower semi-computable.

Let $C_n = \{(Q, x) : d_m^{\cdots}(Q) \le n \text{ and } d_Q(x) \le n\}$. The second projection $\pi_2 : \bigcup_n C_n \to X$ is one-to-one. Indeed, if $(Q_i, x_i) \in \bigcup_n C_n$, $i = 1, 2$ and $\pi_2(Q_1, x_1) = \pi_2(Q_2, x_2)$ then (i) $x_1 = x_2$, (ii) Q_1, Q_2 are m-random hence ergodic, (iii) x_i is Q_i-random so $Q_{x_i} = Q_i$; as a result, $Q_1 = Q_{x_1} = Q_{x_2} = Q_2$. C_n is effectively compact so π_2^{-1} is computable on each $\pi_2(C_n)$ (uniformly in n) by Fact 2.

We know from the proof of Theorem 1 that there exists a constant c such that for all n and all $x \in \mathcal{R}_P^n$, $(Q_x, x) \in C_{n+c}$, hence $\mathcal{R}_P^n \subseteq \pi_2(C_{n+c})$. It implies that π_2^{-1} is computable on each $\mathcal{R}_P^n$, uniformly in n, i.e. π_2^{-1} is P-layerwise computable. Finally, $\pi_1 \circ \pi_2^{-1}$, which maps $x \in \mathcal{R}_P$ to Q_x is P-layerwise computable.

$2 \Rightarrow 1$. Conversely, if $\psi : x \mapsto Q_x$ is P-layerwise computable, then $m = \psi_* P$ is the push-forward of P under ψ, so it is computable by Lemma 1, item 1. $\square$

Remark 2. For $f \in L^1(X, P)$, we denote by f^* the limit of the Birkhoff averages of f. One can also prove that if P is computable then P is effectively decomposable if and only if the function

$$L^1(X, P) \to L^1(X, P)$$
$$f \mapsto f^*$$

is computable.

The effectivity of the ergodic decomposition enables one to extend results from ergodic systems to non-ergodic ones. Let us illustrate it. It was proved in [1] that when P is an ergodic measure, every P-random sequence eventually visits every effective compact set of positive measure under shift iterations. When the decomposition is effective, this theorem can be generalized to non-ergodic measures, giving a version of Poincaré recurrence theorem for random sequences.

Corollary 2. *Let P be a stationary measure that is effectively decomposable. Let F be an effective compact set such that $P(F) > 0$. Every P-random $x \in F$ falls infinitely often in F under shift iterations.*

Proof. x is Q_x-random and Q_x is ergodic. As all random sequences belong to effective open sets of measure one and $x \in F$, $Q_x(F) > 0$. Hence we can apply the result in [1] to the ergodic measure Q_x (strictly speaking their result was proved for *computable* ergodic measures, but it can be relativized without difficulty).

The result actually holds as soon as for every P-random x, Q_x is ergodic and x is Q_x-random.

4.4   V'yugin's Example

In [12], V'yugin constructed a computable stationary measure for which the convergence of Birkhoff's average is not effective. We give a simpler construction and show that this measure is not effectively decomposable.

Let M_i be some effective enumeration of the Turing machines. For each i, let $p_i = 2^{-t_i}$ if M_i halts in time t_i, $p_i = 0$ if t_i does not halt. The real numbers p_i are computable uniformly in i (while they are not uniformly computable as rational numbers). Let P_i be the Markovian stationary measure defined by $P_i[0] = P_i[1] = \frac{1}{2}$ and $\frac{P_i[w01]}{P_i[w0]} = \frac{P_i[w10]}{P_i[w1]} = p_i$ for all $w \in \{0,1\}^*$ (the probability of changing between states 0 and 1 is p_i). Let $P = \sum_i 2^{-i} P_i$. P is computable. We now show that $x \mapsto Q_x$ is not P-layerwise computable (which will imply that P is not effectively decomposable by Theorem 2). Let $f = \chi_{[1]}$. Let $\alpha = \sum_{i: M_i \text{ halts}} 2^{-i}$. $f^*(x) = 0$ for $x = 0^{\mathbb{N}}$, $f^*(x) = 1$ for $x = 1^{\mathbb{N}}$ and $f^*(x) = \frac{1}{2}$ for P-almost all $x \notin \{0^{\mathbb{N}}, 1^{\mathbb{N}}\}$. By definition of Q_x, $f^*(x) = Q_x[1]$ for every P-random x. If f^* were P-layerwise computable, then $P(f^{*-1}[0, 1/4)) = (1 - \alpha)/2$ would be lower semi-computable by basic properties of layerwise computable functions (see [6]).

While P is not effectively decomposable, we can still get a result about random elements.

Proposition 1. *For every P-random x, Q_x is ergodic and x is Q_x-random.*

Proof. The decomposition $P = \sum_i 2^{-i} P_i$ is partial in the sense that some P_i are not ergodic (when M_i does not halt). However we can apply Theorem 1 to this decomposition: P is the barycenter of the computable measure $m' = \sum_i 2^{-i} \delta_{P_i}$, so every P-random x is random for some P_i. (i) If M_i halts, then P_i is ergodic. (ii) If M_i does not halt then $P_i = \frac{1}{2}(\delta_0 + \delta_1)$ (where δ_0 is the measure concentrated on $0^{\mathbb{N}}$, δ_1 on $1^{\mathbb{N}}$). In turn, P_i, which is non-ergodic is effectively decomposable. Hence as x is P_i-random, $Q_x = \delta_0$ or δ_1 and x is Q_x-random.

As a result, Corollary 2 also holds for the measure P.

4.5 A Particular Case: Effective Compact Classes of Ergodic Measures

Proposition 2. *Let P be a computable stationary probability measure. If m_P is supported on an effective compact class of ergodic measures, then P is effectively decomposable.*

Proof. Let $\mathscr{C}$ be an effective compact class of stationary ergodic probability measures. Let $\mathcal{P}(\mathscr{C})$ be the set of probability measures m over $\mathcal{P}(X)$ such that $m(\mathscr{C}) = 1$. $\mathcal{P}(\mathscr{C})$ is an effective compact subset of $\mathcal{P}(X)$: indeed, it is the pre-image of $[1, +\infty]$ under the upper semi-computable function $m \mapsto m(\mathscr{C})$. If $m \in \mathcal{P}(\mathscr{C})$, the barycenter P of m is defined by $P(A) = \int Q(A)\, dm(Q)$ for every measurable set A. The function ψ which maps m to P is computable. Let $\mathcal{I}_{\mathscr{C}}$ be the class of invariant measures that are barycenters of $\mathscr{C}$, i.e. the image of $\mathcal{P}(\mathscr{C})$ under ψ: $\mathcal{I}_{\mathscr{C}}$ is an effective compact class too. By existence and uniqueness of the ergodic decomposition, $\psi : \mathcal{P}(\mathscr{C}) \to \mathcal{I}_{\mathscr{C}}$ is onto and one-to-one; as it is computable and $\mathcal{P}(\mathscr{C})$ is an effective compact set, its inverse is also computable by Fact 2.

The above proposition implies the computability of De Finetti measures on the Cantor space (see [2]).

Example 1. Let m be a computable probability measure over the real interval $[0, 1]$. Pick a real number p at random according to m, and then generate an infinite sequence of $0, 1$ tossing a coin with probability of heads p. As an application of the preceding proposition, we get that the function which maps a random sequence generated by the process to the number p that was picked is P-layerwise computable: it can be computed from the observed outcomes with high probability.

We also learn that the algorithmic theory of randomness fits well with this example: obviously, we expect a random sequence for the whole process to be random for some Bernoulli measure B_p, which is not immediate.

In Section 2.1, we define P-layerwise computable function when P is a computable probability measure. This can be extended straightforwardly to any effective compact class of measures $\mathscr{C}$. The universal $\mathscr{C}$-test induces a decomposition $\bigcup_n \mathcal{R}_n^{\mathscr{C}}$ of the sequences that are random for some measure in $\mathscr{C}$. A function $f : X \to Y$ is $\mathscr{C}$-layerwise computable if it is computable on each $\mathcal{R}_n^{\mathscr{C}}$, uniformly in n. It means that one can compute $f(x)$ if x is random for some measure $P \in \mathscr{C}$, with probability of error bounded by 2^{-n}, whatever P is (as long as it is in $\mathscr{C}$), and for any n.

From Proposition 2 and Corollary 1 we know that for every $P \in \mathcal{I}_{\mathscr{C}}$ and every $x \in \mathcal{R}_P$, Q_x is m-random, hence ergodic and x is Q_x-random. We also prove a quantitative version of this fact. We recall that if $\mathcal{A}$ is an effective compact class of measures, $t_{\mathcal{A}} := \inf_{P \in \mathcal{A}} t_P$ is a universal $\mathcal{A}$-test, i.e. (i) it is lower semi-computable, (ii) $\int t_A \, dP \le 1$ for every $P \in A$ and (iii) t_A multiplicatively dominates every function satisfying (i) and (ii) (see [4] for more details about such class tests). We will consider the class tests $t_{\mathscr{C}}$ and $t_{\mathfrak{I}_{\mathscr{C}}}$.

Theorem 3. *Let $\mathscr{C}$ be an effective compact class of stationary ergodic probability measures. One has:*

1. $t_{\mathscr{C}}(x) \stackrel{*}{=} t_{\mathfrak{I}_{\mathscr{C}}}(x)$
2. *The function $x \mapsto Q_x$ is $\mathfrak{I}_{\mathscr{C}}$-layerwise computable and $\mathscr{C}$-layerwise computable.*

Proof. 1. Of course, $t_{\mathfrak{I}_{\mathscr{C}}} \stackrel{*}{<} t_{\mathscr{C}}$ as $\mathscr{C} \subseteq \mathfrak{I}_{\mathscr{C}}$. Conversely, the $P \in \mathfrak{I}_{\mathscr{C}}$: $\int t_{\mathscr{C}} \, dP = \int (\int t_{\mathscr{C}} \, dQ) \, dm(Q) \le 1$ as m is supported on measures in $Q \in \mathscr{C}$, and $\int t_{\mathscr{C}} \, dQ \le 1$ for such measures. As a result, $t_{\mathscr{C}}$ is a $\mathfrak{I}_{\mathscr{C}}$-test, so $t_{\mathscr{C}} \stackrel{*}{<} t_{\mathfrak{I}_{\mathscr{C}}}$.

2. The proof is similar to the proof of Theorem 2. As $t_{\mathscr{C}}(x) = \inf_{Q \in \mathscr{C}} t_Q(x)$, if $x \in \mathcal{R}^n_{\mathscr{C}}$ then $Q_x \in \mathcal{R}^{n+c}_m$. Again, $\pi_2 : \mathcal{R}^{n+c}_m \times \mathcal{R}^n_{\mathscr{C}} \to \mathcal{R}^n_{\mathscr{C}}$ is computable and bijective so its inverse is computable and maps x to (Q_x, x). Hence $\pi_1 \circ \pi_2^{-1}$ is computable on $\mathcal{R}^n_{\mathscr{C}}$, uniformly in n, i.e. it is $\mathscr{C}$-layerwise computable. As $t_{\mathscr{C}} \stackrel{*}{<} t_{\mathfrak{I}_{\mathscr{C}}}$, it is also $\mathfrak{I}_{\mathscr{C}}$-layerwise computable.

Observe that for generic sequences x, $t_{\mathscr{C}}(x) \stackrel{*}{=} t_{Q_x}(x)$. Indeed, $t_{\mathscr{C}}(x) = \inf_{P \in \mathscr{C}} t_P(x) = t_{Q_x}(x)$ as $t_P(x) = +\infty$ for every $P \in \mathscr{C} \setminus \{Q_x\}$.

4.6   A Weaker Answer: Finitely Decomposable Measures

In the two preceding results, we need the effectivity of the ergodic decomposition. In particular situations, we still get a (weaker) result without this assumption.

Proposition 3. *Let P be a stationary measure such that m_P is supported on a closed set $\mathscr{C}$ of stationary ergodic measures. For every P-random x, Q_x is ergodic.*

To prove it we use the following lemma.

Lemma 4. *Let X, Y be computable metric spaces. Let $f_n : X \to Y$ be uniformly computable functions that converge P-a.e. to a function f. Let $A \subseteq Y$ be a closed set such that $f(x) \in A$ for P-a.e. x. For every P-random x, $\lim f_n(x) \in A$.*

Proof. It is already known if f is constant P-almost everywhere. Let x_0 be a P-random point such that $\lim f_n(x_0) \notin A$. Let $B(y, r)$ be a ball with computable center and radius, containing $\lim f_n(x_0)$ and disjoint from A. Let $g_n(x) = \max(0, r - d(f_n(x), y))$. For P-almost every x, the sequence $g_n(x)$ converges to 0, but $\lim g_n(x_0) = r - d(\lim f_n(x_0), y) > 0$.

Proof (Proof of Proposition 3). For every n, define $Q_n : X \to \mathcal{P}(X)$ by $Q_n(x) = \frac{1}{n}(\delta_x + \ldots + \delta_{T^{n-1}x})$. A sequence x is generic if and only if $Q_n(x)$ is weakly convergent, and in that case Q_x is the limit of $Q_n(x)$. The functions Q_n are uniformly computable. As $Q_x \in \mathscr{C}$ for P-almost every x, $Q_x \in \mathscr{C}$ for every P-random x by Lemma 4.

For instance, if P has a finite decomposition, i.e. if $P = \sum_{i=1}^{n} \alpha_i P_i$ where $\alpha_i \in [0, 1]$, $\sum_i \alpha_i = 1$ and all P_i are ergodic, then regardless of the computability of P, α_i, P_i, for every P-random x, $Q_x \in \{P_1, \ldots, P_n\}$ as the latter set is closed. In this particular case, Q_x is always m-random.

We do not know whether every finitely decomposable measure is effectively decomposable. For instance, are there distinct non-computable ergodic measures P_1, P_2 such that $P := \frac{1}{2}(P_1 + P_2)$ is computable? Such a measure P would be a finitely, non-effectively decomposable measure.

If a finitely, but non-effectively, decomposable measure P exists, and if x is P-random, we do not know whether x is Q_x-random and we do not know whether Q_x is m_P-random. We only know that Q_x is ergodic.

References

1. Bienvenu, L., Day, A., Mezhirov, I., Shen, A.: Ergodic-type characterizations of algorithmic randomness. In: Ferreira, F., Löwe, B., Mayordomo, E., Mendes Gomes, L. (eds.) CiE 2010. LNCS, vol. 6158, pp. 49–58. Springer, Heidelberg (2010)
2. Freer, C.E., Roy, D.M.: Computable exchangeable sequences have computable de finetti measures. In: Ambos-Spies, K., Löwe, B., Merkle, W. (eds.) CiE 2009. LNCS, vol. 5635, pp. 218–231. Springer, Heidelberg (2009)
3. Gács, P.: Uniform test of algorithmic randomness over a general space. Theoretical Computer Science 341, 91–137 (2005)
4. Gács, P.: Lecture notes on descriptional complexity and randomness. Tech. rep., Boston University (2008)
5. Hertling, P., Weihrauch, K.: Random elements in effective topological spaces with measure. Information and Computation 181(1), 32–56 (2003)
6. Hoyrup, M., Rojas, C.: Applications of effective probability theory to martin-löf randomness. In: Albers, S., Marchetti-Spaccamela, A., Matias, Y., Nikoletseas, S., Thomas, W. (eds.) ICALP 2009. LNCS, vol. 5555, pp. 549–561. Springer, Heidelberg (2009)
7. Hoyrup, M., Rojas, C.: Computability of probability measures and Martin-Löf randomness over metric spaces. Inf. Comput. 207(7), 830–847 (2009)
8. Levin, L.A.: On the notion of a random sequence. Soviet Mathematics Doklady 14, 1413–1416 (1973)
9. Martin-Löf, P.: The definition of random sequences. Information and Control 9(6), 602–619 (1966)
10. Nandakumar, S.: An effective ergodic theorem and some applications. In: STOC 2008: Proceedings of the 40th Annual ACM Symposium on Theory of Computing, pp. 39–44. ACM, New York (2008)
11. Phelps, R.R.: Lectures on Choquet's Theorem, 2nd edn. Springer, Berlin (2001)
12. V'yugin, V.V.: Ergodic theorems for individual random sequences. Theoretical Computer Science 207(4), 343–361 (1998)
13. Weihrauch, K.: Computable Analysis. Springer, Berlin (2000)

Computability of the Radon-Nikodym Derivative

Mathieu Hoyrup[1], Cristóbal Rojas[2], and Klaus Weihrauch[3]

[1] LORIA, INRIA Nancy-Grand Est, 615 rue du jardin botanique, BP 239,
54506 Vandœuvre-lès-Nancy, France
mathieu.hoyrup@loria.fr
[2] Department of Mathematics, University of Toronto, 40 St. George Street,
Toronto ON M5S 2E4, Canada
cristobal.rojas@utoronto.ca
[3] Fakultät für Mathematik und Informatik, FernUniversität Hagen,
58084 Hagen, Germany
klaus.weihrauch@fernuni-hagen.de

Abstract. We show that a single application of the non-computable operator EC, which transforms enumerations of sets (in $\mathbb{N}$) to their characteristic functions, suffices to compute the Radon-Nikodym derivative $d\mu/d\lambda$ of a finite measure μ, which is absolutely continuous w.r.t. the σ-finite measure λ. We also give a condition on the two measures (in terms of computability of the norm of a certain linear operator involving the two measures) which is sufficient to compute the derivative.

1 Introduction

Theorem 1 (Radon-Nikodym). *Let $(\Omega, \mathcal{A}, \lambda)$ be a measured space where λ is σ-finite. Let μ be a finite measure that is absolutely continuous w.r.t. λ. There exists a unique function $h \in L^1(\lambda)$ such that for all $f \in L^1(\mu)$,*

$$\int f \, d\mu = \int fh \, d\lambda.$$

The function h is called the **Radon-Nikodym derivative,** *or* **density** *of μ w.r.t. λ, and deonted by $\frac{d\mu}{d\lambda}$.*

Is this theorem computable? Can h be computed from μ and λ? In [11] a negative answer was given.

In this paper we investigate to what extent this theorem is non-computable. We first give an upper bound for its non-computability, showing that it can be computed using a single application of the operator EC (which transforms enumerations of sets of natural numbers into their characteristic functions). In proving this result we use two classical theorems: Levy's zero-one law and Radon-Nikodym Theorem itself. We then give a sufficient condition on the measures to entail the computability of the RN derivative: this condition is the computability of the norm of a certain integral operator associated to the measures.

B. Löwe et al. (Eds.): CiE 2011, LNCS 6735, pp. 132–141, 2011.

2   Preliminaries

2.1   Little bit of Computability via Representations

To carry out computations on infinite objects we encode those objects into infinite symbolic sequences, using representations (see [19] for a complete development). Let $\Sigma = \{0,1\}$. A **represented space** is a pair (X,δ) where X is a set and $\delta \subset \Sigma^{\mathbb{N}} \to X$ is an onto partial map. Every $p \in \mathrm{dom}(\delta)$ such that $\delta(p) = x$ is called a δ-**name** of x (or **name** of x when δ is clear from the context).

Let (X,δ_X) be a represented space. An element $x \in X$ is **computable** if it has a computable name. Let (Y,δ_Y) be another represented space. A realizer for a function $f : X \to Y$ is a (partial) function $F : \Sigma^{\mathbb{N}} \to \Sigma^{\mathbb{N}}$ such that $f \circ \delta_X = \delta_Y \circ F$ (with the expected compatibilities between domains). f is **computable** if it has a computable realizer. Of course, the image of a computable element by a computable function is computable.

Many spaces in Analysis have "canonical" or "natural" representations. For every theorem of the form $(\forall x)(\exists y)Q(x,y)$ we can ask whether there is a computable (w.r.t. the given representations) function (or multi-function) mapping every x to some y such that $Q(x,y)$. Very often problems of this kind have no computable solution. But sometimes a solution of one problem can help to solve another one. Such "helping" between in general non-computable problems can be formalized by a reducibility relation as follows: For functions f, g between represented spaces define $f \leq_W g$ if there are computable (partial) functions K, H om $\Sigma^{\mathbb{N}}$ such that for every realizer G of g, $p \mapsto K(p, G \circ H(p))$ is a realizer of f ([2,3], where $\leq_W$ is called Weihrauch reducibility). In other words, $f \leq_W g$ if f can be computed using one single application of g (provided by an oracle) in the computation. This reducibility has already been used for comparing a number of non-computable mathematical theorems, for example, in [16,17,5,9,2,3,1]. The analogy of this project to reverse mathematics [15] has been discussed in [9,2].

An important non-computable operator is EC, the operator transforming every enumeration of a set of natural numbers into its characteristic function. Define representations En and Cf of $2^{\mathbb{N}}$ by

$$\mathrm{En}(p) = \{n \in \mathbb{N} : 100^n 1 \text{ is a subword of } p\}$$
$$\mathrm{Cf}(p) = \{n \in \mathbb{N} : p_n = 1\}$$

where $p \in \Sigma^{\mathbb{N}}$ and p_n is the n-th symbol in p. Then the En-computable sets are the r.e. sets and the Cf-computable sets are the recursive sets. We define the operator **EC** as the identity from the represented space $(2^{\mathbb{N}}, \mathrm{En})$ to the represented space $(2^{\mathbb{N}}, \mathrm{Cf})$. It transforms every enumeration of a set into its characteristic function. EC is not computable. In [5] it is proved that EC is equivalent to the ordinary limit map of any (sufficiently rich) computable metric space X. Furthermore, EC is complete for effectively Σ_2^0-measurable functions (in the Borel hierarchy) with respect to $\leq_W$ [5, Theorem 7.6]. Many non-computable problems from Analysis are equivalent to EC [2,3].

In this article we introduce a computational background for for the Radon-Nikodym theorem from measure theory and prove that in this setting $RN \leq_W$ EC where RN maps the two measures in Theorem 1 to the function h.

2.2   Computable Measurable Spaces

We start by briefly recalling some basic definitions from measure theory. See for example [12,6,4] for a complete treatment. A **ring** $\mathcal{R}$ over a set Ω is a collection of subsets of Ω which contains the empty set and is closed under finite unions and relative complementation ($B \setminus A \in \mathcal{R}$, for $A, B \in \mathcal{R}$). A σ-**algebra** $\mathcal{A}$ (over the set Ω) is a collection of subsets of Ω which contains Ω and is closed under complementation and countable unions (and therefore also closed under countable intersections). A ring $\mathcal{R}$ generates a unique σ-algebra, denoted by $\sigma(\mathcal{R})$, and defined as the smallest σ-algebra containing $\mathcal{R}$.

In this paper we will work with the **measurable space** $(\Omega, \mathcal{A})$, where $\mathcal{A}$ is a σ-algebra generated by a countable ring $\mathcal{R}$. Members of $\mathcal{A} = \sigma(\mathcal{R})$ will be referred to as **measurable sets**.

A **measure** over a collection $\mathcal{C}$ (which is at least closed by finite unions) of subsets of Ω is a function $\mu : \mathcal{C} \to \mathbb{R}^\infty$ ($= [0, \infty]$) such that i) $\mu(\emptyset) = 0$, $\mu(E) \geq 0$ for all $E \in \mathcal{C}$, and ii) $\mu(\bigcup_i E_i) = \sum_i \mu(E_i)$ for pairwise disjoint sets $E_0, E_1, \ldots \in \mathcal{C}$ such that $\bigcup_i E_i \in \mathcal{C}$. A measure μ over a collection $\mathcal{C}$ is said to be σ-**finite**, if there are sets $E_0, E_1, \ldots \in \mathcal{C}$ such that $\mu(E_i) < \infty$ for all i and $\Omega = \bigcup_i E_i$.

It is well known that a σ-finite measure over ring $\mathcal{R}$ has a unique extension to a measure over the σ-algebra $\sigma(\mathcal{R})$. For measures μ and λ, we say that μ is **absolutely continuous** w.r.t. λ, and write $\boldsymbol{\mu \ll \lambda}$, if $(\lambda(A) = 0 \implies \mu(A) = 0)$ for all measurable sets A.

We now introduce the effective counterparts. In [20], computable measured spaces are introduced, and computability properties of measures are studied. Since we want to formulate our theorem not only for computable measures, we first introduce the computable version of a measurable space as a "computable measure space from [20] without measure".

Definition 1. *A* **computable measurable space** *is a tuple* $(\Omega, \mathcal{A}, \mathcal{R}, \alpha)$ *where*

1. $(\Omega, \mathcal{A})$ *is a measurable space,* $\mathcal{R}$ *is a countable ring such that* $\cup \mathcal{R} = \Omega$ *and* $\mathcal{A} = \sigma(\mathcal{R})$,
2. $\alpha : \mathbb{N} \to \mathcal{R}$ *is a computable enumeration such that the operations* $(A, B) \to A \cup B$ *and* $(A, B) \to A \setminus B$ *are computable w.r.t.* α.

In the following $(\Omega, \mathcal{A}, \mathcal{R}, \alpha)$ will be a computable measurable space. We will consider only measures $\mu : \mathcal{A} \to [0, \infty]$ such that $\mu(E) < \infty$ for every $E \in \mathcal{R}$. Observe that such a measure μ is σ-finite, and therefore well-defined by its values on the ring $\mathcal{R}$. Conversely if a measure μ over $\mathcal{A}$ is σ-finite, one can choose a countable generating $\mathcal{R}$ such that $\mu(E) < \infty$ for all $E \in \mathcal{R}$.

Computability on the space of measures over $(\Omega, \mathcal{A}, \mathcal{R})$ will be expressed via representations.

Definition 2. *Let $\mathcal{M}$ be the set of measures μ such that $\mu(E) < \infty$ for all $E \in \mathcal{R}$ and let $\mathcal{M}_{<\infty}$ be the set of all finite measures. Define representations $\delta_{\mathcal{M}} : \Sigma^{\mathbb{N}} \to \mathcal{M}$ and $\delta_{\mathcal{M}_{<\infty}} : \Sigma^{\mathbb{N}} \to \mathcal{M}_{<\infty}$ as follows:*

1. *$\delta_{\mathcal{M}}(p) = \mu$, iff p is (more precisely, encodes) the list of all $(l, n, u) \in \mathbb{Q}^3$ such that $l < \mu(E_n) < u$, for every $E_n \in \mathcal{R}$.*
2. *$\delta_{\mathcal{M}_{<\infty}}(p) = \mu$, iff $p = \langle p_1, p_2 \rangle$ such that $\delta_{\mathcal{M}}(p_1) = \mu$ and p_2 is (more precisely, encodes) the list of all $(l, u) \in \mathbb{Q}^2$ such that $l < \mu(\Omega) < u$.*

Thus, a $\delta_{\mathcal{M}}$-name p allows to compute $\mu(A)$ for every ring element A with arbitrary precision. A $\delta_{\mathcal{M}_{<\infty}}$-name allows additionally to compute $\mu(\Omega)$. Obviously, $\delta_{\mathcal{M}_{<\infty}} \le \delta_{\mathcal{M}}$ and $\delta_{\mathcal{M}_{<\infty}} \equiv \delta_{\mathcal{M}}$ if $\Omega \in \mathcal{R}$. But in general, not even the restriction of $\delta_{\mathcal{M}}$ to the finite measures is reducible to $\delta_{\mathcal{M}_{<\infty}}$. A computable measurable space with $\delta_{\mathcal{M}}$-computable measure μ is the "computable measure space" from [20]. Definitions 1 and 2 generalize the definitions from [18] (probability measures on Borel subsets of $[0, 1]$), [7,14,13,8,10] (computable finite measures on the Borel subsets of a computable metric space) and [20] (computable measure on a computable σ-algebra).

We will also work with the spaces L^1 of integrable functions and L^2 of square-integrable functions (w.r.t. some measure). A ***rational step function*** is a finite sum

$$s = \sum_{k=1}^{p} \mathbf{1}(E_{i_k})q_{j_k},$$

where $E_{i_k} \in \mathcal{R}$, $q_{j_k} \in \mathbb{Q}$, and $\mathbf{1}(A)$ denotes the characteristic function of A.

The computable numberings of the ring $\mathcal{R} = (E_0, E_1, ...)$ and of the rational numbers $\mathbb{Q} = (q_0, q_1, ...)$ induce a canonical numbering of the the collection $\mathcal{RSF} = (s_0, s_1, ...)$ of rational step functions. Since the collection $\mathcal{RSF}$ is dense in the spaces L^1 and L^2, we can use it to handle computability over these spaces via the following representations:

Definition 3. *For every $\mu \in \mathcal{M}$ define Cauchy representations $\delta_\mu^k : \Sigma^{\mathbb{N}} \to L^k(\mu)$ of $L^k(\mu)$ ($k = 1, 2$) by: $\delta_\mu^k(p) = f$ iff p is (encodes) a sequence $(s_{n_0}, s_{n_1}, ...)$ of rational step functions such that $\|s_{n_i} - f\|_\mu^k \le 2^{-i}$ for all $i \in \mathbb{N}$.*

In the definition above, $\| \cdot \|_\mu^k$ denotes the norm in $L^k(\mu)$.

3   Effective Radon-Nikodym Theorem

3.1   An Upper Bound

In the following, we present our first main result which, in words, says that computability of the Radon-Nikodym derivative is reducible to a single application of the (non-computable) operator EC.

Theorem 2. *The function mapping every σ-finite measure $\lambda \in \mathcal{M}$ and every finite measure μ such that $\mu \ll \lambda$ to the function $h \in L^1(\lambda)$ such that $\mu(E) = \int_E h \, d\lambda$ for all $E \in \sigma(\mathcal{R})$ is computable via the representations $\delta_{\mathcal{M}}$, $\delta_{\mathcal{M}_{<\infty}}$ and δ_λ^1 with a single application of the operator EC.*

For the proof, we will use the following classical result on convergence of conditional expectations, which is a consequence of the more general Doob's martingale convergence theorems (see for example [6], Section 10.5). We recall that a filtration $(\mathcal{F}_n)_n$ is an increasing sequence of σ-algebras on a measurable space. In some sense, $\mathcal{F}_n$ represents the information available at time n. In words, the following result says that if we are learning gradually the information that determines the outcome of an event, then we will become gradually certain what the outcome will be.

Theorem 3 (Levy's zero-one law). *Consider a finite measure λ over the measurable space $(\Omega, \sigma(\mathcal{R}))$. Let $h \in L^1(\lambda)$. Let $(\mathcal{F}_n)_n$ be any filtration such that $\sigma(\mathcal{R}) = \sigma(\bigcup_n \mathcal{F}_n)$. Then*

$$\mathbb{E}(h|\mathcal{F}_n) \xrightarrow[n\to\infty]{} h$$

both λ-almost everywhere and in $L^1(\lambda)$.

Proof (of Theorem 2). We start by proving the result for finite measures.

Lemma 1. *From descriptions of finite measures $\mu \ll \lambda$, one can compute a sequence $\{h_n\}_{n\in\mathbb{N}}$ of $L^1(\lambda)$-computable functions which converges in $L^1(\lambda)$ to the Radon-Nikodym derivative $h = \frac{d\mu}{d\lambda} \in L^1(\lambda)$.*

Proof. Consider the computable enumeration $E_0, E_1, \ldots$ of $\mathcal{R}$. For every n, consider the partition $\mathcal{P}_n$ of the space given by the cells:

– all the possible $E_0^* \cap E_1^* \cap \ldots \cap E_n^*$ where E^* is either E or $(E_0 \cup \ldots \cup E_n) \setminus E$,
– all the $E_{k+1} \setminus (E_0 \cup \ldots \cup E_k)$ for every $k \geq n$.

Observe that all the cells are elements of the ring $\mathcal{R}$. The partitions $\mathcal{P}_n$ induce a filtration which generates the σ-algebra $\sigma(\mathcal{R})$. Let h be the RN-derivative[1] of μ w.r.t. λ. Then $h \in L^1(\lambda)$, so by Levy's zero-one law,

$$\mathbb{E}(h|\mathcal{P}_n) \xrightarrow[n\to\infty]{} h$$

both λ-almost everywhere and in $L^1(\lambda)$.

Now, $h_n := \mathbb{E}(h|\mathcal{P}_n)$ is constant on each cell C of $\mathcal{P}_n$, with value $\frac{\mu(C)}{\lambda(C)}$ if $\lambda(C) > 0$ and 0 otherwise. We now show that the functions h_n are (uniformly) computable elements of $L^1(\lambda)$. For a given ϵ, one can find cells $C_1, \ldots, C_k$ in $\mathcal{P}_n$ such that $\lambda(C_i) > 0$ for all $i = 1, \ldots, k$ and $\sum_{i=1}^k \mu(C_i) > \mu(\Omega) - \epsilon$. Define h_n^ϵ to be constant on C_i with value $\frac{\mu(C_i)}{\lambda(C_i)}$ for $i = 1, \ldots, k$ and 0 on the other cells. As these values are uniformly computable, the functions h_n^ϵ are uniformly $L^1(\lambda)$-computable. Moreover

$$\int |h_n - h_n^\epsilon| \, d\lambda = \int_{\Omega \setminus \bigcup_{i=1}^k C_i} h_n \, d\lambda < \epsilon$$

and thus, the functions h_n are (uniformly) computable elements of $L^1(\lambda)$, and converge to h in $L^1(\lambda)$. The lemma is proved.

---

[1] At this point we use the classical RN theorem. This is where the hypothesis on absolute continuity is used.

Assume now that $\mu \in \mathcal{M}_{<\infty}$ and $\lambda \in \mathcal{M}$. Let once again $(E_0, E_1, \ldots)$ be the computable numbering of the ring $\mathcal{R}$. Let $F_0 := E_0$ and $F_{n+1} := E_{n+1} \setminus \cup_{j=0}^{n} F_j$. Then $(F_i)_i$ is a computable numbering of ring elements such that $F_i \cap F_j = \emptyset$ for $i \neq j$ and $\Omega = \bigcup_i F_i$.

Since $i \mapsto \mu(F_i)$ is computable, there is a computable function $d : \mathbb{N} \to \mathbb{N}$ such that $d(i) > \mu(F_i) \cdot 2^i$. Define a function $w : \Omega \to \mathbb{R}$ by $w(x) := 1/d(i)$ if $x \in F_i$. Then

$$\int w \, \mathrm{d}\lambda = \sum_i \mu(F_i)/d(i) < 2.$$

Define a new measure ν by

$$\nu(E) := \int_E w \, \mathrm{d}\lambda \tag{1}$$

for all $E \in \mathcal{R}$. Then ν is a finite measure, which is equivalent to λ (i.e. $\nu \ll \lambda \ll \nu$) and such that a $\delta_{\mathcal{M}}$-name of ν can be computed from a $\delta_{\mathcal{M}}$-name of λ. Apply Lemma 1 to μ and ν: one can compute a sequence of $L^1(\nu)$-computable functions $\{h'_n\}$ whose limit (in $L^1(\nu)$) is the density $h' = \frac{\mathrm{d}\mu}{\mathrm{d}\nu}$. The sequence $\{h'_n \cdot w\}$ is computable and converges (in $L^1(\lambda)$) to $h = \frac{\mathrm{d}\mu}{\mathrm{d}\lambda} = w \cdot h'$. Since the set $\{i : \exists m > i, \|h'_i \cdot w - h'_m \cdot w\|^1_\lambda > 2^{-i}\}$ is r.e., we can compute its complement using a single application of the operator EC. This gives the indexes of a fast Cauchy subsequence, and hence allow to compute (a δ^1_λ-name of) h. The proof is complete.

The above theorem shows that the Radon-Nikodym theorem is reducible to the (non-computable) operator EC. On the other hand, it was shown in [11] that there exists a computable measure on the unit interval, absolutely continuous w.r.t. Lebesgue measure, and such that the operator EC can be reduced to the computation of the density (which is therefore not L^1-computable). This give us the following corollary.

Corollary 1. *For nontrivial computable measurable spaces, the Radon-Nikodym operator and* EC *are equivalent:* RN $\equiv_W$ EC.

This result characterizes the extent to which the Radon-Nikodym theorem is non-computable. However, it doesn't give us much information on "where" is the non-effectivity. In what follows we present a result which gives an explicit condition (in terms of the computability of the norm of a certain linear operator involving the two measures) allowing to compute the Radon-Nikodym derivative. The proof of this result is somewhat more involved, and some preparation will be required.

3.2   Locating the Non-computability

Let $\mu \in \mathcal{M}_{<\infty}$ be a finite measure over $(\Omega, \mathcal{A}, \mathcal{R})$. Let $u : L^2(\mu) \to \mathbb{R}$ be a linear functional. Classically we have that the following are equivalent:

1. u is continuous,
2. u is uniformly continuous,
3. there exists c (a **bound for** u) such that for every $f \in L^2(\mu)$, $|u(f)| \le c\|f\|_2$.

The smallest bound for u is called **the norm** of u and is denoted by $\|u\|$. That is,

$$\|u\| := \sup\{c \in \mathbb{R} : |u(f)| \le c\|f\|_2\} = \sup_{\|f\|_2 = 1} |u(f)|.$$

Suppose that the operator u is computable in the sense that one can compute the real number $u(f)$ from (a $\delta_{\mathcal{M}}$-name of) μ and (a δ_μ^2-name of) f. Consider now the numbered collection $\mathcal{RSF} = \{r_n\}_{n \in \mathbb{N}}$. Since the sequence $s_i := \frac{r_i}{\|r_i\|_2}$ is uniformly computable (from μ) and dense in $\{f \in L^2(\mu) : \|f\|_2 = 1\}$, it follows that the norm $\|u\|$ of a computable operator u is always a lower-computable (from μ) number. It is not, in general, computable. In case it is computable, we will say that the operator u is **computably normable**.

The following result is an effective version of Riesz-Fréchet Representation Theorem. For simplicity, we state the result in the particular Hilbert space $L^2(\mu)$, but the same proof works for any Hilbert Space provided that the inner product is computable.

Theorem 4. Let u be a non-zero computably normable (from μ) linear functional over $L^2(\mu)$. Then from μ one can compute a name of (the unique) $g \in L^2(\mu)$ such that

$$u(f) = \int fg \, d\mu$$

for all $f \in L^2(\mu)$.

Proof. In the following, all the computations we will be done from a $\delta_{\mathcal{M}}$-name of μ. Chose any computable $x \in L^2(\mu)$ out of $\ker(u)$. As $\|u\|$ is computable, from the classical formula $d(x, \ker(u)) = \frac{|u(x)|}{\|u\|}$, it follows that $d(x, \ker(u))$ is computable too. We can then enumerate a sequence of points $y_n \in E = \ker(u) + x$ which is dense in E. Note that $d(0, E) = d(x, \ker(u))$. Let $z_0 \in E$ be such that $\|z_0\| = d(0, E)$. Since

$$\|y_n - z_0\|^2 = \|y_n\|^2 - \|z_0\|^2$$

we can compute a subsequence y_{n_i} converging effectively to z_0 which is therefore computable. Put $z = \frac{z_0}{\|z_0\|}$ so that $\|z\| = 1$. All what remains is to show that $g = u(z)z$ (which is computable) satisfies the required property. This is done as in the classical proof, namely: z has the property of being orthogonal to $\ker(u)$. That is, $\int zf \, d\mu = 0$ for any $f \in \ker(u)$. Put

$$r := u(f)z - u(z)f.$$

We have $u(r) = u(f)u(z) - u(z)u(f) = 0$ so that $r \in \ker(u)$ and then $\int rz \, d\mu = 0$. This gives,

$$u(f) = u(f) \int z^2 \, d\mu = \int u(f)z^2 \, d\mu = \int (r + u(z)f)z \, d\mu = \int fu(z)z \, d\mu$$

and hence $g = u(z)z$, as was to be shown.

Remark 1. Let μ be a finite measure. For f, g in $L^2(\mu)$, Hölder's inequality implies that $fg \in L^1(\mu)$ and:

$$\|fg\|_1 \leq \|f\|_2 \|g\|_2 \tag{2}$$

so that, in particular, since μ is finite, if $f \in L^2$ then $f \in L^1$ (taking $g \equiv 1$). Moreover, from a δ_μ^2-name of f one can compute a δ_μ^1-name.

Now, let μ and λ be finite measures and consider a new measure $\varphi := \mu + \lambda$. Let $L_\mu : L^2(\varphi) \to \mathbb{R}$ be defined by $L_\mu(f) := \int f \, d\mu$. This is a bounded operator and it is easy to see that from $\delta_{\mathcal{M}}$-names of μ and λ, and from a δ_φ^2-name of f, one can compute the value $L_\mu(f)$.

Definition 4. *A finite measure μ is said to be* **computably normable relative to** *some other finite measure λ, if the norm of the operator L_μ (as defined above) is computable from μ and λ.*

At this point, we are ready to state our second main result.

Theorem 5. *Let $\mu, \lambda \in \mathcal{M}_{<\infty}$ be such that:*

(i) $\mu \ll \lambda$,
(ii) μ is computably normable relative to λ.

Then the Radon-Nikodym derivative $\frac{d\mu}{d\lambda}$ can be computed as an element of $L^1(\lambda)$, from μ and λ.

Proof. We follow Von Neumann's proof. The measure $\varphi := \mu + \lambda$ is computable and by hypothesis the operator $L_\mu : L^2(\varphi) \to \mathbb{R}$ defined by $L_\mu(f) := \int f \, d\mu$ is computably normable. Hence, by Theorem 4 one can compute a name of $g \in L^2(\varphi)$ (and hence a name of g as a point in $L^1(\lambda)$) such that for all $f \in L^2(\varphi)$ the equality:

$$\int f \, d\mu = \int fg \, d\varphi = \int fg \, d\lambda + \int fg \, d\mu \tag{3}$$

holds. This relation can be rewritten as:

$$\int fg \, d\lambda = \int f(1 - g) \, d\mu. \tag{4}$$

Note that (4) holds for any $f \geq 0$ (take $f_n = f\mathbf{1}_{\{f<n\}}$ and apply the Monotone Convergence Theorem). Taking $f = \mathbf{1}_{\{g=1\}}$ in (4) we see that $\lambda(\{g = 1\}) = 0$. Hence the following function is defined λ-almost everywhere:

$$h := \frac{g}{1 - g}.$$

Taking $f = \mathbf{1}_{\{g<0\}}$ and $\mathbf{1}_{\{g>1\}}$ in (4) we see that $0 \leq g \leq 1$ λ-a.e., so $h \geq 0$ λ-a.e. Therefore,

$$\int fh \, d\lambda = \int \left(\frac{f}{1-g}\right) g \, d\lambda = \int \left(\frac{f}{1-g}\right) (1-g) \, d\mu = \int f \, d\mu$$

(by 4). Now, taking f to be the constant function equal to 1, we conclude that $\int h \, d\lambda = 1$ and then it is in $L^1(\lambda)$. This shows that h is the Radon-Nikodym derivative.

It remains to show that the function $h = \frac{g}{1-g}$ is $L^1(\lambda)$-computable.

As g is $L^1(\lambda)$-computable, we can effectively produce a sequence u_i of rational step functions such that $\|u_i - g\|_\lambda < 2^{-i}$. As $g \geq 0$ λ-a.e. we can assume w.l.o.g. that $u_i \geq 0$ (otherwise replace u_i with $\max(u_i, 0)$).

For $n \in \mathbb{N}$ let

$$g_n := \min(g, 1 - 2^{-n}) \qquad h_n := g_n/(1 - g_n)$$
$$u_{in} := \min(u_i, 1 - 2^{-n}) \qquad v_{in} := u_{in}/(1 - u_{in})$$

Since the function $x \mapsto x/(1-x)$ is nondecreasing over $(0, +\infty)$,

$$g_n \leq g_{n+1} \text{ and } \sup_n g_n = g,$$
$$h_n \leq h_{n+1} \text{ and } \sup_n h_n = h.$$

Given a rational number $\epsilon > 0$ we show how to compute n and i such that

$$\|h - v_{in}\|_\lambda < \epsilon.$$

v_{in} will then be a rational step function approximating h up to ϵ, in $L^1(\lambda)$. As a result, it will enable us to compute a δ_μ^1-name of h.

To find n and i, we use the following inequality

$$\|h - v_{in}\|_\lambda \leq \|h - h_n\|_\lambda + \|h_n - v_{in}\|_\lambda \tag{5}$$

We first make the first term small. As for all n, $0 \leq h_n \leq h$, one has $\|h - h_n\|_\lambda = \|h\|_\lambda - \|h_n\|_\lambda$. As $\|h\|_\lambda = \mu(\Omega)$ is given as input and $\|h_n\|_\lambda$ can be computed from n, one can effectively find n such that $\|h\|_\lambda - \|h_n\|_\lambda < \epsilon/2$.

We then make the second term in (5) small:

$$
\begin{aligned}
\|h_n - v_{in}\|_\lambda &= \left\| \frac{g_n}{1 - g_n} - \frac{u_{in}}{1 - u_{in}} \right\|_\lambda \\
&= \left\| \frac{g_n - u_{in}}{(1 - g_n)(1 - u_{in})} \right\|_\lambda \\
&\leq \|g_n - u_{in}\|_\lambda \cdot 2^{2n} \\
&\leq \|g_n - u_{in}\|_\varphi^1 \cdot 2^{2n} \\
&\leq \|g - u_i\|_\varphi^1 \cdot 2^{2n} \\
&\leq 2^{-i} \cdot 2^{2n}.
\end{aligned}
$$

We then compute i such that $2^{-i} \cdot 2^{2n} < \epsilon/2$. One finally gets the expected inequality

$$\|h - v_{in}\|_\lambda \leq \|h - h_n\|_\lambda + \|h_n - v_{in}\|_\lambda < \epsilon$$

and the result follows.

References

1. Brattka, V., de Brecht, M., Pauly, A.: Closed choice and a uniform low basis theorem. arXiv:1002.2800v1 [math.LO] (February 14. 2010)
2. Brattka, V., Gherardi, G.: Effective choice and boundedness principles in computable analysis. Bulletin of Symbolic Logic 17(1), 73–117 (2011)
3. Brattka, V., Gherardi, G.: Weihrauch degrees, omniscience principles and weak computability. Journal of Symbolic Logic 76, 143–176 (2011)
4. Bogachev, V.I.: Measure Theory. Springer, Heidelberg (2006)
5. Brattka, V.: Effective Borel measurability and reducibility of functions. Mathematical Logic Quarterly 51(1), 19–44 (2005)
6. Dudley, R.: Real Analysis and Probability, 2nd edn. Cambridge University Press, Cambridge (2002)
7. Gács, P.: Uniform test of algorithmic randomness over a general space. Theoretical Computer Science 341, 91–137 (2005)
8. Galatolo, S., Hoyrup, M., Rojas, C.: A constructive Borel-Cantelli lemma. constructing orbits with required statistical properties. Theoretical Computer Science 410(21-23), 2207–2222 (2009)
9. Gherardi, G., Marcone, A.: How incomputable is the separable Hahn-Banach theorem? Notre Dame Journal of Formal Logic 50(4), 293–425 (2009)
10. Hoyrup, M., Rojas, C.: Computability of probability measures and martin-löf randomness over metric spaces. Information and Computation 207(7), 830–847 (2009)
11. Hoyrup, M., Rojas, C., Weihrauch, K.: The Radon-Nikodym operator is not computable. In: CCA 2011 (2011)
12. Rudin, W.: Real and Complex Analysis, 3rd edn. McGraw-Hill, New York (1987)
13. Schröder, M.: Admissible representations for probability measures. Mathematical Logic Quarterly 53(4–5), 431–445 (2007)
14. Schröder, M., Simpson, A.: Representing probability measures using probabilistic processes. Journal of Complexity 22(6), 768–782 (2006)
15. Simpson, S.G.: Subsystems of Second Order Arithmetic. Perspectives in Mathematical Logic. Springer, Berlin (1999)
16. Weihrauch, K.: The degrees of discontinuity of some translators between representations of the real numbers. Technical Report TR-92-050, International Computer Science Institute, Berkeley (July 1992)
17. Weihrauch, K.: The degrees of discontinuity of some translators between representations of the real numbers. Informatik Berichte 129, FernUniversität Hagen, Hagen (July 1992)
18. Weihrauch, K.: Computability on the probability measures on the Borel sets of the unit interval. Theoretical Computer Science 219, 421–437 (1999)
19. Weihrauch, K.: Computable Analysis. Springer, Berlin (2000)
20. Wu, Y., Weihrauch, K.: A computable version of the Daniell-Stone theorem on integration and linear functionals. Theoretical Computer Science 359(1–3), 28–42 (2006)

Extracting Winning Strategies in Update Games

Imran Khaliq[1], Bakhadyr Khoussainov[1], and Jiamou Liu[2]

[1] Department of Computer Science, University of Auckland, New Zealand
ikha020@aucklanduni.ac.nz, bmk@cs.auckland.ac.nz
[2] School of Computing and Mathematical Sciences, Auckland University of
Technology, New Zealand
jiamou.liu@aut.ac.nz

Abstract. This paper investigates algorithms for extracting winning strategies in two-player games played on finite graphs. We focus on a special class of games called update games. We present a procedure for extracting winning strategies in update games by constructing strategies explicitly. This is based on an algorithm that solves update games in quadratic time. We also show that solving update games with a bounded number of nonkdeterministic nodes takes linear time.

1 Introduction

Games played on finite graphs have received increasing interest over the last few years. These games are natural models for reactive systems [7], concurrent and communication networks [10], and have close interactions with model checking, verification, automata and logic [5,4,6,13]. Such a game is played between two players, Player 0 and Player 1, over a finite directed graph. The players play the game by moving a token along the edges of the graph. We assume that both players have perfect information, the game is turn-based and each move is deterministic.

There are two interrelated algorithmic problems in the investigation of games played on finite graphs. The first is concerned with finding algorithms that determine the winner of the game. The second is concerned with extracting a winning strategy for the winner. These two problems have been investigated for games with Muller winning conditions. We refer to these games as McNaughton games as McNaughton was the first who studied algorithmic properties of these games [8]. In his paper [8] McNaughton provided an algorithm that detects the winners of these games. Nerode, Remmel and Yakhnis analyzed and improved McNaughton's algorithm in [9]. A. Dawar and P. Hunter showed that detecting the winners in McNaughton games is PSPACE complete problem [11]. In addition, McNaughton in [8] proves that the winners possess winning strategies that can be simulated by finite state automata. The upper bound for the number of states of the automata is $n!$, where n is the number of nodes in the game graph. Dziembowski, Jurdzinski and Walukiewicz showed that $n!$ bound is sharp [3].

A natural question arises to single out special classes of McNaughton games for which there are efficient algorithms that (1) find the winners of the games,

B. Löwe et al. (Eds.): CiE 2011, LNCS 6735, pp. 142–151, 2011.

and (2) extract finite state winning strategies. In [2], Dinneen and Khoussainov investigated a subclass of McNaughton games called update games. They proved that the winner of a given update game can be found in quadratic time on the size of the input game. Various generalizations of update games have been studied in [1], all these generalizations are solved in polynomial time. We also point out that, using straightforward transformations, Büchi and reachability games can be turned into McNaughton games. It is a well known fact that these games can be solved in linear and quadratic time, respectively, and that the winners have memoryless winning strategies [5].

In this paper we investigate update games. In these games Player 0 aims to visit every node of the game graph infinitely often. Update games can be viewed as models of various types of networks in which a message needs to be passed among all members of the network. We pursue several goals: In Section 4 we provide a procedure that generates all the update games in which Player 0 is the winner (Theorem 2). This will be based on structural properties of the underlying game graphs. In Section 5, we present a procedure that, given an update game, explicitly constructs a winning strategy for the winner (Theorem 3). We will provide such an algorithm that runs in quadratic time in the worst case. The algorithm is based on the contraction operator first introduced in [2]. Finally, in the last section we describe an algorithm that detects the winner of update games in linear time given that the number of nondeterministic nodes of Player 1 is bounded by a constant (Theorem 4). We note this requires a new technique since the algorithm based on the contraction operator does not make use of non-deterministic nodes of Player 1.

2   Preliminaries

The games under study are of the form (V_0, V_1, E, Ω) where $(V_0 \cup V_1, E)$ is a finite directed bipartite graph, V_0 and V_1 partition the set $V = V_0 \cup V_1$, $E \subseteq V_0 \times V_1 \cup V_1 \times V_0$, and $\Omega \subseteq 2^V$ is a set of *winning positions*. For each $v \in V$, let vE denote the set $\{v' \in V_0 \cup V_1 \mid (v, v') \in E\}$, the *successors* of v. For this paper we stipulate that $vE \neq \emptyset$ for all $v \in V$. For $\sigma \in \{0, 1\}$, we call nodes in V_σ *Player σ's nodes*.

Intuitively, the players play the game by moving a token along the edges of the graph. The token is initially placed on a node $v_0 \in V$. The play proceeds in rounds. At any round of the play, if the token is placed on a Player σ's node v, then Player σ chooses $u \in vE$, moves the token to u and the play continues on to the next round. Formally, a *play (starting from v_0)* is a sequence $\rho = v_0, v_1, v_2, \ldots$ such that $v_{i+1} \in v_i E$ for all $i \in \mathbb{N}$. We set $\mathsf{Inf}(\rho) = \{v \in V \mid \forall i \exists j > i : v_i = v\}$. Thus, any play is infinite and $\mathsf{Inf}(\rho) \neq \emptyset$. We say Player 0 *wins the play ρ* if $\mathsf{Inf}(\rho) \in \Omega$; otherwise, Player 1 wins the play. These games are called *McNaughton games* or games with Muller winning conditions. We simply refer to them as games.

A *strategy* for Player σ is a function that takes as input initial segments of plays $v_0, v_1, \ldots v_k$ where $v_k \in V_\sigma$ and outputs some v_{k+1} such that $v_{k+1} \in v_k E$.

We will concentrate on finite state strategies, which is realized by a finite I/O (Mealy) automaton $\mathcal{S} = (Q, q_0, \delta)$ where Q is the finite set of states, $q_0 \in Q$ is the initial state, and $\delta : Q \times V_\sigma \to Q \times V_{1-\sigma}$ is the *transition function*. The strategy $\mathcal{S}$ is a *k-state strategy* if $|Q| = k$. One state strategies are also called *memoryless strategies*. Thus, a memoryless strategy for Player σ is simply a function $\mathcal{S} : V_\sigma \to V_{1-\sigma}$. A play $\rho = v_0, v_1, v_2, \ldots$ is *consistent* with $\mathcal{S}$ if there exists a sequence of states $q_0, q_1, q_2, \ldots$ such that for all $i \in \mathbb{N}$ we have the following: If $v_i \in V_\sigma$, then $\delta(q_i, v_i) = (q_{i+1}, v_{i+1})$; If $v_i \in V_{1-\sigma}$, then $q_{i+1} = q_i$. Thus, the strategy does not change its state when Player $(1 - \sigma)$ makes moves. A strategy for Player σ is *winning* from node v_0 if assuming Player σ always acts according to the strategy, all plays starting from v_0 generated by the players are winning for Player σ. A game Γ is *determined* if one of the players has a winning strategy starting at any given node v of the game. *To solve a game* means to find all positions from which a given player wins.

3   Update Games and their Basic Properties

A game (V_0, V_1, E, Ω) is an *update game* if $\Omega = \{V\}$. An *update network* is an update game where Player 0 wins the game from some node. we denote the update game by (V_0, V_1, E). Dinneen and Khoussainov in [2] discussed basic properties of these games. Our focus will be to give a refined analysis of the winning strategies in update games.

Let Γ be a game. The *0-attractor* set of X, denoted $\mathsf{Attr}_0(X)$, is the set of all nodes $v \in V$ that satisfy the following: Player 0 has a memoryless strategy $\mathcal{S}$ such that every play consistent with $\mathcal{S}$ that begins from v eventually reach some node in X. It is well-known that the set $\mathsf{Attr}_0(X)$ for any $X \subseteq V$ can be computed in time $O(|V| + |E|)$ [9].

Proposition 1. *A game Γ is an update network if and only if $\mathsf{Attr}_0(\{v\}) = V$ for all $v \in V$.*

Proof. Let $V = \{v_1, v_2, \ldots, v_n\}$. If $\mathsf{Attr}_0(\{v_i\}) = V$ for all $v \in \{1, \ldots, n\}$, then Player 0 wins as follows: When v_i is visited, Player 0 applies the memoryless "attractor" strategy to visit v_{i+1} (here we let $n + 1 = 1$).   □

It is clear that the winning strategy described in the above proof requires $|V|$ states. Proposition 1 gives us a simple algorithm that solves update games:

Corollary 1. *There exists an algorithm that solves update games $\Gamma = (V_0, V_1, E)$ in time $O(|V| \cdot (|V| + |E|))$. Furthermore, if Γ is an update network, then Player 0 wins with an $|V|$-state strategy. Otherwise, Player 1 wins with a memoryless strategy.*   □

A *t-star network* is a game where $|V_0| = 1$, $|V_1| = t$ and the only node in V_0 is linked to every node in V_1 via an edge.

Lemma 1. *A game Γ where $|V_0| = 1$ is an update network if and only if it is a t-star network for some t.*   □

Let Γ be an update game. By Corollary 1, if Γ is an update network then Player 0 has an n-state winning strategy. Our goal is to build more sophisticated finite state wining strategies for update games. To do this we recast some of the results from [2]. One of the important concepts in the analysis is the notion of a forced cycle defined below.

Definition 1. *Let Γ be an update game.*

1. *For a Player 0's node v define the set $\mathsf{Forced}(v) = \{u \mid v \in uE, |uE| = 1\}$. We say that Player 1 is* forced *to move from u to v if $u \in \mathsf{Forced}(v)$.*
2. *A* forced cycle *is a sequence of pairwise distinct nodes $x_1, y_1, x_2, y_2, \ldots, x_k, y_k$ such that $x_i \in V_0$, $y_i \in \mathsf{Forced}(x_{i+1})$ and $y_k \in \mathsf{Forced}(x_1)$ for all $i = 1, 2, \ldots, k$. A forced cycle of length 2 is called a* spike. *If x, y form a spike we denote it by $x \leftrightarrow y$.*

We would sometimes abuse the notation by referring to a forced cycle as a set of nodes. We now state several properties (without proofs) of an update network Γ with $|V_0| > 1$.

Lemma 2.   *1. For all $v \in V_0$, $\mathsf{Forced}(v) \neq \emptyset$.*
2. *For every $v \in V_0$, there exists $w \in \mathsf{Forced}(v)$ and $u \in V_0$ such that $(u, w) \in E$ and $u \neq v$.*
3. *The game Γ contains a forced cycle which is not a spike.* $\square$

Thus, every update game without forced cycles can not be an update network. Note that existence of forced cycles does not guarantee that the game is an update network.

4   Contraction and Unfolding Operators

Our goal is to introduce an operator that reduces the sizes of update games. This will be important for building finite state winning strategies in update games. Our definition follows [2].

Definition 2. *Let $\Gamma = (V_0 \cup V_1, E)$ be an update game and C be a forced cycle in Γ. The* contraction operator, *when applied to Γ and C, produces a new update game $\Gamma(C)$ as follows (See Fig. 1 for an example):*

1. *Contract all nodes in $C \cap V_0$ (resp. $C \cap V_1$) to a new node x(resp. y) of Player 0 (resp 1).*
2. *Put directed edges between x and y.*
3. *Replace all edges (u, v), where $u \in V \setminus C$ and $v \in C$, by the edges (u, x) if $v \in V_0$, otherwise replace the edge (u, v) with (u, y).*
4. *Replace all edges (u, v), where $u \in C$ and $v \in V \setminus C$, by the edge (x, v) (Note that $u \in V_0$ as C is a forced cycle).*
5. *Keep all other nodes and edges in $V \setminus C$ intact.*

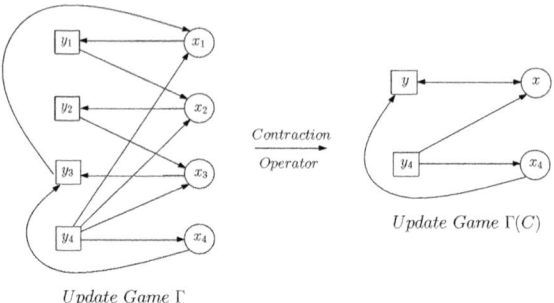

Fig. 1. In the update game Γ the forced cycle C is $\{x_1, y_1, x_2, y_2, x_3, y_3\}$. In the figure circles represent Player 0's nodes and squares represent Player 1's nodes.

Lemma 3. *Let Γ be an update game with a forced cycle C. Then Γ is an update network if and only if $\Gamma(C)$ is an update network.*

Proof. Assume Player 0 wins the game $\Gamma(C)$ with winning strategy $\mathcal{S}$. In the following, we intuitively describe a winning strategy for Player 0 in Γ: Player 0 plays the game Γ by simulating a play π that is consistent with $\mathcal{S}$ in $\Gamma(C)$. If the play π visits nodes other than x and y, then Player 0 copies the movements of π in Γ. Once π moves from a node $v \notin \{x, y\}$ to x or y, then Player 0 in Γ moves to any node in $C \cap vE$. Then Player 0 first force the play to go around the cycle C and then when π leaves $\{x, y\}$ to some node u, Player 0 will move to a node $w \in V_0 \cap C$ such that $u \in wE$ and move to u. Since $\mathcal{S}$ is a winning strategy, it is easy to see that Player 0 also wins in the game Γ.

Conversely, assume Player 0 wins the game Γ with winning strategy $\mathcal{S}'$. Then Player 0 also wins $\Gamma(C)$ by simulating a play in Γ that is consistent with $\mathcal{S}'$. $\quad\square$

Suppose $\Gamma = (V_0, V_1, E)$ is an update game. By iteratively applying the contraction operator on forced cycles that are not spikes, we obtain a sequence of games $\Gamma_0, \Gamma_1, \ldots, \Gamma_k$ where $\Gamma_0 = \Gamma$, and $\Gamma_{i+1} = \Gamma_i(C_i)$ for all $i \in \{0, \ldots, k-1\}$, where C_i is a non-spike forced cycle in Γ_i. The last game Γ_k does not have a non-spike forced cycle. We call this sequence a *maximal contraction sequence of Γ*. Given Γ, finding a non-spike forced cycle C (or detecting such forced cycle does not exist), and constructing $\Gamma(C)$ take time $O(|V| + |E|)$. Therefore a maximal contraction sequence can be built in time $O(k \cdot (|V| + |E|))$. Combining Lemma 1, Lemma 2 and Lemma 3, we have:

Theorem 1 (Deciding Update Games). *There exists an algorithm that given an update game Γ constructs a maximal contraction sequence in time $O(k \cdot (|V| + |E|))$, where $k \leq |V|$. Moreover, Γ is an update network if and only if the last game in the sequence is a star network.* $\quad\square$

Our next goal is to extract a finite state winning strategies for update networks Γ using Theorem 1. For this we define an operator that, in some sense, inverses the contraction operator:

Definition 3. *Let Γ be an update game and $x \leftrightarrow y$ be a spike in Γ. Let C be any non-spike forced cycle that is disjoint from the graph of Γ. The unfolding operator, applied to $\Gamma, x \leftrightarrow y$ and C, proceeds in two steps as follows:*

1. **Replacement**: *Replace the spike $x \leftrightarrow y$ with a new forced cycle C of length at least 4.*
2. **Enrichment**: *First, associate edges of the forms (v, y) and (v, x) in Γ, where $v \notin \{x, y\}$, with sets $M_v^1 \subset C$ and $M_v^0 \subset C$ of Player 1 and Player 0 nodes, respectively. In addition, associate with each edge of the form (x, v), where $v \in V_1$, a set $M_v \subset C$ of Player 0 nodes. Now the enrichment operation proceeds as follows:*
   (a) *Replace each edge (v, y) as above with the edges (v, u), where $u \in M_v^1$.*
   (b) *Replace each edge (v, x) as above with the edges (v, u), where $u \in M_v^0$.*
   (c) *Replace each edge (x, v) as above with the edges (u, v), where $u \in M_v$.*

This resulting game is not uniquely determined and depends on the sets M_v^0, M_v^1, M_v. For notational convenience we suppress the parameters C, M_v^0, M_v^1 and M_v and denote the resulting game by $\Gamma(x \leftrightarrow y)$. We call it an unfolded game *of Γ.*

The proof of the following lemma follows from the definitions:

Lemma 4. *Let $x \leftrightarrow y$ be a spike in $\Gamma(C)$ obtained from contracting C. Then the following hold:*

1. *The unfolding operator applied to the update game $\Gamma(C)$, the spike $x \leftrightarrow y$, and forced cycle C produces the original game Γ, that is $\Gamma = \Gamma(C)(x \leftrightarrow y)$.*
2. *Γ is an update network if and only if $\Gamma(C)(x \leftrightarrow y)$ is an update network.* □

Theorem 1 and Lemma 4 allow us to construct update networks. Namely, start with a star network and consecutively apply the unfolding operation. All update games obtained in this way are update networks. Conversely, for every update network Γ there exists a sequence of unfolded games that starts from a star network such that the sequence produces Γ. We single out this observation as the following theorem:

Theorem 2 (Building Update Networks). *All update networks can be obtained by consecutively applying the unfolding operation to star-networks.* □

5   Extracting Winning Strategies

Let Γ be an update game and let $\Gamma_0, \ldots, \Gamma_k$ be a maximal contraction sequence of Γ. By Lemma 4, each Γ_{i+1} is an unfolding of Γ_i through a spike $x_i \leftrightarrow y_i$ in the game Γ_i. Assume that the game Γ_k is a t-star network. Our goal is to explicitly construct a finite state winning strategy for Player 0 by using this sequence.

Let $\mathcal{S} = (Q, q_0, \delta)$ be a finite state strategy in game Γ. We say that the spike $x \leftrightarrow y$ is *used by $\mathcal{S}$* if for all state $q \in Q$ we have $\delta(q, x) = (q', y)$ for some $q' \in Q$. Otherwise, we say that the spike is *unused by $\mathcal{S}$*. For instance, when Γ is a t-star network, Player 0 has an obvious t-state winning strategy that visits all the Player 1 nodes in cyclic order. Every spike in Γ is used by the strategy. When the strategy $\mathcal{S}$ is clear, we drop the reference to $\mathcal{S}$ and simply say the spike is used or unused.

Lemma 5. *Assume that Player 0 has a t-state winning strategy S in the update game Γ. If the spike $x \leftrightarrow y$ is used by S then Player 0 has a t-state winning strategy in $\Gamma(x \leftrightarrow y)$. Otherwise Player 0 has a $(t+1)$-state winning strategy in $\Gamma(x \leftrightarrow y)$.*

Proof. We prove the lemma by explicitly constructing the strategy for Player 0 in $\Gamma(x \leftrightarrow y)$. The formal construction of the strategy is quite technical, thus we only provide a rough outline. Intuitively, Player 0 plays the game $\Gamma(x \leftrightarrow y)$ by simulating a play π in game Γ that is consistent with the strategy S. Suppose the spike $x \leftrightarrow y$ is used by S. When the play π visits x and S indicates Player 0 to visit y, Player 0 will visit a node $u_0 \in C$ and start to go around the cycle C while remaining in the same state. Upon returning to u_0, Player 0 changes its state and resume the simulation of π from the point where π has gone from x to y and back to x. This strategy of Player 0 still has t states as no new state is created for it to go around the cycle C. Suppose the spike $x \leftrightarrow y$ is not used by S. Then the winning strategy of Player 0 in $\Gamma(x \leftrightarrow y)$ is defined similarly. However, we need an extra state so that once π visits x, Player 0 can "branch-off" the play in $\Gamma(x \leftrightarrow y)$ to go around the cycle C. Hence the strategy has $t+1$ states   □

Consider the maximal contraction sequence $\Gamma_0 \ldots, \Gamma_k$ on Γ. Assume that the last game Γ_k is not a star-network. Our goal is to build a winning strategy for Player 1 in the original game. The strategy we construct will be memoryless as one would expect from Corollary 1.

Lemma 6. *Let Γ be an update game.*

1. *If there exists a node $v \in V_1$ without incoming edges then any memoryless strategy for Player 1 in Γ and in all unfolded games of Γ is a winning strategy.*
2. *Suppose that there exists a node $v \in V_0$ such that $\mathsf{Forced}(v) = \emptyset$. Player 1 has a memoryless winning strategy in Γ and all unfolded games of Γ.*
3. *Suppose that $|V_0| > 1$ and there exists a node $v \in V_0$ such that for all $w \in \mathsf{Forced}(v)$ no $u \neq v$ with $(u, w) \in E$ exists. Player 1 has a memoryless winning strategy in Γ and all unfolded games of Γ.*   □

The two lemmas above and Theorem 1 give us the following theorem.

Theorem 3 (Extracting Winning Strategies). *Suppose that we are given a maximal contraction sequence $\Gamma_0, \ldots, \Gamma_k$ of an update game Γ.*

1. *Assume that Γ_k is a t-star network for some $t \in \mathbb{N}$. Player 0 has a $(t+m)$-state winning strategy in game Γ, where m is the total number of unused spikes that are unfolded in the sequence $\Gamma_0, \ldots, \Gamma_k$. Moreover, the strategy can be built in time proportional to m.*
2. *Assume that Γ_k is not a t-star network for any $t \in \mathbb{N}$. Player 1 has a memoryless winning strategy in game Γ. Moreover, the strategy can be built in time proportional to k.*   □

Note that the winning strategies extracted by the procedure above depend on the number of contractions of forced cycles, and the number of states in this strategy may not be minimal. We remark that the problem of computing the minimal state winning strategies in update networks is NP-complete. For instance the Hamiltonian cycle problem can be reduced to finding a memoryless winning strategy problem in update networks. Indeed, given a directed graph G, subdivide every edge (u, v) into two edges (u, x) and (x, v) where x is a new node. This new graph is now the underlying graph for an update game $\Gamma(G)$. In $\Gamma(G)$, Player 0's nodes are the original nodes in G and Player 1's nodes are the new nodes. Clearly, the graph G has a Hamiltonian cycle if and only if Player 0 has a one-state winning strategy in the game $\Gamma(G)$.

6 Update Games with a Fixed Number of Nondeterministic Nodes

A natural question arises if there exists a better algorithm that solves update games than the algorithm described in Theorem 1. For instance, one would like to know if it is possible to solve update games in linear time on the size of the graph. In a game Γ, we say a node u is *nondeterministic* if $|uE| > 1$. In this section we provide a linear-time algorithm to solve update games where there are at most k nondeterministic nodes of Player 1, where $k \geq 1$ is fixed. We denote the class of all such games by $\mathcal{U}_k$. Our algorithm takes a game Γ from the class $\mathcal{U}_k$, and reduces the game to an equivalent game from the class $\mathcal{U}_{k-1}$. The process may eventually produce a game from the class $\mathcal{U}_0$. The following is an obvious lemma that characterizes all the update networks from the class $\mathcal{U}_0$.

Lemma 7. *Let Γ be an update game from the class $\mathcal{U}_0$. Then Γ is an update network if and only if the underlying graph G of Γ is strongly connected.* □

Let Γ be an update game from the class $\mathcal{U}_k$, and let $b_0, b_1, \ldots, b_{k-1}$ be all nondeterministic nodes of Player 1 in Γ. Consider the graph H obtained by removing all these nondeterministic nodes from the underlying graph G. Let $C_0, \ldots, C_{t-1}$ be all the strongly connected components of H each of cardinality at least 2. We define a new update game Γ' called the *derivative* of Γ:

Definition 4. *For every strongly connected component C_i, collapse all Player 0 nodes in C_i into one node denoted by x_i, collapse all Player 1 nodes in C_i into one node denoted by y_i. Keep the other nodes of G intact. Note that some of the edges of G might collapse into one edge. The resulting graph is the underlying graph of the derivative game Γ'.(See Figure 2 for an example)*

It is obvious that for each component C_i in the game Γ, Player 0 has a strategy f_i such that Player 0 using this strategy can stay in C_i forever and visit all the nodes of C_i infinitely often. We prove the next lemma using these strategies and a similar argument as the proof of Lemma 3.

Lemma 8. *The game Γ is an update network if and only if Γ' is an update network.* □

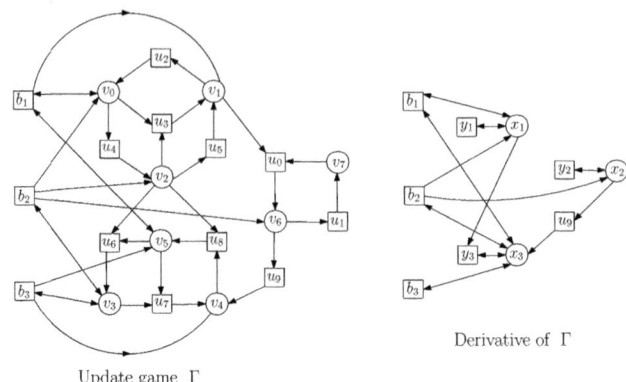

Update game Γ

Derivative of Γ

Fig. 2. Transformation of update game Γ to its derivative

The next lemma gives us a sufficient condition for Player 1 to win the update game Γ.

Lemma 9. *If Γ is an update network then at least one of the nondeterministic nodes b_0, ..., b_{k-1} becomes deterministic in the derivative game Γ'.*

Proof. Suppose no b_i where $i \in \{0, \ldots, k-1\}$ is deterministic in Γ'. We show that Player 1 has a memoryless winning strategy in Γ'. Consider the natural partial order $\leq$ defined on the spikes $x_1 \leftrightarrow y_1$, ..., $x_{t-1} \leftrightarrow y_{t-1}$ as follows. Say that $x_i \leftrightarrow y_i \leq x_j \leftrightarrow y_j$ if and only if there exists a path from C_j to C_i such that the path does not use the nondeterministic nodes $b_0, b_1, \ldots, b_{k-1}$.

We now define a strategy for Player 1 in Γ' as follows. Let $x \leftrightarrow y$ be a maximal spike in Γ'. Note that for any Player 1's node w with (w, x) there exists $w_x \neq x$ such that $(w, w_x) \in E'$. Define the strategy g_x so that $g_x(w) = w_x$. At all other nodes $u \neq w$, the strategy g_x is defined arbitrarily. This strategy is clearly a winning strategy for Player 1.   □

Now consider the sequence of update games obtained by taking derivatives:

$$\Gamma_0 = \Gamma, \ \Gamma_1 = \Gamma_0', \ \Gamma_2 = \Gamma_1', \ldots$$

This process stops at stage s if either Γ_s has no nondeterministic nodes or Γ_s has the same number of nondeterministic nodes as Γ_{s-1}. We call the sequence $\Gamma_0, \Gamma_1, \ldots, \Gamma_s$ the *maximal derivative sequence* of Γ. From the three lemmas above we have that Γ is an update network if and only if $\Gamma_s \in \mathcal{U}_0$ and the underlying graph G_s of Γ_s is strongly connected. Constructing Γ_{i+1} from Γ_i takes time proportional to $|V| + |E|$. This can be done using Tarjan's algorithm that computes strongly connected components[12]. Thus, we have the following theorem.

Theorem 4 (Deciding Update Games). *There exists an algorithm that given an update game Γ constructs the maximal derivative sequence in time*

$O(k \cdot (|V| + |E|))$, where k is the number of nondeterministic nodes of Player 1 in Γ. Moreover, the game Γ is an update network if and only if $\Gamma_s \in \mathcal{U}_0$ and G_s is a strongly connected component. □

The following is an obvious corollary:

Corollary 2. *There is a linear time algorithm that solves update games from the class U_k.* □

References

1. Bodlaender, H.L., Dinneen, M.J., Khoussainov, B.: On Game-Theoretic Models of Networks. In: Eades, P., Takaoka, T. (eds.) ISAAC 2001. LNCS, vol. 2223, pp. 550–561. Springer, Heidelberg (2001)
2. Dinneen, M.J., Khoussainov, B.: Update Games and Update Networks. Journal of Discrete Algorithms 1(1) (2003)
3. Dziembowski, S., Jurdzinski, M., Walukiewicz, I.: How Much Memory is Needed to Win Infinite Games. In: Proceedings of 12th Annual IEEE Symposium on Logic in Computer Science, Warsaw, Poland, pp. 99–110 (1997)
4. Emerson, E.A., Jutla, C.S., Sistla, A.P.: On Model Checking for Fragments of μ-calculus. In: Courcoubetis, C. (ed.) CAV 1993. LNCS, vol. 697, pp. 385–396. Springer, Heidelberg (1993)
5. Grädel, E., Thomas, W., Wilke, T. (eds.): Automata, Logics, and Infinite Games. LNCS, vol. 2500. Springer, Heidelberg (2002)
6. Ishihara, H., Khoussainov, B.: Complexity of Some Infinite Games Played on Finite Graphs. In: Kučera, L. (ed.) WG 2002. LNCS, vol. 2573, pp. 270–281. Springer, Heidelberg (2002)
7. Mang, Y.C.: Games in open systems verification and synthesis. PhD Thesis, University of California at Berkeley (2002)
8. McNaughton, R.: Infinite Games Played on Finite Graphs. Annals of Pure and Applied Logic 65, 149–184 (1993)
9. Nerode, A., Remmel, J., Yakhnis, A.: McNaughton Games and Extracting Strategies for Concurrent Programs. Annals of Pure and Applied Logic 78, 203–242 (1996)
10. Nerode, A., Yakhnis, A., Yakhnis, V.: Distributed Concurrent Programs as Strategies in Games. In: Logical Methods, pp. 624–653 (1992)
11. Hunter, P., Dawar, A.: Complexity Bounds for Regular Games. In: Jedrzejowicz, J., Szepietowski, A. (eds.) MFCS 2005. LNCS, vol. 3618, pp. 495–506. Springer, Heidelberg (2005)
12. Tarjan, R.: Depth-first Search and Linear Graph Algorithms. SIAM Journal on Computing 1, 146–160 (1972)
13. Thomas, W.: Infinite games and verification (Extended abstract of a tutorial). In: Brinksma, E., Larsen, K.G. (eds.) CAV 2002. LNCS, vol. 2404, pp. 58–64. Springer, Heidelberg (2002)

A Generalised Dynamical System, Infinite Time Register Machines, and Π_1^1-CA$_0$

Peter Koepke[1] and Philip D. Welch[2]

[1] Mathematisches Institut, Rheinische Friedrich-Wilhelms-Universität Bonn,
Endenicher Allee 60, 53115 Bonn, Germany
koepke@math.uni-bonn.de
[2] School of Mathematics, University of Bristol, Clifton, Bristol, BS8 1TW,
United Kingdom
p.welch@bristol.ac.uk

Abstract. We identify a number of theories of strength that of Π_1^1-CA$_0$. In particular: (a) the theory that the set of points attracted to the origin in a generalised transfinite dynamical system of any n-dimensional integer torus exists; (b) the theory asserting that for any $Z \subseteq \omega$ and n, the halting set H_n^Z of infinite time n-register machine with oracle Z exists.

Suppose $f : \mathbb{N}^n \longrightarrow \mathbb{N}^n$. We are going to consider transfinite iterations of such $f : \mathbb{N}^n \longrightarrow \mathbb{N}^n$ as a *generalised dynamical system*. If one wishes, one may think of f acting on the points of an n-dimensional lattice torus where we identify ∞ with 0. We set this up as follows. Given a point $r = (r_1, \ldots, r_n) \in \mathbb{N}^n$ set:

$$r^0 = (r_1^0, \ldots, r_n^0) = (r_1, \ldots, r_n);$$
$$r^{\alpha+1} = (r_1^{\alpha+1}, \ldots, r_n^{\alpha+1}) = f((r_1^\alpha, \ldots, r_n^\alpha));$$
$$r^\lambda = (r_1^\lambda, \ldots, r_n^\lambda) = (\text{Liminf}^*_{\alpha \to \lambda} r_1^\alpha, \text{Liminf}^*_{\alpha \to \lambda} r_2^\alpha, \ldots, \text{Liminf}^*_{\alpha \to \lambda} r_n^\alpha)$$

where we define $\text{Liminf}^*_{\alpha \to \lambda} r_1^\alpha = \text{Liminf}_{\alpha \to \lambda} r_1^\alpha$ if the latter is $< \omega$, and set it to 0 otherwise, thus:

$$r_i^\lambda = \begin{cases} \text{Liminf}_{\alpha \to \lambda} r_i^\alpha & \text{if the latter is } < \omega \\ 0 & \text{otherwise.} \end{cases}$$

We may wonder about the behaviour of points under this dynamic. For example which points ultimately end up at the origin O? As a more amusing example let $p = (p_0, p_1, p_2) \in (\mathbb{N}^n)^3$ be a triple of three points on the n-dimensional lattice. In general they thus form a proper triangle. Then define:

$$T_f = \{(p_0, p_1, p_2) \in \mathbb{N}^{n3} \mid \exists \alpha \; p_0^\alpha = p_1^\alpha = p_2^\alpha\}.$$

T_f is thus the set of possible starting triangles, which at some point collapse and become coincident after iteration of their vertices (and remain collapsed of course from some point α_0 onwards).

Definition 1. *Let* GDS$_T$ *be the statement: "$\forall n \forall f : \mathbb{N}^n \longrightarrow \mathbb{N}^n (T_f$ exists)."*

B. Löwe et al. (Eds.): CiE 2011, LNCS 6735, pp. 152–159, 2011.

Clearly a certain amount of analysis is needed to show that GDS_T holds. But how much? We use the common nomenclature for *subsystems of second order number theory* as taken, for example from [6]. Here "$\mathrm{ACA_0}$" and "$\Pi_1^1\text{-}\mathrm{CA_0}$" stand for arithmetical and Π_1^1-Comprehension axioms respectively. "$\mathrm{ATR_0}$" is arithmetical transfinite recursion. Recall that these are boldface theories allowing real (*i.e.* set of integer) parameters (and that $\mathrm{ACA_0}$ is included in $\mathrm{ATR_0}$).

Theorem 1. *Over* $\mathrm{ATR_0}$ *the statements* GDS_T *and* $\Pi_1^1\text{-}\mathrm{CA_0}$ *are equivalent.*

Similar results hold for starting triangles which in particular collapse to the origin, or triangles which become collinear at some stage. As mentioned above, we could define

$$Z_f =_{\mathrm{df}} \{p \in \mathbb{N}^n \mid \exists \alpha\, p^\alpha = O\}.$$

The appropriate statement GDS_Z would be "$\forall n \forall f : \mathbb{N}^n \longrightarrow \mathbb{N}^n (Z_f \text{ exists})$."

As [6, VI.1] shows, a typical theory of the same strength as $\Pi_1^1\text{-}\mathrm{CA_0}$ is: *"For every tree* $T \subseteq \omega^{<\omega}$ *there is a perfect subtree* $P \subseteq T$ *such that the set of paths through* T *not a path through* P *forms a countable set."* This is thus Cantor's result that any closed set in $\mathbb{R}$ can decomposed as a countable set together with a perfect set. There are many other statements known to be equivalent to $\Pi_1^1\text{-}\mathrm{CA_0}$.

A feature of our dynamical systems is the use of the liminf process on the lattice of some arbitrary but finite dimension. The device used is to then redefine such a coordinate at a limit stage which has been sent out to ∞ to be zero. This method has also been employed in the *Infinite Time Register Machines* of Koepke and Miller [4] which we shall use to prove the theorem. We assume the notation from this paper or from [1].

Briefly such a machine consists of a standard Shepherdson-Sturgis Register Machine [5] (or see Cutland [2]), with a standard program for such a device. Such a program is a finite set of instructions, to do some basic register contents manipulation of the finitely many registers $R_0, \ldots, R_{N-1}$ of the machine. The infinitary behaviour is defined by letting register values at limit stages be liminf*s of previous values.

Definition 2. *Let* N *be a natural number. An* N-register machine *has registers* $R_0, R_1, \ldots, R_{N-1}$ *which can hold natural numbers. An* N-register program *is a finite list* $P = I_0, I_1, \ldots, I_{s-1}$ *of instructions, each of which may be of one of five kinds where* m, n *range over the numbers* $0, 1, \ldots, N-1$:

1. *the* zero instruction $Z(n)$ *changes the contents of* R_n *to* 0;
2. *the* successor instruction $S(n)$ *increases the natural number contained in* R_n *by* 1;
3. *the* oracle *instruction* $O(n)$ *replaces the content of the register* R_n *by the number* 1 *if the content is an element of the oracle, and by* 0 *otherwise;*
4. *the* transfer instruction $T(m, n)$ *replaces the contents of* R_n *by the natural number contained in* R_m;
5. *the* jump instruction $J(m, n, q)$ *is carried out as follows: the contents* r_m *and* r_n *of the registers* R_m *and* R_n *are compared; then, if* $r_m = r_n$, *the machine proceeds to the* qth *instruction of* P; *if* $r_m \neq r_n$, *the machine proceeds to the next instruction in* P.

The instructions of the program can be addressed by their indices which are called program states. *At ordinal time τ the machine will be in a* configuration *consisting of a program state $I(\tau) \in \omega$ and the register contents which can be viewed as a function $R(\tau) : N \to \omega$. $R(\tau)(n)$ is the content of register R_n at time τ. We also write $R_n(\tau)$ instead of $R(\tau)(n)$.*

Definition 3. *Let P be an N-register program. program. Let $R_0(0), \ldots, R_{N-1}(0)$ be natural numbers and $Z \subseteq \omega$ be an oracle. These data determine the* infinite time register computation

$$I : \theta \to \omega, R : \theta \to ({}^N \omega)$$

with program P, input $R_0(0), \ldots, R_{N-1}(0)$ and oracle Z by recursion:

1. *θ is an ordinal or $\theta = \mathrm{Ord}$; θ is the* length *of the computation;*
2. *$I(0) = 0$; the machine starts in state 0;*
3. *If $\tau < \theta$ and $I(\tau) \notin s = \{0, 1, \ldots, s-1\}$ then $\theta = \tau + 1$; the machine halts if the machine state is not a program state of P;*
4. *If $\tau < \theta$ and $I(\tau) \in s$ then $\tau + 1 < \theta$; the next configuration is determined by the instruction $I_{I(\tau)}$ according to the previous definition;*
5. *If $\tau < \theta$ is a limit ordinal, then $I(\tau) = \liminf_{\sigma \to \tau} I(\sigma)$ and for all $k < \omega$*

$$R_k(\tau) = \liminf{}^*_{\sigma \to \tau} R_k(\sigma).$$

So the register R_k is reset in case $\liminf_{\sigma \to \tau} R_k(\sigma) = \omega$.

If the computation halts then $\theta = \beta + 1$ is a successor ordinal and $R(\beta)$ is the final register content. In this case we say that P computes $R(\beta)(0)$ from $R(0)$ and the oracle Z, and we write $P : R(0), Z \mapsto R(\beta)(0)$.

Definition 4. *A partial function $F : \omega^n \rightharpoonup \omega$ is* computable *if there is some N-register program P such that for every n-tuple $(a_0, \ldots, a_{n-1}) \in \mathrm{dom}(F)$,*

$$P : (a_0, \ldots, a_{n-1}, 0, 0, \ldots, 0), \emptyset \mapsto F(a_0, \ldots, a_{n-1}).$$

Using the liminf process to define an instruction at a limit stage of time λ, rather neatly sets the machine to perform the beginning instruction of the outermost loop, (which we can think of as a 'subroutine') entered unboundly often below λ. After executing the instruction $I(\alpha)$ the program may move on to execute the next line of instructions ($I(\alpha + 1)$ is the line following $I(\alpha)$), or else if a jump is involved, $I(\alpha + 1)$ may be elsewhere in the program. If the liminf of the contents of a register becomes infinite, then we reset the contents value to 0: this allows the machine to continue.

Elementary arguments show that any register machine, working on any program, with any initial distribution of register contents, will either halt, or enter a permanently looping cycle by some countable ordinal stage. It is useful to add the capability of a machine to consult an *oracle* $Z \subseteq \omega$, and receive a 0/1 as to whether $R_i(\alpha)$ is in Z or not.

Again, however, immediately there is a question of, *how long* must one wait in order to see that the given machine with its program, oracle, and starting register values, is indeed looping?

Koepke and Miller give a criterion for a machine to be in a looping cycle. They dub a pair (I, R) of an instruction number (about to be performed) and a list of the current register contents R a *constellation*:

Lemma 1. *Let* $I : \theta \longrightarrow \omega, R : \theta \longrightarrow {}^n\omega$ *be a computation of the n register machine with program P and with oracle Z for order type θ many stages. Then if this computation has not halted by stage θ, then it will never do so if θ is sufficiently large so that there is some constellation (I', R') so that*

$$\mathrm{otp}(\{\beta \mid I(\beta) = I' \wedge R(\beta) = R'\} \geq \omega^\omega.$$

This gives then a simple criterion to check that we have gone far enough to test whether the computation will halt. Immediately:

Corollary 1. *Any such computation either halts, or cycles after a countable ordinal number of stages.*

It is also easy to sketch an argument that shows how *admissible ordinals* play a role. We let $\omega_k =_{\mathrm{df}} \omega_k^{\mathrm{CK}}$ be the the the kth *admissible ordinal* for $k < \omega$. (An ordinal α is *admissible* if $L_\alpha \models \mathrm{KP}$ - the latter denotes Kripke-Platek set theory; ω_1^{CK} is thus the least non-recursive ordinal.) It is well known that any admissible ordinal is Π_2-*reflecting*: that is if $\varphi(\boldsymbol{y})$ is any Π_2 formula of the language of set theory, with $\boldsymbol{y} \in L_\alpha$, then if $L_\alpha \models \varphi(\boldsymbol{y})$ then there is $\beta < \alpha$ with $L_\beta \models \varphi(\boldsymbol{y})$. Since the operations of a register machine are Δ_1 and so *absolute* to any transitive admissible set (containing the oracle set Z if there is one: note that a program P of length m is essentially an element of $\mathrm{HF} = L_\omega$), it is clear that we may define the operations of any ITRM inside any admissible set. Imagine we have a virtually trivial machine with $Z = \varnothing$, and only one register R_0! As we run the computation, suppose that the program does not halt on its starting input by stage $\eta =_{\mathrm{df}} \omega_1^{\mathrm{CK}}$. Let $\delta_0 < \eta$ be any ordinal. Then trivially:

$$R_0(\eta) = \mathrm{Liminf}^*_{\delta_0 < \beta \to \eta} R_0(\beta) = k \leq \omega$$

and

$$I(\eta) = \mathrm{Liminf}_{\delta_0 < \beta \to \eta} I(\beta) = I_l \text{ where } l < m.$$

Let us first consider the case that $\mathrm{Liminf}_{\delta_0 < \beta \to \eta} R_0(\beta) < \omega$. These values can be expressed by a conjunction of a Σ_2 and a Π_2 statement about δ_0. Hence by Π_2 reflection, there is some $\delta_1 \in (\delta_0, \eta)$ so that

$$R_0(\delta_1) = R_0(\eta) = \mathrm{Liminf}_{\delta_0 < \beta \to \delta_1} R_0(\beta) = k \wedge I(\delta_1) = I_l.$$

But as δ_0 was arbitrary, this can be repeated; in short we may define by a Σ_1 recursion a sequence of ordinals δ_ι for $0 < \iota < \eta$ with $R_0(\delta_\iota) = R' \wedge I(\delta_\iota) = I'$, with the constellation $(I', R') = (I_l, k)$ ready to be used in Lemma 1. Hence the computation never halts.

Now, suppose the other case holds, and that $\mathrm{Liminf}_{\beta \to \eta} R_0(\beta) = \omega$; hence by fiat $R_0(\eta)$ is reset to 0. Suppose again $I_l = I(\eta)$. Now, suppose the computation continues to $\eta + \eta$ without halting. The segment of computation $[\eta, \eta + \eta]$ is *precisely* the same as that of the machine starting instead on the instruction I_l 'at time zero' with register content $R_0(\eta)$ as initial value, and performing η many steps. We conclude again by the same Lemma 1 that either the machine is looping, or that $R_0(\eta + \eta) = 0$ due to a resetting of the register contents to 0 at time $\eta + \eta$. However now we may have that $I(\eta + \eta)$ is different from $I(\eta)$. Nevertheless we may iterate the argument. Let $\eta' =_{\mathrm{df}} \omega_2^{\mathrm{CK}}$ the second admissible ordinal. Suppose the computation continues as far as η' without halting. Recalling that admissible ordinals are multiplicatively closed, we conclude that for every ordinal $\gamma = \eta \cdot (\delta + 1)$ for some $\delta < \eta'$, that if the machine is not looping by stage γ then $R_0(\gamma) = 0$ again by the 'resetting rule', and the machine then will perform an instruction $I(\gamma)$. We claim that for some instruction number I_q for $q < m$ that the constellation $(I_q, 0)$ occurs with order type at least ω^ω before stage η': for some q $(I_q, 0)$ occurs unboundedly in η' and if this only occurs with order type some $\beta < \omega^\omega$ this would afford a map of β cofinally into η' enumerating stages where this constellation occurs, all this in a Σ_1-definable fashion over L_β. This would contradict the latter's admissibility. The conclusion is that the computation either halts or enters an indefinite loop by the second admissible ordinal.

It is then a matter of induction, to use this argument again on machines with n registers, to see that they either halt or enter an infinite loop by stage $\omega_{n+1}^{\mathrm{CK}}$ (the details of the induction are in Theorem 9 of [3]).

We define:

1. The n-register halting set $H_n =_{\mathrm{df}} \{\langle e, r_0, \ldots, r_{n-1}\rangle \mid P_e(r_0, \ldots, r_{n-1}) \downarrow\}$,
2. the assertion ITRM$_n$: "The *n-register halting set* H_n exists",
3. and the similar relativized statement **ITRM**: "For any $Z \subseteq \omega$, for any $n < \omega$ the *n-halting set* H_n^Z exists."

Theorem 2

1. ITRM$_n$ *can be proven in* KP + *"there exist n admissible ordinals* $> \omega$."
2. **ITRM** *can be proven in* Π_1^1-CA$_0$.

Proof (1.) has essentially been outlined above where ITRM$_1$ was done in detail.

For (2.), Π_1^1-CA$_0$ proves that for any $Z \subseteq \omega$ the *hyperjump* HJ(Z) of Z exists (in Simpson [6] VII.1.16). The hyperjump (as defined in [6, VII.1.5]) is a complete $\Sigma_1^{1,Z}$-set of integers. Now Π_1^1-CA$_0^{\mathrm{set}}$ (recall that this is also a boldface theory as outlined in VII.3 *op. cit.*, and is, for the language of second order arithmetic, L^2, a conservative extension of Π_1^1-CA$_0$) proves that for every $Z \subseteq \omega$ there is a class of constructible sets L^Z constructed from Z, which is a model of the Axiom Beta. As L^Z is a model of Beta, we have that HJ(Z) is a set in L^Z. Repeating this argument, we have that for every $k < \omega$ that $L^Z \models$ HJ(k, Z) exists (where HJ(k, Z) is the k'th hyperjump of Z). Then we may conclude that $L^Z \models \omega_k^{Z,\mathrm{CK}}$ exists. But the arguments deployed above show then that for any k $L^Z \models H_k^Z$ exists. By absoluteness $H_k^Z = (H_k^Z)^{L^Z}$. This suffices.

Now for the converse:

Theorem 3

1. $\text{ATR}_0 + \textbf{ITRM} \vdash \Pi_1^1\text{-CA}_0$.
2. There is a fixed $k < \omega$ so that for any $n < \omega$

$$\text{ATR}_0 + \text{ITRM}_{n \cdot k} \vdash \text{``HJ}(n, \varnothing) \text{ exists.''}.$$

Proof (Sketch of (1.)) Define $T \subseteq^{<\omega} \omega$ to be a *tree* in the usual way. A *path through* T is then a function $h : \omega \longrightarrow T$ with $\forall i(h(i)$ is T-extended by $h(i+1))$. By Simpson [6] VI.1.1 it suffices to show that if $\langle T_k \mid k < \omega \rangle$ is a sequence of trees, that there exists the set $X = \{k \mid T_k \text{ has a path}\}$. Let then $\langle T_k \mid k < \omega \rangle$ be such a sequence. We may assume that it is coded in some recursive manner by $Z \subseteq \omega$. Modify the program of Lemma 1 of [1], to search for paths through trees coded into Z: this is straightforward, as that program checks for ill-foundedness anyway: all that is needed to do is to incorporate the recursive function decoding a T_k from Z. Suppose then that $P_f^Z(x)$ is the program searching for the answer to the ill-foundedness of T_x. Thus for any k $P_f^Z(k) \downarrow 1$ iff T_k is ill-founded, and $P_f^Z(k) \downarrow 0$ otherwise. Adjusting the program slightly just to diverge in the latter case, there is then a program $P_{f'}^Z$ so that

$$T_k \text{ is ill-founded iff } \langle f', k, 0, \ldots, 0 \rangle \in H_m^Z$$

(where we have a sequence of $m - 2$ zeroes, where $P_{f'}$ uses $m - 1$ registers; the number of registers the latter actually uses will depend also on the number of registers used in the implementation of the code of Lemma 1 of [1]). **ITRM** tells us that H_m^Z exists as a set. Hence by ACA_0 so does $X = \{k \mid T_k \text{ has a path}\}$.

By, e.g., [6, VII.1.16], we may also show that

$$\text{ATR}_0 + \textbf{ITRM} \vdash \text{``For every } Z \subseteq \omega(\text{HJ}(Z) \text{ exists)''}$$

to achieve the same conclusion. We first sketch this before detailing (2.).

Fix Z. Using Z as an oracle, we can, using the properties of ITRMs test for each given index e in turn, whether $\{e\}^Z$ is in WF or not. This again incorporates the argument of Theorem 1 of [1], which shows how given $Y \subseteq \omega$ there is a program P_e^Y (uniform in Y) that decides with a 0/1 output if $Y \in \text{WF}$. We wish to use this for each Y of the form $\{e\}^Z$. We thus extend the 'pseudo-code' of the program of Lemma 1 of [1], so that on input e with data Z it may compute membership questions about $\{e\}^Z$, and then run P_e^Y on $Y = \{e\}^Z$ as a sub-program. However this extension is straightforward (and also uniform in Z). This provides an index f so that $P_f^Z(e) \downarrow 1$ iff $\{e\}^Z \in \text{WF}$, and $P_f^Z(e) \downarrow 0$ otherwise. Adjusting this program to one with an index f' so that $P_{f'}^Z(e) \uparrow$ whenever $\{e\}^Z \notin \text{WF}$, we shall have $\{e\}^Z \in \text{WF}$ iff $\langle f', e, 0, \ldots, 0 \rangle \in H_m^Z$ (where we have a sequence of $m - 2$ zeroes, where $P_{f'}$ uses $m - 1$ registers; again the number of registers the latter actually uses will depend also on the number of registers used in the implementation of the code of [1, Lemma 1]).

ITRM tells us that H_m^Z exists as a set. Hence by ACA_0 so does $\mathcal{O}^Z$. However, there is (uniformly in Z) a (1-1) recursive function $h : \mathbb{N} \longrightarrow \mathbb{N}$ with $m \in \mathrm{HJ}(Z) \leftrightarrow h(m) \notin \mathcal{O}^Z$. (The complement of $\mathcal{O}^Z$ and $\mathrm{HJ}(Z)$ are both complete Σ_1^1 sets of integers and hence recursively isomorphic.) Hence $\mathrm{HJ}(Z)$ exists. Hence assuming that we used overall k registers, what we actually have shown in the above is that $\mathrm{ATR}_0 + \mathrm{ITRM}_k \vdash$ "$\mathrm{HJ}(1, Z) = \mathrm{HJ}(Z)$ *exists.*"

This argument can be repeated to get any n^{th} hyperjump of Z. Let $Z^{h(1)} =_{\mathrm{df}}$ $\mathrm{HJ}(1, Z)$ as above. We then have argued that there is a k-register machine (for some $k < \omega$) so that it gives 0/1 answers to queries as to whether $n \in Z^{h(1)}$. In order to compute $Z^{h(2)} =_{\mathrm{df}} \mathrm{HJ}(2, Z)$ it suffices then to compute $\mathcal{O}^{Z^{h(1)}}$. We need thus to check answers to queries such as whether $\{e\}^{Z^{h(1)}} \in \mathrm{WF}$ or not. In order to do this we employ again the routine to check for the eth recursive function in a set Y and ask whether it is in WF; here whenever we have to query $?n \in Y?$ since we shall take $Y = Z^{h(1)}$, we must run (as a subroutine) the procedure to test if $?n \in Z^{h(1)}?$. We may reserve the first k registers for this work (as described above), and the next k registers for computing the 0/1 answer to $?\{e\}^{Z^{h(1)}} \in \mathrm{WF}?$ In this fashion we may compute $\mathcal{O}^{Z^{h(1)}}$ and hence $Z^{h(2)}$. Inductively this means that the nth hyperjump can be computed using $n \cdot k$ registers.

However, applying this to $Z = \varnothing$ is our (2.).

Using the registers in the algorithm for the iterated hyperjump more efficiently one could probably obtain a tighter bound for the number of registers needed, like $n + k$ instead of $n \cdot k$. The paper [7] also contains results on certain infinite time computabilities and reverse mathematics.

We shall now prove Theorem 1:

Proof Assume GDS_T. We define a function $f : \mathbb{N}^m \longrightarrow \mathbb{N}^m$ so that for any Z, $H_n^Z \leq_T T_{f'}$ for some $m > n$. We use the fact that the n-register machine halting set H_n^Z is in fact decidable by some n'-register machine with program $P_{e(n)}^Z$. This is one of the basic properties of ITRMs (*cf.* Theorem 4 of either [4] or [1]); recall that in particular this will require more registers than the n under observation. Let us assume $e(n)$ chosen so that $\langle e, r_0, \ldots, r_{n-1} \rangle \in H_n^Z$ iff $P_{e(n)}^Z(e, r_0, \ldots, r_{n-1}, 0, \ldots, 0) \downarrow 1$. We'll make some further harmless modifications to $P_{e(n)}^Z$; this will describe a a program $P_{e(n)'}^Z$: firstly, that if $P_{e(n)}^Z$ uses n_0 registers, then $P_{e(n)'}^Z$ uses a final extra register R_{n_0} which however is not addressed before the halting procedure shortly to be described.

Secondly that before halting, if $P_{e(n)}^Z$ is to set to halt with a 1 in R_0, $P_{e(n)'}^Z$ sets R_{n_0} to 0; otherwise it leaves R_{n_0} unaltered. Now consider the register contents as the coordinates of points $p_i \in \mathbb{N}^{n_0+1}$ for $i < 3$ by letting the j'th coordinate of p_i be R_j, for $j \leq n_0$. The action of the program $P_{e(n)'}^Z$ defines a function $f : \mathbb{N}^{n_0+1} \longrightarrow \mathbb{N}^{n_0+1}$ with each machine step corresponding to another iterative application of f. Suppose we start $P_{e(n)'}^Z$ with the register configuration $(e, r_0, \ldots, r_{n-1}, 0, \ldots, i)$; we set $p_i = p_i^0$ to have the coordinates of this starting configuration. Then the register configuration at stage α of the computational process yields the coordinates of p_i^α.

Then we shall have $\langle e, r_0, \ldots, r_{n-1} \rangle \in H_n^Z$ iff $(p_0^0, p_1^0, p_2^0) \in T_f$. By ACA$_0$ we thus have that H_n^Z is a set.

Conversely, assume **ITRM**. Let $f : \mathbb{N}^n \longrightarrow \mathbb{N}^n$. By using repeated tupling functions, prime power coding and the usual devices, we may assume that f is coded by some $Z = Z_f \subseteq \omega$. In particular, as an oracle we may 'query' of Z given $(r_0, \ldots, r_{n-1})$, the value of $f(r_0, \ldots, r_{n-1})$. We may thus define ITRM machines P^f that incorporate the function f. However then it is an elementary argument to write down an algorithm using such P^f on input a triple of points p_i each in $\mathbb{N}^n$, to compute the iterates of f on each p_i and see if they become coincident. If this algorithm is exemplified by P_e^f on n_0 registers then the halting set $H_{n_0}^{Z_f}$ tells us the answer to this question. We thus have for the triangle set T_f that $T_f \leq_T H_{n_0}^{Z_f}$. As **ITRM** asserts the existence of $H_{n_0}^{Z_f}$, by ACA$_0$ T_f exists.

References

1. Carl, M., Fischbach, T., Koepke, P., Miller, R., Nasfi, M., Weckbecker, G.: The basic theory of infinite time register machines. Archive for Mathematical Logic 49(2), 249–273 (2010)
2. Cutland, N.: Computability: an Introduction to Recursive Function Theory. CUP, Cambridge (1980)
3. Koepke, P.: Ordinal computability. In: Ambos-Spies, K., Löwe, B., Merkle, W. (eds.) CiE 2009. LNCS, vol. 5635, pp. 280–289. Springer, Heidelberg (2009)
4. Koepke, P., Miller, R.: An enhanced theory of infinite time register machines. In: Beckmann, A., Dimitracopoulos, C., Löwe, B. (eds.) CiE 2008. LNCS, vol. 5028, pp. 306–315. Springer, Heidelberg (2008)
5. Shepherdson, J., Sturgis, H.: Computability of recursive functionals. Journal of the Association of Computing Machinery 10, 217–255 (1963)
6. Simpson, S.: Subsystems of second order arithmetic. Perspectives in Mathematical Logic. Springer, Heidelberg (1999)
7. Welch, P.D.: Weak systems of determinacy and arithmetical quasi-inductive definitions. Journal of Symbolic Logic 76 (2011)(to appear)

Computability Power of Mobility in Enhanced Mobile Membranes

Shankara Narayanan Krishna[1] and Gabriel Ciobanu[2,3]

[1] Department of Computer Science and Engineering, Indian Institute of Technology
Bombay, Powai, Mumbai, India 400 076
krishnas@cse.iitb.ac.in
[2] Institute of Computer Science, Romanian Academy, Str. Gh. Asachi, Nr. 2, Iasi,
700 481, Romania
[3] Facultatea de Informatică, Universitatea "Al. I. Cuza", General Berthelot, 16,
Iaşi 700483, Romania
gabriel@info.uaic.ro

Abstract. We explore the computability power of the various forms of mobility in membrane computing. First we improve a previous computational completeness result by showing that 5 membranes together with $endo, exo, fendo$ and $fexo$ operations are enough for Turing completeness (it is known that 3 membranes do not suffice). Then we show that 10 membranes along with $fendo, fexo$; 9 membranes with $endo, exo, pendo$; 8 membranes with $fendo, fexo, pendo$ and 12 membranes with $endo, exo, fendo$ give computational completeness. When we look at restricted mobility described by $rendo, rexo, rfendo$ and $rfexo$, we show that they do not give computational completeness; moreover, adding pure mobility to the restricted operations does not lead to the level of RE.

1 Introduction

Membrane Computing [6] is a branch of natural computing motivated by the structure and functioning of the living cell. All the computing devices considered in membrane computing are called *P Systems* in honour of Păun, the father of the area. Initial models were based on a cell-like hierarchical arrangement of membranes delimiting compartments where multisets of objects evolve according to rewriting rules. Several biological operations were modelled in the context of membrane computing : communication of objects based on symport, antiport, concentration difference among membranes; object evolution based on catalysts, splicing, insertion-deletion; system evolution based on mitosis, autopoiesis, endocytosis, exocytosis, phagocytosis, pinocytosis to name a few. Passing from the classical hierarchical arrangement of membranes, systems where the membranes are arranged in the nodes of an arbitrary graph, as well as systems inspired by the network of neurons and their spiking action were investigated. The link between membrane computing and other related areas like brane calculi, Petri

B. Löwe et al. (Eds.): CiE 2011, LNCS 6735, pp. 160–170, 2011.

nets as well as the applications of membrane computing in areas like computer graphics, cryptography and economics have been explored. The state of the art is given in [7].

Inspired by the operations of endocytosis and exocytosis, we use a variant of membrane systems called enhanced mobile membranes. They were previously studied in [4], and used to describe some biological mechanisms of the immune system in [1]. The operations governing the mobility of the enhanced mobile membranes are endocytosis (endo), exocytosis (exo), enhanced endocytosis (fendo) and enhanced exocytosis (fexo). The interplay between these four operations is quite powerful, and the computational power of a Turing machine is obtained using twelve membranes without using the context-free evolution of objects [4]. In this paper we improve the previous computational completeness results by showing that five membranes are enough, and then investigate the computability aspects of the different forms of mobility in enhanced mobile membranes. We obtain several combinations of operations that provide Turing completeness. A form of restricted mobility described by some new rules do not lead to computational completeness.

Prerequisites. We refer to [2,8] for the elements of formal language theory used in this paper. Here $\mathbf{N}$ denotes the set of natural numbers; V denotes a finite alphabet; V^* is the free monoid generated by V under the operation of concatenation and the empty string denoted by λ, as unit element. We denote by $\mathbf{NRE}$ recursively enumerable sets of natural numbers.

A **matrix grammar** is a quadruple $G = (N, T, M, S)$ where N, T are sets of non-terminals and terminals respectively, S is the start symbol, and M is a finite set of matrices of the form $(r_1, \ldots, r_n), n \geq 1$, with context-free rewriting rules $r_i : \alpha_i \to \beta_i, \alpha_i \in N, \beta_i \in (N \cup T)^*$. For two strings x, y we say that $x \Rightarrow y$ iff there are strings $x_0, \ldots, x_m$ and a matrix $(r_1, \ldots, r_n) \in M$ such that $x_0 = x, x_m = y$, and $x_{i-1} = x'_{i-1}\alpha_i x''_{i-1}, x_i = x'_{i-1}\beta_i x''_{i-1}$ for some $x'_{i-1}, x''_{i-1} \in (N \cup T)^*$, for all $1 \leq i \leq m$. In other words, a direct derivation in the matrix grammar G corresponds to applying the rules of a matrix, in order. We denote by MAT^λ the families of languages generated by matrix grammars without appearance checking.

If we allow appearance checking, then the matrix grammar is the tuple $G = (N, T, M, S, F)$ where F is a subset of the totality of occurrences of rules in the matrices of M. For any two strings x, y we say that $x \Rightarrow_{ac} y$ iff there are strings $x_0, \ldots, x_m$ and a matrix $(r_1, \ldots, r_n) \in M$ such that $x_0 = x, x_m = y$, and $x_{i-1} = x'_{i-1}\alpha_i x''_{i-1}, x_i = x'_{i-1}\beta_i x''_{i-1}$ for some $x'_{i-1}, x''_{i-1} \in (N \cup T)^*$, for all $1 \leq i \leq m$, or else, the rule $\alpha_i \to \beta_i \in F$, α_i is not a subword of x_{i-1}, and $x_{i-1} = x_i$. We denote by $\mathrm{MAT}^\lambda_{ac}$ the families of languages generated by matrix grammars with appearance checking. In case all rules are λ-free, we remove the superscript λ from the notation.

A **matrix grammar in the strong binary normal form** is a construct $G = (N, T, S, M, F)$, where $N = N_1 \cup N_2 \cup \{S, \dagger\}$, with these three sets mutually disjoint, two distinguished symbols $B^{(1)}, B^{(2)} \in N_2$, and the matrices in M of one of the following forms:

1. $(S \rightarrow XA)$, with $X \in N_1, A \in N_2$,
2. $(X \rightarrow Y, A \rightarrow x)$, with $X, Y \in N_1, A \in N_2, x \in (N_2 \cup T)^*, |x| \leq 2$,
3. $(X \rightarrow Y, B^{(j)} \rightarrow \dagger)$, with $X, Y \in N_1, j = 1, 2$,
4. $(X \rightarrow \lambda, A \rightarrow x)$, with $X \in N_1, A \in N_2, x \in T^*, |x| \leq 2$.

There is only one matrix of type 1, and F consists of the rules $B^{(j)} \rightarrow \dagger$ with $j = 1, 2$ appearing in matrices of type 3 ($\dagger$ is a trap-symbol, once introduced it is never removed). A matrix of type 4 is used only once, in the last step of a derivation. It is proved in [3] that each recursively enumerable language can be generated by a matrix grammar in the strong binary normal form.

Register Machines. The proof of Theorem 1 is based on the concept of Minsky's register machine [5]. An *n-register machine* is a construct $M = (n, P, i, h)$, where (i) n is the number of registers, (ii) P is a set of labeled instructions of the form $j : (op(r), k, l)$, where $op(r)$ is an operation on register r of M, j, k, l are labels from the set $Lab(M)$ which identify the instructions in a one-to-one manner, (iii) i is the initial label, and (iv) h is the final label. A register machine is capable of the following instructions:

1. $(\text{add}(r), k, l)$: Add one to the contents of register r and proceed to instruction k or to instruction l; in the deterministic variants usually considered in the literature we demand $k = l$.
2. $(\text{sub}(r), k, l)$: If register r is not empty, then subtract one from its contents and go to instruction k, otherwise proceed to instruction l.
3. *Halt*: Stop the machine; this additional instruction can only be assigned to the final label h.

In their *deterministic variant*, such n-register machines can be used to compute any partial recursive function $f : \mathbf{N}^\alpha \rightarrow \mathbf{N}^\beta$; starting with $(n_1, \ldots, n_\alpha) \in \mathbf{N}^\alpha$ in registers 1 to α, M computes $f(n_1, \ldots, n_\alpha) = (r_1, \ldots, r_\beta)$ if it halts in the final label with registers 1 to β containing r_1 to r_β. If the final label cannot be reached, then $f(n_1, \ldots, n_\alpha)$ remains undefined. In their *non-deterministic variant*, n-register machines can compute any recursively enumerable set of non-negative integers (or of vectors of non-negative integers). Starting with all registers being empty, we consider a computation of the n-register machine to be successful, if it halts with the result being contained in the first (β) register(s) and with all other registers being empty.

Proposition 1. *If $L \subseteq V^*$, $\text{card}(V) = k, L \in \text{RE}$, then a 3-register machine M exists such that for every $w \in V^*$ we have $w \in L$ if and only if M halts when starting with $val_{k+1}(w)$ in its first register; in the halting step, all registers of the machine are empty.*

2   Enhanced Mobile Membranes

We assume the reader is familiar with membrane computing; for the state of the art, see [7]. We follow the notations used in [4] for our class of membrane systems called *P systems with enhanced mobile membranes*. A *P system with enhanced mobile membranes* is a construct $\Pi = (V, H, \mu, w_1, \ldots, w_n, R, i)$, where: $n \geq 1$ (the initial *degree* of the system); V is an alphabet (its elements are called *objects*); H is a finite set of *labels* for membranes; μ is a *membrane structure*, consisting of n membranes, labelled with elements of H; $w_1, w_2, \ldots, w_n$ are strings over V, describing the initial *multisets of objects* placed in the n regions of μ, i is the output membrane of the system, and R is a finite set of *developmental rules* of the following forms:

(a) $[a]_h[\ ]_m \rightarrow [[w]_h]_m$, for $h, m \in H, a \in V, w \in V^*$ *endocytosis*
an elementary membrane labeled h enters the adjacent membrane labeled m (m is not necessarily an elementary membrane) under the control of object a; the labels h and m remain unchanged during this process; however, the object a is modified to w during the operation.

(b) $[[a]_h]_m \rightarrow [w]_h[\ ]_m$, for $h, m \in H, a \in V, w \in V^*$ *exocytosis*
an elementary membrane labeled h is sent out of a membrane labeled m (membrane m is not necessarily elementary) under the control of object a; the labels of the two membranes remain unchanged; however the object a from membrane h is modified during this operation;

(c) $[\ ]_h[a]_m \rightarrow [[\ ]_h w]_m$, for $h, m \in H, a \in V, w \in V^*$ *forced endocytosis*
an elementary membrane labeled h enters the adjacent membrane labeled m (m is not necessarily an elementary membrane) under the control of object a of m; the labels h and m remain unchanged during this process; however, the object a is modified to w during the operation.

(d) $[a[\ ]_h]_m \rightarrow [\ ]_h[w]_m$, for $h, m \in H, a \in V, w \in V^*$ *forced exocytosis*
an elementary membrane labeled h is sent out of a membrane labeled m (membrane m is not necessarily elementary) under the control of object a of m; the labels of the two membranes remain unchanged; however the object a of membrane m is modified to w during the operation.

(e) $[u]_h[v]_m \rightarrow [[u]_h v]_m$, for $h, m \in H, u, v \in V^*$ *pure endocytosis*
an elementary membrane labeled h containing u enters the adjacent membrane containing v; the objects do not evolve in the process. u and v cannot both be the empty string simultaneously.

(f) $[u[v]_h]_m \rightarrow [u]_m[v]_h$, for $h, m \in H, u, v \in V^*$ *pure exocytosis*
an elementary membrane labeled m containing v comes out of the membrane labeled m containing u. The objects do not evolve in the process. u and v cannot both be the empty string simultaneously.

The rules are applied according to the following principles:

1. All the rules are applied in parallel, non-deterministically choosing the rules, the membranes, and the objects, but in such a way that the parallelism is maximal; this means that in each step we apply a multiset of rules such that no further rule can be added to the multiset, no further membranes and objects can evolve at the same time.

2. The membrane m from each type (a) – (f) of rules as above is said to be *passive*, while the membrane h is said to be *active*. In any step of a computation, any object and any active membrane can be involved in at most one rule, but the passive membranes are not considered involved in the use of rules (hence they can be used by several rules at the same time as passive membranes);

3. The evolution of objects and membranes takes place in a bottom-up manner. After having a (maximal) multiset of rules chosen, they are applied starting from the innermost membranes, level by level, up to the skin membrane (all these sub-steps form a unique evolution step, called a *transition* step).

4. When a membrane is moved across another membrane by rules (a)-(f), its whole contents (its objects) are moved.

5. All objects and membranes which do not evolve at a given step (for a given choice of rules which is maximal) are passed unchanged to the next configuration of the system.

By using the rules in this way, we get transitions among the configurations of the system. A sequence of transitions is a computation, and a computation is successful if it halts (it reaches a configuration where no rule can be applied). At the end of a halting computation, the number of objects from a special membrane called output membrane is considered as a the result of the computation. A non-halting computation provides no output.

The family of all sets of numbers $\mathbf{N}(\Pi)$ which are obtained as a result of a halting computation by a P system Π with enhanced mobile membranes of degree at most n using rules $\alpha \subseteq \{exo, endo, fendo, fexo, pendo, pexo\}$, is denoted by $\mathbf{NEM}_n(\alpha)$. Here $endo$ and exo represent endocytosis and exocytosis, $fendo$ and $fexo$ represent forced endocytosis and forced exocytosis, while $pendo, pexo$ represent pure endocytosis and pure exocytosis. When we restrict $|w| = 1$ in rules (a) – (d), we call the operations $rendo, rexo, rfendo$ and $rfexo$ where r stands for "restricted". The mobile membranes considered in [4] differs from the above definition in the following respects: (i) allows object evolution rules of the kind $[a \rightarrow v]_h$, $a \in V, v \in V^*$ and contextual evolution rules of the kind $[[a]_j [b]_h]_k \rightarrow [[w]_j [b]_h]_k$ for $h, j, k \in H, a, b \in V, w \in V^*$, and (ii) it does not exist in [4] the pure mobility rules $pendo, pexo$, nor their restricted versions.

3 Computability Power of Mobility

In this section, we explore the expressive power of the various forms of mobility. In [4] there are presented some preliminary investigations in this direction, and a computational completeness result has been obtained using the operations $endo, exo, fendo$ and $fexo$ and 12 membranes. Nothing is known about the expressiveness of $endo$ and exo (similarly $fendo$ and $fexo$) other than the fact that for membrane systems with 3 membranes, the operations $endo$ and exo together are equivalent to the operations of $fendo$ and $fexo$. We have the following results:

1. We improve the computational completeness result of [4] by showing that 5 membranes along with $endo, exo, fendo$ and $fexo$ suffices (it is known [4] that 3 membranes do not suffice);
2. we show that 10 membranes along with $fendo, fexo$ gives computational completeness;
3. we show that 9 membranes along with $endo, exo, pendo$ give computational completeness;
4. we show that 8 membranes along with $fendo, fexo, pendo$ give computational completeness;
5. we show that 12 membranes along with $endo, exo, fendo$ give computational completeness;
6. when we look at restricted mobility, $rendo, rexo, rfendo$ and $rfexo$ does not give computational completeness;
7. Adding pure mobility to restricted operations does not take us to the level of RE.

Note that in the following proofs, we denote by μ the membrane structure as well as the initial contents of all membranes.

Theorem 1. $\text{NEM}_5(endo, exo, fendo, fexo) = \textbf{NRE}$.

Proof. We only prove the assertion $\textbf{NRE} \subseteq \text{NEM}_5(endo, exo, fendo, fexo)$, and infer the other inclusion from the Church-Turing thesis. The proof is based on the observation that each set from **NRE** is the range of a recursive function. Thus, we will prove the following assertion. For each recursively enumerable function $f : \textbf{N} \to \textbf{N}$, there is a Π with 5 membranes satisfying the following condition: For any arbitrary $x \in \textbf{N}$, the system Π first "generates" a multiset of the form c^x and halts if and only if $f(x)$ is defined, and, if so, the result of the computation is $f(x)$.

In order to prove this assertion, we consider a register machine $\mathcal{M}$ with 3 registers, the last one being a special output register which is never decremented. Let there be a program P consisting of h instructions $l_1, \ldots, l_h$ which computes f. Let l_h correspond to the instruction HALT and l_1 be the first instruction. The input value x is expected to be in register 1 and the output value in register 3. Without loss of generality, we can assume that all registers other than the first one are empty at the beginning of a computation. We construct the P system $\Pi = (V, \{0, 1, 2, 3, 4\}, \mu, \emptyset, \{a_0\}, \{a_1\}, \{K_0\}, \emptyset, R, 3)$ with $\mu = [[a_0]_1 [a_1]_2 [K_0]_3 [ ]_4]_0$. $V = \{l_i, l'_i, l''_i, L_i, L'_i \mid 1 \le i \le h\} \cup \{K_0, a_0, a_1, c\}$. The rules R are:

1. $[K_0]_3 [ ]_1 \to [[K_0]_3]_1, [[ ]_3 a_0]_1 \to [ ]_3 [a_0 c]_1, [[K_0]_3]_1 \to [l_1]_3 [ ]_1$
   (Generation of c^x, the initial contents of register 1: Membrane 3 with K_0 enters membrane 1, and comes out each time adding a c to membrane 1. To terminate, K_0 is changed to l_1)

2. $[l_i]_3[\ ]_1 \to [[l_j]_3]_1, [[\ ]_3 a_0]_1 \to [\ ]_3[a_0 c]_1$
3. $[l_i]_3[\ ]_2 \to [\ [l_j]_3]_2$ and $[[\ ]_3 a_1]_2 \to [\ ]_3[a_1 c]_2,$
4. $[l_i]_3[\ ]_4 \to [[l_j c]_3]_4, [[l_j]_3]_4 \to [l_j]_3[\ ]_4$

   (Simulation of an increment instruction: $l_i : (inc(k), l_j)$. An endo and fexo rule given by rule 2 is used to increment counter 1: membrane 3 enters membrane 1 changing the instruction label, and coming out after adding a c in membrane 1. Similar rule (rule 3) between membranes 2, 3 for incrementing counter 2, with a_1 playing the role of a_0 for increment. To increment counter 3, we use the rules (rule 4) between membranes 3 and 4)

5. $[\ ]_1[l_i]_3 \to [[\ ]_1 L_i]_3,$
6. $[[c]_1]_3 \to [\ ]_1[\ ]_3, [\ ]_4[L_i]_3 \to [[\ ]_4 L'_i]_3,$
7. $[[\ ]_1 L'_i]_3 \to [\ ]_1[l''_i]_3,$
8. $[[\ ]_4 l''_i]_3 \to [\ ]_4[l''_i]_3,$
9. $[[\ ]_4 L'_i]_3 \to [\ ]_4[L'_i]_3,$
10. $[L'_i]_3[\ ]_4 \to [[l''_i]_3]_4,$
11. $[[l''_i]_3]_4 \to [l''_i]_3[\ ]_4$

    (Simulation of a decrement instruction $l_i : (dec(1), l'_i, l''_i)$. The simulation is initiated by rule 5, when membrane 1 enters membrane 3 by an fendo rule, l_i is replaced with L_i. If there is a c in membrane 1, then membrane 1 exits membrane 3 using rule 6; in parallel, membrane 4 enters membrane 3 using an fendo rule, replacing L_i with L'_i. If there were no c's in membrane 1, then membrane 1 will still be inside membrane 3, hence rule 7 is used, replacing L'_i with l''_i, an fexo rule. Membrane 4 exits membrane 3 irrespective of when membrane 1 exits membrane 3. If the symbol L'_i is present in membrane 3 after both membranes 1,4 exit it, then it means that there was a c in membrane 1; this L'_i is now replaced with l'_i using the endo, exo rules 10,11. Rules for decrementing counter 2 are similar, with membrane 2 playing the role of membrane 1.)

If $\mathcal{M}$ halts, then eventually, we will have the instruction l_h in membrane 3 and membranes 1,2 will have the final contents of counters 1,2. Using the rule $[l_h]_3[\ ]_4 \to [[\ ]_3]_4$, the label l_h is erased. If we assign 3 as the output membrane, then its contents will be same as the contents of the output counter 3 at the end of a halting computation.   □

Theorem 2. $\text{NEM}_{10}(fendo, fexo) = \text{NRE}.$

Proof. The proof is done by simulating a matrix grammar $G = (N, T, S, M, F)$ with appearance checking in the strong binary normal form. We construct the P system $\Pi = (V, \{0, 1, 1', 2, 2', 3, 4, 5, 6, 7\}, \mu, \{XA\}, \emptyset, \ldots, \emptyset, R, 0)$ with $\mu = [\ [XA]_0[\ ]_3[\ ]_4[\ ]_5[\ ]_1[\ ]_{1'}[\ ]_2[\ ]_{2'}[\ ]_6]_7$. Here, XA in membrane 0 corresponds to the initial matrix $(S \to XA)$. Membrane 0 is the output membrane. Let there be n_1 matrices of types 2,4 in G labeled $1, \ldots n_1$ and n_2 matrices of type 3 in G labeled $n_1+1, \ldots, n_1+n_2$. $V = N \cup T \cup \{X_{ij}, A_{ij} \mid 0 \le i, j \le n_1, X \in N_1, A \in N_2\} \cup \{X'_j, X''_j \mid X \in N_1, n_1 + 1 \le j \le n_1 + n_2\} \cup \{Z, \dagger\}$. The rules are

1. $[X]_0[\ ]_3 \to [X_{ii}[\ ]_3]_0, [A[\ ]_3]_0 \to [A_{jj}]_0[\ ]_3$
2. $[X_{ki}]_0[\ ]_4 \to [X_{k-1i}[\ ]_4]_0, [A_{kj}[\ ]_4]_0 \to [A_{k-1j}]_0[\ ]_4, \ k > 0$
3. $[X_{0,i}]_0[\ ]_5 \to [Y[\ ]_5]_0, [A_{0,j}[\ ]_5]_0 \to [x]_0[\ ]_5$
4. $[A_{kj}[\ ]_5]_0 \to [\ \dagger\ ]_0[\ ]_5, \ k > 0$
5. $[A_{0j}[\ ]_4]_0 \to [\ \dagger\ ]_0[\ ]_4,$
6. $[\ \dagger\ ]_0[\ ]_4 \to [\ \dagger[\ ]_4]_0, [\ \dagger[\ ]_4]_0 \to [\ \dagger\ ]_0[\ ]_4$

(Simulation of a type 2 matrix $m_i : (X \to Y, A \to x)$. Rules 1 are used to remember the matrix m_i to be simulated. If X, A belong to the same matrix, then we obtain X_{ii} and A_{ii}. Rules 2 are then used to check if both X, A belong to the same matrix. If yes, then A_{0i} is generated in membrane 0 in the immediate next step after X_{0i}. This is followed by rule 3, by which X_{0i} and A_{0i} are replaced. In case rule 1 gives rise to X_{ii} and A_{jj} with $i \neq j$, then an infinite computation is triggered by rules 4, 5 and 6.

For $i \in \{1, 2\}$, and a matrix $m_j : (X \to Y, B^{(i)} \to \dagger)$ of type 3,

7. $[X]_0[\ ]_i \to [X'_j[\ ]_i]_0,$
8. $[X'_j]_0[\ ]_{i'} \to [X''_j[\ ]_{i'}]_0, [B^{(i)}[\ ]_i]_0 \to [\ \dagger\ ]_0[\ ]_i,$
9. $[X''_j[\ ]_{i'}]_0 \to [Y]_0[\ ]_{i'}$
10. $[Y[\ ]_i]_0 \to [Y]_0[\ ]_i$
11. $[\ \dagger\ ]_0[\ ]_i \to [\ \dagger[\ ]_i]_0, [\ \dagger[\ ]_i]_0 \to [\ \dagger\ ]_0[\ ]_i$

(Simulation of a type 3 matrix $m_j : (X \to Y, B^{(i)} \to \dagger)$. The membrane labeled i enters membrane 0 replacing X with X'_j. This is followed by 2 parallel rules: membrane i' entering membrane 0 replacing X'_j with X''_j, and membrane i exiting membrane 0 in the presence of $B^{(i)}$. If $B^{(i)}$ is present, an infinite computation is triggered by rule 11. Membrane i' exits membrane 0 replacing X''_j with Y. If $B^{(i)}$ is absent, then membrane i will be inside membrane 0. In this case, it exits membrane 0 replacing Y with Y.)

12. $[Z]_0[\ ]_6 \to [[\ ]_6]_0, [A[\ ]_6]_0 \to [\ \dagger\ ]_0[\ ]_6, A \in N_2$

(Simulation of a type 4 matrix $m_j : (X \to \lambda, A \to x)$. This is done using rules 1-6, replacing X with a new symbol Z. After this, we check if membrane 0 contains any non-terminals, and if so, an infinite computation is triggered by rule 6. Otherwise, a halting computation is obtained, with membrane 0 containing the output.) □

Theorem 3. $\mathrm{NEM}_9(endo, exo, pendo) = \mathrm{NRE}$.

Proof. The proof is done by simulating a matrix grammar $G = (N, T, S, M, F)$ with appearance checking in the strong binary normal form. As in Theorem 2, let there be n_1 matrices of types 2,4 and n_2 matrices of type 3. We construct the P system $\Pi = (V, \{0, 1, 2, 3, 4, 5, 6, 7, 8\}, \mu, w_0, \ldots, w_8, R, 0)$ with membrane structure $[\ [XA]_0[\ ]_3[\ ]_4[\ ]_5[[B^{(1)}B^{(2)}]_2[\alpha]_1]_6[\ ]_7]_8$; XA in membrane 0 corresponds to the initial matrix $(S \to XA)$. $V = N \cup T \cup \{X_{ij}, A_{ij} \mid 0 \leq i, j \leq n_1, X \in N_1, A \in N_2\} \cup \{X_j \mid n_1 + 1 \leq j \leq n_2\} \cup \{\alpha, \beta\} \cup \{Z, \dagger\}$. Membrane 0 is the output membrane. The rules are

1. $[X]_0[\ ]_3 \to [[X_{ii}]_0]_3, [[A]_0]_3 \to [A_{jj}]_0[\ ]_3,$
2. $[X_{il}]_0[\ ]_4 \to [[X_{i-1l}]_0]_4, [[A_{jk}]_0]_4 \to [A_{j-1k}]_0[\ ]_4$ for $i, j > 0$

3. $[X_{0i}]_0[\ ]_5 \to [[Y]_0]_5, [[A_{0i}]_0]_5 \to [x]_0[\ ]_5$,
4. $[[A_{jk}]_0]_5 \to [\ \dagger\ ]_0[\ ]_5$ if $j > 0$, $[[A_{0k}]_0]_4 \to [\ \dagger\ ]_0[\ ]_4$,
5. $[\ \dagger\ ]_0[\ ]_5 \to [[\ \dagger\ ]_0]_5, [[\ \dagger\ ]_0]_5 \to [\ \dagger\ ]_0[\ ]_5$
   (Simulation of a type 2 matrix $m_i : (X \to Y, A \to x)$. Similar to Theorem 2)
6. $[X]_0[\ ]_6 \to [[X_j]_0]_6, [X_j B^{(i)}]_0[B^{(i)}]_2 \to [[B^{(i)}]_2 X_j B^{(i)}]_0, [\alpha]_1[]_0 \to [[\alpha]_1]_0$,
7. $[[B^{(i)}]_2]_0 \to [\ \dagger\ ]_2[]_0, [[\alpha]_1]_0 \to [\beta]_1[\ ]_0$,
8. $[X_j]_0[\beta]_1 \to [[X_j]_0 \beta]_1, [[X_j]_0]_1 \to [Y]_0[\ ]_1$,
9. $[\beta]_1[Y]_0 \to [[\beta]_1 Y]_0, [[\beta]_1]_0 \to [\alpha]_1[\ ]_0$,
10. $[[Y]_0]_6 \to [Y]_0[\ ]_6$,
11. $[[\beta]_1]_6 \to [\ \dagger\ ]_1[\ ]_6$,
12. $[[\ \dagger\ ]_i]_6 \to [\ \dagger\ ]_i[\ ]_6, [\ \dagger\ ]_i[\ ]_6 \to [[\ \dagger\ ]_i]_6$ for $i = 1, 2$.
    (Simulation of a type 3 matrix $m_j : (X \to Y, B^{(i)} \to \dagger)$. Membrane 0 enters membrane 6 replacing X with X_j. This is followed by the *pendo, endo* rules 6, by which membranes 1,2 enter 0. Of course, membrane 2 enters only if there is a $B^{(i)}$ in membrane 0. The presence of a $B^{(i)}$ in membrane 0 triggers an infinite computation. Membrane 1 exits membrane 0 replacing α with β. This is followed by a *pendo* rule in 8, by which membrane 0 enters membrane 1. This helps in replacing X_j with Y. Next, β is replaced with α by rule 9. If rule 10 is used before rule 9, we get an infinite computation. Membrane 0 comes out using rule 10, and another matrix can be simulated)
13. $[Z]_0[\ ]_7 \to [[\ ]_0]_7, [[A]_0]_7 \to [\ \dagger\ ]_0[\ ]_7, A \in N_2$
    (Simulation of a type 4 matrix $m_j : (X \to \lambda, A \to x)$. This is similar to Theorem 2.)   □

Theorem 4. $NEM_8(fendo, fexo, pendo) = NRE$.

Theorem 5. $NEM_{12}(endo, exo, fendo) = NRE$.

Both Theorems 4, 5 can be proved by simulation of a matrix grammar with appearance checking in the strong binary normal form. The simulation of type-2 and type-4 matrices in Theorem 4 and Theorem 5 are exactly as in Theorems 2 and 3 respectively. We omit the details for simulation of a type-3 matrix here. The following result can be observed from the above Theorems.

Theorem 6. $NMAT \subseteq NEM_5(endo, exo)$.

We now address the question of expressiveness for systems with the operations *rendo, rexo, rfendo* and *rfexo* along with the pure mobility operations *pendo, pexo*.

Theorem 7. *For all $n \in \mathbf{N}$, we have*

1. $NEM_n(rendo, rexo, rfendo, rfexo) \subseteq NMAT_{ac} \subset NRE$,
2. $NEM_n(rendo, rexo, rfendo, rfexo, pendo, pexo) \subseteq NMAT_{ac} \subset NRE$.

Proof. We give the proof idea here. The full proof can be found in a technical report on the first author's webpage. Lets discuss (1). The proof technique is to simulate the given P system Π using a random context matrix grammar with appearance checking but without λ-rules, and use the result $\mathbf{N}RCM(M, CF - \lambda, ac) = \mathbf{NMAT}_{ac}$. Here, $\mathbf{N}RCM(M, CF - \lambda, ac)$ denotes the family of sets of numbers recognized by random context matrix grammars with appearance checking, using context free non erasing rules. The operations in Π are all non-erasing; this helps in choosing non-erasing rules in the grammar as well. Assume that Π has alphabet V, n membranes labeled 1 to n, and i_0 denotes the output membrane. Let membrane i have k_i symbols $1 \le i \le n$ in the initial configuration. The random context matrix grammar G is constructed as follows:

1. The initial matrix is $(S \to CZE_{1,\emptyset} \ldots N_{i,list} \ldots E_{n,\emptyset}w_1 w_j \ldots w_{i_0}, \emptyset, \emptyset)$, where w_i denotes the initial contents of membrane i. The contents of the output membrane w_{i_0} is kept at the end of the string. A symbol $E_{i,\emptyset}$ says that membrane i is elementary in the current configuration, while the symbol $N_{i,list}$ says that membrane i is non-elementary in the current configuration, and $list$ is the list of i's children. The symbol Z initialized to $\emptyset$ keeps track of the pairs of membranes that have been part of a rule in the transition from the current configuration to the next. Each time a rule is used between a pair (i, j) of membranes, (i, j) is added to Z. In case no rule is applicable to a pair (m, k) of membranes, then this fact is verified by appearance checking, and then (m, k) is added to Z. When Z contains all distinct pairs of membranes, one step of Π has been simulated. C is a symbol to choose randomly a pair (i, j) of membranes whose interaction we are trying to simulate. Thus, C is replaced with some $C_{i,j}$, $C_{i,j}$ by some $C_{m,k}$ and so on.

2. Depending on the rules used in a step of Π, the symbols $E_{i,\emptyset}$ and $N_{j,list}$ are updated. When $|Z| = \binom{n}{2} - n$, Z is reinitialized to $\emptyset$ and $C_{l,r}$ (assuming the pair (l, r) got simulated last) is replaced with C.

3. To terminate, we have to make sure that we have reached a halting computation. For this, at some non-deterministic point of time, we replace C with U. Then for all membranes other than the output, we replace the contents with symbols X after making sure that they have no rule to participate in given the current configuration. We replace symbols a_{i_0} of i_0 with a after ensuring the same. We are then left with a string of the form $UZN_{1,list} \ldots E_{n,\emptyset}X \ldots Xw$, where w is the contents of i_0. Then we replace the symbols $N_{j,list}, E_{i,\emptyset}, Z, U$ also with X. Then the number of symbols X we have is $\kappa = n + 2 + (k_1 + \ldots + k_n) - k_{i_0}$ which is finite. Then we get a string of the form $X^\kappa w$, $w \in V^*$. The terminal alphabet T of G is $V \cup \{X\}$.

4. Since the family MAT_{ac} is closed under quotient by letters, we do the quotient operation κ times on $L(G)$. Then we have $\mathbf{N}(\partial_X^\kappa(L)) = \mathbf{N}(\Pi)$.   □

References

1. Aman, B., Ciobanu, G.: Describing the Immune System Using Enhanced Mobile Membranes. Electronic Notes in Theoretical Computer Science 194, 5–18 (2008)
2. Dassow, J., Păun, G.: Regulated Rewriting in Formal Language Theory. Springer, Heidelberg (1989)
3. Freund, R., Păun, G.: On the number of non-terminal symbols in graph-controlled, programmed and matrix grammars. In: Margenstern, M., Rogozhin, Y. (eds.) MCU 2001. LNCS, vol. 2055, pp. 214–225. Springer, Heidelberg (2001)
4. Krishna, S.N., Ciobanu, G.: On the Computational Power of Enhanced Mobile Membranes. In: Beckmann, A., Dimitracopoulos, C., Löwe, B. (eds.) CiE 2008. LNCS, vol. 5028, pp. 326–335. Springer, Heidelberg (2008)
5. Minsky, M.L.: Computation: Finite and Infinite Machines. Prentice-Hall, Englewood Cliffs (1967)
6. Păun, G.: Computing with Membranes. Journal of Computer and System Sciences 61(1), 108–143 (2000)
7. Păun, G., Rozenberg, G., Salomaa, A. (eds.): The Oxford Handbook of Membrane Computing. Oxford University Press, Oxford (2010)
8. Salomaa, A.: Formal Languages. Academic Press, London (1973)

Nature-Based Problems in Cellular Automata

Martin Kutrib

Institut für Informatik, Universität Giessen, Arndtstraße 2, 35392 Giessen, Germany
kutrib@informatik.uni-giessen.de

Abstract. Cellular automata were proposed by John von Neumann in order to solve the logical problem of nontrivial self-reproduction. From this biological point of view he employed a mathematical device which is a multitude of interconnected automata operating in parallel to form a larger automaton. His famous early result reveals that it is logically possible for such nontrivial computing device to replicate itself ad infinitum. Nowadays (artificial) self-reproduction is one of the cornerstones of automata theory, which plays an important role in the field of molecular nanotechnology. We briefly summarize some important developments on cellular automata as model to investigate further nature-based problems. On our short tour on the subject we will address the French Flag Problem, the Firing Squad Problem in growing cellular arrays, oblivious cellular automata, and the fault-tolerant Early Bird Problem.

1 Introduction

More than half a century ago John von Neumann had become interested in the question of whether computing machines can construct copies or variants of themselves, whether artificial self-reproducing structures exist. He started to investigate the logic necessary for replication and found one of the cornerstones in the theory of automata. He employed a mathematical device which is a homogeneously structured multitude of interconnected finite-state machines operating in parallel to form a larger machine, a cellular automaton. John von Neumann showed that there are cellular automata which can replicate themselves ad infinitum. The name of these automata originates from the context in which they were developed. Due to their intuitive and colorful concepts, cellular automata have soon been considered from a computational and modeling point of view. So, from the very beginning they were both, an interesting and challenging model for theoretical computer science and an interesting model for practical applications. Their inherent massive parallelism renders obvious applications as model for systems that are beyond direct measurements. Here, on our short tour on nature-based problems in cellular automata we will address the French Flag Problem that came up with the question of whether one can achieve regulative global polarity in organisms without polarity in individual cells. In connection with the simulation of pigmentation patterns on the shells of sea-snails it is supposed that glands stop their action synchronously at the same time, even though their number was growing during the synchronization

B. Löwe et al. (Eds.): CiE 2011, LNCS 6735, pp. 171–180, 2011.

process. This inspired the investigation of synchronization in growing cellular automata. Another phenomenon occurring in nature is obliviousness. In order to be economic, information that has not been used for a certain time is of little importance and may be forgotten. So, obliviousness can be studied as an additional property of cellular automata. Finally we consider fault-tolerance and reliable computations in cellular automata using the example of the Early Bird Problem.

2   Cellular Automata

A cellular automaton is a linear array of identical deterministic finite state machines, sometimes called cells. Except for the outermost cells each one is connected to its both nearest neighbors. We identify the cells by positive integers. The state transition depends on the current state of a cell itself and the current states of its neighbors, where the two outermost cells receive information associated with a boundary symbol on their free input lines. There exists a special, so-called quiescent state, with the property that if some cell and all of its neighbors are in the quiescent state then the cell remains in the quiescent state. The state changes take place simultaneously at discrete time steps.

Definition 1. *A* cellular automaton *(CA) is a system* $\langle S, \delta, s_0, \# \rangle$, *where*

1. *S is the finite, nonempty set of* cell states,
2. *$s_0 \in S$ is the* quiescent state,
3. *$\# \notin S$ is the permanent* boundary symbol, *and*
4. *$\delta : (S \cup \{\#\}) \times S \times (S \cup \{\#\}) \to S$ is the* local transition function *which satisfies $\delta(p, s_0, q) = s_0$, for all $p, q \in \{s_0, \#\}$.*

In general, the global behavior of a cellular automaton is of interest. It is induced by the local behavior of all cells, that is, by the local transition function. More precisely, a *configuration* of a cellular automaton $\langle S, \delta, s_0, \# \rangle$ at time $t \geq 0$ is a description of its global state, which is formally a mapping $c_t : \{1, \ldots, n\} \to S$, for $n \geq 1$. Configurations may be represented as words over the set of cell states in their natural ordering. For example, the initial configuration consisting of n cells in the quiescent state is represented by s_0^n. Successor configurations are computed according to the global transition function Δ. Let c_t, $t \geq 0$, be a configuration with $n \geq 2$, then its successor c_{t+1} is defined as follows:

$$c_{t+1} = \Delta(c_t) \iff \begin{cases} c_{t+1}(1) = \delta(\#, c_t(1), c_t(2)) \\ c_{t+1}(i) = \delta(c_t(i-1), c_t(i), c_t(i+1)), i \in \{2, \ldots, n-1\} \\ c_{t+1}(n) = \delta(c_t(n-1), c_t(n), \#) \end{cases}$$

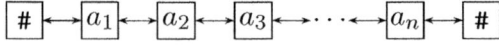

Fig. 1. A (one-dimensional) cellular automaton

For $n = 1$, the next state of the sole cell is $\delta(\#, c_t(1), \#)$. Thus, Δ is induced by δ. A computation can be represented as space-time diagram, where each row is a configuration and the rows appear in chronological ordering.

3 French Flag Problem

The French Flag Problem originates from [30]. Originally, it has been formulated as a problem of regulation, or to be more precise, a problem of pattern formation of simple axial patterns. In order to model the determination of a pattern in a tissue having three regions of cells with discrete properties and sharp bounds between them, the French flag model has been established as a concept of a morphogen. A morphogen is a signaling molecule that regulates the pattern formation. It acts directly on cells dependent on its concentration, where morphogen gradients generate different cell states in distinct spatial order. In particular, in the French flag model, the French flag is used to represent the effect of a morphogen on cell differentiation. The cell states are represented by the different colors of the French flag: high concentrations activate a *blue* gene, lower concentrations activate a *white* gene, where *red* indicates cells below the necessary concentration threshold.

In terms of cellular automata the problem is to construct a CA that when started with all cells in the quiescent state turns into a French flag upon excitation from the outside world at one of the ends. The colors are represented by states and the red region appears at the end excited. Moreover, if the array is cut into two or more pieces, it is required that all pieces turn into French flags with the same orientation as the original. So, if the array is in some unstable equilibrium then pattern formation is initiated by a small disturbance that pushes the array out of this equilibrium. This assumption is reasonable in embryonic morphogenesis (cf. [8]). Since a solution turns the entire array into an asymmetric French flag, the array exhibits a polarity. This global polarity can be achieved by single cells exhibiting a local polarity, that is, by cells that may distinguish between left and right. This local polarity is an unlikely activity for cells of organisms. From this point of view, in [6,8] cellular automata having no polarity on the cellular level, that is, with symmetric transition functions are investigated. Here a transition function is symmetric if a cell still receives the states of its both neighbors but does not know which state comes from the left and which from the right. Formally, we require

$$\delta(p, s, q) = \delta(q, s, p)$$

for all $p, q \in S \cup \{\#\}$ and $s \in S$. It turned out that the French Flag Problem can be solved by such automata. A precise definition of the problem and the solution which is sketched in the following are presented in [6,8]. The solution relies on the concept of signals. Roughly speaking, signals are described as follows: If a cell changes to the state of its neighbor after some k time steps, and if subsequently its neighbors and their neighbors do the same, then the basic signal moves with speed $\frac{1}{k}$ through the array. With the help of auxiliary signals, rather complex

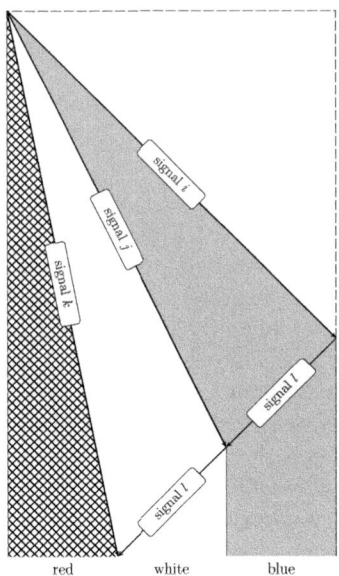

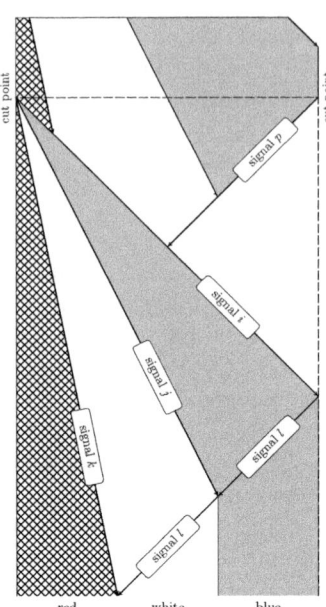

Fig. 2. Basic signals cutting the array into thirds

Fig. 3. Signal p after cutting the array

signals can be established. For a more general treatment of signals and so-called (time-)constructible mappings we refer to [1,3,13,21,27].

The underlying idea for the solution of the French Flag Problem is depicted in Figure 2. After one of the ends is excited from the outside world it emits three signals. Signal i moves with speed 1, signal j with speed $\frac{1}{2}$, and signal k with speed $\frac{1}{5}$ to the opposite end. Signal i bounces at the opposite end and continues to move as signal l with speed 1. Let n denote the length of the array. After $\frac{4}{3}n$ steps signal l arrives at cell $\frac{2}{3}n$. At he same time signal j arrives at that cell and, thus, the boundary between the second and last third is marked. Similarly, after $\frac{5}{3}n$ steps signals l and k meet at cell $\frac{1}{3}n$ which marks the boundary between the first and second third. The detailed solution ensures for arrays whose length is not divisible by three that the length x of each third meets the condition $|3x - n| \leq 2$. In order to obtain the colors, a cell which has been passed through by signal i turns to states indicating blue. When a blue cell is passed through by signal j it turns to white, and when a white cell is passed through by signal k it turns to red.

So far, we have sketched a solution that seems to need polarity of the cells in order to send the signals in the required direction. But, in fact, we can simulate this polarity to some extent by a symmetric transition function with a simple but clever trick. Immediately after emitting signal i the excited cell sends another signal which labels the cells passed through cyclically by 1,2,3,1,... These labels are stored as a part of the state. Now it is easy for a cell to recognize which state received comes from right or left. For example, a cell with label 2 has a right neighbor labeled 3 and a left neighbor labeled 1. Since the transition function is symmetric it is evident that we obtain a French flag with reverse orientation if the right end is initially excited.

So it remains to be explained what happens if the array is cut. Due to the labels, a cell which is excited (set to state i) as a result of a cut knows whether or not it is on the same side as the original excitation was. If it is on the same side it emits signals i, j, and k as before and turns the subarray into a French flag having the same orientation. Otherwise it emits a signal p (cf. Figure 3). This signal tries to get to the opposite end, where it emits signals i, j, and k in order to establish a French flag. On its way it stops possible signals j and k but is stopped by a signal i. In the latter case, the p signal has no purpose. In the former case, the array has been cut at both ends and, thus, even for such situations a new French flag is established.

A further result in [6,8] is a solution of the Firing Squad Synchronization Problem in symmetric cellular automata. Later these results have strongly been generalized in [25], where it is shown that *every* CA can be simulated in real-time by a symmetric CA. A further generalization to CA with a wider class of interconnection schemes can be found in [12].

4   Synchronization of Growing Cellular Automata

The famous *Firing Squad Synchronization Problem* has been raised by Myhill. According to [22] it came up with the problem to turn on all parts of a self-reproducing machine simultaneously. Over the years lots of generalizations and modifications have been investigated and solved. In particular, in connection with the simulation of pigmentation patterns on the shells of sea-snails it is supposed that glands stop their action synchronously at the same time, even though their number was growing during the synchronization process. This inspired the investigation of synchronization in growing cellular automata [7].

In such systems growth takes place at the ends by allowing the end cells to divide into two or more cells. Cell division occurs at a fixed rate, say p cells in q time steps at the right end ($0 \leq p < q$), and r cells in t time steps at the left end ($0 \leq r < t$). In order to model the growth a finite counter is used in the leftmost as well as in the rightmost cell. The counters are initially set to 0. For the rightmost cell, if at some time the counter is u, the cell checks whether $u + p \geq \frac{q}{2}$. If not then the cell does not divide and the counter is set to the value $u + p$. Otherwise, the cell divides into two, the counter moves to the rightmost of these two cells and is set to $u + p - q$. In this way, after q steps there have been exactly p cell divisions and the value of the counter is again 0. Similarly at the leftmost cell. So, after x time steps the length of a configuration having initially l cells is $\gamma(p, q, r, t, l, x) = l + \lceil \frac{xp}{q} \rceil + \lceil \frac{xr}{t} \rceil$.

Roughly speaking, the Firing Squad Synchronization Problem (FSSP) is to construct a cellular automaton M so that if M is started with all cells but the leftmost one quiescent (the soldiers) and the leftmost one in some state g (the general), then eventually M will be in a configuration where all cells are in some state f (the firing state), and no cell has been in the firing state before.

More formally, for growing cellular automata the problem is to construct a growing CA $\langle S, \delta, s_0, \# \rangle$ so that for all $l, x \geq 1$, (i) there is a $k \geq l$ and states $s_1, s_2 \in S$, such that $\Delta^x(s_0^l) = s_1 s_0^k s_2$, (ii) there is a state $g \in S$, such that $|\Delta^x(s_0^l)| = |\Delta^x(g s_0^{l-1})| = \gamma(p, q, r, t, l, x)$, and (iii) there is a state $f \in S$ and an $m \geq 1$ such that for all $0 \leq x < m$ the configuration $\Delta^x(g s_0^{l-1})$ does not contain state f, but for $x \geq m$ the configuration $\Delta^x(g s_0^{l-1})$ is f^l.

Statement (i) says that an initially quiescent array will remain quiescent except for the ends at which the growth takes place. The growth itself is described by (ii), and the synchronization is stated in (iii).

By tricky utilization of signals the problem has been solved in [7]. The main result reads as follows.

Theorem 2. *There exists a solution of the FSSP in growing cellular automata.*

5   Oblivious Cellular Automata

Another phenomenon occurring in nature is obliviousness. In order to be economic, information that has not been used for a certain time is supposed to be of little relevance and therefore may be forgotten. So, obliviousness seems to be an interesting additional property of cellular automata in order to obtain a better model for the study of certain natural phenomena. Moreover, if it is possible to perform any computation of classical cellular automata by oblivious cellular automata, then the constructions can be seen as strategies of self-repair (with respect to the faults caused by obliviousness). In [9] a long construction is given which shows that, in fact, such simulations are possible with respect to so-called *R-simulations (result-simulations)*. That is, for any given classical CA that does a computation on an input and comes to a result, there is an oblivious CA that does a computation on the encoded input and comes to the same but encoded result. Next we present the definition of obliviousness in CA and the main theorem from [9].

Given a cellular automaton $M = \langle S, \delta, s_0, \# \rangle$, let $\tau : \{1, \ldots, n\} \times S \times \mathbb{N} \to \mathbb{N} \cup \{\infty\}$, for $n \geq 1$, be a mapping, where $\tau(i, p, t)$ gives the last time step between time 0 and time t at which cell i was in state p. If cell i was never in that state until time t then $\tau(i, p, t)$ is equal to ∞. In order to introduce obliviousness we now require that the local transition function δ obeys the following rules. Assume that an application of the original (non-oblivious) transition function δ_o at time t causes cell i to enter state s. Then δ also causes cell i to enter state s provided that i has been in state s within the last $\varphi(t)$ time steps. Otherwise state s has been forgotten and δ sends cell i into the quiescent state s_0 instead. The mapping $\varphi : \mathbb{N} \to \mathbb{N}$ is a parameter which controls the degree of obliviousness. Formally, for all $i \in \{1, 2, \ldots, n\}$ and $t \geq 0$, let $c_t(0) = c_t(n + 1) = \#$ and define

$$\delta(c_t(i-1), c_t(i), c_t(i+1)) = \delta_o(c_t(i-1), c_t(i), c_t(i+1))$$
$$\text{if } t - \tau(i, \delta_o(c_t(i-1), c_t(i), c_t(i+1)), t) \leq \varphi(t),$$

and $\delta(c_t(i-1), c_t(i), c_t(i+1)) = s_0$ otherwise.

Clearly, the quiescent state can never be forgotten. In general, φ has to meet further conditions in order to avoid artificial situations. For example, one can require that φ is constant or monotonically increasing and unbounded. If $\varphi(t) \geq t$ then none of the states can be forgotten and we have a classical cellular automaton. The next theorem is a main result in [9].

Theorem 3. *Let* $M = \langle S, \delta, s_0, \# \rangle$ *be a (classical) cellular automaton and* φ *be monotonically increasing and unbounded. Then there is an oblivious cellular automaton which R-simulates* M.

6   The Fault-Tolerant Early Bird Problem

In massively parallel computing systems that consist of hundred thousands of processing elements each single component is subject to failure, such that the probability of misoperations and loss of function of the whole system increases with the number of its elements. It was von Neumann [29] who first stated the problem of building reliable systems out of unreliable components. Biological systems may serve as good examples. Due to the necessity to function normally even in case of certain failures of their components, nature developed mechanisms which invalidate the errors. In other words, they are working with fault tolerance. Error detecting and correcting components should not be global to the whole system because they themselves are subject to failure. Therefore, the fault tolerance has to be a design feature of the single elements.

In [4] reliable arrays are constructed under the assumption that at each time step a cell fails with a constant probability. Moreover, such a failure does not incapacitate the cell permanently, but only violates its rule of operation in the current step. Under the same constraint that cells themselves (and not their links) may fail, fault-tolerant computations have been investigated, for example in [5,23], where encodings are established that allow the correction of so-called K-separated misoperations, in [17,18,26,31], where the firing squad synchronization problem is considered in defective cellular automata, and in terms of interacting automata with nonuniform delay in [10,20], where the synchronization of the networks is the main object, too. Syntactical pattern recognition in cellular automata with static and dynamic defects of the cells and their links is the main contribution in [14,15,16].

Here we focus on reliable solutions of the Early Bird Problem in CA with defective cells. The defects are as follows: It is assumed that each cell has a self-diagnosis circuit which is run once before the actual computation. The result of that diagnosis is stored locally in a special register, respectively. In this way intact cells can detect whether their neighbors are defective or not when they receive the first states. Moreover, it is assumed that during the actual computation no new defects may occur. Otherwise, the whole computation would become invalid. What is the effect of a defective cell? It is reasonable to require that a defective cell cannot modify information, that is, cannot apply the local transition function. On the other hand, it must be able to transmit information in order to avoid the parallel computation being broken into two non-interacting lines and, thus, being impossible at all. This means that a defective cell forwards the states of its neighbors to the opposite neighbors unchanged with a delay of one time step. Therefore, the speed of information transmission, which occurs in both directions simultaneously, is one cell per time step. In order to adapt the quiescent condition of the ordinary transition function, it is assumed that intact cells in quiescent state can not change their state just because defective neighbors are transmitting border and/or quiescent states.

Another point of view on such devices is to define a transmission delay between every two adjacent cells and to allow nonuniform delays [10,20]. Now the number of defective cells between two intact ones determines the corresponding delay.

The *Early Bird Problem* (EBP) is another typical representative of the class of problems whose solutions are defined by the global behavior and are obtained by homogeneous parallel local state transitions of a system. It was defined and named in [24] for elementary cyclic graphs where each vertex is a deterministic finite automaton. In terms of one-dimensional cellular automata we are concerned with initial configurations where all cells are quiescent. As long as a cell is quiescent it may be excited from the outside world. In this case, it changes to the so-called *bird state* (B) (a bird has arrived at the cell). The problem is to construct a CA such that after a while, those cells are designated (by state B) that have been excited first (the early birds). That is, to distinguish between early and late birds. All the other cells must be in another state (∗).

The problem has been solved for cellular automata in [28]. A simplified solution using only five states has been given in [19]. This solution is optimal for the number of states since in [11] it has been proved that the EBP is unsolvable with four states. An example for the five state solution of [19] is depicted in Figure 4. Basically, the idea is that each bird sends out waves to the left and right (states < and >). When wavefronts are colliding, successively two clashing waves are deleted and replaced by two ∗ which continue to move in the same direction but can pass other waves. Similarly when a wave arrives at the border. When a ∗ arrives at a bird, no further waves are emitted. When a wave arrives at a bird, the bird is deleted. In this case another bird has sent more waves and, thus, was the elder one.

As expected, this attractively simple solution does not solve the problem for cellular automata with defective cells. However, the principle to determine the age of a bird by its number of emitted signals can still be used. In [2] a fault-tolerant solution is presented, whose rough idea is as follows.

A basic principle of the solution algorithm is that signals circulate between two bird cells until the next step of the algorithm is completed (cf. [2] for a detailed description and analysis).

1. When a bird arrives, it emits signals to the left and right at each time step until it receives a signal from another bird. Now it bounces arriving signals. The difference of the numbers of signals emitted by the birds is twice the age difference of the birds.
2. The birds convert the number of emitted signals into binary. The binary number is sent as signals which circulate, too.

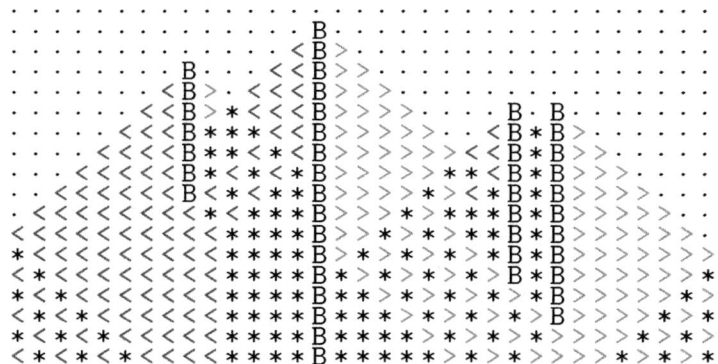

Fig. 4. Example of a state optimal solution of the EBP in classical cellular automata

3. The difference of binary numbers of neighboring birds is computed and divided by two, in order to determine the elder bird. Again, the difference is sent as signals which circulate.
4. All differences in between two farther birds are summed up in a specific way in order to compare their ages. The birds in between them are deleted and behave like defective cells.
5. The last step of the algorithm is repeated until all birds have the same age and, thus, are the global early birds.

References

1. Buchholz, T., Kutrib, M.: Some relations between massively parallel arrays. Parallel Comput. 23, 1643–1662 (1997)
2. Fay, B., Kutrib, M.: The fault-tolerant early bird problem. IEICE Trans. Inf. Syst. E87-D, 687–693 (2004)
3. Fischer, P.C.: Generation of primes by a one-dimensional real-time iterative array. J. ACM 12, 388–394 (1965)
4. Gács, P.: Reliable computation with cellular automata. J. Comput. System Sci. 32, 15–78 (1986)
5. Harao, M., Noguchi, S.: Fault tolerant cellular automata. J. Comput. System Sci. 11, 171–185 (1975)
6. Herman, G.T., Liu, W.H.: The daughter of Celia, the French flag, and the firing squad. Simulation 21, 33–41 (1973)
7. Herman, G.T., Liu, W.H., Rowland, S., Walker, A.: Synchronization of growing cellular arrays. Inform. Control 25, 103–122 (1974)
8. Herman, G.T.: Models for cellular interactions in development without polarity of individual cells II. Int. J. Systems Sci. 3, 149–175 (1972)
9. Höllerer, W.O., Vollmar, R.: On forgetful cellular automata. J. Comput. System Sci. 11, 237–251 (1975)
10. Jiang, T.: The synchronization of nonuniform networks of finite automata. Inform. Comput. 97, 234–261 (1992)

11. Kleine-Büning, H.: The early bird problem is unsolvable in a one-dimensional cellular space with 4 states. Acta Cybernet. 6, 23–31 (1983)
12. Kobuchi, Y.: A note on symmetrical cellular spaces. Inform. Process. Lett. 25, 413–415 (1987)
13. Kutrib, M.: Cellular automata – A computational point of view. In: New Developments in Formal Languages and Applications, pp. 183–227. Springer, Heidelberg (2008)
14. Kutrib, M., Löwe, J.T.: Fault tolerant parallel pattern recognition. In: Theoretical and Practical Issues on Cellular Automata (ACRI 2000), Springer, pp. 72–80. Springer, Heidelberg (2000)
15. Kutrib, M., Löwe, J.-T.: Massively parallel pattern recognition with link failures. In: Jeffery, K., Hlaváč, V., Wiedermann, J. (eds.) SOFSEM 2000. LNCS, vol. 1963, pp. 392–399. Springer, Heidelberg (2000)
16. Kutrib, M., Löwe, J.T.: Massively parallel fault tolerant computations on syntactical patterns. Fut. Gener. Comput. Systems 18, 905–919 (2002)
17. Kutrib, M., Vollmar, R.: Minimal time synchronization in restricted defective cellular automata. J. Inform. Process. Cybern. EIK 27, 179–196 (1991)
18. Kutrib, M., Vollmar, R.: The firing squad synchronization problem in defective cellular automata. IEICE Trans. Inf. Syst. E78-D, 895–900 (1995)
19. Legendi, T., Katona, E.: A 5 state solution of the early bird problem in a one dimensional cellular space. Acta Cybernet. 5, 173–179 (1981)
20. Mazoyer, J.: Synchronization of a line of finite automata with nonuniform delays. Research Report TR 94-49, Ecole Normale Supérieure de Lyon (1994)
21. Mazoyer, J., Terrier, V.: Signals in one-dimensional cellular automata. Theoret. Comput. Sci. 217, 53–80 (1999)
22. Moore, E.F.: The firing squad synchronization problem. In: Sequential Machines – Selected Papers, pp. 213–214. Addison-Wesley, Reading (1964)
23. Nishio, H., Kobuchi, Y.: Fault tolerant cellular spaces. J. Comput. System Sci. 11, 150–170 (1975)
24. Rosenstiehl, P., Fiksel, J.R., Holliger, A.: Intelligent graphs: Networks of finite automata capable of solving graph problems. In: Graph Theory and Computing, pp. 219–265. Academic Press, London (1972)
25. Szwerinski, H.: Symmetrical one-dimensional cellular spaces. Inform. Control 67, 163–172 (1985)
26. Umeo, H.: A fault-tolerant scheme for optimum-time firing squad synchronization. In: Parallel Computing: Trends and Applications North-Holland, pp. 223–230. North-Holland, Amsterdam (1994)
27. Umeo, H., Kamikawa, N.: A design of real-time non-regular sequence generation algorithms and their implementations on cellular automata with 1-bit inter-cell communications. Fund. Inform. 52, 257–275 (2002)
28. Vollmar, R.: On two modified problems of synchronization in cellular automata. Acta Cybernet. 3, 293–300 (1977)
29. von Neumann, J.: Probabilistic logics and the synthesis of reliable organisms from unreliable components. Automata Studies 34, 43–98 (1956)
30. Wolpert, L.: Positional information and the spatial pattern of cellular differentiation. J. Theoret. Biol. 25, 1–47 (1969)
31. Yunès, J.B.: Fault tolerant solutions to the firing squad synchronization problem. Technical Report LITP 96/06, Institut Blaise Pascal, Paris (1996)

Multi-Resolution Cellular Automata for Real Computation*

James I. Lathrop, Jack H. Lutz, and Brian Patterson

Department of Computer Science, Iowa State University, Ames IA 50011,
United States of America
jil,lutz,patterbj@cs.iastate.edu

Abstract. This paper introduces *multi-resolution cellular automata (MRCA)*, a multi-resolution variant of cellular automata. Cells in an MRCA are allowed to "fission" one or more times during the course of execution. At any given time, the MRCA may thus be carrying out computations on a variety of spatial scales. Our main theorem uses the MRCA model to give a natural characterization of the computability of sets in Euclidean space, provided that their boundaries are computably nowhere dense.

1   Introduction

The primary objective of computable analysis is to provide a realistic theoretical foundation for scientific computing, which consists mainly of large-scale, high-precision computations in problem domains involving real and complex numbers, functions on Euclidean spaces, differential equations, and other continuous, as opposed to discrete, mathematical objects. The first task for computable analysis was thus to formulate notions of computability and complexity that are appropriate for such problem domains. This task is well on its way to being achieved. It was begun by Turing [21,22], who defined computable real numbers. It was furthered by Grzegorczyk [8] and Lacombe [13], who used the oracle Turing machine model from Turing's Ph.D. thesis [23] to define the computability of functions from the reals to the reals. After a somewhat slow start, progress accelerated dramatically in the 1980s. Pour-El, Richards, and Weihrauch conducted deep and influential investigations in computable analysis [18,24], and Ko, Friedman, Kreitz, and Weihrauch formulated and investigated useful and informative models of the computational complexity of real-valued functions [10,12,11,25]. Computability and complexity in analysis is now a large and active research area that includes rigorous investigations of computation in Hilbert spaces, Banach spaces, differentiable manifolds, and many of the other mathematical settings of large-scale scientific computing. Braverman and Cook [3] coined the term "bit-computability" for the approach shared by all the above models and argued convincingly that this is indeed a good theoretical foundation for scientific computing.

---

* This research was supported in part by National Science Foundation Grants 0652569 and 0728806.

B. Löwe et al. (Eds.): CiE 2011, LNCS 6735, pp. 181–190, 2011.

The objective of this paper is to introduce a spatial model of computation–a variant of the cellular automaton model–that will contribute to our understanding of computability and complexity in analysis. In the half century since their introduction by Ulam and von Neumann [4], cellular automata have been used in many ways. They have (1) served as modeling tools for the physical, biological, and social sciences; (2) been used to investigate speculative frontiers such as artificial life and hypercomputation; and (3) shed new light on universality, parallelism, communication, fault tolerance, and other fundamental aspects of the theory of computing [4,26,7,20,14,17]. The present paper is an investigation of the third sort. Our primary interest is in using cellular automata to further our knowledge of the foundations of computable analysis.

Our starting point is the computability of subsets of Euclidean space, as defined by Brattka and Weihrauch [1] and also exposited in [25,2,3]. In this conference version of our paper, for ease of exposition, we confine ourselves to subsets of the 2-dimensional Euclidean unit square $[0,1]^2$. A subset X of $[0,1]^2$ is computable if there is an algorithm that turns pixels green when they are clearly in X and red when they are clearly not in X. A key feature of this definition is that the pixels may have arbitrary, but finite, resolution. More formally, for each rational point $q \in (\mathbb{Q} \cap [0,1])^2$ and each $n \in \mathbb{N}$, let $B(q, 2^{-n})$ denote the open ball of radius 2^{-n} centered at q. A set $X \subseteq [0,1]^2$ is computable if there is a computable function $f : (\mathbb{Q} \cap [0,1])^2 \times \mathbb{N} \to \{0,1\}$ such that, for all $q \in (\mathbb{Q} \cap [0,1])^2$ and $n \in \mathbb{N}$, the following two conditions hold.

(i) If $B(q, 2^{-n}) \subseteq X$, then $f(q,n) = 1$ (green).
(ii) If $B(q, 2^{1-n}) \cap X = \emptyset$, then $f(q,n) = 0$ (red).

If the hypotheses of (i) and (ii) are both false, then $f(q, n)$ must still be defined, and it may be either 1 or 0.

Here we want a cellular automaton that achieves the spirit of this definition, with its cells corresponding to the above pixels. The first thing to note is that such a cellular automaton must allow arbitrary, but finite, precision. Accordingly, we allow cells to "fission" zero or more times during the course of execution. When a cell discovers that it is completely in or out of the set X, we want it to turn green or red, respectively and then stay that color. Other cells may of course still be computing and/or fissioning, so we may have cells of many different resolutions (sizes) at any given time. With this motivation, we define our model, the *multi-resolution cellular automaton (MRCA)* model of computation, so that it has these features. Details appear in section 3.

We define a set $X \subseteq [0,1]^2$ to be MRCA-computable if there exist open sets $G \subseteq X$ and $H \subseteq [0,1]^2 \smallsetminus X$, with $G \cup H$ dense in $[0,1]^2$, and an MRCA that, starting with all cells uncolored, achieves the following.

(I) For every $x \in G$, there is some finite time at which x (i.e. some cell containing x) becomes green and stays that way.
(II) For every $x \in H$, there is some finite time at which x (i.e. some cell containing x) becomes red and stays that way.

All other points, including points on the topological boundary of X, remain uncolored throughout the execution of the MRCA. (Note that this definition implies that the sets G and H are, in fact, computably open.)

Our main theorem states that, for sets X with suitably small boundaries, MRCA-computability is equivalent to computability. In this context, "suitably small" means that the boundary is *computably nowhere dense*, i.e., nowhere dense in a sense that is witnessed by an algorithm. This is not a severe restriction, but it seems to be an inevitable consequence of the fact that we have prevented the MRCA from "changing its mind" about the color of a cell, while the Brattka-Weihrauch definition does allow the values of $f(q, n)$, for q near the boundary of X, to oscillate as the precision parameter n increases.

We note that Myrvold [15] and Parker [16] have considered notions somewhat similar to MRCA-computability. (See also [17].) Myrvold's "decidability ignoring boundaries" differs in that there is no small-boundary hypothesis, and the sets G and H are required to be the interiors of X and its complement, respectively. The latter requirement is, for our purposes here, unduly restrictive: We show that there is a computable set with computably nowhere dense boundary whose interior is not computably open. Parker's "decidability up to measure zero" requires the set of points not correctly "decided" to be small in a measure-theoretic sense, while we require it to be small in a topological sense.

Other cellular automaton models use concepts similar to fission to address issues other than real computation. Zeno machines [6] (or Accelerating Turing Machines) are Turing machines that take 2^{-n} steps to complete the nth computational step. Infinite time Turing machines [9] (ITTMs) generalize Zeno machines by allowing a *limit state* wherein every cell is set to the limit supremum of the values previously displayed in that cell. These models subdivide time in a way similar to how MRCAs subdivide space. A Scale-Invariant Cellular Automata (SCA) [19] create a lattice graph of cells where each level is half the width of the preceding level. Cells at depth i perform computation every 2^{-i} time units. In contrast, the MRCA sets cell size more dynamically than by depth in a lattice and all cells update at each time step.

Section 2 below discusses computability in the Euclidean plane, our small-boundary hypothesis, and a theorem needed for our main result. Section 3 presents the MRCA model, and section 4 presents our main theorem. Section 5 gives a brief summary.

2   Real Computation and Small Boundaries

In this section we review some fundamental aspects of the computability of sets in the Euclidean plane. We first fix some useful terminology and notation.

We work in the unit square $[0, 1]^2$. The *open ball* with *center* $x \in [0, 1]^2$ and *radius* $r \geq 0$ is the set $B(x, r) = \{y \in [0, 1]^2 \big| |x - y| < r\}$, where $|x - y|$ denotes the Euclidean distance from x to y. We write $\mathcal{B}$ for the set of all such balls having rational centers ($x \in \mathbb{Q}^2$) and rational radii ($r \in \mathbb{Q}$), recalling that $\mathcal{B}$ is a countable basis for the Euclidean topology on $[0, 1]^2$. For $X \subseteq [0, 1]^2$, we use the standard topological notations $\overline{X}$, X°, and ∂X for the *closure, interior,* and *boundary* of X, respectively. Recall that $\partial X = \overline{X} \cap [0, 1]^2 \smallsetminus X$.

For each $n \in \mathbb{N}$, we write $[n] = \{0, 1, \ldots, n - 1\}$. For each $n \in \mathbb{N}$ and $i, j \in [2^n]$, we define the *closed dyadic square* $Q(n, i, j) = [i \cdot 2^{-n}, (i + 1) \cdot 2^{-n}] \times [j \cdot 2^{-n}, (j + 1) \cdot 2^{-n}]$, and we write $\mathcal{Q}$ for the set of all such squares. Note that $Q(0, 0, 0) = [0, 1]^2$. As we shall see, $\mathcal{Q}$ is the set of all possible cells of an MRCA.

A set $G \subseteq [0, 1]^2$ is *computably open* (or Σ_1^0) if there is a computably enumerable set $\mathcal{A} \subseteq \mathcal{B}$ such that $G = \cup \mathcal{A}$. (Since each $B(x, r) \in \mathcal{B}$ is specified by its rational center and radius, it is clear what it means for a subset of $\mathcal{B}$ to be computably enumerable.) This notion is easily seen to be incomparable with the notion of computability defined in the introduction. For example, if $\alpha \in (0, 1)$ is a real number that is lower semicomputable but not computable, then $[0, \alpha)^2$ is computably open but not computable. Conversely, the square $\left[0, \frac{1}{2}\right]^2$ is computable but not open, hence not computably open.

Recall that a set $D \subseteq [0, 1]^2$ is *dense* if it meets every nonempty open ball. A set $Z \subseteq [0, 1]^2$ is *nowhere dense* if it is not dense in any nonempty open ball, i.e., if, for every $B \in \mathcal{B} \smallsetminus \{\emptyset\}$, then exists $B' \in \mathcal{B} \smallsetminus \{\emptyset\}$ such that $B' \subseteq B \smallsetminus Z$. Effectivizing this, we say that Z is *computably nowhere dense* if there is a computable function $f : \mathcal{B} \smallsetminus \{\emptyset\} \to \mathcal{B} \smallsetminus \{\emptyset\}$ such that, for all $B \in \mathcal{B} \smallsetminus \{\emptyset\}$, $f(B) \subseteq B \smallsetminus Z$.

Nowhere dense sets are very small. For example, the Baire category theorem says that no countable union of nowhere dense sets contains all of $[0, 1]^2$. Computably nowhere dense sets are very small in an even stronger sense.

Observation 1. *A set $Z \subseteq [0, 1]^2$ is computably nowhere dense if and only if there is a computably open dense set $G \subseteq [0, 1]^2$ such that $Z \cap G = \emptyset$.*

Recall that a *separator* of two (disjoint) sets A and B is a set S such that $A \subseteq S$ and $B \cap S = \emptyset$. The following result is crucial to our main theorem in section 4.

Theorem 2. *If $X \subseteq [0, 1]^2$ is a set whose boundary is computably nowhere dense, then the following two conditions are equivalent.*

(1) X is computable.
(2) X is a separator of two computably open sets whose union is dense.

The small-bounary hypothesis is needed here. For example, if $X = \left[0, \frac{1}{2}\right]^2 \cup (\mathbb{Q} \cap [0, 1])^2$, then (1) holds, but (2) fails.

It is tempting to think that the two computably open sets in (2) can be assumed to be interiors of X and its complement. However, we can easily give an example of a computable set X with computably nowhere dense boundary whose interior is not computably open.

3   Multi-Resolution Cellular Automata

We now introduce multi-resolution cellular automata (MRCA), a model of computation that generalizes cellular automata by allowing cells to fission.

Let $U_2 = \{(0,1), (1,0), (0,-1), (-1,0)\}$ be the set of direction vectors and let S be a finite set of states. We write $S_\perp = S \cup \{\perp\}$, where $\perp$ is an "undefined" symbol. We define a *neighborhood status* for S to be a function $\nu : U_2 \times \{0,1\} \to S_\perp$, and we write N_S for the set of all such ν.

Definition. A *multi-resolution cellular automaton (MRCA)* is a triple $A = (S, \delta, s)$ where S is a finite set of states; $s \in S$ is the *start state*; and $\delta : S \times N_S \to S \cup S^4$ is the *transition function*.

Here we give a brief, intuitive summary of the operation (semantics) of an MRCA. The MRCA initially has just one cell, $[0,1]^2$, and it is in the state s. During each clock cycle, all cells of the MRCA are updated simultaneously. The "neighbors" of a cell Q are the eight surrounding cells suggested by the following picture.

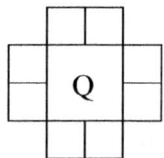

The update of Q depends upon the current states of Q and these neighboring cells. If any of the neighboring cells are outside of $[0,1]^2$ or have already subdivided, their states are, for the purpose of updating Q, undefined ($\perp$). If some existing cell contains two of these neighboring cells, its state is used for both of them in updating Q. If the value of δ, applied to this information is in S, then this is the new state of Q. If the value of δ is $(s_1, s_2, s_3, s_4) \in S^4$, then Q fissions into the four cells

$$\begin{array}{|c|c|} \hline Q_2 & Q_1 \\ \hline Q_3 & Q_4 \\ \hline \end{array}$$

and each cell Q_i is assigned the state s_i.

We have now specified the basic features of the MRCA model. Additional features may be added for specific purposes. For example, in Section 4 we use this model to color portions of $[0,1]^2$ in a monotone way (i.e. without mind changes). For this purpose, we define a *(monotone) k-coloring* MRCA to be a 4-tuple $A = (S, \delta, s, c)$, where

(i) (S, δ, c) is an MRCA;
(ii) $c : S \dashrightarrow [k]$ is the *state-coloring partial function*; and
(iii) for all $q \in dom(c), \nu \in N_S$, and $q' \in S$, if q' is $\delta(q, \nu)$ or a component of $\delta(q, \nu)$, then $q' \in dom(c)$ and $c(q') = c(q)$.

Condition (iii) here ensures that δ never updates an already-colored cell in such a way as to alter or remove the color of any part of the cell.

For each $i \in [k]$, the *set colored i* by a k-coloring MRCA $A = (S, \delta, s, c)$ is the set $S_i(a)$ defined in the now-obvious way.

Example 1. For each $n \in \mathbb{Z}^+$, let $Q_n = Q\left(2n, \dfrac{4^n - 4}{3}, 0\right) = \left[\dfrac{1}{3}(1 - 4^{1-n}), \dfrac{1}{3}(1 - 4^{-n})\right] \times \left[0, 4^{-n}\right]$, and let $X = \bigcup_{n=1}^{\infty} Q_n$. Consider the 1-coloring MRCA $A = (S, \delta, s, c)$, where $S = \{s, t, u, v\}$; $c(v) = 0$; $c(s), c(t)$, and $c(u)$ are undefined; and, for all $\nu \in N_S$,

$$\delta(s, \nu) = (t, t, u, t),$$
$$\delta(t, \nu) = t,$$
$$\delta(u, \nu) = (t, t, v, s),$$
$$\delta(v, \nu) = v.$$

It is easy to see that $S_0(A) = X$, as depicted below with 0 = green (green may appear as dark grey while red appears as light grey in grayscale versions of this paper).

4   MRCA Characterization of Computability

This section presents our main theorem, an MRCA characterization of computability under the small-boundary hypothesis of Section 2.

We define a set $X \subseteq [0,1]^2$ to be *MRCA-computable* if there exist open sets $G \subseteq X$ and $H \subseteq [0,1]^2 \setminus X$, with $G \cup H$ dense on $[0,1]^2$, and a 2-coloring MRCA A such that $S_1(A) = G$ and $S_0(A) = H$. With the convention 0 = red and 1 = green, this is the definition that we gave in the introduction.

The following technical lemma is the central part of our argument for the main theorem of this paper.

Lemma 1. *If χ is a cellular k-coloring that is computable, then there is a k-coloring MRCA A such that, for each $i \in [k]$, $S_i(A) = S_i(\chi)$.*

We now state and prove the main result of this paper, deferring proof of this lemma.

Theorem 3 (Main Theorem). *If $X \subseteq [0,1]^2$ is a set whose boundary is computably nowhere dense, then X is computable if and only if X is MRCA-computable.*

Our proof of Theorem 3 uses the following coloring notion. We define a *(cellular) k-coloring* of $[0,1]^2$ to be a partial function $\chi : \mathcal{Q} \dashrightarrow [k]$ satisfying the consistency condition that, for all $Q_1, Q_2 \in dom(\chi)$, $Q_1 \cap Q_2 \neq \emptyset \Rightarrow \chi(Q_1) = \chi(Q_2)$.

We call $\chi(Q)$ the *color* of cell Q in the coloring χ. For each $i \in [k]$, the *set colored i by χ* is then $S_i(\chi) = \cup \{Q \mid \chi(Q) = i\}$.

Proof. Let $X \subseteq [0,1]^2$ be a set whose boundary is computably nowhere dense.

Assume that X is computable. Then, by Theorem 2, there exist computably open sets $G, H \subseteq [0,1]^2$ such that $G \cup H$ is dense on $[0,1]^2$ and X is a separator of G and H. Since G and H are computably open, there exist c.e. sets $\mathcal{A}_G, \mathcal{A}_H \subseteq \mathcal{B}$ such that $G = \bigcup \mathcal{A}_G$ and $H = \bigcup \mathcal{A}_H$. By standard techniques, each open ball $B \in \mathcal{B}$ can be written as a countable union of dyadic squares $Q \in \mathcal{Q}$. Since this process is effective, it follows that there exist c.e. sets $\mathcal{R}_G \subseteq \mathcal{Q}$ and $\mathcal{R}_H \subseteq \mathcal{Q}$ such that $G = \bigcup \mathcal{R}_G$ and $H = \bigcup \mathcal{R}_H$. Since G and H are disjoint open sets, every cell in $\mathcal{R}_G$ must be disjoint from every cell in $\mathcal{R}_H$. Hence the partial function $\chi : \mathcal{Q} \dashrightarrow [2]$ defined by

$$\chi(Q) = \begin{cases} 1 & \text{if } Q \in \mathcal{R}_G \\ 0 & \text{if } Q \in \mathcal{R}_H \\ \text{undefined} & \text{otherwise.} \end{cases}$$

is a well-defined cellular 2-coloring of $[0,1]^2$. Clearly, $S_1(\chi) = \bigcup \mathcal{R}_G = G$ and $S_0(\chi) = \bigcup \mathcal{R}_H = H$. Moreover, since $\mathcal{R}_G$ and $\mathcal{R}_H$ are c.e., χ is computable. It follows by Lemma 1 that there is a 2-coloring MRCA A such that $S_1(A) = S_1(\chi) = G$ and $S_0(A) = S_0(\chi) = H$.

Conversely, assume that X is MRCA computable. Then there exists computably open sets $G \subseteq X$ and $H \subseteq [0,1]^2 \setminus X$, with $G \cup H$ dense on $[0,1]^2$. Then X is a separator of G and H, so X is computable by Theorem 2. □

We now sketch the proof of Lemma 1.

Proof Sketch of Lemma 1. It is well-known that a Turing machine may be simulated by a one-dimensional cellular automaton. We extend this result to an MRCA by utilizing cell divisions to create space when necessary. The result is a Turing machine computation that "falls" towards a single point as more space is needed by the computation. With the slight modification described below, the entire computation can be made to fall toward a point, thus giving us a way to spatially dove-tail MRCA computations. We call each instance of this computation on a given input a *computational unit*.

As the Turing machine simulation proceeds, additional space may be required to continue the simulation. Also, our construction requires that the Turing machine simulation permanently moves out of the right-most cell. This is easily accomplished by periodically adding a tape cell when not required by the Turing machine simulation and shifting the entire simulation to the left one cell. This allows the space on the right to be vacated. These are both accomplished by dividing the left-most cell and sliding the entire computation to the left, causing the the entire computation to fall down the "rabbit hole" [5] as it becomes smaller and smaller. This procedure is as shown in Figure 1.

For each set of points enclosed in a dyadic-width square, we create four rotated copies of the computational unit as described above with input describing this square of points. The rotated copies are labelled in Figure 2 according to their

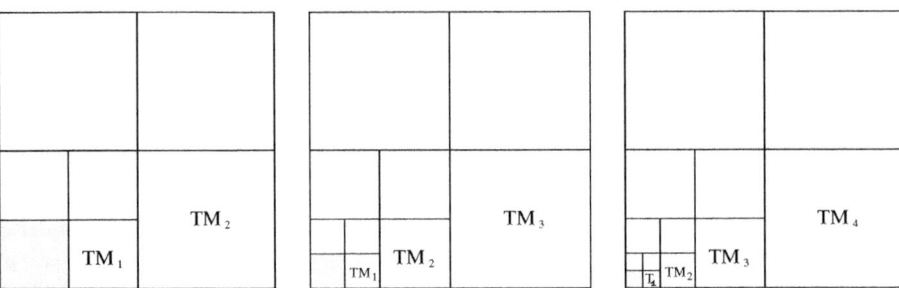

Fig. 1. Creation of space for Turing machine computation by "falling down a rabbit hole." TM_i represents the ith cell of computation. The "rabbit hole" is colored pink and is the cell that divides.

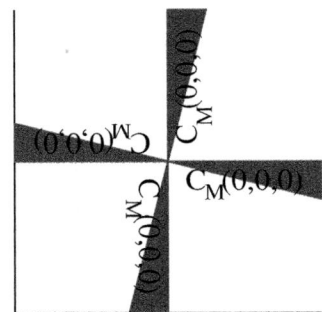

Fig. 2. Initial state of each C_M with initial state $C_M(0,0,0)$ rotated to show the direction of computation in each computational unit. The area colored green is the space reserved for that unit's computation.

rotation. We create these duplicate computational units to simplify coloring sets of points. When a computational unit concludes that its input should be colored, it first colors the cells it used to compute. Because the four computational units for a given square are identical, they will eventually color a "pinwheel" pattern as depicted in green in Figure 2. To color the rest of the square, we institute a general MRCA rule: If a north or south neighbor and an east or west neighbor is set to a color, then the current cell changes to that color.

Note that later, "child" computational units for coloring smaller areas will not intersect earlier, "parent" computational units because the computational unit of the parent will have compressed to avoid any such interference.

In order to guarantee every dyadic square has its color computed, there are rules for periodic fission to create four computational units for each dyadic square in the same "pinwheel" pattern as the parent. Address information is then passed from each "parent" computational unit and updated to reflect the "child" dyadic subsquare. These computational units then begin in the same initial state using as input the correct address for the space the child computation would color. The end result of this process is depicted in Figure 3 for computational unit

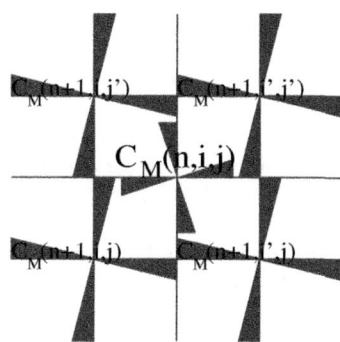

Fig. 3. Shows how the child computational units of $C_M(n,i,j)$ are laid out where $i' = i + 2^{-n-1}$ and $j' = j + 2^{-n-1}$. The green lines along each axis eminating from the parent computational units represent "cookies states" that the parent computational unit can follow to color the entire depicted square. Although these lines appear to cross the child units, the child units will preserve cookies of their ancestors.

$C_M(n,i,j)$ with parent computational units in green and child computational units in red.

5   Conclusion

We have presented a multi-resolution variant of cellular automata that gives an intuitive, spatial characterization of the computability of plane sets with small boundaries. The notion of cells that fission may at first seem fanciful, but the MRCA model is motivated by a concrete objective (real computation), and it can be implemented in software. It will be interesting to see what other aspects of computable analysis can be illuminated by the MRCA model.

References

1. Brattka, V., Weihrauch, K.: Computability on subsets of Euclidean space I: Closed and compact subsets. Theoretical Computer Science 219, 65–93 (1999)
2. Braverman, M.: On the complexity of real functions. In: Forty-Sixth Annual IEEE Symposium on Foundations of Computer Science, pp. 155–164 (2005)
3. Braverman, M., Cook, S.: Computing over the reals: Foundations for scientific computing. Notices of the AMS 53(3), 318–329 (2006)
4. Burks, A.: Essays on Cellular Automata. Univeristy of Illinois Press, US (1970)
5. Carroll, L.: Alice's Adventures in Wonderland. Project Gutenberg (2008)
6. Copeland, B.J.: Accelerating Turing machines. Minds and Machines 12, 281–301 (2002)
7. Griffeath, D., Moore, C.: New Constructions in Cellular Automata. Oxford University Press, USA (2003)
8. Grzegorczyk, A.: Computable functionals. Fundamenta Mathematicae 42, 168–202 (1955)

9. Hamkins, J.D., Lewis, A.: Infinite time Turing machines. Journal of Symbolic Logic 65(2), 567–604 (2000)
10. Ko, K., Friedman, H.: Computational complexity of real functions. Theoretical Computer Science 20, 323–352 (1982)
11. Ko, K.-I.: Complexity Theory of Real Functions. Birkhäuser, Boston (1991)
12. Kreitz, C., Weihrauch, K.: Complexity theory on real numbers and functions. In: Cremers, A.B., Kriegel, H.-P. (eds.) GI-TCS 1983. LNCS, vol. 145, pp. 165–174. Springer, Heidelberg (1982)
13. Lacombe, D.: Extension de la notion de fonction recursive aux fonctions d'une ow plusiers variables reelles, and other notes. Comptes Rendus (1955), 240:2478-2480; 241:13-14, 151-153, 1250-1252
14. McIntosh, H.V.: One Dimensional Cellular Automata. Luniver Press, United Kingdom (2009)
15. Myrvold, W.C.: The decision problem for entanglement. In: Cohen, R.S., Horne, M.S.J. (eds.) Potentiality, Entanglement and Passion-at-a-Distance: Quantum Mechanical Studies for Abner Shimony, pp. 177–190. Kluwer Academic Publishers, Dordrecht and Boston (1997)
16. Parker, M.W.: Undecidability in $\mathbb{R}^n$: Riddled basins, the KAM tori, and the stability of the solar system. Philosophy of Science 70, 359–382 (2003)
17. Parker, M.W.: Three concepts of decidability for general subsets of uncountable spaces. Theoretical Computer Science 351, 2–13 (2006)
18. Pour-El, M.B., Richards, J.I.: Computability in Analysis and Physics. Springer, Heidelberg (1989)
19. Schaller, M., Svozil, K.: Scale-invariant cellular automata and self-similar Petri nets. The European Physical Journal B 69, 297–311 (2009)
20. Schiff, J.: Cellular Automata: A Discrete View of the World. Wiley Interscience, USA (2008)
21. Turing, A.M.: On computable numbers with an application to the Entscheidungsproblem. Proc. London Math. Soc. 42(2), 230–265 (1936)
22. Turing, A.M.: On computable numbers with an application to the Entscheidungsproblem. A correction. Proc. London Math. Soc. 43(2), 544–546 (1936)
23. Turing, A.M.: Systems of Logic Based on Ordinals. Ph.D. thesis, Princeton, New Jersey, USA (1938)
24. Weihrauch, K.: Computability. Springer, New York (1987)
25. Weihrauch, K.: Computable Analysis. An Introduction. Springer, Heidelberg (2000)
26. Wolfram, S.: A New Kind of Science. Wolfram Media, Champaign, Illinois, USA (2002)

Deciding According to the Shortest Computations

Florin Manea*

[1] Otto-von-Guericke-Universität Magdeburg, Fakultät für Informatik
PSF 4120, D-39016 Magdeburg, Germany
[2] Faculty of Mathematics and Computer Science, University of Bucharest,
Str. Academiei 14, 010014 Bucharest, Romania
flmanea@fmi.unibuc.ro

Abstract. In this paper we propose, and analyze from the computational complexity point of view, a new variant of nondeterministic Turing machines. Such a machine accepts a given input word if and only if one of its shortest possible computations on that word is accepting; on the other hand, the machine rejects the input word when all the shortest computations performed by the machine on that word are rejecting. Our main results are two new characterizations of $\mathbf{P}^{\mathrm{NP}[\log]}$ and $\mathbf{P}^{\mathrm{NP}}$ in terms of the time complexity classes defined for such machines.

1 Preliminaries

The computation of a nondeterministic Turing machine (and, in fact, any computation of a nondeterministic machine, that consists in a sequence of moves) can be represented as a (potentially infinite) tree. Each node of this tree is an instantaneous description (ID for short, a string encoding the configuration of the machine at a given moment: the content of the tapes and the state), and its children are the IDs encoding the possible configurations in which the machine can be found after a (nondeterministic) move is performed. If the computation is finite then the tree is also finite and each leaf of the tree encodes a final ID: an ID in which the state is either accepting or rejecting. The machine accepts if and only if one of the leaves encodes the accepting state (also in the case of infinite trees), and rejects if the tree is finite and all the leaves encode the rejecting state.

Therefore, in the case of finite computations, one can check if a word is accepted/rejected by a machine by searching in the computation-tree for a leaf that encodes an accepting ID. Theoretically, this is done by a simultaneous traversal of all the possible paths in the tree (as we can deduce, for instance, from the definition of the time complexity of a nondeterministic computation). However, in practice, it is done by traversing each path at a time, until an accepting ID is found, or until the whole tree was traversed. Unfortunately, this may be a very time consuming task. Consequently, one may be interested in methods of using nondeterministic machines in a more efficient manner.

---

* The work of Florin Manea is supported by the *Alexander von Humboldt Foundation*.

B. Löwe et al. (Eds.): CiE 2011, LNCS 6735, pp. 191–200, 2011.
© Springer-Verlag Berlin Heidelberg 2011

Our paper proposes such a method: the machine accepts (rejects) a word if and only if one of the shortest paths in the computation-tree ends (respectively, all the shortest paths end) with an accepting ID (with rejecting IDs). Intuitively, we traverse the computations-tree on levels and, as soon as we reach a level containing a leaf, we look if there is a leaf encoding an accepting ID on that level, and accept, or if all the leaves on that level are rejecting IDs, and, consequently, reject. We show that the class of languages that are decided according to this strategy by Turing machines, whose shortest computations have a polynomial number of steps, equals the class $\mathbf{P}^{\mathbf{NP}[\log]}$. As a consequence of this result we can also show that the class of languages that are decided by Turing machines, working in nondeterministic polynomial time on any input but deciding according to the computations that have a minimal number of nondeterministic moves, also equals the class $\mathbf{P}^{\mathbf{NP}[\log]}$. These results continue a series of characterizations of $\mathbf{P}^{\mathbf{NP}[\log]}$, started in [9]. Then, we propose another method: the machine accepts (rejects) a word if and only if the the first leaf that we meet in a breadth-first-traversal of the computations-tree encodes an accepting ID (respectively, encodes a rejecting ID); note that in this case, one must define first an order between the sons of a node in the computations-tree. We show that, in the case of ordering the tree lexicographically, the class of languages that are decided, according to this new strategy, by Turing machines whose shortest computations have a polynomial number of steps equals the class $\mathbf{P}^{\mathbf{NP}}$.

The research presented in this paper is related to a series of papers presenting variants of nondeterministic Turing machines, working in polynomial time, that accept (or reject) a word if and only if a specific property is (respectively, is not) verified by the possible computations of the machine on that word. We recall, for instance: polynomial machines which accept if at least $1/2$ of their computations are accepting, and reject if at least $1/2$ of their computations are rejecting (the class $\mathbf{PP}$), or polynomial machines that accept if at least $2/3$ of the computation paths accept and reject if at most $1/3$ of the computation paths accept (the class of bounded-error probabilistic polynomial time $\mathbf{BPP_{path}}$); several other examples can be found on the Complexity Zoo web page[1] or [6]. However, instead of looking at all the computations, we look just at the shortest ones, and instead of asking questions regarding the number of accepting/rejecting computations, we just ask existential questions about the shortest computations.

Our work finds motivations also in the area of nature-inspired supercomputing models. Some of these models (see [5,7], for instance) were shown to be complete by simulating, in a massively parallel manner, all the possible computations of a nondeterministic Turing machine; characterizations of several complexity classes, like $\mathbf{NP}, \mathbf{P}$ and $\mathbf{PSPACE}$, were obtained in this framework. However, these machines were, generally, used to accept languages, not to decide them; in the case when a deciding model was considered ([5]), the rejecting condition was just a mimic of the rejecting condition from classical computing models. Modifying such nature-inspired machines in order to decide as soon as a possible accepting/rejecting configuration is obtained, in one of the computations simulated in

---

[1] www.complexityzoo.com, a web page constructed and managed by Scott Aaronson.

parallel, seems to be worth analyzing: such a halting condition looks closer to what really happens in nature, and it leads to a reduced consume of resources, comparing to the case when the machine kept on computing until all the possibilities were explored. Also, from a theoretical point of view, considering such halting conditions could lead to novel characterizations of a series of complexity classes (like the ones discussed in this paper) by means of nature-inspired computational models, as they seem quite close to the idea of deciding with respect to the shortest computations.

Further we review a seris of basic notions. By reasons of space we omit, here and in the rest of the paper, some classical definitions and give only sketches of the proofs. All these can be found in the technical report [4].

The reader is referred to [1,3,6] for the basic definitions regarding Turing machines, oracle Turing machines, complexity classes and complete problems.

We denote by $\mathbf{P^{NP}}$ the class of languages decided by deterministic Turing machines, that work in polynomial time, with oracles from $\mathbf{NP}$. We denote by $\mathbf{P^{NP[log]}}$ the class of languages decided by deterministic Turing machines, that work in polynomial time, with oracles from $\mathbf{NP}$, and which can enter the query state at most $\mathcal{O}(\log n)$ times in a computation on a input word of length n.

The following problem is complete for $\mathbf{P^{NP}}$, with respect to polynomial time reductions (see [8] for a proof):

Problem 1. (Odd - Traveling Salesman Problem, TSP_{odd}) Let n be a natural number, and d be a function $d : \{1, \ldots, n\} \times \{1, \ldots, n\} \to I\!N$. Decide if the minimum value of $I = \{\sum_{i=1}^{n} d(\pi(i), \pi(i+1)) \mid \pi$ is a permutation of $\{1, \ldots, n\}$, and $\pi(n+1) = \pi(1)\}$ is odd.

We assume that the input of this problem is given as the natural number n, and n^2 numbers representing the values $d(i, j)$, for all i and j. The size of the input is the number of bits needed to represent the values of d times n^2.

Next we describe a $\mathbf{P^{NP[log]}}$-complete problem; however, we need a few preliminary notions (see [2] for a detailed presentation). Let n be a natural number and let $C = \{c_1, \ldots, c_n\}$ be a set of n candidates. A preference order on C is an ordered list $\langle c_{\pi(1)} < c_{\pi(2)} < \ldots < c_{\pi(n)} \rangle$, where π is a permutation of $\{1, \ldots, n\}$; if c_i appears before c_j in the list we say that the candidate c_i is preferred to the candidate c_j in this order. Given a multiset V of preference orders on a set of n candidates C (usually V is given as a list of preference orders) we say that the candidate c_i is a Condorcet winner, with respect to the preferences orders of V, if c_i is preferred to each other candidate in strictly more than half of the preference orders. We define the Dodgson score of a candidate c, with respect to V, as the smallest number of exchanges of two adjacent elements in the preference orders from V (switches, for short) needed to make c a Condorcet winner; we denote this score with $Score(C, c, V)$. In [2] it was shown that the following problem is $\mathbf{P^{NP[log]}}$-complete, with respect to polynomial time reductions:

Problem 2. (Dodgson Ranking, DodRank) Let n be a natural number, let C be a set of n candidates, and c and d two candidates from C. Let V be a multiset of preference orders on C. Decide if $Score(C, c, V) \leq Score(C, d, V)$.

We assume that the input of this problem is given as the natural number n, two numbers c and d less or equal to n, and a list of preference orders V, encoded as permutations of the set $\{1, \ldots, n\}$. If we denote by $\#(V)$ the number of preference orders in V, then the size of the input is $\mathcal{O}(\#(V)n \log n)$.

The connection between decision problems and languages is discussed in [3]. When we say that a decision problem is solved by a Turing machine, of certain type, we actually mean that the language corresponding to that decision problem is decided by that machine.

2   Shortest Computations

In this section we propose a modification of the way Turing machines decide an input word. Then we propose a series of results on the computational power of these machines and the computational complexity classes defined by them.

Definition 1. *Let M be a Turing machine and w be a word over the input alphabet of M. We say that w is accepted by M with respect to shortest computations if there exists at least one finite possible computation of M on w, and one of the shortest computations of M on w is accepting; w is rejected by M w.r.t. shortest computations if there exists at least one finite computation of M on w, and all the shortest computations of M on w are rejecting. We denote by $L_{sc}(M)$ the language accepted by M w.r.t. shortest computations, i.e., the set of all words accepted by M, w.r.t. shortest computations. We say that the language $L_{sc}(M)$ is decided by M w.r.t. shortest computations if all the words not accepted by M, w.r.t. shortest computations, are rejected w.r.t. shortest computations.*

The following remark shows that the computational power of the newly defined machines coincides with that of classic Turing machines.

Remark 1. The class of languages accepted by Turing machines w.r.t. shortest computations equals **RE**, while the class of languages decided by Turing machines w.r.t. shortest computations equals **REC**.

Next we define a computational complexity measure for the Turing machines that decide w.r.t. shortest computations.

Definition 2. *Let M be a Turing machine, and w be a word over the input alphabet of M. The time complexity of the computation of M on w, measured w.r.t. shortest computations, is the length of the shortest possible computation of M on w. A language L is said to be decided in polynomial time w.r.t. shortest computations if there exists a Turing M machine and a polynomial f such that the time complexity of a computation of M on each word of length n, measured w.r.t. shortest computations, is less than $f(n)$, and $L_{sc}(M) = L$. We denote by $PTime_{sc}$ the class of languages decided by Turing machines in polynomial time w.r.t. shortest computations.*

The main result of this section is the following:

Theorem 1. $PTime_{sc} = \mathbf{P}^{\mathbf{NP}[\log]}$.

Proof. The proof has two parts. First we show the *upper bound* $PTime_{sc} \subseteq$ $\mathbf{P^{NP[log]}}$, and, then we show the *lower bound* $PTime_{sc} \supseteq \mathbf{P^{NP[log]}}$.

For the first part of the proof let $L \subseteq V^*$ be a language in $PTime_{sc}$ and let M be a Turing machine that decides L in polynomial time w.r.t. shortest computations. Also, let f be a polynomial such that the time complexity of the computation of M on each word of length n, measured w.r.t. shortest computations, is less than $f(n)$. A machine M', with access to an **NP**-oracle, accepting L works as follows on an input word w: it searches (by binary search) the minimum length of an accepting computation of M on w, with length less or equal to $f(|w|)$, and then it verifies, by another oracle query, if there exists a shorter rejecting computation of M on w; if the answer to the last query is negative, the machine accepts, otherwise it rejects. It is clear that M' works in polynomial time and makes at most $\mathcal{O}(\log n)$ queries to the oracle; therefore, $L \in \mathbf{P^{NP[log]}}$.

For the second inclusion, the class $PTime_{sc}$ is closed under polynomial-time reductions, so it suffices to show that the $\mathbf{P^{NP[log]}}$-complete problem *DodRan* can be solved in polynomial time by a Turing machine M, deciding w.r.t. shortest computations. The input of M consists in the number n, the set C of n candidates, c and d two candidates from C, and V the multiset of preference orders on C (encoded as explained in the previous section). First, M chooses nondeterministically some switches in V, that are supposed to make c and d Condorcet winners, respectively; the length of a possible computation performed in this step depends on the choice of the numbers k_1 and k_2 of switches: if these numbers are smaller, then the computation is shorter. Then the machine makes (deterministically) the switches chosen in the previous step. The length of a possible computation, until this moment, is still determined by the choice of k_1 and k_2. Further, the machine verifies if those switches make indeed c and d winners, according to the orders modified by the previously chosen moves. If they were both transformed in winners by the chosen switches, the computation continues; otherwise, the machine makes a sequence of dummy steps, long enough to make that computation irrelevant for the final answer of the machine on the given input, and rejects. Note that at least one choice of the switches, in the first step, makes both c and d winners. Now, the shortest computations are those ones in which both c and d were transformed into winners and the chosen numbers k_1 and k_2 are minimal. But this is exactly the case when $k_1 = Score(C, c, V)$ and $k_2 = Score(C, d, V)$. Finally, all the computations in which c and d were transformed into winners are completed by a deterministic comparison between k_1 and k_2. The decision of such a computation is to accept, if $k_1 \leq k_2$, or to reject, otherwise; clearly, all the shortest computations return the same answer. Consequently, M accepts if and only if $Score(C, c, V) \leq Score(C, d, V)$, and reject otherwise. It is rather easy to see that M works in polynomial time, since each of the 5 steps described above can be completed in polynomial time.

In conclusion we showed that *DodRan* can be solved in polynomial time by a Turing machine that decides w.r.t. shortest computations. As a consequence of the above, it follows that $PTime_{sc} \supseteq \mathbf{P^{NP[log]}}$, and this ends our proof. $\square$

The technique used in the previous proof to show that $\mathbf{P}^{\mathbf{NP}[\log]}$-complete problems can be solved by in polynomial time by Turing machines that decide w.r.t. shortest computations suggests another characterization of $\mathbf{P}^{\mathbf{NP}[\log]}$: it also equals the class of languages decided by Turing machines that work in polynomial time and accept, resspectively reject, an input if and only if the computation that has the minimal number of non-deterministic moves, on that input, is accepting, respectively rejecting.

3   The First Shortest Computation

In the previous section we have proposed an decision mechanism of Turing machines that basically consisted in identifying the shortest computations of a machine on an input word, and checking if one of these computations is an accepting one, or not. Now we analyze how the properties of the model are changed if we order the computations of a machine and the decision is made according to the first shortest computation, in the defined order.

Let $M = (Q, V, U, q_0, acc, rej, B, \delta)$ be a t-tape Turing machine, and assume that $\delta(q, a_1, \ldots, a_t)$ is a totally ordered set, for all $a_i \in U$, $i \in \{1, \ldots, t\}$, and $q \in Q$; we call such a machine an *ordered Turing machine*. Let w be a word over the input alphabet of M. Assume s_1 and s_2 are two (potentially infinite) sequences describing two possible computations of M on w. We say that s_1 is lexicographically smaller than s_2 if s_1 has fewer moves than s_2, or they have the same number of steps (potentially infinite), the first k IDs of the two computations coincide and the transition that transforms the kth ID of s_1 into the $k + 1$th ID of s_1 is smaller than the transition that transforms the kth ID of s_2 into the $k + 1$th ID of s_2, with respect to the predefined order of the transitions. It is not hard to see that this is a total order on the computations of M on w. Therefore, given a finite set of computations of M on w one can define the lexicographically first computation of the set as that one which is lexicographically smaller than all the others.

Definition 3. *Let M be an ordered Turing machine, and w be a word over the input alphabet of M. We say that w is accepted by M with respect to the lexicographically first computation if there exists at least one finite possible computation of M on w, and the lexicographically first computation of M on w is accepting; w is rejected by M w.r.t. the lexicographically first computation if the lexicographically first computation of M on w is rejecting. We denote by $L_{lex}(M)$ the language accepted by M w.r.t. the lexicographically first computation. We say that the language $L_{lex}(M)$ is decided by M w.r.t. the lexicographically first computation if all the words not contained in $L_{lex}(M)$ are rejected by M.*

As in the case of Turing machines that decide w.r.t. shortest computations, the class of languages accepted by Turing machines w.r.t. the lexicographically first computation equals **RE**, while the class of languages decided by Turing machines w.r.t. the lexicographically first computation equals **REC**. The time complexity of the computations of Turing machines that decide w.r.t. the lexico-graphically first computation is defined exactly as in the case of machines that decide w.r.t. shortest computations. We denote by $PTime_{lex}$ *the class of languages decided by Turing machines in polynomial time w.r.t. the lexicographically first computation.* In this context, we are able to show the following theorem.

Theorem 2. $PTime_{lex} = \mathbf{P^{NP}}$.

Proof. In the first part of the proof we show that $PTime_{lex} \subseteq \mathbf{P^{NP}}$. Let L be a language in $PTime_{lex}$ and let M be a Turing machine that decides L in polynomial time w.r.t. the lexicographically first computation. Also, let f be a polynomial such that the time complexity of the computation of M on each word of length n, measured w.r.t. the lexicographically first computation, is less than $f(n)$. A deterministic Turing machine M', with $\mathbf{NP}$ oracle, accepting L implements the following strategy, on an input word w: it searches (by binary search) the minimum length of a computation of M on w, with length less or equal to $f(|w|)$, and, then, by querying the oracle, it tries to construct, ID by ID, the lexicographically first computation of M, having that length; then, it decides according to the decision that M took on that computation.

To show the second inclusion, note that, similar to the case of machines deciding w.r.t. shortest computations, the class $PTime_{lex}$ is closed to polynomial-time reductions. Thus, it is sufficient to show that the $\mathbf{P^{NP}}$-complete problem TSP_{odd} can be solved in polynomial time by a Turing machine M that decides w.r.t. the lexicographically first computation. Therefore we construct a Turing machine M that solves TSP_{odd} w.r.t. the lexicographically first computation. The input of this machine consists in a natural number n, and n^2 natural numbers, encoding the values of the function $d : \{1, \ldots, n\} \times \{1, \ldots, n\} \rightarrow I\!N$. First, we choose a possible permutation π of $\{1, \ldots, n\}$ and compute the sum $S = \sum_{i=1}^{n} d(\pi(i), \pi(i+1))$; in this way we have computed an upper bound for the solution of the Traveling Salesman Problem, defined by the function d. Then we try to find another permutation π' that leads to a smaller sum. For this we choose first a number S_0 that has as many digits as S (of course, it may have several leading zeros); however, the computations are ordered in such a manner that a computation in which smaller numbers are chosen comes before (lexicographically) a computation in which a greater number is chosen. Then M verifies if S' can be equal to the sum $\sum_{i=1}^{n} d(\pi'(i), \pi'(i+1))$, for a permutation π' nondeterministically chosen. If the answer is yes then it means that S_0 is also a possible solution of the problem; otherwise we conclude that the nondeterministic choices made so far were not really the good ones, so we reject after we make a long-enough sequence of dummy steps, in order not to influence the decision of the machine. Finally, we verify if S_0 is odd, and accept if and only if this condition holds. By the considerations made above, it is clear that in all the shortest computations we identified some numbers that may be solutions of the Traveling Salesman Problem; moreover, in the first of the shortest computations we have identified the smallest such number, i.e., the real solution of the problem. Consequently, the decision of the machine is to accept or to reject the input according to the parity of the solution identified in the first shortest computation, which is correct. Also, it is not hard to see that every possible computation of M has polynomial length, and always ends with a decision.

Summarizing, we showed that TSP_{odd} can be solved in polynomial time by a Turing machine that decides w.r.t. the lexicographically first computation. To conclude, we have $PTime_{lex} \supseteq \mathbf{P^{NP}}$, and this concludes our proof.   □

Remark 2. Note that the proof of Theorem 1 shows that $\mathbf{P^{NP[\log]}}$ can be also characterized as the class of languages that can be decided in polynomial time w.r.t. shortest computations by nondeterministic Turing machines whose shortest computations are either all accepting or all rejecting. On the other hand, in the proof of Theorem 2, the machine that we construct to solve w.r.t. the lexicographically first computation the TSP_{odd} problem may have both accepting and rejecting shortest computations on the same input. This shows that $\mathbf{P^{NP[\log]}} = \mathbf{P^{NP}}$ if and only if all the languages in $\mathbf{P^{NP}}$ can be decided w.r.t. shortest computations by nondeterministic Turing machines whose shortest computations on a given input are either all accepting or all rejecting.

There is a point where the definition of the ordered Turing machine doesn't seem satisfactory: each time a machine has to execute a nondeterministic move, for a certain state and a tuple of scanned symbols, the order of the possible moves is the same, regardless of the input word and the computation performed until that moment. Therefore, we consider another variant of ordered Turing machines, in which such informations are considered:

Let M be a Turing machine. We denote by $\langle M \rangle$ a binary encoding of this machine (see, for instance, [3]). It is clear that the length of the string $\langle M \rangle$ is a polynomial with respect to the number of states and the working alphabet of the machine M. Let $g : \{0,1,\#\}^* \to \{0,1,\#\}^*$ be a function such that $g(\langle M \rangle \# w_1 \# w_2 \# \ldots \# w_k) = w_1' \# w_2' \# \ldots \# w_p'$, given that $w_1, \ldots, w_k$ are binary encodings of the IDs that appear in a computation of length k of M (we assume that they appear in this order, and that w_1 is an initial configuration), and $w_1', \ldots, w_p'$ are the IDs that can be obtained in one move from w_k. Clearly, this function induces canonically an ordering on the computations of a Turing machine. Assume s_1 and s_2 are two (potentially infinite) sequences describing two possible computations of M on w. We say that s_1 is g-smaller than s_2 if the first k IDs of the two computations, which can be encoded by the strings $w_1, \ldots, w_k$, coincide, and $g(\langle M \rangle \# w_1 \# w_2 \# \ldots \# w_k) = w_1' \# w_2' \# \ldots \# w_p'$, the $k+1$th ID of s_1 is encoded by w_i', the $k+1$th ID of s_2 is encoded by w_j', and $i < j$. It is not hard to see that g induces a total order on the computations of M on w; thus we will call such a function *an ordering function*. Therefore, given a finite set of computations of M on w we can define the g-first computation of the set as the one that is g-smaller than all the others.

Definition 4. *Let M be a Turing machine, and $g : \{0,1,\#\}^* \to \{0,1,\#\}^*$ be an ordering function. We say that w is accepted by M with respect to the g-first shortest computation if there exists at least one finite possible computation of M on w, and the g-first of the shortest computations of M on w is an accepting* *one; w is rejected by M w.r.t. the lexicographically first computation if the g-first shortest computation of M on w is a rejecting computation. We denote by* $L^g_{fsc}(M)$ *the language accepted by M w.r.t. the g-first shortest computation, i.e., the set of all words accepted by M, w.r.t. the g-first shortest computation. As in the case of regular Turing machines, we say that the language $L^g_{fsc}(M)$ is decided by M w.r.t. the g-first shortest computation if all the words not contained in $L^g_{fsc}(M)$ are rejected by that machine, w.r.t. the g-first shortest computation.*

It is not surprising that, if g is Turing computable, the class of languages accepted by Turing machines w.r.t. the g-first shortest computation equals **RE**, while the class of languages decided by Turing machines w.r.t. the lexicographically first computation equals **REC**. The time complexity of the computations of Turing machines that decide w.r.t. the g-first shortest computation is defined exactly as in the case of machines that decide w.r.t. shortest computations. We denote by $PTime^g_{fsc}$ *the class of languages decided by Turing machines in polynomial time w.r.t. the g-first shortest computation.* Also, we denote by $PTime_{ofsc}$ the union of all the classes $PTime^g_{fsc}$, where the ordering function g can be computed in polynomial deterministic time. We are now able to show the following theorem.

Theorem 3. $PTime_{ofsc} = \mathbf{P^{NP}}$.

Notice that $\mathbf{P^{NP[\log]}} \subseteq PTime^g_{fsc} \subseteq \mathbf{P^{NP}}$, for all the ordering functions g which can be computed in polynomial deterministic time. The second inclusion is immediate from the previous Theorem, while the first one follows from the fact that any language in $\mathbf{P^{NP[\log]}}$ is accepted w.r.t. shortest computations, in polynomial time, by a nondeterministic Turing machine whose shortest computations are either all accepting or all rejecting; clearly, the same machine can be used to show that the given language is in $PTime^g_{fsc}$.

It is interesting to see that for some particular ordering functions, as for instance the one that defines the lexicographical order discussed previously, a stronger result holds: $PTime^g_{fsc} = \mathbf{P^{NP}}$ (where g is the ordering function). We leave as an open problem to see if this relation holds for all the ordering functions, or, if not, to see when it hold.

4   Conclusions and Further Work

In this paper we have shown that considering a variant of Turing machine, that decides an input word according to the decisions of the shortest computations of the machine on that word, leads to new characterizations of two well studied complexity classes $\mathbf{P^{NP[\log]}}$ and $\mathbf{P^{NP}}$. These results seem interesting since they provide alternative definitions of these two classes, that do not make use of any other notion than the Turing machine (like oracles, reductions, etc.). From a theoretical point of view, an attractive continuation of the present work would be to analyze if the equality results in Theorems 1 and 2 relativize. It is not hard to see that the upper bounds shown in these proofs are true even if we allow all the machines to have access to an arbitrary oracle. It remains to be settled if a similar result holds in the case of the lower bounds.

Nevertheless, other accepting/rejecting conditions related to the shortest computations could be investigated. As we mentioned in the Introduction, several variants of Turing machines that decide a word according to the number of accepting, or rejecting, computations were already studied. We intend to analyze what happens if we use similar conditions for the shortest computations of a Turing machine. In this respect, using the ideas of the proof of Theorem 2, one can show that:

Theorem 4. *Given a nondeterministic polynomial Turing machine M_1, one can construct a nondeterministic polynomial Turing machine, with access to NP-oracle, M_2, whose computations on an input word correspond bijectively to the short computations of M_1 on the same word, such that two corresponding computations are both either accepting, or rejecting.*

This Theorem is useful to show upper bounds on the complexity classes defined by counting the accepting/rejecting shortest computations, e.g., $\mathbf{PP_{sc}} \subseteq \mathbf{PP^{NP}}$ (where $\mathbf{PP_{sc}}$ is the class of decision problems solvable by a nondeterministic polynomial Turing machine which accepts if and only if at least $1/2$ of the shortest computations are accepting, and rejects otherwise) or $\mathbf{BPP_{sc}} \subseteq \mathbf{BPP^{NP}_{path}}$ (where $\mathbf{BPP_{sc}}$ is the class of decision problems solvable by an nondeterministic polynomial Turing machine which accepts if at least $2/3$ of the shortest computations are accepting, and rejects if at least $2/3$ of the shortest computations are rejecting). However, in some cases one can show stronger upper bounds; for instance, $\mathbf{PP_{sc}} \subseteq \mathbf{PP^{NP[log]}_{ctree}}$ (where $\mathbf{PP^{NP[log]}_{ctree}}$ is the class of decision problems solvable by a $\mathbf{PP}$-machine which can make a total number of $\mathcal{O}(\log n)$ queries to an $\mathbf{NP}$-language in its entire computation tree, on an input of length n).

References

1. Hartmanis, J., Stearns, R.E.: On the Computational Complexity of Algorithms. Trans. Amer. Math. Soc. 117, 533–546 (1965)
2. Hemaspaandra, E., Hemaspaandra, L.A., Rothe, J.: Exact analysis of Dodgson elections: Lewis Carroll's 1876 voting system is complete for parallel access to NP. J. ACM 44(6), 806–825 (1997)
3. Hopcroft, J.E., Ullman, J.D.: Introduction to Automata Theory, Languages and Computation. Addison-Wesley (1979)
4. Manea, F.: Deciding According to the Shortest Computations. Tech. rep., Otto-von-Guericke-Universität Magdeburg, Fakultät für Informatik (2011), http://theo.cs.uni-magdeburg.de/pubs/preprints/pp-afl-2011-02.pdf
5. Manea, F., Margenstern, M., Mitrana, V., Pérez-Jiménez, M.J.: A New Characterization of NP, P, and PSPACE with Accepting Hybrid Networks of Evolutionary Processors. Theory Comput. Syst. 46(2), 174–192 (2010)
6. Papadimitriou, C.M.: Computational complexity. Addison-Wesley (1994)
7. Pérez-Jiménez, M.J.: A Computational Complexity Theory in Membrane Computing. In: W. Membrane Computing. LNCS, vol. 5957, pp. 125–148. Springer (2009)
8. Wagner, K.W.: More Complicated Questions About Maxima and Minima, and Some Closures of NP. Theor. Comput. Sci. 51, 53–80 (1987)
9. Wagner, K.W.: Bounded Query Classes. SIAM J. Comput. 19(5), 833–846 (1990)

Computation of Similarity—Similarity Search as Computation

Stoyan Mihov[1] and Klaus U. Schulz[2,*]

[1] Institute for Parallel Processing, Bulgarian Academy of Sciences, 25A,
Acad. G. Bonchev St, 1113 Sofia, Bulgaria
[2] Centrum für Informations- und Sprachverarbeitung,
Ludwig-Maximilians-Universität München, Oettingenstr. 67, 80538 München,
Germany

Abstract. We present a number of applications in Natural Language Processing where the main computation consists of a similarity search for an input pattern in a large database. Afterwards we describe some efficient methods and algorithms for solving this computational challenge. We discuss the view of the similarity search as a special kind of computation, which is remarkably common in applications of Computational Linguistics.

1 Introduction and Road Map

In current reviews on work in Computational Linguistics and its application areas two general paradigms are often compared. *Rule-based approaches* collect linguistic knowledge in hand-crafted rule sets and manually created resources [4]. *Statistical approaches* on the other hand rely on tagged training data and apply Machine Leaning techniques to address classification, labeling of data, and related problems [1,7]. Since both approaches come with characteristic strengths and weaknesses, it is important to look at suitable combinations of both lines. In this paper we address one particular form of such a combination, which from our point of view should be considered as a computational paradigm of its own. We use "similarity computation" as a preliminary name.

Looking at methods and resources, similarity computation is based on rules plus probabilities (or some form of costs) and a set of correct objects. Problems to be solved are typically described in terms of some form of similarity search. A given input object, the pattern, is transformed into correct objects using the rules while minimizing costs. Concrete instances of similarity computation can be found in many areas of Natural Language Processing.

- **Statistical machine translation.** The problem is to find the sequence of words in the target language, which is most similar to a sentence in the source language in respect to a translation model and a language model.

---

* Funded by EU FP7 Project IMPACT (IMProving ACcess to Text).

B. Löwe et al. (Eds.): CiE 2011, LNCS 6735, pp. 201–210, 2011.

- **Speech recognition.** The problem is to find a sequence of words where the pronunciation matches a given speech audio signal. The match is in respect to the acoustic similarity and a given language model.
- **Text correction.** A kernel problem in text correction is the detection and correction of erroneous words based on a background lexicon. For a given erroneous word, the computation of suitable correction candidates in the lexicon is a form of similarity computation.
- **Interpreting historical texts.** Due to missing normalization of language, words in historical texts come in many distinct spelling variants. In the context of Information Retrieval, non-expert users ask queries in modern spelling. Hence, in order to properly access historical documents we need to find modern pendants for words in historical spelling. In practice, solutions to this problem represent some special forms of similarity computation.

From a bird's eye perspective, most research on similarity computation is centered around two orthogonal questions. The **first core problem** is to find an adaquate and optimal notion of similarity for a given application. The formal notion of similarity used in the search - in terms of rules and costs - should coincide with linguistic intuition and the specific requirements of each application. The **second core problem** is to realize some form of efficient search for similar objects in large databases for a given notion of similarity. In the long run, the **synthesis of the two directions** can be considered as the ultimate goal. Relevant research questions as to a synthesis are, e.g.:

- How to combine high efficiency of the search with appropriateness of notions of similarity? How to adapt a known efficient algorithm for some given similarity measure to more general or other similarity measures?
- Self (and online) adaption: in a given (online) application, use intermediate results obtained from fast similarity computation to automatically improve rules and/or costs.

For all these lines of research, concrete examples will be given below. Note that the above form of self adaption is directed towards finding the right notion of similarity, the first core problem mentioned above. An interesting question to be raised on this context is whether self-adaption can also be used to improve efficiency of search (second core problem).

In the remainder of the paper, we first give concrete examples for the orientation scheme described above. In Section 2 we describe the aforementioned four instances of similarity computation in Natural Language Processing in more detail. In each case we also comment on adequate notions of similarity for the given application (first core problem). In Section 3 we describe efficient methods for similarity search, illuminating work on the second core problem. Section 4 is devoted to the synthesis problem, in the sense described above. In Section 5 we give experimental results which show that similarity computation can be surprisingly fast. In the Conclusion (Section 6) we add some general considerations, looking at similarity search as a special form of computation.

2   Similarity Search in Natural Language Processing

2.1   Speech Recognition

For solving the speech recognition task the state-of-the-art approach [5] is first to convert the audio signal into a sequence of feature vectors. In a typical implementation [11] this is done by slicing the signal into a sequence of overlapping frames. For each frame a feature vector is produced, which usually consists of the frame's energy and cepstrum coefficients (cf. Figure 1, top left).

The feature vectors are classified using Bayesian classifiers implemented as Gaussian mixtures, which define the probability for a feature vector to belong to a given class (cf. Figure 1, top middle). For each of the phonemes a model is built based on a continuous left-to-right 3-state Hidden Markov Model (HMM) (cf. Figure 1, top right). The training of the Gaussian mixtures and the HMMs is done using variants of the Baum-Welch algorithm [15].

For building the speech recognition model (Figure 1 bottom right) first, a deterministic finite-state automaton recognizing all the phonetized words from the vocabulary is constructed. In the next step the phonemes in the network are substituted with the corresponding acoustic models – 3-state left-to-right HMMs. At the end, bigram-weighted transitions are added from the final states to the starting state. The probabilities of those transitions are estimates of the likelihood of co-occurrence of the two consecutive words in a sequence (cf. Figure 1, bottom left).

The recognition process is reduced to finding the most probable path in the recognition model for the given sequence of feature vectors.

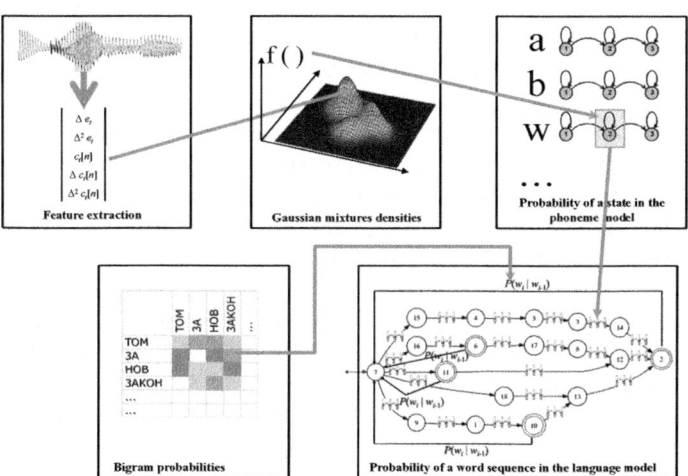

Fig. 1. General scheme of the speech recognition process

2.2   Statistical Machine Translation

The main idea behind statistical machine translation is first to derive a translation model and a language model from large corpora.

Typically [13] the language model is a bigram or trigram Markov model trained from large monolingual corpus. The language model is used to represent the language relevancy of a sequence of words in the target language.

The translation model has to evaluate the probability that a given source sentence is a translation of a sequence of words or phrases in the target language. The translation model is built by estimating the alignment probabilities of individual pairs of words or phrases from an aligned bilingual corpus. Those alignment probabilities are estimated using the expectation maximization algorithm.

The translation process is reduced to the search of a sequence of words or phrases in the target language, which is most probable in respect to the translation and the language model. A full search would be exponential in the length of the source sentence. Because of that in practice a beam search variant is used.

2.3   Text Correction

Text correction algorithms look at the tokens of a given text and apply some form of error detection and error correction [6]. Typically, the detection of errors is based on a large background lexicon for the given language. Tokens V that do not appear in the lexicon are marked as possible errors. For error correction, entries W of the lexicon similar to V are selected.

For text correction, adequate notions of similarity are refinements of the Levenshtein distance. The (standard) *Levenshtein distance* between two words V, W is the minimal number of (standard) edit operations needed to transform V into W [8]. Standard edit operations are the deletion of a symbol, the insertion of a symbol, and the substitution of a symbols by another symbol. Optimal refined notions of similarity depend on the application subarea.

Spell checking. Spell checkers for typed texts have to be aware of (mainly) two types of errors. *Cognitive errors* are caused by missing competence of the writer. In the literature, there are many lists on typical cognitive spelling errors. *Typing errors* are a result of missing care when typing texts. Typical examples are transpositons of letters, substitutions of letters corresponding to neighboured keys etc.

OCR correction. Typical OCR errors depend both on the OCR engine and the given input document (font, paper quality,..). Merges and splits of symbols should be used as additional edit operations.

In both cases, optimal refinements of the Levenshtein distance use new edit operations which come with specific costs depending on the symbols used.

2.4   Interpreting Historical Texts

When going back to early history, historical texts can only be interpreted by experts. Texts from the younger history (printing period) in most cases contain words which are similar to modern pendants. In German historical texts, the letter t of a modern word is often written th in the old spelling, which points to a patterns $t \mapsto th$. For machine interpretation of historical texts (e.g., indexing), we need to find the modern pendants for historical spellings [2,3]. We can use a similarity search in a modern lexicon, where the notion of similarity is based on the given set of patterns and their likelihood. The problem of how to find an optimal notion of similarity is difficult: particularities of the spelling variants found in input texts depends on the temporal period, the geographic place and even the author.

3 Approximate Search Methods

3.1 Approximate Search in Large Lexica

We face a scenario where for an input pattern (string) P, distance measure d and similarity bound b we want to retrieve all lexicon words W where the distance $d(P, W)$ does not exceed b.

Lexicon representation and search space. For the design of efficient search procedures, the internal representation of lexica is important. The following methods assume that the lexicon is implemented as a deterministic finite-state automaton $\mathcal{A}_L$. The automaton may be used to enumerate all words in the lexicon in a systematic way, using a complete traversal of the automaton. During this traversal, common prefixes of lexicon words are generated only once. At each point of the computation we face a unique prefix. We extend the prefix to look for further correction suggestions. In the automaton picture we choose a transition from the current state reached. Backtracking always returns to the last state visited where at least one transition has not been considered so far. The traversal of the automaton defines a huge search space. The challenge is to prune this search space by recognizing situations as soon as possible where the current prefix/path cannot lead to a word W satisfying the given similarity bound. In what follows, by a *filter mechanism* we mean a method which, given the pattern P and a prefix U_1 of a lexicon word, decides (i) if there exists a string U_2 such that $d(P, U_1 U_2) \leq b$, and (ii) if $d(P, U_1) \leq b$. Note that such a filter may be used to prune a maximal part of the aforementioned search space.

Dynamic programming. A first filter mechanism was described in [14], where also the above idea of a filtered traversal of the lexicon automaton was introduced. Oflazer uses the dynamic programming matrix, well-known as a standard procedure for computing the Levenshtein distance [20]. In what follows, if not mentioned otherwise, d denotes the Levenshtein distance. In Oflazer's approach, lines of the matrix correspond to the prefixes of the pattern P. Columns correspond to the prefixes of the current prefix visited during the lexicon traversal. Each cell $M[i, j]$ of the matrix stores the Levenshtein distance between the prefixes of length i and j of the pattern resp. lexicon prefix. Each transition step in the lexicon adds a new column to the matrix. Obviously, we may stop to extend the current path once all entries of the last column are larger than b.

When reaching a final state of the lexicon automaton, the bottom cell of the last column shows if we have found a word W such that $d(P, W) \leq b$.

Universal Levenshtein automata.   As a second, more efficient filter the authors [19,9] introduced the concept of a universal Levenshtein automaton. The automaton for bound b does not depend on the pattern P. The symbols σ consumed during lexicon traversal are translated into bitvectors that represent the distribution of the symbol σ in characteristic substrings of P. The bitvectors obtained represent the input of the automaton. The automaton for bound b fails (no transition possible) as soon as the sequence $U_1 = \sigma_1 \cdots \sigma_k$ (which is translated into bitvectors) cannot be extended to a string $U_1 U_2$ such that $d(P, U_1 U_2) \leq b$. The automaton accepts the translation of the symbols of U_1 (reaching a final state) iff $d(P, U_1) \leq b$. Note that the universal Levenshtein automaton for a given bound b needs to be computed only once in an offline step. The automaton then can be used for any patterns P.

Forward-backward method.   A remaining point for possible optimizations of the above solutions is the so-called "wall effect": if we tolerate b errors and start searching in the lexicon from left to right, then in the first b steps we have to consider *all* prefixes of lexicon words although only a tiny fraction of these prefixes will lead to a useful lexicon word. The forward-backward method [9] is one approach for overcoming the wall effect. Two subcases are distinguished: (i) most of the discrepancies between the pattern and the lexicon word are in the first half of the strings; and (ii) most of the discrepancies are in the second half. We apply two subsearches. For subsearch (i) during traversal of the first half of the pattern we tolerate at most $b/2$ errors. In this way, the inital search space is pruned. Then search proceeds by tolerating up to b errors. At this point, the branching of search paths is usually low. For subsearch (ii) the traversal is performed on the reversed automaton and the reversed pattern in a similar way – in the first half starting from the back only $b/2$ errors are allowed, afterwards the traversal to the beginning tolerates b errors. In [9] it has been shown that the performance gain compared to the classical solution is enormous.

3.2   Viterbi Search in Huge HMMs

As discussed in the previous section, the speech recognition process is reduced to finding the most probable path in a huge HMM for the given sequence of feature vectors derived from the audio signal.

The computational complexity of the naive approach, which considers all possible paths, would be $O(M^T)$, where M is the number of states in the language model HMM and T is the length of the sequence of feature vectors. The Viterbi algorithm [15] can be used for reducing the complexity to $O(M^2 T)$. The Viterbi algorithm is a dynamic programming algorithm, which assumes that the most likely path up to a certain moment t only depends on the emission at the moment t and the most likely path at the moment $t - 1$.

In practice M is in the order of millions and T is in the order of a thousand. Therefore a full Viterby search can hardly be performed in real time. To reduce the computational complexity a common approach is to employ beam search. In the Viterby algorithm at each step only a limited number of paths - the best scoring paths - are considered for continuation.

4 Synthesis

Adapting Levenshtein Automata and Forward-Backword Method to more general distance measures. The forward-backward method (cf. Section 3.1) based on universal Levenstein automata as a control mechanism for search was initially introduced for the standard Levenshtein distance. Already in [9] generalizations have been described for variants of the Levenshtein distance where transpositions are used as additional edit operations.

As a next level of generalization, the situation has been considered where edit operations (e.g. substitutions) are restricted to certain combinations of characters. This again needs new modifications of the universal automata used for control [18]. Interestingly, we may use the same automaton for two given distances as long as they use the same general type of edit operations (e.g., deletions, insertions, and substitutions), but for distinct combinations of symbols. Is is only necessary to adapt the translation of symbols into bitvectors.

Recently, in [10] the concept of a universal Levenshtein automaton has been further generalized to a very general family of distances between strings that includes all the above cases as special examples. Necessary and sufficient conditions for the existence of "universal neighbourhood automata" are given.

Online adaption of similarity measures. When we look at the application areas described in Section 2 we see that for each application domain each single application scenario typically comes with its own best notion of similarity. This is easily seen, e.g., for OCR correction, where the most natural similarity measure always takes the typical OCR errors into account. In the absence of good training data we need to analyze the OCR output using fast correction technology. As a result we can make a hypothesis on the errors that occurred and use this statistics to fine-tune the costs for specific edit operations and combinations of symbols. Using then this information, a second run of the correction system may lead to improved results. In [18] we showed that the method may be used to improve the selection of correction candidates from the lexicon for OCR errors.

Similar principles for adapting statistical parameters of an initial model are often used in speech recognition. The Baum-Welch algorithm [15], as a realization of the Expectation Maximization principle [16], starts with a hypothesis on paths visited in a Hidden Markov Model for a given input sequence. This hypothesis is used for a reestimation of transition and emission probabilities, which results in an improved model.

We believe that the general idea behind the Expectation Maximization principle can be applied in most situations where we consider some form of similarity computation. The central question is how to obtain a reliable hypothesis on the rules that occurred for the input data. If we succeed in deriving a realistic estimate for the number each rule has been used, an improved model is obtained.

Note that for the parameter re-estimation it is not essential that we correctly predict the treatment (application of rules/transitions) for each *single part* of the input data (local analysis). In fact, as long as the *general statistics* derived for all input data (global analysis) is a good approximation, an improved model is obtained which then gives a better picture of local phenomena.

5   Some Experimental Results

We have done empirical experiments using the Levenshtein distance to extract all the words from a wordlist within a given distance to a pattern. Three different methods are compared: (i) the "forward method" described in [14]; (ii) the forward-backward method [9]; and (iii) the "ideal" method, which explicitly maps each possible garbled word to the set of all close words from the dictionary. Theoretically the ideal method can be implemented as a trie (a finte-state automaton with a tree-like structure) for all possible garbled words with the explicit set of the neighbor dictionary words directly linked to the trie's leaves. In practice however, this simple construction leads to astronomical memory requirements. In order to have the comparison of each of the other methods against the theoretically most efficient method, we have constructed the trie for the garbled words only, for which we have performed the test.

Table 1 presents the time needed for retrieving all dictionary words within a given distance to a pattern. The dictionary used is a Bulgarian wordform wordlist containing half a million entries with an average length of 10 characters. The times given are in milliseconds measured on a 2.66 GHz Intel Core 2 machine running Linux. Table 2 presents the results of the experiments on a 1.2 million entries database of bibliographic entries with an average length of 50 characters. Both tables show that the Forward-backward algorithm dramatically increases the search efficiency by reducing the wall effect.

As discussed in Section 3 a full Viterbi search is practically impossible for HMMs with millions of states. Therefore the so-called beam-search is used in practice. The beam-search explores the HMM network by examining the most

Table 1. Experimental results on a 0.5 M words Bulgarian dictionary

Distance	2	3	4	5
Forward	3.958	21.21	—	—
Forward-Backward	0.412	1.145	4.723	10.992
Ideal	0.002	0.003	0.003	0.003

Table 2. Experimental results on a 1.2 M entries bibliography

Distance	2	3	4	5
Forward	17.059	89.398	—	—
Forward-Backward	1.478	3.063	18.432	38.51
Ideal	0.009	0.005	0.006	0.008

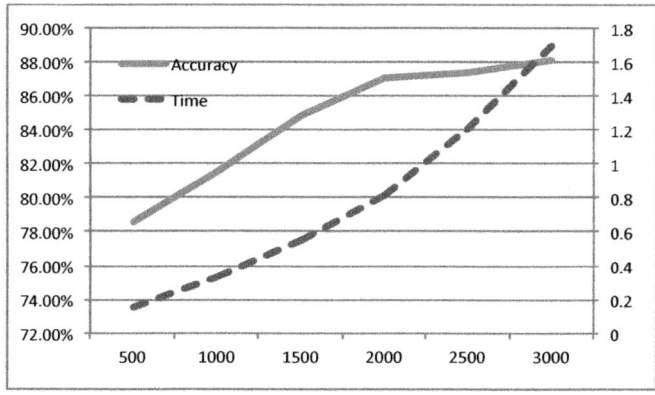

Fig. 2. Beam search accuracy and speed [11]

promising states only. On each step the number of active states is reduced to the given limit by eliminating the states with the lowest probabilities so far. In this way the time complexity is $O(L^2T)$, where L is the limit for the active states and T is the length of the sequence of feature vectors.

On Figure 2 the accuracy of the beam search in respect to the active states (on the x axis) is shown with a solid line. The corresponding time required is given with a dashed line. On the right axis the ratio of the processing time requirement with respect to real time is given. The figure shows that after a certain point increasing the number of active states does not significantly increase the accuracy of the recognition process.

6   Conclusion

We illuminated concrete instances of and challenges for "similarity computation", which is considered as a kind of programming paradigm. Similarity computation combines rules and statistics. A characteristic feature is the interdependence between two problems: first, in each application scenario we need to find the optimal notion of similarity - a kind of optimal balance. Second, on this basis, given an input pattern P the search for similar objects in the background resource of "correct objects" should be very fast. A successful search should come with insights on the rules used, the form of the derivation. In this way, efficient search procedures support the task to adapt the notion of similarity to a given input problem. Similarity computation shares some ideas with optimization theory. In both cases we consider input problems, a notion of "correct solution" and costs. However, in typical applications of optimization theory, the cost function (and the set of rules) is stable. There is no need to find an "optimal balance".

References

1. Burges, C.J.: A tutorial on support vector machines for pattern recognition. Data Mining and Knowledge Discovery 2(2), 121–167 (1998)
2. Ernst-Gerlach, A., Fuhr, N.: Generating search term variants for text collections with historic spellings. In: Lalmas, M., MacFarlane, A., Rüger, S.M., Tombros, A., Tsikrika, T., Yavlinsky, A. (eds.) ECIR 2006. LNCS, vol. 3936, pp. 49–60. Springer, Heidelberg (2006)
3. Gotscharek, A., Neumann, A., Reffle, U., Ringlstetter, C., Schulz, K.U.: Enabling information retrieval on historical document collections: the role of matching procedures and special lexica. In: AND 2009: Proceedings of The Third Workshop on Analytics for Noisy Unstructured Text Data, pp. 69–76. ACM, New York (2009), doi:10.1145/1568296.1568309
4. Gross, M.: The Construction of Local Grammars. In: Finite-State Language Processing, pp. 329–352. The MIT Press, Cambridge (1997)
5. Huang, X., Acero, A., Hon, H.W.: Spoken Language Processing. Prentice Hall, New Jersey (2001)
6. Kukich, K.: Techniques for automatically correcting words in texts. ACM Computing Surveys, 377–439 (1992)
7. Lafferty, J., McCallum, A., Pereira, F.: Conditional random fields: Probabilistic models for segmenting and labeling sequence data. In: Proc. 18th International Conf. on Machine Learning, pp. 282–289. Morgan Kaufmann, San Francisco (2001)
8. Levenshtein, V.: Binary codes capable of correcting deletions, insertions, and reversals. Sov. Phys. Dokl. (1966)
9. Mihov, S., Schulz, K.U.: Fast approximate search in large dictionaries. Computational Linguistics 30(4), 451–477 (2004)
10. Mitankin, P., Mihov, S., Schulz, K.U.: Deciding word neighborhood with universal neighborhood automata. Theoretical Computer Science (in Press)
11. Mitankin, P., Mihov, S., Tinchev, T.: Large vocabulary continuous speech recognition for Bulgarian. In: Proceedings of the RANLP 2009 (2009)
12. Ney, H., Essen, U., Kneser, R.: On structuring probabilistic dependencies in stochastic language modelling. Computer Speech & Language 8, 1–38 (1994)
13. Och, F.J., Ney, H.: A systematic comparison of various statistical alignment models. Computational Linguistics 29(1), 19–51 (2003)
14. Oflazer, K.: Error-tolerant finite-state recognition with applications to morphological analysis and spelling correction. Computational Linguistics 22(1), 73–89 (1996)
15. Rabiner, L.: A tutorial on hidden Markov models and selected applications in speech recognition. Proc. of IEEE 77 (1989)
16. Redner, R., Walker, H.: Mixture densities, maximum likelihood and the EM algorithm. SIAM Review 26(2), 195–239 (1984)
17. Ringlstetter, C., Schulz, K.U., Mihov, S.: Adaptive text correction with webcrawled domain-dependent dictionaries. ACM Trans. Speech Lang. Process. 4(4), 9 (2007)
18. Schulz, K., Mihov, S., Mitankin, P.: Fast selection of small and precise candidate sets from dictionaries for text correction tasks. In: ICDAR 2007: Proceedings of the Ninth International Conference on Document Analysis and Recognition, pp. 471–475. IEEE Computer Society Press, Washington, DC, USA (2007)
19. Schulz, K.U., Mihov, S.: Fast string correction with Levenshtein automata. International Journal of Document Analysis and Recognition 5(1), 67–85 (2002)
20. Wagner, R., Fisher, M.: The string-to-string correction problem. Journal of the ACM (1974)

Adapting Rabin's Theorem for Differential Fields

Russell Miller[1,2,*] and Alexey Ovchinnikov[1,**]

[1] Department of Mathematics, Queens College, City University of New York,
65-30 Kissena Boulevard, Flushing NY 11367, United States of America
Russell.Miller@qc.cuny.edu, Alexey.Ovchinnikov@qc.cuny.edu
[2] Department of Mathematics, CUNY Graduate Center, 365 Fifth Avenue,
New York NY 10016, United States of America

Abstract. Harrington extended the first half of Rabin's Theorem to differential fields, proving that every computable differential field can be viewed as a computably enumerable subfield of a computable presentation of its differential closure. For fields F, the second half of Rabin's Theorem says that this subfield is Turing-equivalent to the set of irreducible polynomials in $F[X]$. We investigate possible extensions of this second half, asking both about the degree of the differential field K within its differential closure and about the degree of the set of constraints for K, which forms the closest analogue to the set of irreducible polynomials.

1 Introduction

Rabin's Theorem is fundamental to the study of computability theory and fields. Proven in [12] in 1960, it gives an effective construction of the algebraic closure of a computable field F around the field itself, and describes the exact conditions necessary for the original field to be computable (as a subfield of the algebraic closure), namely the decidability of the set of reducible polynomials in $F[X]$. As the notion of algebraic closure is essential to modern field theory, the question it addresses is absolutely natural, and it answers that question convincingly.

The practice of closing a field algebraically foreshadowed the construction of the differential closure of a differential field, and the stream of results which flowed from this notion with the work of Kolchin and many others, beginning in the mid-twentieth century, emphasized its importance. Moreover, starting later in that century, differentially closed fields have become the focus of a great deal of work in model theory. The theory **DCF₀** of differentially closed fields of

* The corresponding author was partially supported by Grant # DMS-1001306 from the National Science Foundation and by grants numbered 62632-00 40 and 63286-00 41 from The City University of New York PSC-CUNY Research Award Program.
** The second author was partially supported by Grant # CCF-0952591 from the National Science Foundation and by Grant # 60001-40 41 from The City University of New York PSC-CUNY Research Award Program. Both authors wish to acknowledge useful conversations with Dave Marker and Michael Singer.

B. Löwe et al. (Eds.): CiE 2011, LNCS 6735, pp. 211–220, 2011.

characteristic 0 is in many ways even more interesting to model theorists than the corresponding theory $\mathbf{ACF_0}$ for algebraically closed fields: both are ω-stable, but $\mathbf{ACF_0}$ is strongly minmal, whereas $\mathbf{DCF_0}$ has infinite Morley rank. Today differentially closed fields are widely studied by both algebraists and logicians.

Therefore it is natural to attempt to replicate Rabin's Theorem in the context of computable differential fields and their differential closures. Harrington took a significant step in this direction in [5] in 1974, proving that every computable differential field does indeed have a computable differential closure, and can be enumerated inside that closure, just as Rabin showed can be done for a computable field inside its algebraic closure. All these results, and the terms used in them, are defined fully in the next section. However, Harrington's theorem mirrors only the first half of Rabin's Theorem: it remains to determine what conditions would guarantee – or better yet, would be equivalent to – decidability of the original differential field within its differential closure. With this abstract we begin these efforts, giving the current state of knowledge in Sections 4 and 5. Sections 2 and 3 sketch most of the necessary background. For further questions we suggest [14] for general background in computability, [4] for computable model theory, [2] for field arithmetic, [8,11,10] for introductions to computable fields, and [1,7] for differential fields in the context of model theory.

2   Computable Differential Fields

Differential fields are a generalization of fields, in which the field elements are often viewed as functions. The elements are not treated as functions, but the differential operator(s) on them are modeled on the usual notion of differentiation of functions.

Definition 1. A *differential field* is a field K with one or more additional unary functions δ_i satisfying the following two axioms for all $x, y \in K$:

$$\delta_i(x + y) = \delta_i x + \delta_i y \qquad \delta_i(x \cdot y) = (x \cdot \delta_i y) + (y \cdot \delta_i x).$$

The *constants* of K are those x such that, for all i, $\delta_i x = 0$. They form a differential subfield C_K of K.

So every field can be made into a differential field by adjoining the zero operator $\delta x = 0$. For a more common example, consider the field $F(X_1, \ldots, X_n)$ of rational functions over a field F, with the partial derivative operators $\delta_i = \frac{\partial}{\partial X_i}$. We will be concerned only with *ordinary differential fields*, i.e. those with a single differential operator δ.

The next definitions arise from the standard notion of a computable structure. To avoid confusion, we use the domain $\{x_0, x_1, \ldots\}$ in place of ω.

Definition 2. A *computable field* F consists of a set $\{x_i : i \in I\}$, where I is an initial segment of ω, such that these elements form a field with the operations given by Turing-computable functions f and g:

$$x_i + x_j = x_{f(i,j)} \qquad x_i \cdot x_j = x_{g(i,j)}.$$

A *computable differential field* is a computable field with one or more differential operators δ as in Definition 1, each of which is likewise given by some Turing-computable function h with $\delta(x_i) = x_{h(i)}$.

Fröhlich and Shepherdson were the first to consider computable algebraically closed fields, in [3]. However, the definitive result on the effectiveness of algebraic closure is Rabin's Theorem. To state it, we need the natural notions of the root set and the splitting set.

Definition 3. Let F be any computable field. The *root set* R_F of F is the set of all polynomials in $F[X]$ having roots in F, and the *splitting set* S_F is the set of all polynomials in $F[X]$ which are reducible there. That is,

$$R_F = \{p(X) \in F[X] : (\exists a \in F) \; p(a) = 0\}$$
$$S_F = \{p(X) \in F[X] : (\exists \text{ nonconstant } p_0, p_1 \in F[X]) \; p = p_0 \cdot p_1\}.$$

Both these sets are computably enumerable. They are computable whenever F is isomorphic to a prime field $\mathbb{Q}$ or $\mathbb{F}_p$, or to any finitely generated extension of these. At the other extreme, if F is algebraically closed, then clearly both R_F and S_F are computable. However, there are many computable fields F for which neither R_F nor S_F is computable; see the expository article [8, Lemma 7] for a simple example. Fröhlich and Shepherdson showed that R_F is computable iff S_F is, and Rabin's Theorem then related them both to a third natural c.e. set related to F, namely its image inside its algebraic closure. (Rabin's work actually ignored Turing degrees, and focused on S_F rather than R_F, but the theorem stated here follows readily from Rabin's proof.) More recent works [9,15] have compared these three sets under stronger reducibilities, but here, following Rabin, we consider only Turing reducibility, denoted by $\leq_T$, and Turing equivalence $\equiv_T$.

Theorem 4 (Rabin's Theorem, in [12]). *For every computable field F, there exist an algebraically closed computable field E and a computable field homomorphism $g : F \to E$ such that E is algebraic over the image $g(F)$. Moreover, for every embedding g satisfying these conditions, the image $g(F)$ is Turing-equivalent to both the root set R_F and the splitting set S_F of the field F.*

We will refer to any embedding $g : F \to E$ satisfying the conditions from Rabin's Theorem as a *Rabin embedding* of F. Since this implicitly includes the presentation of E (which is required by the conditions to be algebraically closed), a Rabin embedding is essentially a presentation of the algebraic closure of F, with F as a specific, but perhaps undecidable, subfield.

As we shift to consideration of differential fields, we must first consider the analogy between algebraic closures of fields and differential closures of differential fields. The theory **DCF$_0$** of differentially closed fields K of characteristic 0 is a complete theory, and was axiomatized by Blum (see e.g. [1]) using the axioms for differential fields of characteristic 0, along with axioms stating that, for every pair of nonzero differential polynomials $p, q \in K\{Y\}$ with $\text{ord}(p) > \text{ord}(q)$, there exists some $y \in K$ with $p(y) = 0 \neq q(y)$. (The *differential polynomial ring*

$K\{Y\}$ is the ring $K[Y, \delta Y, \delta^2 Y, \ldots]$ of all algebraic polynomials in Y and its derivatives. The *order* of $p \in K\{Y\}$ is the greatest $r \geq 0$ such that $\delta^r Y$ appears nontrivially in $p(Y)$. By convention the zero polynomial has order $-\infty$, and all other constant polynomials have order -1. Blum's axioms therefore include formulas saying that all nonconstant algebraic polynomials $p \in K[Y]$ have roots in $\hat{K}$, by taking $q = 1$.)

For a differential field K with extensions containing elements x_0 and x_1, we will write $x_0 \cong_K x_1$ to denote that $K\langle x_0 \rangle \cong K\langle x_1 \rangle$ via an isomorphism fixing K pointwise and sending x_0 to x_1. This is equivalent to the property that, for all $h \in K\{Y\}$, $h(x_0) = 0$ iff $h(x_1) = 0$; a model theorist would say that x_0 and x_1 realize the same atomic type over K. The same notation $x_0 \cong_F x_1$ could apply to elements of field extensions of a field F, for which the equivalent property would involve only algebraic polynomials $h \in K[Y]$.

Let $K \subseteq L$ be an extension of differential fields. An element $x \in L$ is *constrained over* K if x satisfies some *constraint* over K, as defined here.

Definition 5. Let K be a differential field. A *constraint* for K is a pair (p, q) of monic differential polynomials in $K\{Y\}$ with the properties that $\mathrm{ord}(p) > \mathrm{ord}(q)$, that p is irreducible as a polynomial in $K[Y, \delta Y, \ldots]$, and that for all differential field extensions L_0 and L_1 of K and all $x_i \in L_i$ such that $p(x_i) = 0 \neq q(x_i)$, we have $x_0 \cong_K x_1$. Such elements x_0 and x_1 are said to *satisfy the constraint* (p, q). We denote the complement by

$$T_K = \{(p, q) \in (K\{Y\})^2 : (p, q) \text{ is } not \text{ a constraint}\},$$

and refer to $\overline{T_K}$ as the *constraint set* for K. If T_K is computable, we say that K has a *constraint algorithm*.

The notation T_K is intended to parallel the notation R_F and S_F. (Also, recall that C_K already denotes the constant subfield of K). Definition 5 parallels the definition of the splitting set S_F in function if not in form. For fields F, irreducible polynomials $p(X)$ have exactly the same property: if $p(x_0) = p(x_1) = 0$ (for x_0 and x_1 in any algebraic field extensions of F), then $x_0 \cong_F x_1$ (that is, $F(x_0) \cong F(x_1)$ via an F-isomorphism mapping x_0 to x_1). So T_K is indeed the analogue of S_F: both are Σ_1^0 sets, given that K and F are both computable, and both are the negations of the properties we need to produce isomorphic extensions. To see that T_K is Σ_1^0, note that $(p, q) \in \overline{T_K}$ iff all $x_0, x_1 \in \hat{K}$ and all $h \in K\{Y\}$ satisfy:

$$[p(x_0) = p(x_1) = 0 \ \& \ q(x_0) \neq 0 \neq q(x_1)] \implies (h(x_0) = 0 \iff h(x_1) = 0),$$

the latter condition (over all h) being equivalent to $K\langle x_0 \rangle \cong K\langle x_1 \rangle$.

If $x \in L$ is constrained over K by (p, q), then there exists a differential subfield of L, extending K and containing x, whose transcendence degree as a field extension of K is finite. Indeed, writing $K\langle x \rangle$ for the smallest differential subfield of L containing x and all of K, we see that the transcendence degree of $K\langle x \rangle$ over K is the smallest order of any nonzero element of $K\{Y\}$ with root x. This will be seen below to be exactly the order of p. The elements of L which are constrained over K turn out to form a differential field in their own right. If this subfield is all of L, then L itself is said to be a *constrained extension* of K.

An algebraic closure $\overline{F}$ of a field F is an algebraically closed field which extends F and is algebraic over it. Of course, one soon proves that this field is unique up to isomorphism over F (that is, up to isomorphisms which restrict to the identity on the common subfield F). On the other hand, each F has many algebraically closed extensions; the algebraic closure is just the smallest of them. Likewise, each differential field K has many differentially closed field extensions; a *differential closure* of K is such an extension which is constrained over K. As with fields, the differential closure of K turns out to be unique up to isomorphism over K, although the proof for differential fields is significantly more difficult and was first accomplished by Shelah in [13] using the notion of ω-stability. On the other hand, the differential closure $\hat{K}$ of K is generally not *minimal*: there exist differential field embeddings of $\hat{K}$ into itself over K whose images are proper differential subfields of $\hat{K}$. This provides a first contrast between $\mathbf{DCF_0}$ and $\mathbf{ACF_0}$, since the corresponding statement about algebraic closures is false.

With this much information in hand, we can now state the parallel to the first half of Theorem 4.

Theorem 6 (Harrington [5], Corollary 3). *For every computable differential field K, there exists a differentially closed computable differential field L and a computable differential field homomorphism $g : K \to L$ such that L is constrained over the image $g(K)$.*

For the sake of uniform terminology, we continue to refer to a computable function g as in Theorem 6 as a *Rabin embedding* for the differential field K.

Harrington actually proves the existence of a computable structure L which is the prime model of the theory T generated by $\mathbf{DCF_0}$ and the atomic diagram of K. Thus L is a computable structure in the language $\mathcal{L}'$ in which the language of differential fields is augmented by constants for each element of K. The embedding of K into L is accomplished by finding, for any given $x \in K$, the unique element $y \in L$ that is equal to the constant symbol for x. Clearly this process is computable, since L is a computable $\mathcal{L}'$-structure, and so we have our embedding of K into L. Since L is the prime model of T, it must be constrained over K: otherwise it could not embed into the constrained closure, which is another model of T. So L satisfies the definition of the differential closure of K, modulo the computable embedding. The same holds in positive characteristic.

The root set and splitting set of a differential field K are still defined, of course, just as for any other field. However, with the differential operator δ now in the language, several other sets can be defined along the same lines and are of potential use as we attempt to adapt Rabin's Theorem. The most important of these is the constraint set, which is analogous in several ways to the splitting set and will be discussed in the next section.

We will also need a version of the Theorem of the Primitive Element for differential fields. This was provided long ago by Kolchin.

Theorem 7 (Kolchin; see [6], p. 728). *Assume that an ordinary differential field F contains an element x with $\delta x \neq 0$. If E is a differential subfield of* *the differential closure $\overline{F}$ and E is generated (as a differential field over F) by finitely many elements, then there is a single element of E which generates all of E as a differential field over F.*   □

Kolchin gave counterexamples in the case where δ is the zero derivation on F, and also extended this theorem to partial differential fields with m derivations: the generalized condition there is the existence of m elements whose Jacobian is nonzero.

3   Constraints

Proposition 8. *Let K be a differential field. Then for each $x \in \hat{K}$, there is exactly one $p \in K\{Y\}$ such that x satisfies a constraint of the form $(p, q) \in \overline{T_K}$. Moreover, $\mathrm{ord}(p)$ is least among the orders of all nonzero differential polynomials in the radical differential ideal $I_K(x)$ of x within $K\{Y\}$:*

$$I_K(x) = \{p \in K\{Y\} : p(x) = 0\},$$

and $\deg(p)$ is the least degree of $\delta^{\mathrm{ord}(p)}Y$ in any polynomial in $K\{Y\}$ of order $\mathrm{ord}(p)$ with root x.

Proof. Since $\hat{K}$ is constrained over K, each $x \in \hat{K}$ satisfies at least one constraint $(p, q) \in \overline{T_K}$. Set $r = \mathrm{ord}(p)$, and suppose there were a nonzero $\tilde{p}(Y) \in I_K(x)$ with $\mathrm{ord}(\tilde{p}) < r$. By Blum's axioms for **DCF$_0$**, there would exist $y \in \hat{K}$ with $p(y) = 0 \neq q(y) \cdot \tilde{p}(y)$, since the product $(q \cdot \tilde{p})$ has order $< r$. But then y also satisfies (p, q), yet $\tilde{p}(y) \neq 0 = \tilde{p}(x)$, so that $K\langle x\rangle \not\cong K\langle y\rangle$. This would contradict Definition 5. Hence r is the least order of any nonzero differential polynomial with root x.

It follows from minimality of r that $\{x, \delta x, \ldots, \delta^{r-1}x\}$ is algebraically independent over K. The polynomial $p(x, \delta x, \ldots, \delta^{r-1}x, Y)$ must then be the minimal polynomial of $\delta^r x$ over the field $K(x, \ldots, \delta^{r-1}x)$, since $p(Y, \delta Y, \ldots, \delta^r Y)$ is irreducible in $K[Y, \delta Y, \ldots, \delta^r Y]$. This implies the claim in Proposition 8 about $\deg(p)$, and also shows that every constraint satisfied by x must have first component p.   □

In fact, the irreducibility of $p(Y)$ is barely necessary in Definition 5. The condition that $K\langle x\rangle \cong K\langle y\rangle$ for all x, y satisfying the constraint shows that $p(Y)$ cannot factor as the product of two distinct differential polynomials. The only reason for requiring irreducibility of p is to rule out the possibility that p is a perfect square, cube, etc. in $K\{Y\}$. If these were allowed, the uniqueness in Proposition 8 would no longer hold.

It is quickly seen that if $(p, q) \in \overline{T_K}$, then also $(p, q \cdot h) \in \overline{T_K}$ for every $h \in K\{Y\}$. So the constraint satisfied by an $x \in \hat{K}$ is not unique. However, Proposition 8 does show that the following definition makes sense.

Definition 9. If $K \subseteq L$ is an extension of differential fields, then for each $x \in L$, we define $\mathrm{ord}_K(x) = \mathrm{ord}(p)$, where (p, q) is the constraint in $\overline{T_K}$ satisfied by x. If no such constraint exists, then $\mathrm{ord}_K(x) = \infty$. Notice that in the constrained closure of K, every element has finite order over K.

4   Decidability in the Constraint Set

We now present our principal result thus far on constraint sets and Rabin embeddings for differential fields. Of course, we hope to establish more results, and perhaps the converse, in the near future.

Theorem 10. *Let K be any computable differential field with a single nonzero derivation δ, and $g : K \rightarrow \hat{K}$ a Rabin embedding of K. Then all of the following are computable in an oracle for the constraint set $\overline{T_K}$: the Rabin image $g(K)$, the set A of finite subsets of $\hat{K}$ algebraically independent over $g(K)$, and the order function ord_K on $\hat{K}$.*

Proof. First we show that ord_K is computable from a $\overline{T_K}$-oracle. Given any $x \in \hat{K}$, we use the oracle to find some $(p, q) \in \overline{T_K}$ satisfied by x. Since $\hat{K}$ is the constrained closure of K, such a constraint must exist, and when we find it, we know by Proposition 8 that $\mathrm{ord}_K(x) = \mathrm{ord}(p)$.

Next we show that $g(K) \leq_T \overline{T_K}$. Our procedure accepts an arbitrary $x \in \hat{K}$ as input, and searches through $\overline{T_K}$, using its oracle, for a constraint (p, q) with $p^g(x) = 0 \neq q^g(x)$. (Here p^g represents the polynomial in $\hat{K}\{Y\}$ whose coefficients are the images under g of the coefficients of p.) Since $\hat{K}$ is constrained over K, it must eventually find such a constraint (p, q), and it concludes that $x \in g(K)$ iff $\mathrm{ord}(p) = 0$ and $p(Y)$ is linear in Y. Of course, if $p(Y) = Y - b$ (hence is of order 0, when viewed as a differential polynomial), then $x = g(b) \in g(K)$. Conversely, if $x \in g(K)$, then $(Y - g^{-1}(x), \ 1)$ is readily seen to lie in $\overline{T_K}$, since it is satisfied by no element of $\hat{K}$ except x. Proposition 8 shows that our algorithm will find a constraint with first coordinate $(Y - g^{-1}(x))$, hence will conclude correctly that $x \in g(K)$.

It remains to show that $A \leq_T T_K$. Given a finite subset $S = \{b_1, \ldots, b_k\} \subseteq \hat{K}$, we decide whether $S \in A$ as follows. First, we search for a nonzero $h \in g(K)[X_1, \ldots, X_k]$ with $h(b_1, \ldots, b_k) = 0$. If we find such an h, we conclude that $s \notin A$. Simultaneously, we search for a constraint $(p, q) \in \overline{T_K}$ with $r = \mathrm{ord}(p) \geq k$, an $x \in \hat{K}$ satisfying (p, q), and elements $y_{k+1}, \ldots y_r$ of $\hat{K}$ such that:

- each of $x, \delta x, \ldots, \delta^{r-1}x$ is algebraic over the field generated over $g(K)$ by the set $S' = \{b_1, \ldots, b_k, y_{k+1}, \ldots, y_r\}$; and
- each element of S' is algebraic over the subfield $g(K)(x, \delta x, \ldots, \delta^{r-1}x)$.

Of course, being algebraic over a c.e. subfield of $\hat{K}$ is itself a Σ_1^0 property, so all of this involves a large search. If we do find all these items, we conclude that $s \in B$.

Now a polynomial h as described above exists iff S is algebraically dependent over $g(K)$. We must show that S is algebraically independent iff the second alternative holds. The backwards direction is quick: if we find all the required elements, then $\{x, \delta x, \ldots, \delta^{r-1} x\}$ is algebraically independent over $g(K)$ (since x satisfies the constraint (p, q) and $\mathrm{ord}(p) = r$) and algebraic over $g(K)(S')$, yet S' has only r elements itself, hence must be algebraically independent over $g(K)$. In particular, its subset S is algebraically independent, as required. It remains to show the forwards direction.

Now S generates a differential subfield F of $\hat{K}$, and by Theorem 7, since δ is assumed to be nonzero, F is generated as a differential field over $g(K)$ by a single element $x \in F$. Let (p, q) be any constraint satisfied by x. If S is indeed algebraically independent over $g(K)$, then F has finite transcendence degree $\geq k$ over $g(K)$, and since $\{x, \delta x, \ldots, \delta^{\mathrm{ord}(p)-1} x\}$ forms a transcendence basis for F (as a field) over $g(K)$, we know that $\mathrm{ord}(p) \geq k$. Moreover, S must extend to a transcendence basis $S \cup \{y_{k+1}, \ldots, y_{\mathrm{ord}(p)}\}$ of F over $g(K)$. This yields all the elements needed for the second alternative to hold. $\square$

We note that the existence of a nonzero derivation δ was used only in the proof that $A \leq_T T_K$. In particular, the Rabin image $g(K)$ is computable in a T_K-oracle, regardless of the derivation. This means that $(g(K) \cap C_{\hat{K}})$ is a T_K-computable subfield of the constant field $C_{\hat{K}}$, which in turn is a computable subfield of $\hat{K}$. Indeed, the restriction of g to C_K is a Rabin embedding of the computable field C_K into its algebraic closure $C_{\hat{K}}$, in the sense of Theorem 4, the original theorem of Rabin for fields.

Therefore, if C is any computable field without a splitting algorithm, we can set $K = C$ to be a differential field with $C_K = C$ (by using the zero derivation). Theorem 6 gives a Rabin embedding g of this differential field K into a computable presentation of $\hat{K}$. Theorem 4 shows that $g(K) = g(C_K)$ is noncomputable within the computable subfield $C_{\hat{K}}$, and therefore must be noncomputable within $\hat{K}$ itself. Finally, Theorem 10 shows that the constraint set $\overline{T_K}$ of this differential field K was noncomputable.

So there do exist computable differential fields, even with the simplest possible derivation, for which the constraint set is noncomputable. In the opposite direction, it is certainly true that if K itself is already differentially closed, then its constraint set is computable, since the constraints are exactly those (p, q) with $p(Y)$ of order 0 and linear in Y. (Such a pair is satisfied by exactly one $x \in K$, hence by exactly one $x \in \hat{K} = g(K)$, using the identity function as the Rabin embedding g. Thus it trivially satisfies Definition 5.) We do not yet know any examples of computable differential fields which have computable constraint set, yet are not differentially closed. The decidability of the constraint set is a significant open problem for computable differential fields in general. So likewise is the decidability of constrainability: for which $p \in K\{Y\}$ does there exist a q with $(p, q) \in \overline{T_K}$? The comments in the proof of Theorem 10 make it clear that $p(Y) = \delta Y$ is an example of a differential polynomial which is not constrainable.

5 Decidability in the Rabin Image

Rabin's Theorem for fields, stated above as Theorem 4, gave the Turing equivalence of the Rabin image $g(F)$ and the splitting set S_F. Our principal analogue of S_F for differential fields K is the set T_K, and Theorem 10 makes some headway in identifying sets, including $g(K)$ but not only that set, whose join is Turing-equivalent to T_K. It is also natural to ask about Rabin's Theorem from the other side: what set (or what join of sets) must be Turing-equivalent to $g(K)$? We now present one step towards an answer to that question, using the notion of a *linear* differential polynomial in $K\{Y\}$. Recall that "linear" here is used in exactly the sense of field theory: the polynomial has a constant term, and every other term is of the form $a\delta^i Y$, for some $a \in K$ and $i \in \omega$. If the constant term is 0, then the polynomial is *homogeneous* as well, every term having degree 1. The solutions in $\hat{K}$ of a homogeneous linear polynomial $p(Y)$ of order r are well known to form an r-dimensional vector space over the constant field $C_{\hat{K}}$. By additivity, the solutions in $\hat{K}$ to any linear polynomial of order r then form the translation of such a vector space by a single root x of $p(Y)$. Of course, not all of the solutions in $\hat{K}$ need lie in K: the solutions to $p(Y) = 0$ in K (if any exist) form the translation of a vector space over C_K of dimension $\leq r$.

Proposition 11. *In a computable differential field K whose field C_K of constants is algebraically closed, the full linear root set FR_K:*

$$\{linear\ p(Y) \in K\{Y\} : p(Y) = 0\ has\ solution\ space\ in\ K\ of\ dim\ ord(p)\},$$

is computable from an oracle for the image $g(K)$ of K in any computable differential closure $\hat{K}$ of K under any Rabin embedding g. Moreover, the Turing reduction is uniform in indices for $\hat{K}$ and g.

Proof. To begin with, suppose that $x \in \hat{K}$ is a constant. Now x satisfies some constraint $(p, q) \in \overline{T_K}$. Since $\delta x = 0$, $p(Y)$ either is the polynomial δY or else has order 0. In the latter case, x is algebraic over K, and indeed turns out to be algebraic over the constant subfield C_K, hence lies in C_K. But we cannot have $p(Y) = \delta Y$: there is no $q(Y)$ such that $(\delta Y, q) \in \overline{T_K}$. (The infinitely many elements of C_K all are roots of δY, yet all satisfy distinct types over K, so in order for $(\delta Y, q)$ to lie in $\overline{T_K}$, all those elements would have to be roots of $q(Y)$. However, q must be nonzero with order < 1.) So every constant in $\hat{K}$ lies in the image $g(C_K)$. It follows that if a linear $p(Y) \in K\{Y\}$ lies in FR_K, then the entire solution space of $p(Y) = 0$ in $\hat{K}$ is contained within K.

Now, given as input a linear $p \in K\{Y\}$, say of order r and with constant term $z \in K$, write $p_0(Y) = p(Y) - z$, so $p_0(Y)$ is homogeneous. We search through $\hat{K}$ until we find:

- an $x \notin g(K)$ with $p(x) = 0$; or
- an $x \in g(K)$ with $p(x) = 0$ *and* $ord(p)$-many solutions $x_0, \ldots, x_{r-1}$ in $g(K)$ to the homogeneous equation $p_0(Y) = 0$, such that $\{x_0, \ldots, x_{r-1}\}$ is linearly independent over $C_{\hat{K}}$.

Deciding linear independence over $C_{\hat{K}}$ is not difficult: one simply checks whether the Wronskian matrix $(\delta^i x_j)_{i,j<r}$ has determinant 0.

Recall that $C_{\hat{K}} = C_K$. In the former case, therefore, K cannot contain a space of solutions to $p(Y) = 0$ of dimension ord(p). In the latter case, K must have a solution space of dimension ord(p), which is to say, a complete solution space. □

It would be of interest to try to extend this result to the case where C_K need not be algebraically closed, and/or to the situation involving the differential closure of an extension of K by finitely many (or possibly infinitely many) algebraically independent constants.

References

1. Blum, L.: Differentially closed fields: a model theoretic tour. In: Bass, H., Cassidy, P., Kovacic, J. (eds.) Contributions to Algebra. Academic Press, New York (1977)
2. Fried, M.D., Jarden, M.: Field Arithmetic. Springer, Berlin (1986)
3. Frohlich, A., Shepherdson, J.C.: Effective procedures in field theory. Phil. Trans. Royal Soc. London, Series A 950, 407–432 (1956)
4. Harizanov, V.S.: Pure computable model theory. In: Handbook of Recursive Mathematics, vol. 1, pp. 3–114. Elsevier, Amsterdam (1998)
5. Harrington, L.: Recursively presentable prime models. The Journal of Symbolic Logic 39(2), 305–309 (1974)
6. Kolchin, E.R.: Extensions of Differential Fields, I. Annals of Mathematics 43(4), 724–729 (1942)
7. Marker, D.: Model Theory: An Introduction. Springer, Heidelberg (2002)
8. Miller, R.G.: Computable fields and Galois theory. Notices of the American Mathematical Society 55(7), 798–807 (2008)
9. Miller, R.G.: Is it harder to factor a polynomial or to find a root? Transactions of the American Mathematical Society 362(10), 5261–5281 (2010)
10. Miller, R.G.: An Introduction to Computable Model Theory on Groups and Fields. To appear in Groups, Complexity and Cryptology (May 2011)
11. Miller, R.G.: Computability and differential fields: a tutorial. In: Guo, L., Sit, W. (eds.) Differential Algebra and Related Topics: Proceedings of the Second International Workshop, World Scientific, Singapore (2011); ISBN 978-981-283-371-6 (to appear)
12. Rabin, M.: Computable algebra, general theory, and theory of computable fields. Transactions of the American Mathematical Society 95, 341–360 (1960)
13. Shelah, S.: Uniqueness and characterization of prime models over sets for totally transcendental first-order theories. Journal of Symbolic Logic 37, 107–113 (1972)
14. Soare, R.I.: Recursively Enumerable Sets and Degrees. Springer, New York (1987)
15. Steiner, R.M.: Computable fields and weak truth-table reducibility. In: Ferreira, F., Löwe, B., Mayordomo, E., Mendes Gomes, L. (eds.) CiE 2010. LNCS, vol. 6158, pp. 394–405. Springer, Heidelberg (2010)

Quantum Information Channels in Curved Spacetime

Prakash Panangaden

[1] School of Computer Science, McGill University, 3480 University Street,
Montreal QC H3A 2A7, Canada
[2] Oxford University Computing Laboratory, Wolfson Building, Parks Road, Oxford,
OX1 3QD, United Kingdom

Abstract. Quantum field theory in curved spacetime reveals a fundamental ambiguity in the quantization procedure: the notion of vacuum, and hence of particles, is observer dependent. A state that an inertial observer in Minkowski space perceives to be the vacuum will appear to an accelerating observer to be a thermal bath of radiation. The impact of this Davies-Fulling-Unruh noise on quantum communication has been explored in a recent paper by Bradler, Hayden and the author.

I will review the results of that paper. The problem of quantum communication from an inertial sender to an accelerating observer and private communication between two inertial observers in the presence of an accelerating eavesdropper was studied there. In both cases, they were able to establish compact, tractable formulas for the associated communication capacities assuming encodings that allow a single excitation in one of a fixed number of modes per use of the communications channel. Group theoretical ideas play a key role in the calculation.

I close with a discussion of some issues of quantum communication in curved spacetime that have yet to be understood.

1 Quantum Communication in Rindler Spacetime

There are three key ingredients in the formulation of a quantum field theory: the algebra of operators, the decomposition of the field into positive and negative frequency components which determines the Hilbert space of one-particle states and the Fock space and finally the representation of the $*$-algebra of operators on this Fock space. In ordinary flat spacetime one uses Poincaré invariance to define the vacuum. Equivalently, one can use the Fourier transform to determine the positive and negative frequency decomposition and then the vacuum. It is possible to prove that this is the unique Poincaré invariant state. The key point is that the notion of particle is determined by the decomposition into positive and negative frequencies.

A well-known feature of quantum field theory in curved spacetimes is the creation of particles from a vacuum [1], which points to a fundamental ambiguity: the notion of particle is not an absolute one in the absence of Poincaré invariance. Even in flat spacetimes one has the Davies-Fulling-Unruh effect [2,3,4,5] whereby a uniformly accelerating observer in Minkowski space detects a thermal bath

B. Löwe et al. (Eds.): CiE 2011, LNCS 6735, pp. 221–229, 2011.
© Springer-Verlag Berlin Heidelberg 2011

of radiation in a state that an inertial observer perceives as a vacuum. The point here is that an accelerating observer uses a different time coordinate—the parameter along the integral curves of the boost Killing vector field—in order to determine the decomposition into positive and negative frequencies.

In quantum information theory, on the other hand, one typically treats the notion of particle as canonical and concepts like "pure state" and "mixed state" are taken to have absolute meaning. In [6,7] the consequences for quantum information theory of this ambiguity in the definition of vacuum (and particle) states is studied.

Previous work had studied how relativistic effects impact entanglement and quantum communication [8,9,10,11,12,13,14,15,16]. Most such work studied the degradation caused when protocols not designed for relativistic situations are employed in situations where relativistic effects are significant and cause degradation of entanglement. The approach in [6,7] was to design protocols specifically with relativistic effects in mind, in the spirit of [17,18,19]. In particular, instead of just looking at measures of entanglement one has to consider the effect of error correction and quantify this through the notion of channel capacity.

We focus on two scenarios. In the first, an inertial observer, Alice, attempts to send quantum information to an accelerating receiver, Bob, by physically transmitting scalar "photons" of chosen modes. Owing to the thermal noise perceived by the receiver, quantum error correcting codes are required to protect the quantum information.

The second scenario is more elaborate. Two inertial observers—again call them Alice and Bob—communicate by exchanging scalar "photons" of chosen modes, while an accelerating observer—traditionally called Eve—attempts to eavesdrop or wiretap their communication channel. This time, it is Eve who detects thermal noise and therefore cannot perfectly decode the communications between Alice and Bob, thus allowing the possibility of private communication between them. Of course, we are not proposing this as a practical scheme for cryptography but, rather, as an exploration of the impact of relativistic quantum field theory on quantum information theory.

The concept of private capacity in the classical setting is due to Maurer [20] and independently Ahlswede and Csiszar [21]. The private capacity of a quantum channel was first studied by Cai et al. and Devetak [22,23]. These capacities measure the optimal rate at which Alice can transmit classical bits to Bob that remain secret from Eve, in the limit of many uses of the channel. In the present paper we introduce the private *quantum* capacity of a quantum channel, which measures the usefulness of the channel for sending private quantum mechanical data (qubits) instead of bits.

As mentioned above, the standard approach to quantum field theory in flat spacetime is to decompose the field into "positive" and "negative" frequency modes as defined by the Fourier transform. One then defines creation and annihilation operators that correspond to these modes and the vacuum state is defined to be the state killed by all the annihilation operators. The Poincaré invariance of Minkowski spacetimes means that the vacuum state is the unique state that is invariant under the action of the Poincaré group. In Rindler space, it is natural for the accelerating observer to use his or her own timelike Killing field (the boost) to define the notion of positive and negative frequency. This means that there will be a mismatch between Alice's notion of vacuum state and that of the accelerating observer. The transformation between the creation and annihilation operators of the different (and inequivalent) quantum field theories is given by a linear map, called a *Bogoliubov transformation* [1], between the creation and annihilation operators of the two quantum field theories.

The explicit form of the Bogoliubov transformation is well known and we use it to define a *channel* which we call the Unruh channel. In quantum information theory, a channel is simply any physically realizable transformation of a quantum state. The idea is that the process of transmission may introduce noise and loss of information. Thus, an initially pure quantum state may become mixed.

In the Unruh channel, Alice prepares some state in her chosen d-dimensional space encoded in terms of Minkowski modes. An accelerating observer (Bob or Eve depending on the scenario) intercepts this, but using an apparatus that detects excitations of the quantum field defined according to the prescription of the Rindler quantum field theory. So the state that she detects will be described by some infinite-dimensional density matrix. A detailed analysis of this density matrix makes it possible to extract quantitative information about the private and quantum capacities. We evaluate both the quantum capacity from Alice to an accelerating Bob *and* the private capacity for inertial Alice and Bob trying to exchange quantum information while simultaneously confounding an accelerating eavesdropper. Figure 1 contains spacetime diagrams illustrating the two communication scenarios.

Both quantities exhibit surprising behavior. The quantum capacity, the optimal rate at which a sender can transmit qubits to a receiver through some noisy channel, usually exhibits a threshold behavior; channels below some quality threshold have quantum capacity exactly zero. For the Unruh channels, however, we find that the quantum capacity is strictly positive for all accelerations, reaching zero only in the limit of infinite acceleration. It is therefore always possible to transmit quantum data to an accelerating receiver provided the sender is not behind the receiver's horizon. Careful choices of encoding can therefore eliminate the degradation in fidelity known to occur if one uses a naive teleportation protocol to communicate with an accelerating receiver [9] (see also [24]). In addition to characterizing quantum transmission to an accelerating receiver, our analysis applies equally well to the study of quantum data transmission through an optical amplifier, which may well be its more important application.

The private quantum capacity is likewise positive for all nonzero eavesdropper accelerations. Thus, in principle, any eavesdropper acceleration, no matter how small, can be exploited to safeguard transmissions of quantum data between two inertial observers. Curiously, the private quantum capacity has a simple formula when the channel between the inertial observers is noiseless; the formula reveals that in this case the private quantum capacity is exactly equal to

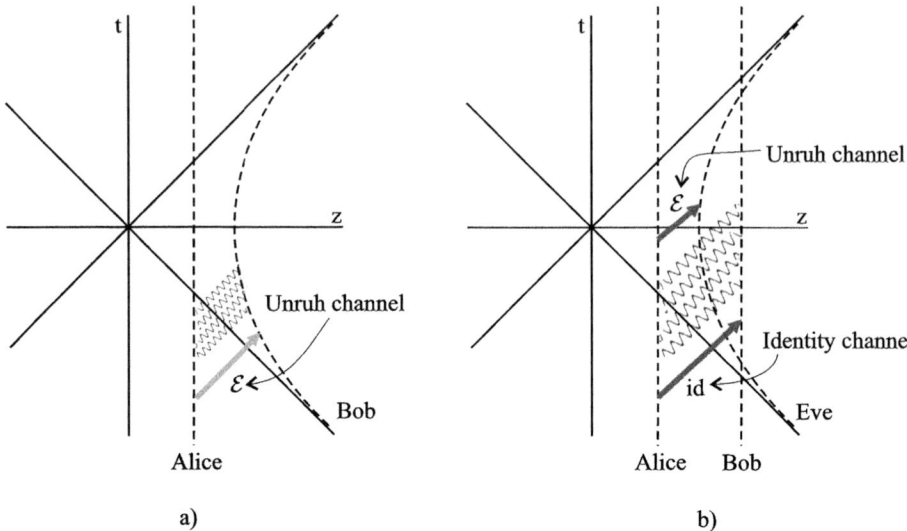

Fig. 1. Spacetime diagrams for the two communication scenarios. (a) Alice is an inertial observer try to send quantum information to the uniformly accelerated Bob. The wavy lines indicate transmission via wave packets and the d-rail qudit encoding. (b) In the second diagram, Alice and the intended receiver, Bob, are both inertial observers. In our idealized scenario, they are assumed to share an noiseless quantum channel. A uniformly accelerated eavesdropper, Eve, attempts to wiretap Alice's message to Bob.

the entanglement-assisted quantum capacity to the eavesdropper's environment, despite the absence of any entanglement assistance in the problem.

2   Quantum Capacity and Private Quantum Capacity

The ability of a quantum channel to transmit quantum information is measured by its quantum capacity, the optimal rate at which qubits can be reliably transmitted in the limit of many uses of the channel and vanishing error. There are many equivalent ways to define the quantum capacity [25]. Here we use a version which focuses on the transmission of halves of maximally entangled states across the noisy channel. Recall that $|\Phi_{2^k}\rangle$ represents the maximally entangled state on k pairs of qubits.

Definition 1. *An* (n, k, δ) *entanglement transmission code from Alice to Bob consists of an encoding channel* $\mathcal{A}$ *taking a k-qubit system R' into the input of $\mathcal{N}^{\otimes n}$ and a decoding channel $\mathcal{B}$ taking the output of $\mathcal{N}^{\otimes n}$ to a k-qubit system $C \cong R'$ satisfying*

$$\left\| (\mathrm{id} \otimes \mathcal{B} \circ \mathcal{N}^{\otimes n} \circ \mathcal{A})(\Phi_{2^k}) - \Phi_{2^k} \right\|_1 \leq \delta. \tag{1}$$

A rate Q is an achievable rate *for entanglement transmission if for all $\delta > 0$ and sufficiently large n there exist $(n, \lfloor nQ \rfloor, \delta)$ entanglement transmission codes. The* quantum capacity $Q(\mathcal{N})$ *is the supremum of all the achievable rates.*

In any capacity problem, the objective is to understand the structure of the optimal codes. Doing so normally results in a theorem characterizing the capacity in terms of simple entropic functions optimized over a single use of the channel, a so-called "single-letter formula." In general, the structure of the optimal codes is still unknown for the quantum capacity problem, however, they can be characterized in the case of qudit Unruh channels.

2.1   Private Quantum Capacity: General Case

The private quantum capacity is the optimal rate at which a sender (Alice) can send qubits to a receiver (Bob) while simultaneously ensuring that those qubits remain encrypted from the eavesdropper's (Eve's) point of view. At first glance, this would not seem to be a very interesting concept. The impossibility of measuring quantum information without disturbing it would seem to ensure that successful transmission of quantum information would make it automatically private. One can imagine a passive eavesdropper, however, who *could* have nontrivial access to the qubits should she choose to exercise it. The setting we will ultimately be primarily concerned with here is a relativistic version of that passive eavesdropper, in particular, the case in which the eavesdropper is uniformly accelerated.

Definition 2. *A* quantum wiretap channel *consists of a pair of quantum channels*
$(\mathcal{N}_{A \to B}, \mathcal{E}_{A \to E})$ *taking the density operators on A to those on B and E, respectively.*

$\mathcal{N}$ should be interpreted as the channel from Alice to Bob and $\mathcal{E}$ the channel from Alice to Eve. Let $U_{\mathcal{N}} : A \to B \otimes B_c$ and $U_{\mathcal{E}} : A \to E \otimes E_c$ be isometric extensions of the channels $\mathcal{N}$ and $\mathcal{E}$. In particular, $\mathcal{N}(\cdot) = \mathrm{Tr}_{B_c} U_{\mathcal{N}} \cdot U_{\mathcal{N}}^{\dagger}$ and $\mathcal{E}(\cdot) = \mathrm{Tr}_{E_c} U_{\mathcal{E}} \cdot U_{\mathcal{E}}^{\dagger}$. In many circumstances, $\mathcal{E}$ will be a degraded version of the "environment" of the Alice-Bob channel, meaning that there exists a channel $\mathcal{D}$ such that $\mathcal{E}(\cdot) = \mathcal{D} \circ \mathrm{Tr}_B U_{\mathcal{N}} \cdot U_{\mathcal{N}}^{\dagger}$. For the uniformly accelerated eavesdropper, however, this needn't be the case so we don't require *a priori* that there be a particular relationship between $\mathcal{N}$ and $\mathcal{E}$.

Recall that $\pi_{2^k} = \mathbb{I}/2^k$, is the maximally mixed state on k qubits.

Definition 3. *An (n, k, δ, ϵ) private entanglement transmission code from Alice to Bob consists of an encoding channel $\mathcal{A}$ taking a k-qubit system R' into the input of $\mathcal{N}^{\otimes n}$ and a decoding channel $\mathcal{B}$ taking the output of $\mathcal{N}^{\otimes n}$ to a k-qubit system $C \cong R'$ satisfying*
 1. Transmission: $\|(\mathrm{id} \otimes \mathcal{B} \circ \mathcal{N}^{\otimes n} \circ \mathcal{A})(\Phi_{2^k}) - \Phi_{2^k}\|_1 \leq \delta.$
 2. Privacy: $\|(\mathrm{id} \otimes \mathcal{E}^{\otimes n} \circ \mathcal{A})(\Phi_{2^k}) - \pi_{2^k} \otimes (\mathcal{E}^{\otimes n} \circ \mathcal{A})(\pi_{2^k})\|_1 \leq \epsilon.$
A rate Q is an achievable rate *for private entanglement transmission if for all $\delta, \epsilon > 0$ and sufficiently large n there exist $(n, \lfloor nQ \rfloor, \delta, \epsilon)$ private entanglement transmission codes. The* private quantum capacity $Q_p(\mathcal{N}, \mathcal{E})$ *is the supremum of all the achievable rates.*

The transmission criterion states that halves of EPR pairs encoded by $\mathcal{A}$, sent through the channel and then decoded by $\mathcal{B}$ will be preserved by the communications system with high fidelity. Alternatively, one could ask that arbitrary pure states or even arbitrary states entangled with a reference sent through $\mathcal{B} \circ \mathcal{N}^{\otimes n} \circ \mathcal{A}$ be preserved with high fidelity. The different definitions are equivalent for the standard quantum capacity $Q(\mathcal{N}) = Q_p(\mathcal{N}, \mathrm{Tr})$, which is defined with no privacy requirement [25]. The equivalence extends straightforwardly to the private quantum capacity.

3   Capacity Results

In the paper [7] we explicitly describe the Unruh channel. One gets an infinite dimensional density matrix which may look intractable at first glance. However, by using the symmetry of the problem and some representation theory results we were able to show that channel transforms nicely and furthermore that it has a special property called conjugate degradability.

The Fock spaces for the incoming and outgoing particles carry representations of $SU(d)$ for the following reason. Since we are dealing with bosons we get completely symmetric states so they are representations of the symmetric group. Because of the well known intertwining of the symmetric group and the groups $SU(d)$ (called Schur-Weyl duality) they are also representations of $SU(d)$. It can be shown that the transformations effected by the Unruh channel mesh properly with the group actions on the state spaces: the qudit Unruh channel is $SU(d)$-*covariant*. We recall the relevant notation and give a formal definition of covariance. Recall that $\mathcal{DM}(\mathcal{H})$ stands for the space of density matrices on the Hilbert space $\mathcal{H}$.

Definition 4. *Let G be a group, $\mathcal{H}_{in}, \mathcal{H}_{out}$ be Hilbert spaces and let $r_1 : G \to GL(\mathcal{H}_{in}), r_2 : G \to GL(\mathcal{H}_{out})$ be two unitary representations of the group G. Let $\mathcal{K} : \mathcal{DM}(\mathcal{H}_{in}) \to \mathcal{DM}(\mathcal{H}_{out})$ be a channel. We say that $\mathcal{K}$ is **covariant with respect to** G and the representations r_1, r_2, if*

$$\mathcal{K}\left(r_1(g)\rho r_1(g)^{\dagger}\right) = r_2(g)\mathcal{K}(\rho)r_2(g)^{\dagger} \tag{2}$$

holds for all $g \in G, \rho \in \mathcal{DM}(\mathcal{H}_{in})$.

Usually covariance means that the property holds for all possible pairs of representations, but in our case we are only interested in the representations that are carried by the input and output spaces of the channel. What covariance really means is that certain equations hold for "Lie algebraic reasons" and not because of some special property of a particular representation.

The following result is established in [7].

Theorem 1. *The qudit Unruh channel is $SU(d)$-covariant for the fundamental representation in the input space and any completely symmetric representation carried by the output space of the channel.*

Using this result one can calculate the channel capacity and the private channel capacity explicitly. The main result is that the capacity has an explicit form and that there are no threshold phenomena where the capacity jumps. The private quantum capacity turns out to be equal to the entanglement-assisted capacity to the eavesdropper's environment. Applied to the qudit Unruh channels, we find the private quantum capacity is positive for all non-zero eavesdropper accelerations, no matter how small. It is strange that these two quantities should be the same since there is no relation between the communication tasks. In the private capacity problem there is no entanglement assistance and in the latter there is no question of privacy.

4   Where Do We Go from Here?

There are a number of interesting questions that are raised. First of all one can contemplate carrying out these calculations in black hole geometries where the Hawking radiation [26] would have an effect on the communication capacity. Similarly one can contemplate these same calculations in cosmological space-times especially in the de Sitter universe [27] where there are cosmological horizons.

A number of interesting questions can be raised in the general area of black holes and information [28,29,30]. The most fundamental is of course the question of whether information is lost. There are however, other interesting questions from the perspective of modern quantum information theory. For example, it is possible that black holes are "almost" optimal cloners [31].

Another basic question is "can an observer in the Rindler spacetime prepare a pure state?" If this is not possible then one has to rethink basics of quantum information theory in curved spacetime. The fact that there is a thermal bath for an accelerating observer suggests that it would be difficult to cancel out all this radiation in all modes. Perhaps one will have to prepare states that are pure when restricted to some sector of Fock space.

Acknowledgments

I would like to thank McGill University and EPSRC for their generous support during my sabbatical leave, NSERC (Canada) and the Office of Naval Research (USA) for funding this research and the Computing Laboratory of the University of Oxford for its hospitality. I thank my colleagues Kamil Bradler and Patrick Hayden for a very interesting collaboration. Finally, I am grateful to Viv Kendon and Sonja Smets for inviting me to speak in the special session of CiE 2011.

References

1. Parker, L.: Particle creation in expanding universes. Phys. Rev. Lett. 21, 562–564 (1968)
2. Fulling, S.A.: Nonuniqueness of canonical field quantization in riemannian space-time. Phys. Rev. D 7(10), 2850–2862 (1973)

3. Davies, P.C.W.: Scalar particle production in schwarzschild and rindler metrics. J. Phys. A 8(4), 609–616 (1975)
4. Unruh, W.G.: Notes on black hole evaporation. Phys. Rev. D 14, 870–892 (1976)
5. Unruh, W.G., Wald, R.M.: What happens when an accelerating observer detects a rindler particle. Phys. Rev. D 29(6), 1047–1056 (1984)
6. Bradler, K., Hayden, P., Panangaden, P.: Private communication via the Unruh effect. Journal of High Energy Physics JHEP08(074) (August 2009), doi:10.1088/1126-6708/2009/08/074
7. Bradler, K., Hayden, P., Panangaden, P.: Quantum communication in Rindler spacetime. Arxiv quant-ph 1007.0997 (July 2010)
8. Alsing, P.M., Milburn, G.J.: Lorentz invariance of entanglement. Quantum Information and Computation 2, 487 (2002)
9. Alsing, P.M., Milburn, G.J.: Teleportation with a uniformly accelerated partner. Phys. Rev. Lett. 91(18), 180404 (2003)
10. Peres, A., Terno, D.R.: Quantum information and relativity theory. Rev. Mod. Phys. 76(1), 93–123 (2004)
11. Gingrich, R.M., Adami, C.: Quantum entanglement of moving bodies. Phys. Rev. Lett. 89(27), 270402 (2002)
12. Caban, P., Rembieliński, J.: Lorentz-covariant reduced spin density matrix and Einstein-Podolsky-Rosen Bohm correlations. Physical Review A 72, 12103 (2005)
13. Doukas, J., Carson, B.: Entanglement of two qubits in a relativistic orbit. Physical Review A 81(6), 62320 (2010)
14. Fuentes-Schuller, I., Mann, R.B.: Alice Falls into a Black Hole: Entanglement in Noninertial Frames. Physical Review Letters 95, 120404 (2005)
15. Datta, A.: Quantum discord between relatively accelerated observers. Physical Review A **80**(5) 80(5), 52304 (2009)
16. Martin-Martinez, E., León, J.: Quantum correlations through event horizons: Fermionic versus bosonic entanglement. Physical Review A 81(3), 32320 (2010)
17. Kent, A.: Unconditionally secure bit commitment. Phys. Rev. Lett. 83(7), 1447–1450 (1999)
18. Czachor, M., Wilczewski, M.: Relativistic Bennett-Brassard cryptographic scheme, relativistic errors, and how to correct them. Physical Review A 68(1), 10302 (2003)
19. Cliche, M., Kempf, A.: Relativistic quantum channel of communication through field quanta. Physical Review A 81(1), 12330 (2010)
20. Maurer, U.M.: The strong secret key rate of discrete random triples. In: Communication and Cryptography – Two Sides of One Tapestry, pp. 271–284. Kluwer Academic Publishers, Dordrecht (1994)
21. Ahlswede, R., Csiszar, I.: Common randomness in information theory and cryptography. IEEE Transactions on Information Theory 39, 1121–1132 (1993)
22. Cai, N., Winter, A., Yeung, R.W.: Quantum privacy and quantum wiretap channels. Problems of Information Transmission 40(4), 318–336 (2005)
23. Devetak, I.: The private classical capacity and quantum capacity of a quantum channel. IEEE Transactions on Information Theory 51(1), 44–55 (2005)
24. Schützhold, R., Unruh, W.G.: Comment on Teleportation with a uniformly accelerated partner. arXiv:quant-ph/0506028 (2005)
25. Kretschmann, D., Werner, R.F.: Tema con variazioni: quantum channel capacity. New Journal of Physics 6, 26-+ (2004)
26. Hawking, S.W.: Particle creation by black holes. Comm. Math. Phys. 43(3), 199–220 (1975)

27. Hawking, S., Ellis, G.: The large scale structure of space-time. Cambridge Monographs on Mathematical Physics. Cambridge University Press, Cambridge (1973)
28. Hawking, S.W.: Is information lost in black holes? In: Wald, R.M. (ed.) Black Holes and Relativistic Stars, pp. 221–240. University of Chicago Press, Chicago (1998)
29. Hayden, P., Preskill, J.: Black holes as mirrors: Quantum information in random subsystems. Journal of High Energy Physics 0709(120) (2007)
30. Page, D.: Black hole information. Available on ArXiv hep-th/9305040 (May 1993)
31. Adami, C., Steeg, G.L.V.: Black holes are almost optimal quantum cloners. arXiv:quant-ph/0601065v1 (January 2006)

Recognizing Synchronizing Automata with Finitely Many Minimal Synchronizing Words is PSPACE-Complete

Elena V. Pribavkina[1] and Emanuele Rodaro[2]

[1] Ural State University, 620083, Ekaterinburg, Russia
[2] Centro de Matematica, Faculdade de Ciências
Universidade do Porto, 4169-007 Porto, Portugal
elena.pribavkina@usu.ru, emanuele.rodaro@fc.up.pt

Abstract. A deterministic finite-state automaton $\mathscr{A}$ is said to be *synchronizing* if there is a *synchronizing* word, i.e. a word that takes all the states of the automaton $\mathscr{A}$ to a particular one. We consider synchronizing automata whose language of synchronizing words is finitely generated as a two-sided ideal in Σ^*. Answering a question stated in [1], here we prove that recognizing such automata is a PSPACE-complete problem.

Let $\mathscr{A} = \langle Q, \Sigma, \delta \rangle$ be a deterministic and complete finite-state automaton (DFA). The action of the transition function δ can naturally be extended to the free monoid Σ^*. This extension will still be denoted by δ. For convenience for each $v \in \Sigma^*$ and $q \in Q$ we will write $q \cdot v = \delta(q, v)$ and put $S \cdot v = \{q \cdot v \mid q \in S\}$ for any $S \subseteq Q$.

An automaton $\mathscr{A} = \langle Q, \Sigma, \delta \rangle$ is called *synchronizing* if it possesses a *synchronizing* word, that is a word w which takes all states of $\mathscr{A}$ to a particular one: $q \cdot w = q' \cdot w$ for all $q, q' \in Q$.

Over the past forty years synchronizing automata and especially shortest synchronizing words have been widely studied, motivated mostly by the famous Černý conjecture [2] which states that any n-state synchronizing automaton possesses a synchronizing word of length at most $(n-1)^2$. This conjecture has been proved for a large number of classes of synchronizing automata, nevertheless in general it remains one of the most longstanding open problems in automata theory. For more details see the surveys [3,4,5].

By $\mathrm{Syn}_{\mathscr{A}}$ we denote the language of all words synchronizing a given automaton $\mathscr{A}$. This language is well-known to be regular. Moreover, it is easy to see that this language is a two-sided ideal in Σ^*, i.e. $\mathrm{Syn}_{\mathscr{A}} = \Sigma^* \mathrm{Syn}_{\mathscr{A}}^{min} \Sigma^*$. The set of generators of this ideal, denoted by $\mathrm{Syn}_{\mathscr{A}}^{min}$, consists of all synchronizing words such that none of its proper prefixes nor suffixes is synchronizing. We refer to such words as *minimal synchronizing words*. The class of automata possessing only finitely many minimal synchronizing words was studied in [1], where such automata were called *finitely generated synchronizing automata*. Here we preserve from [1] the denotation **FG** for this class. In particular, in [1] a linear upper bound on the length of shortest synchronizing words for the class **FG** is proved, thus the Černý conjecture is true for this class of automata.

B. Löwe et al. (Eds.): CiE 2011, LNCS 6735, pp. 230–238, 2011.
© Springer-Verlag Berlin Heidelberg 2011

In addition in [1] the problem of checking whether a given synchronizing automaton belongs to **FG** is considered. In the sequel we refer to this problem as the FINITENESS problem:

- *Input*: A synchronizing DFA $\mathscr{A} = \langle Q, \Sigma, \delta \rangle$.
- *Question*: Is the language $\mathrm{Syn}_{\mathscr{A}}^{min}$ finite?

Since $\mathrm{Syn}_{\mathscr{A}}$ is a regular language and

$$\mathrm{Syn}_{\mathscr{A}}^{min} = \mathrm{Syn}_{\mathscr{A}} \setminus (\Sigma \, \mathrm{Syn}_{\mathscr{A}} \cup \mathrm{Syn}_{\mathscr{A}} \, \Sigma)$$

clearly FINITENESS is a decidable problem. Whereas testing if a given DFA is synchronizing is a polynomial task [2,6], in [1] it is shown that the FINITENESS problem is unlikely to have a polynomial algorithm:

Theorem 1 ([1]). *The problem* FINITENESS *is co-NP-hard.*

However this theorem does not frame this problem precisely between the known complexity classes, for instance a natural question is whether or not FINITENESS is in co-NP. Answering an open question posed in [1] here we prove that FINITENESS is a PSPACE-complete problem.

The paper is organized as follows. In Section 1 we introduce some definitions and recall some properties of the class **FG** from [1]. In Section 2 we prove the main result.

1   Preliminaries

A subset S of Q is called *reachable* if there is a word $v \in \Sigma^*$ with $S = Q.v$. Given a subset S of Q by $\mathrm{Fix}(S)$ we denote the set of all words *fixing* S:

$$\mathrm{Fix}(S) = \{w \in \Sigma^* \mid S.w = S\}.$$

By $\mathrm{Syn}(S)$ we denote the set of all words bringing S to a singleton:

$$\mathrm{Syn}(S) = \{w \in \Sigma^* \mid |S.w| = 1\}.$$

In this notations we have $\mathrm{Syn}_{\mathscr{A}} = \mathrm{Syn}(Q)$. Besides $\mathrm{Syn}_{\mathscr{A}}$ is contained in $\mathrm{Syn}(S)$ for any S. The following Lemma holds.

Lemma 1 ([1]). *Given a word* $w \in \Sigma^*$ *there exists an integer* $\beta \geq 0$ *such that the set* $m(w) = Q.w^{\beta}$ *is fixed by* w. *Moreover* $m(w)$ *is the largest subset of* Q *with this property and for any word* $w \in \Sigma^*$ *and* $\alpha \in \mathbb{N}$

$$m(w^{\alpha}) = m(w).$$

The subset $m(w)$ of Q from the previous lemma is called *the maximal fixed set with respect to* w. The following theorem gives a combinatorial characterization of the class **FG**.

Theorem 2 ([1]). *Given a synchronizing DFA $\mathscr{A} = \langle Q, \Sigma, \delta \rangle$ the following are equivalent:*

*(i) $\mathscr{A}$ is in **FG**;*
(ii) for any reachable subset $S \subseteq Q$ such that $1 < |S| < |Q|$, for each $w \in \mathrm{Fix}(S)$

$$\mathrm{Syn}(S) = \mathrm{Syn}(m(w)).$$

This theorem gives rise to the following algorithm FINCHECK [1] for the FINITENESS problem.

FINCHECK:

```
1 From the synchronizing DFA 𝒜 = ⟨Q, Σ, δ⟩ construct its power
  automaton 𝒫(𝒜) consisting only of subsets reachable from Q.
2 For each state S of 𝒫(𝒜) do:
  2.1 For each state T of 𝒫(𝒜) with S ⊆ T do:
    2.2 If Fix(T) ∩ Fix(S) ≠ ∅, then
      2.3 If Syn(T) ≠ Syn(S), then exit and return FALSE
3 Otherwise exit and return TRUE
```

It is easy to see that the language $\mathrm{Syn}(S)$ is recognized by the automaton $\mathscr{A}_{\mathrm{Syn}(S)} = \langle Q_{\mathrm{Syn}(S)}, \Sigma, \delta, S, \sigma \rangle$ with the state set $Q_{\mathrm{Syn}(S)}$ consisting of all subsets S' of Q with $|S'| \leq |S|$, with S as the initial state, and the set of final states σ consisting of singletons. The language $\mathrm{Fix}(S)$ is recognized by the automaton $\mathscr{A}_{\mathrm{Fix}(S)} = \langle Q_{\mathrm{Fix}(S)}, \Sigma, \delta, S, \{S\} \rangle$, where $Q_{\mathrm{Fix}(S)}$ consists of all $S' \subseteq Q$ with $|S'| = |S|$. We have the following theorem.

Theorem 3. FINITENESS *is in PSPACE.*

Proof. We prove that FINCHECK runs in polynomial space. Indeed, we can give some linear order (for instance, lexicographical) on the power set of Q so that we can try a pair T, S with $S \subseteq T$. Thus, we are reduced to prove that conditions 2.2 and 2.3 run in polynomial space. The language $\mathrm{Fix}(T) \cap \mathrm{Fix}(S)$ is recognized by the product of automata $\mathscr{A}_{T,S} = \langle Q_{\mathrm{Fix}(T)} \times Q_{\mathrm{Fix}(S)}, \Sigma, \delta_{T,S}, (T, S), \{(T, S)\} \rangle$ where

$$\delta_{T,S}(H_1, H_2) = (\delta(H_1), \delta(H_2)).$$

Obviously, we cannot afford to construct $\mathscr{A}_{T,S}$ directly. Instead we check the emptiness of the language $\mathrm{Fix}(T) \cap \mathrm{Fix}(S)$ by checking if there is a loop at the state (T, S) in $\mathscr{A}_{T,S}$. Observe that we can look for a loop of length at most $|Q_{\mathrm{Fix}(T)} \times Q_{\mathrm{Fix}(S)}| = 2^{|T|+|S|}$. To check if there is a loop of at most such length we run SEARCH$((T, S), (T, S), |T| + |S|)$. In general SEARCH$(u, v, N)$ checks in $\mathscr{A}_{T,S}$ if there is a path connecting u to v of length at most 2^N:

SEARCH(u, v, N)

```
1 if N = 0 then return u == v or δ_{T,S}(u, a) == v for some a ∈ Σ
2 For each q ∈ Q_{Fix(T)} × Q_{Fix(S)}
  2.1 If SEARCH(u, q, N − 1) and SEARCH(q, v, N − 1) then return TRUE
3 return FALSE
```

The space used to run SEARCH$((T, S), (T, S), |T| + |S|)$ is obviously polynomial. Indeed the depth of the recursion is at most $|T| + |S|$ and on each level we need only $O(1)$ state descriptions. Calculating the transition function also requires only polynomial space. Therefore checking $\text{Fix}(T) \cap \text{Fix}(S) \neq \varnothing$ is a polynomial space task.

Checking $\text{Syn}(T) \neq \text{Syn}(S)$ is equivalent to check $\text{Syn}(T) \cap \text{Syn}(S)^c \neq \varnothing$ or $\text{Syn}(T)^c \cap \text{Syn}(S)^c \neq \varnothing$ where $\text{Syn}(H)^c$ is the complement language of $\text{Syn}(H)$. An analogous argument used previously to prove that checking that $\text{Fix}(T) \cap \text{Fix}(S) \neq \varnothing$ is a polynomial space task can be easily adapted to prove that checking both conditions $\text{Syn}(T) \cap \text{Syn}(S)^c \neq \varnothing$ and $\text{Syn}(T)^c \cap \text{Syn}(S)^c \neq \varnothing$ are also polynomial space tasks.   □

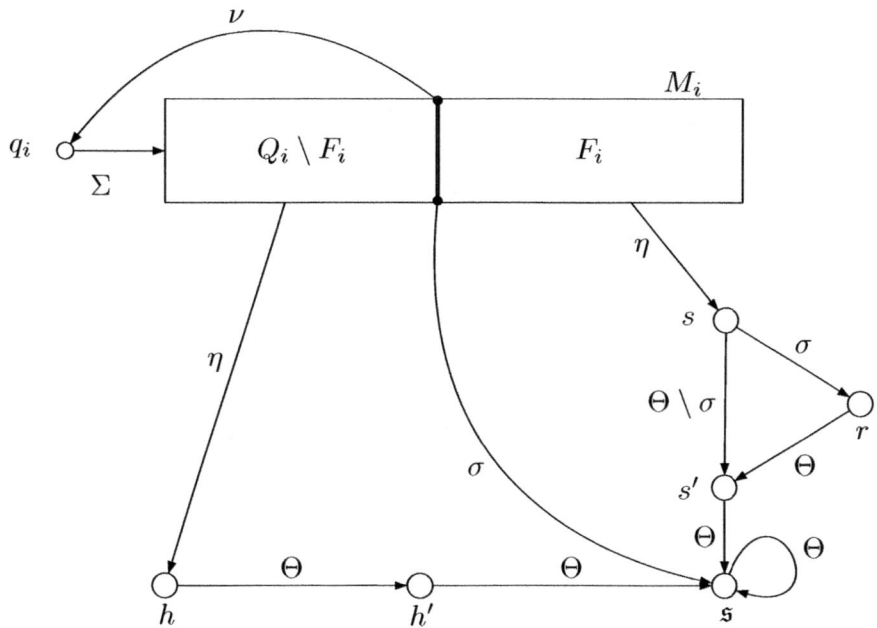

Fig. 1. Automaton $\mathcal{M}$

2   PSPACE-Completeness of FINITENESS

To prove that **FINITENESS** is a PSPACE-complete problem we reduce the following well-known PSPACE-complete problem [7] to the complement of **FINITENESS**. This problem deals with checking emptiness of intersection of a collection of DFAs:

NON-EMPTINESS DFA

- *Input*: Given n DFA's $M_i = \langle Q_i, \Sigma, \delta_i, q_i, F_i \rangle$ for $i = 1, \dots, n$.
- *Question*: $\bigcap_i L[M_i] \neq \varnothing$?

Given an instance of NON-EMPTINESS DFA problem, we can suppose without loss of generality that each initial state q_i has no incoming edges and $q_i \notin F_i$. Indeed, excluding the case for which the empty word ε belongs to $\bigcap_i L[M_i]$, we can always build a DFA $M_i' = \langle Q_i', \Sigma, \delta_i', q_i', F_i \rangle$ which recognizes the same language of M_i such that the initial state q_i' has no incoming edges. This can be easily achieved adding a new initial state q_i' to Q_i and extending δ_i to a function δ_i' with $\delta_i'(q_i', a) = \delta_i(q_i, a)$ for all $a \in \Sigma$. Moreover, we can assume that the sets Q_i for $i = 1, \ldots, n$ are disjoint.

For our reduction, from DFAs M_i, $i = 1, \ldots, n$, we build a special DFA $\mathcal{M} = \langle Q, \Theta, \phi \rangle$ with $Q = \cup_{i=1}^n Q_i \cup \{\eth, \eth', \hbar, \hbar', \mathfrak{r}, \mathfrak{s}\}$ where $\eth, \eth', \hbar, \hbar', \mathfrak{r}, \mathfrak{s}$ are new states not belonging to any Q_i. We add three new letters to Σ getting $\Theta = \Sigma \cup \{\sigma, \nu, \eta\}$ and the transition function ϕ is defined by the following rules (Fig. 1):

for all $i = 1, \ldots, n$, for all $q \in Q_i$ and $\alpha \in \Sigma$, $\phi(q, \alpha) = \delta_i(q, \alpha)$;
for all $i = 1, \ldots, n$ and for all $q \in Q_i$, $\phi(q, \nu) = q_i$;
for all $i = 1, \ldots, n$ and for all $q \in F_i$, $\phi(q, \eta) = \eth$;
for all $i = 1, \ldots, n$ and for all $q \in (Q_i \setminus F_i)$, $\phi(q, \eta) = \hbar$;
for all $i = 1, \ldots, n$ and for all $q \in Q_i$ $\phi(q, \sigma) = \mathfrak{s}$;
for all $\alpha \in \Theta \setminus \{\sigma\}$, $\phi(\eth, \alpha) = \eth'$ and $\phi(\eth, \sigma) = \mathfrak{r}$;
for all $\alpha \in \Theta$, $\phi(\hbar, \alpha) = \hbar'$, $\phi(\hbar', \alpha) = \mathfrak{s}$, $\phi(\eth', \alpha) = \mathfrak{s}$, $\phi(\mathfrak{r}, \alpha) = \mathfrak{s}$, $\phi(\mathfrak{s}, \alpha) = \mathfrak{s}$.

Remark 1. By the definition of the transition function ϕ we get $\phi(Q, w) \cap Q_i \neq \varnothing$ if and only if $w \in (\Sigma \cup \{\nu\})^*$.

From this remark and the definition of ϕ we easily derive the following lemma.

Lemma 2. *For any $w \in \Theta^*$ we have $\phi(Q, w) \cap Q_i \neq \varnothing$ for all $i = 1, \ldots, n$ if and only if there is some $j \in \{1, \ldots, n\}$ such that $\phi(Q, w) \cap Q_j \neq \varnothing$.*

The following lemma reduces NON-EMPTINESS DFA to a reachability problem in automaton $\mathcal{M}$.

Lemma 3. *There is a word $w \in \Theta^+$ with $|w| \geq 2$ such that $\phi(Q, w) = \{\mathfrak{s}, \mathfrak{r}\}$ if and only if $\bigcap_{i=0}^n L[M_i] \neq \varnothing$.*

Proof. Let $w' \in \bigcap_{i=0}^n L[M_i]$. Then $\phi(q_i, w') \in F_i$ for all $i = 1, \ldots, n$. Consider the word $w = \nu w' \eta \sigma$. Obviously, $|w| \geq 2$. Moreover, $\phi(Q, \nu) = \{q_1, \ldots, q_n, \mathfrak{s}, \hbar', \eth'\}$, $\phi(Q, \nu w') = \{q_1', \ldots, q_n', \mathfrak{s}\}$ with $q_i' \in F_i$ for all $i = 1, \ldots, n$, hence $\phi(Q, \nu w' \eta) = \{\eth, \mathfrak{s}\}$ and so $\phi(Q, \nu w' \eta \sigma) = \{\mathfrak{r}, \mathfrak{s}\}$.

Conversely, suppose that there is a word $w \in \Theta^+$ such that $\phi(Q, w) = \{\mathfrak{s}, \mathfrak{r}\}$. If $w \in (\Sigma \cup \{\nu\})^+$, then $\phi(Q, w) \cap Q_i \neq \varnothing$, which is a contradiction. Therefore w contains some factor belonging to $\{\sigma, \eta\}^+$. Thus we can factorize $w = uy\rho$ where u is the maximal prefix of w belonging to $(\Sigma \cup \{\nu\})^*$, $y \in \{\sigma, \eta\}$ and $\rho \in \Theta^*$. We have two cases: either $u \in \Sigma^*$ or u contains some factor of ν^+. We treat these two cases considering $u = u'\nu^k x$ where x is the maximum suffix of u belonging to Σ^*, $u' \in (\Sigma \cup \{\nu\})^*$ and $k \in \{0, 1\}$ depending whether u contains some factor of ν^+ or not. Consider $S = \phi(Q, u'\nu^k)$. If $k = 0$ then $u' = \varepsilon$ and so $\{q_1, \ldots, q_n\} \subseteq Q = S$, otherwise if $k = 1$ an easy calculation shows that also in this case $\{q_1, \ldots, q_n\} \subseteq S$ holds. We now consider the action of the word $xy\rho$ on S. We have two cases.

- *Case 1: $y = \sigma$.* If $k = 0$, then $S = Q$. Thus if $|x| > 0$ then $\phi(S, x\sigma) = \{\mathfrak{s}\}$, but then we have $\phi(Q, w) = \{\mathfrak{s}\}$, a contradiction. Therefore we can suppose $x = \varepsilon$ and so $\phi(S, x\sigma) = \{\mathfrak{h}', \mathfrak{r}, \mathfrak{s}\}$. However, in order to bring $\mathfrak{h}'$ inside the set $\{\mathfrak{s}, \mathfrak{r}\}$ we necessarily have $\rho \in \Theta^+$. Then again we get a contradiction $\phi(Q, w) = \{\mathfrak{s}\}$. Therefore we can suppose $k = 1$ from which we get $\phi(S, x) \subseteq \bigcup_{i=1}^n Q_i \cup \{\mathfrak{h}', \mathfrak{d}', \mathfrak{s}\}$, whence $\phi(S, x\sigma) = \{\mathfrak{s}\}$ and $\phi(Q, w) = \{\mathfrak{s}\}$, a contradiction.
- *Case 2: $y = \eta$.* Since $\{q_1, \ldots, q_n\} \subseteq S$, by Lemma 2 we have that $S' = \phi(S, x)$ is a subset of Q with $S' \cap Q_i \neq \varnothing$ for all $i = 1, \ldots, n$. We consider two cases: either $S' \cap Q_i \subseteq F_i$ for all $i = 1, \ldots, n$ or not.
  In the first case, since Σ acts on each Q_i and $\{q_1, \ldots, q_n\} \subseteq S$, we get $x \in \Sigma^+$ and $x \in \bigcap_{i=0}^n L[M_i]$, i.e. $\bigcap_{i=0}^n L[M_i] \neq \varnothing$.
  Otherwise there is some positive integer j such that $(S' \cap Q_j) \setminus F_j \neq \varnothing$. Hence $\mathfrak{h} \in \phi(S', \eta)$ and to bring $\mathfrak{h}$ inside the set $\{\mathfrak{s}, \mathfrak{r}\}$ we have to apply a word ρ of length at least two. Since $\phi(S', \eta) \subseteq \{\mathfrak{d}, \mathfrak{d}', \mathfrak{h}, \mathfrak{h}', \mathfrak{s}\}$, we get a contradiction $\phi(Q, w) = \{\mathfrak{s}\}$.   □

Now we prove the following lemma.

Lemma 4. *The automaton $\mathcal{M} = \langle Q, \Theta, \phi \rangle$ previously defined belongs to the class **FG**. Moreover for any $h \in \Theta^+$, $m(h) \subseteq \bigcup_{i=1}^n Q_i \cup \{\mathfrak{s}\}$.*

Proof. Clearly $\mathcal{M}$ is a synchronizing automaton with the sink state $\mathfrak{s}$. We observe that for any word $h \in \Theta^+$ the maximal subset of Q fixed by h contains the sink state $\mathfrak{s}$. From Lemma 1, $m(h) = m(h^\alpha)$ for any $\alpha \in \mathbb{N}$, therefore since $\mathfrak{s} \in m(h)$, if $m(h) \cap \{\mathfrak{d}, \mathfrak{d}', \mathfrak{h}, \mathfrak{h}', \mathfrak{r}\} \neq \varnothing$, then for $\alpha \geq 2$ we would have the contradiction $|m(h^\alpha)| < |m(h)|$, whence $m(h) \subseteq \bigcup_{i=1}^n Q_i \cup \{\mathfrak{s}\}$.

Arguing by contradiction suppose that $\mathcal{M} \notin \mathbf{FG}$. By Theorem 2 there is a reachable set S with $1 < |S| < |Q|$ and a non-empty word $h \in \mathrm{Fix}(S)$ such that $\mathrm{Syn}(m(h)) \subsetneq \mathrm{Syn}(S)$. Let $u \in \mathrm{Syn}(S) \setminus \mathrm{Syn}(m(h))$. Since $|S| > 1$, $S \subseteq m(h) \subseteq \bigcup_{i=1}^n Q_i \cup \{\mathfrak{s}\}$ and $\mathfrak{s} \in S$ ($\mathfrak{s}$ is the sink state, and S is a reachable subset different from $\mathfrak{s}$) we deduce that $S \cap \bigcup_{i=1}^n Q_i \neq \varnothing$.

Suppose that $u \in (\Sigma \cup \nu)^*$. Since any word from $(\Sigma \cup \{\nu\})^*$ leaves $\bigcup_{i=1}^n Q_i$ inside $\bigcup_{i=1}^n Q_i$, and $S \cap \bigcup_{i=1}^n Q_i \neq \varnothing$ we have $\phi(S, u) \cap (\bigcup_{i=1}^n Q_i) \neq \varnothing$. Since $\{\mathfrak{s}\} = \phi(S, u)$ we get $\mathfrak{s} \in \bigcup_{i=1}^n Q_i$, which is a contradiction. Thus we can suppose that u contains some factor of $\{\eta, \sigma\}^+$ and so we can suppose $u = u'yu''$ where $u' \in (\Sigma \cup \nu)^*$, $y \in \{\eta, \sigma\}$ and $u'' \in \Theta^*$. We consider two cases.

- *Case 1: $y = \sigma$.* Since $S \subseteq m(h) \subseteq \bigcup_{i=1}^n Q_i \cup \{\mathfrak{s}\}$, we have $\phi(S, u'\sigma) = \{\mathfrak{s}\}$, $\phi(m(h), u'\sigma) = \{\mathfrak{s}\}$ which contradicts the condition $u \in \mathrm{Syn}(S) \setminus \mathrm{Syn}(m(h))$.
- *Case 2: $y = \eta$.* Since $\phi(S, u') \subseteq \bigcup_i Q_i \cup \{\mathfrak{s}\}$, we can have either $\phi(S, u'\eta) = \{\mathfrak{d}, \mathfrak{s}\}$ or $\phi(S, u'\eta) = \{\mathfrak{h}, \mathfrak{s}\}$, or $\phi(S, u'\eta) = \{\mathfrak{h}, \mathfrak{d}, \mathfrak{s}\}$. By definition of the transition function ϕ it is straightforward to check that

$$\mathrm{Syn}(\{\mathfrak{h}, \mathfrak{s}\}) = \mathrm{Syn}(\{\mathfrak{d}, \mathfrak{s}\}) = \mathrm{Syn}(\{\mathfrak{h}, \mathfrak{d}, \mathfrak{s}\}) = \Theta^2.$$

Hence $u'' \in \Theta^2$. Since $m(h) \subseteq \bigcup_{i=1}^n Q_i \cup \{\mathfrak{s}\}$ we get $\phi(m(h), u'\eta) \subseteq \{\mathfrak{h}, \mathfrak{d}, \mathfrak{s}\}$, whence $\phi(m(h), u'\eta u'') = \{\mathfrak{s}\}$ which again contradicts $u \in \mathrm{Syn}(S) \setminus \mathrm{Syn}(m(h))$.   □

The following lemma reduces the reachability problem stated in lemma 3 to the problem co-FINITENESS.

Lemma 5. *Let* $\mathcal{M} = \langle Q, \Theta, \phi \rangle$ *be the previously defined automaton. Then there is a synchronizing automaton* $\mathcal{M}'$ *such that the language* $\mathrm{Syn}_{\mathcal{M}'}^{min}$ *is infinite if and only if there exists* $w \in \Theta^+$ *with* $|w| \geq 2$ *such that* $\phi(Q, w) = \{\mathfrak{r}, \mathfrak{s}\}$.

Proof. We modify automaton $\mathcal{M} = \langle Q, \Theta, \phi \rangle$ in order to obtain a new automaton $\mathcal{M}' = \langle Q', \Theta', \phi' \rangle$ with $Q' = Q \cup \{\mathfrak{p}, \mathfrak{p}'\}$, $\Theta' = \Theta \cup \{\tau\}$, where τ is a new symbol, and the following transition function (Fig. 2):
$\phi'(q, x) = \phi(q, x)$ for all $x \in \Theta$ and $q \in Q$,
let $G = Q \setminus \{\mathfrak{s}, \mathfrak{r}\}$, then $\phi'(G, \tau) = \{\mathfrak{p}\}$,
$\phi'(\mathfrak{p}, \tau) = \mathfrak{p}$, $\quad \phi'(\mathfrak{p}, x) = \mathfrak{p}'$ for all $x \in \Theta$,
$\phi'(\mathfrak{p}', y) = \mathfrak{s}$ for all $y \in \Theta'$, and
$\phi'(\mathfrak{r}, \tau) = \mathfrak{r}$, $\phi'(\mathfrak{s}, \tau) = \mathfrak{s}$.
Suppose there exists $w \in \Theta^+$ with $|w| \geq 2$ such that $\phi(Q, w) = \{\mathfrak{s}, \mathfrak{r}\}$. The image $\phi'(Q', w)$ is equal to $\{\mathfrak{s}, \mathfrak{r}\}$, and so this set is reachable also in $\mathcal{M}'$. Since $\phi'(\{\mathfrak{s}, \mathfrak{r}\}, \tau) = \{\mathfrak{s}, \mathfrak{r}\}$, then $\tau \in \mathrm{Fix}(\{\mathfrak{s}, \mathfrak{r}\})$ and clearly $m(\tau) = \{\mathfrak{p}, \mathfrak{s}, \mathfrak{r}\}$. On the other hand, if we take any $a \in \Theta$, then $\phi(\{\mathfrak{s}, \mathfrak{r}\}, a) = \{\mathfrak{s}\}$ but $\phi(\{\mathfrak{p}, \mathfrak{s}, \mathfrak{r}\}, a) = \{\mathfrak{p}', \mathfrak{s}\}$. Thus $\mathrm{Syn}(\{\mathfrak{s}, \mathfrak{r}\}) \neq \mathrm{Syn}(\{\mathfrak{p}, \mathfrak{s}, \mathfrak{r}\})$, and so by the Theorem 2, $\mathrm{Syn}_{\mathcal{M}'}^{min}$ is infinite.
Conversely, if the language $\mathrm{Syn}_{\mathcal{M}'}^{min}$ is infinite, then by Theorem 2, there is a reachable subset $S \subseteq Q'$ with $1 < |S| < |Q'|$ and a word $h \in \mathrm{Fix}(S)$ over $\Theta \cup \{\tau\}$ such that $\mathrm{Syn}(m(h)) \subsetneq \mathrm{Syn}(S)$. Let $u \in \mathrm{Syn}(S) \setminus \mathrm{Syn}(m(h))$. We consider the following two cases.

– *Case 1.* Suppose $h \in \Theta^+$. By Lemma 1 $m(h) = m(h^\alpha)$ for any $\alpha \in \mathbb{N}$. Thus $m(h) \subseteq Q$. Moreover, since $h \in \Theta^+$ by Lemma 4 we get $m(h) \subseteq \bigcup_{i=1}^n Q_i \cup \{\mathfrak{s}\}$. Therefore $S \subseteq m(h) \subseteq \bigcup_{i=1}^n Q_i \cup \{\mathfrak{s}\}$. We prove that the set S is also a reachable subset in $\mathcal{M}$. Let $v \in \Theta'^*$ be a word such that

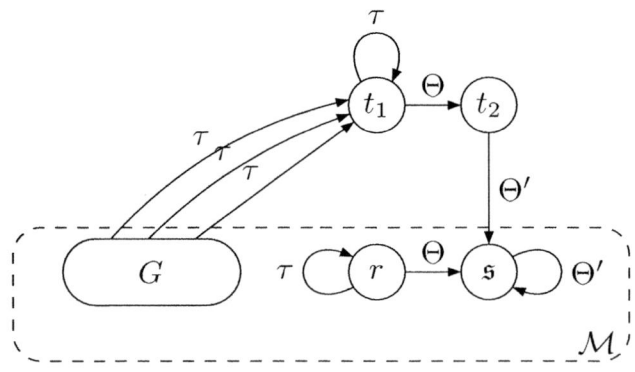

Fig. 2. Automaton $\mathcal{M}'$

$\phi'(Q', v) = S$. Suppose that v contains a factor of τ^+, since $S \subseteq \bigcup_{i=1}^n Q_i \cup \{\mathfrak{s}\}$ then necessarily $S = \{\mathfrak{s}\}$ which contradicts $|S| > 1$. Thus we can suppose $v \in \Theta^*$ and in this case it is easy to check that $\phi(Q, v) = S$.

Since Θ acts inside $\mathcal{M}$, then by Lemma 4 $(\mathrm{Syn}(m(h)) \cap \Theta^*) = (\mathrm{Syn}(S) \cap \Theta^*)$. Therefore any word $u \in \mathrm{Syn}(S) \setminus \mathrm{Syn}(m(h))$ has a factor in τ^+ and so we can factorize $u = u'\tau u''$ with $u' \in \Theta^*$ and $u'' \in \Theta'^*$. We consider two further subcases.

i) $u' \in (\Sigma \cup \nu)^*$. Then we have $\phi'(S, u') \subseteq \bigcup_{i=1}^n Q_i \cup \{\mathfrak{s}\}$ and $\phi'(m(h), u') \subseteq \bigcup_{i=1}^n Q_i \cup \{\mathfrak{s}\}$, and so $\phi'(S, u'\tau) = \phi'(m(h), u'\tau) = \{\mathfrak{p}, \mathfrak{s}\}$ which is a contradiction.

ii) u' contains a factor of $\{\sigma, \eta\}^+$. Therefore either $u' = t\sigma t'$ or $u' = t\eta t'$ with $t \in (\Sigma \cup \{\nu\})^*$. If $u' = t\sigma t'$ with $t \in (\Sigma \cup \{\nu\})^*$, then $\phi'(S, t\sigma) = \phi'(m(h), t\sigma) = \{\mathfrak{s}\}$. This is again a contradiction. Thus we may assume $u' = t\eta t'$ with $t \in (\Sigma \cup \{\nu\})^*$. We recall that $S = \phi'(Q', v) = \phi(Q, v)$ with $v \in \Theta^*$ and $S \subseteq \bigcup_{i=1}^n Q_i \cup \{\mathfrak{s}\}$. Let $S' = \phi'(Q', vt) = \phi(Q, vt)$. Clearly $S' \subseteq \bigcup_i Q_i \cup \{\mathfrak{s}\}$. Suppose $S' \subseteq \bigcup_i F_i \cup \{\mathfrak{s}\}$. Consider the word $w = vt\eta\sigma$, then it is straightforward to check that it is a word with $|w| \geq 2$ such that $\phi(Q, vt\eta\sigma) = \{\mathfrak{s}, \mathfrak{r}\}$ and so we are done. Therefore we can suppose that $S' \setminus (F_j \cup \{\mathfrak{s}\}) \neq \varnothing$ for some $j \in \{1, \ldots, n\}$. Since $S' \subseteq \bigcup_i Q_i \cup \{\mathfrak{s}\}$, we get $\mathfrak{h} \in \phi'(Q', vt\eta) \subseteq \{\mathfrak{h}, \mathfrak{d}, \mathfrak{s}\}$. Since $\phi'(S, u) = \{\mathfrak{s}\}$ and $u = t\eta t'\tau u''$, then $\phi'(\mathfrak{h}, t'\tau u'') = \mathfrak{s}$. However, a simple inspection on automaton $\mathcal{M}'$ reveals that any word $x \in \Theta'^*$ bringing $\mathfrak{h}$ to $\mathfrak{s}$ brings also $\mathfrak{d}$ to $\mathfrak{s}$, i.e. if $\phi'(\mathfrak{h}, x) = \mathfrak{s}$ then $\phi'(\mathfrak{d}, x) = \mathfrak{s}$. Therefore $\phi'(\mathfrak{d}, t'\tau u'') = \mathfrak{s}$ and so since $\phi'(m(h), t\eta) \subseteq \{\mathfrak{h}, \mathfrak{d}, \mathfrak{s}\}$ we get also $\phi'(m(h), t\eta t'\tau u'') = \phi'(m(h), u) = \{\mathfrak{s}\}$, whence $u \in \mathrm{Syn}(m(h))$, a contradiction.

- *Case 2.* Suppose that h contains a factor of τ^+. Since $\phi'(Q', \tau) = \{\mathfrak{p}, \mathfrak{r}, \mathfrak{s}\}$, then if h contains some factors of Θ^+, it is not difficult to see that $\phi'(Q', h^2) \subseteq \{\mathfrak{s}, \mathfrak{p}'\}$. So $\phi'(Q', h^3) = \{\mathfrak{s}\}$, which leads to a contradiction $S \subseteq m(h) = \{\mathfrak{s}\}$. Thus we have $h = \tau^\beta$ for some positive integer $\beta > 0$, and so $m(h) = \{\mathfrak{p}, \mathfrak{s}, \mathfrak{r}\}$. Observe that, since $\mathfrak{s}$ is a sink state, then any reachable subset $S \subseteq Q'$ contains $\mathfrak{s}$. Therefore, since $|S| \geq 2$, $S \neq m(h)$ (otherwise $\mathrm{Syn}(m(h)) = \mathrm{Syn}(S)$) and $\mathfrak{s} \in S$, we have only two possibilities for S: either $S = \{\mathfrak{s}, \mathfrak{r}\}$ or $S = \{\mathfrak{s}, \mathfrak{p}\}$.

  i) Let $S = \{\mathfrak{s}, \mathfrak{r}\}$. Since S is reachable in $\mathcal{M}'$, let $v \in \Theta'^+$ be a word such that $\phi'(Q', v) = \{\mathfrak{s}, \mathfrak{r}\}$. If $v \in \Theta^+$, then it is easy to see that $\phi'(Q', v) = \{\mathfrak{s}, \mathfrak{r}\}$ implies also $\phi(Q, v) = \{\mathfrak{s}, \mathfrak{r}\}$, and we are done. Thus we can assume that v contains a factor of τ^+, so we can factorize $v = w\tau^\alpha z$ with $w \in \Theta^*$ and some positive integer $\alpha > 0$. Let us first suppose that $w \neq \varepsilon$ and $\phi(Q, w) \neq \{\mathfrak{s}, \mathfrak{r}\}$. Then either $\mathfrak{r}$ is not in $\phi(Q, w)$ or $\phi(Q, w)$ contains a state different from both $\mathfrak{s}$ and $\mathfrak{r}$. In the first case by the construction of $\mathcal{M}'$ we get that $\mathfrak{r} \notin \phi'(Q', v)$, which is impossible. In the other case we get $\phi'(Q', w\tau^\alpha) = \{\mathfrak{p}, \mathfrak{s}, \mathfrak{r}\}$, however there is no $z \in \Theta'^+$ such that $\phi'(\{\mathfrak{p}, \mathfrak{s}, \mathfrak{r}\}, z) = \{\mathfrak{s}, \mathfrak{r}\}$, which is again impossible, whence $\phi(Q, w) = \{\mathfrak{s}, \mathfrak{r}\}$. If $w = \varepsilon$, then $\phi'(Q', \tau^\alpha) = \{\mathfrak{p}, \mathfrak{s}, \mathfrak{r}\}$, and by the same reason we get a contradiction with $\phi'(Q', v) = \{\mathfrak{s}, \mathfrak{r}\}$.

ii) Let $S = \{\mathfrak{s}, \mathfrak{p}\}$. We prove that in this case $\mathrm{Syn}(m(h)) = \mathrm{Syn}(\{\mathfrak{p}, \mathfrak{s}, \mathfrak{r}\}) = \mathrm{Syn}(\{\mathfrak{p}, \mathfrak{s}\}) = \mathrm{Syn}(S)$ which contradicts the initial assumption. Clearly it is enough to show that $\mathrm{Syn}(\{\mathfrak{p}, \mathfrak{s}\}) \subseteq \mathrm{Syn}(\{\mathfrak{p}, \mathfrak{s}, \mathfrak{r}\})$. Consider a word $g \in \mathrm{Syn}(\{\mathfrak{p}, \mathfrak{s}\})$. Since τ fixes both $\{\mathfrak{p}, \mathfrak{s}\}$ and $\{\mathfrak{p}, \mathfrak{s}, \mathfrak{r}\}$, we can assume that g starts with a letter $a \in \Theta$, so $g = ag'$. Then $\phi'(\{\mathfrak{s}, \mathfrak{p}\}, a) = \{\mathfrak{s}, \mathfrak{p}'\} = \phi'(\{\mathfrak{p}, \mathfrak{s}, \mathfrak{r}\}, a)$, thus $\phi'(\{\mathfrak{p}, \mathfrak{s}\}, g) = \phi'(\{\mathfrak{p}, \mathfrak{s}, \mathfrak{r}\}, g)$, i.e. $g \in \mathrm{Syn}(\{\mathfrak{p}, \mathfrak{s}, \mathfrak{r}\})$.
□

We are now in position to prove our main theorem.

Theorem 4. FINITENESS *is PSPACE-complete.*

Proof. By Theorem 3 FINITENESS is in PSPACE. Since the construction of the automata $\mathcal{M}, \mathcal{M}'$ can be performed in polynomial time from the automata M_i, $i = 1, \ldots, n$, by Lemmas 3 and 5 we can reduce NON–EMPTINESS DFA to co-FINITENESS.
□

Acknowledgements

The first-named author acknowledges support from the Federal Education Agency of Russia, grant 2.1.1/3537, and from the Russian Foundation for Basic Research, grants 09-01-12142 and 10-01-00793. The second-named author acknowledges the support of the Centro de Matemâtica da Universidade do Porto financed by FCT through the programmes POCTI and POSI, with Portuguese and European Community structural funds, as well as the support of the FCT project SFRH/BPD/65428/2009.

References

1. Pribavkina, E.V., Rodaro, E.: Synchronizing automata with finitely many minimal synchronizing words. Inf. Comput. 209, 568–579 (2011)
2. Černý, J.: Poznámka k homogénnym eksperimentom s konečnými automatami [in slovak]. Mat.-Fyz. Čas. Slovensk. Akad. Vied. 14, 208–216 (1964)
3. Mateescu, A., Salomaa, A.: Many-valued truth functions, Černý's conjecture and road coloring. EATCS Bull 68, 134–150 (1999)
4. Sandberg, S.: Homing and synchronizing sequences. In: Broy, M., Jonsson, B., Katoen, J.-P., Leucker, M., Pretschner, A. (eds.) Model-Based Testing of Reactive Systems. LNCS, vol. 3472, pp. 5–33. Springer, Heidelberg (2005)
5. Volkov, M.V.: Synchronizing automata and the Černý conjecture. In: Martín-Vide, C., Otto, F., Fernau, H. (eds.) LATA 2008. LNCS, vol. 5196, pp. 11–27. Springer, Heidelberg (2008)
6. Eppstein, D.: Reset sequences for monotonic automata. SIAM J. Comput. 19, 500–510 (1990)
7. Kozen, D.: Lower bounds for natural proof systems. In: Proc. of 18th FOGS, pp. 254–266 (1977)

Consecutive Ones Property Testing: Cut or Swap

Mathieu Raffinot[*]

LIAFA, Université Paris Diderot—Paris 7, 75205 Paris Cedex 13, France
raffinot@liafa.jussieu.fr

Abstract. Let C be a finite set of n elements and $\mathcal{R} = \{r_1, r_2, \ldots, r_m\}$ a family of m subsets of C. The family $\mathcal{R}$ verifies the consecutive ones property if there exists a permutation P of C such that each r_i in $\mathcal{R}$ is an interval of P. Several algorithms have been proposed to test this property in time $O(\sum_{i=1}^{m} |r_i|)$, all being involved. We present a simpler algorithm, based on a new partitioning scheme.

1 Introduction

Let $C = \{c_1, \ldots, c_n\}$ be a finite set of n elements and $\mathcal{R} = \{r_1, r_2, \ldots, r_m\}$ a family of m subsets of C. Those sets can be seen as a 0-1 matrix, where set C represents the columns and each r_i the ones of row i. Figure 1 shows such a matrix.

The family $\mathcal{R}$ verifies the consecutive ones property (C1P) if there exists a permutation P of C such that each r_i in $\mathcal{R}$ is an interval of P. For instance, the family given by the matrix of 1 verifies C1P. Efficiently testing C1P has received a lot of attention in the literature for this problem to be strongly related to the recognition of interval graphs, the recognition of planar graphs, modular decomposition and others graph decompositions. The consecutive ones property is the core or many other algorithms that have applications in a wide range of domains, from VLSI circuit conception through planar embeddings [11] to computational biology for the reconstruction of ancestral genomes [1,4].

We denote $|\mathcal{R}| = \sum_{i=1}^{m} |r_i|$. Several $O(|\mathcal{R}|)$ time algorithms have been proposed to test this property, following five main approaches.

The first approach and still the most well known one is the use of PQ-tree structure [2]. A PQ-tree represents a set of permutations defined by the possible orders of its leaves obtained by changing the order of the children of any node depending of its type. The nodes are either P or Q nodes. For a P node, any order of its children is valid, while for a Q node only the complete reversal of its children is accepted. For instance, in Figure 1, the PQ-tree represents the order $c_4c_2c_6c_1c_3c_7c_5c_8$, but also $c_4c_2c_6c_7c_3c_1c_5c_8$, $c_4c_6c_2c_7c_3c_1c_8c_5$, and so on. If a family verifies C1P, then one can build a PQ-tree representing all column

---

[*] This work was partially supported by the french ANR-2010-COSI-004 MAPPI Project.

B. Löwe et al. (Eds.): CiE 2011, LNCS 6735, pp. 239–249, 2011.

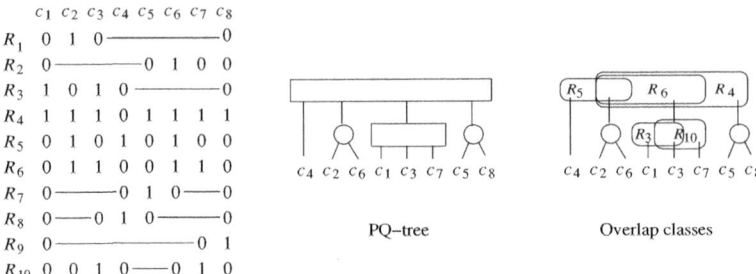

Fig. 1. A matrix verifying the consecutive ones property, its associated PQ-tree and the information contained in overlap classes. In the PQ-tree, Q nodes are represented by boxes, while P nodes by circles.

orders for which the C1P will be verified. For instance, the PQ-tree in Figure 1 represents all orders for which the family given by the matrix at its rights verifies C1P. If a family does not verify C1P, its associated PQ-tree is said empty.

Given a family, in order to build its associated PQ-tree, each row is inserted one after the other in the tree while the PQ-tree is not empty. This update is done through a procedure called Refine which complexity is amortized on the size of the tree. The main drawback of this approach is that the implementation of Refine in its right complexity is still a challenge. It uses a series of 11 templates depending on the form of the tree and choosing which to use in constant time is a huge programming difficulty, that has only slightly been reduced by Young [13] using a recursive Refine that allows us to reduce the number of templates. Moreover, extracting a certificate that the family does really not verify C1P from this approach is hard. Therefore, given a PQ-tree implementation, one can hardly be confident neither in its validity nor in its time complexity. This is the reason why many other algorithmic approaches have been tempted to test C1P using simpler and/or certified algorithms.

One of those attempt consists in first transforming the C1P testing problem to interval graph recognition by adding fake rows and then use a special LexBFS traversal that produces a first order on C that has some special properties [6]. A recursive partitioning phase is then necessary following both this LexBFS order and an order on the rows derived from a clique tree built from the LexBFS traversal. This approach is also complex, both to understand and to program, and surprisingly the links between these two first approaches are not that clear.

A third approach was to try to design the PC-tree [9], an easiest structure to refine than the PQ-tree. However as Haeupler and Tarjan noticed in [7], the authors did not consider "implementations issues" (sic) than lead to incorrect algorithms for the recognition of planar graphs.

A fourth approach appeared in [8] with the idea of simplifying the C1P test by avoiding PQ-tree. However, the algorithm remains very involved.

A last and more recent approach has been presented by R. McConnell in [10]. This approach is a breakthrough in the understanding of the intrinsic constraints of C1P and the real nature of the PQ-tree. We describe this approach in details since our method is a tricky simplification of it. McConnell shows that each Q node of the PQ-tree represents in fact an overlap class of a subset of the rows. Two rows r_i and r_j of $\mathcal{R}$ overlap if $r_i \cap r_j \neq \emptyset$, $r_i \setminus r_j \neq \emptyset$, and $r_j \setminus r_i \neq \emptyset$. An overlap class is a equivalence class of the overlap relation, that is, two rows r_i and r_j are in the same class if there is a chain of overlaps from r_i to r_j. For instance, the two non trivial overlap classes of the family example given by the matrix of Figure 1 are shown on the same figure on the right.

Overlap classes partition the set of rows and form a laminar family, and thus they can be organized in an inclusion forest which is the skeleton of the PQ-tree. All remaining P nodes might also been derived from the overlap classes. However, for an overlap class to be a node of the PQ-tree, it also has to verify the consecutive one property. Thus, where is the gain ? The trick used by McConnell is that verifying the C1P of an overlap class is independent of the other overlap classes and somehow easier provided a spanning tree of the overlap graph of the class. Using a partitioning approach guided by this tree, it is linear in the total size of the rows in an overlap class to test if this overlap class verifies C1P. Consequently, by testing overlap classes one after the other, one can verify if the whole set $\mathcal{R}$ fulfills C1P in $O(|\mathcal{R}|)$ time. The technical complexity of the approach is twofold: (a) compute overlap classes and (b) a spanning tree of each class. Point (a) is performed in [10] through an algorithm of Dahlhaus published as a routine of [5] used for undirected graph split decomposition. It is considered by McConnell as a black box that takes as input $\mathcal{R}$ and returns a list of overlap classes and for each overlap class the list of rows that belongs to. Point (b) is then computed in [10] for each overlap class by a complex add-on from the list of rows in the class.

In this article we present a simplification of McConnell's approach by introducing a new partitioning scheme. It should be noted first that McConnell's approach can already be very slightly simplified using existing tools. Indeed, the algorithm of Dahlhaus for computing overlap classes is an algorithmic pearl that has been recently simplified and made computationable in the sense that the original version uses an LCA while the simplified version presented in [3] only uses partitioning. Moreover, a modification of Dahlhaus's approach allows us to extract a spanning tree of each overlap class. This modification is not obvious but remains simpler than the add-on of [10]. However, building a spanning tree from Dahlhaus is intrinsically difficult, because the two concepts are somehow antinomic: Dahlhaus's approach maintains some ambiguities in the row overlaps that permit to gain on the overall computation, while computing a spanning tree requires solving most of these ambiguities, which is sometimes difficult. In this paper, we successfully maintain these ambiguities even in the partitioning phase, avoiding the spanning tree.

To clearly present our approach let us consider the difference between the PQ-tree approach and that of McConnell in terms of partitioning. In the first, we maintain and update a partition even if rows can be included one in the other (that does not *cut* any class of the partition). This is where the difficulty arises: in the same time we manage both a partition and an inclusion tree. In the second approach the idea is to impose that each row added surely overlaps a previous one, which simplifies the partitioning and insures the linear time complexity, but at the cost of the computation of a spanning tree of each overlap class.

Our approach lies in between. For each overlap call we update a partition, but we also allow some fail and swap in the partitioning scheme. We compute an order that guaranties that when adding a new row r_1, if it does not overlap any row already considered, then the row following r_2 will, and moreover r_1 overlaps r_2 and will be considered next. We thus swap r_1 and r_2 in the order we update the partition if r_1 does not cut. We call this order a "swap overlap order". This order could of course be obtained from a spanning tree, but we explain below how we can compute such an order at a very small computational price by entering deeper in Dahlhaus's algorithm, that we also slightly simplify for our needs. Our algorithm thus runs in 3 main steps: (1) the computation of each overlap class using an algorithm close to that of Dahlhaus, (2) for each class the computation of a swap overlap order, and (3) a partitioning of each class guided by this order using a new partitioning scheme. If the partitioning fails on a class, the C1P is not verified. Steps 1 and 2 are performed in the same time, but for clarity we present them in two distinct steps.

This article is organized as follows. In the following Section 2 we present a variation of Dahlhaus's algorithm for computing overlap classes. In Section 3 we explain our main notion of *swap overlap order* and explain how to slightly modify Dahlhaus's algorithm to generate such an order for each overlap class. In Section 4 we eventually explain how to test C1P on each overlap class using the swap overlap order associated to.

2   Computing Overlap Classes

In this section we recall and slightly modify the algorithm of Dahlhaus for computing overlap classes already simplified and presented in [3]. The computational problem to efficiently compute the overlap classes comes from the fact that the underlying overlap graph, where r_i are the vertices and (r_i, r_j) is an edge if r_i overlaps r_j, might have $\Theta(|\mathcal{R}|^2)$ edges and thus be quadratic in $O(|\mathcal{R}|)$. An *overlap class* is a connected component of this graph.

Let LR be the list of all $r \in \mathcal{R}$ sorted in decreasing size order. The ordering of sets of equal size is arbitrarily fixed, and thus LR is a total order. Given $r \in \mathcal{R}$, we denote Max(r) as the largest row $X \in \mathcal{R}$ taken in LR order such that $X <_{LR} r$ and X overlaps r. This definition is modified from that in [3] to consider the order LR in the definition of Max(r).

Note that Max(r) might be undefined for some sets of $\mathcal{R}$. In this latter case, in order to simplify the presentation of some technical points, we write Max(r) = $\emptyset$. Dahlhaus's algorithm is based on the following observation:

Lemma 1 ([5,3]). *Let $r \in \mathcal{R}$ such that Max(r) $\neq \emptyset$. Then for all $X \in \mathcal{R}$ such that $X \cap r \neq \emptyset$ and $|r| \leq |X| \leq |Max(r)|$, X overlaps r or Max(r).*

The trick we propose below for computing the overlap order of each overlap class is also based on lemma 1.

Let us assume first that we already computed all $\text{Max}(r)$. For each column $c \in \mathcal{C}$ we compute the list $SL(c)$ of all sets $r \in \mathcal{R}$ to which c belongs. This list is sorted in increasing order of the sizes of the sets respecting LR, thus in decreasing order in LR. Computing and sorting all lists for all $c \in \mathcal{C}$ can be done in $O(|\mathcal{R}|)$ time using a stable bucket sort.

Dahlhaus's overlap class identification is built on those lists. For all $c \in \mathcal{C}$, Let r be a set containing c such that $\text{Max}(r) \neq \emptyset$. We define a new interval on $SL(c)$ beginning in r, continuing from r in the order of $SL(c)$ and finishing by the greatest row in $SL(c)$ such that $|Y| \leq |\text{Max}(r)|$. Notice that this greatest row Y is not necessarily equal to $\text{Max}(r)$. If it is the case, the interval is said of type M (for Max included), of type E (for External) otherwise. Given an interval I, $\text{First}(I)$ is the first row of the interval, thus the row which generates the interval.

We "bucket" sort the intervals in a table $TI[1..m]$ of m entries the following way. For an interval $I = [r_{i_1} \ldots r_{i_k}]$, I is added to all $T[i_j]$, $1 \geq j \geq k$. We refer the reader to [12] for an example of a family and the intervals associated to.

To compute overlap classes, we mark them one after the other, keeping the numbering of the overlap class each row belongs to in a table $NC[1..m]$ all initialized to 0.

Algorithm A: computing all overlap classes

1. Initialize the counter $nc = 1$ to count the overlap class we are tagging;
2. Choose an arbitrary $l, 1 \leq l \leq m$ such that there exist at least on interval in $TI[l]$ of type M;
3. For all interval(s) $I = [r_{i_1} \ldots r_{i_k}]$ of type M in $TI[l]$,
   (a) remove all occurrences of I out of TI;
   (b) mark each row in I to belong to overlap class nc, thus $NC[i_j] = nc$, $1 \leq j \leq k$;
   (c) recurse this algorithm from step 3 on all i_j, $1 \leq j \leq k$, such that $TI[i_j]$ is not empty;
4. For all interval(s) $J = [r_{i_1} \ldots r_{i_k}]$ of type E in $TI[l]$,
   (a) remove all occurrences of J out of TI;
   (b) mark each row in J to belong to overlap class nc, thus $NC[i_j] = nc$, $1 \leq j \leq k$;
   (c) recurse this algorithm from step 3 on all i_j, $1 \leq j \leq k$, such that $TI[i_j]$ is not empty;
   (d) end the recursive procedure;
5. Increment nc and apply step 2 while $TI[l]$ is not empty.

Rows that are not marked during this algorithm are themselves an overlap class of a single element that it is not necessary to consider further for testing C1P. We focus below on overlap classes that contain at least 2 rows. Algorithm A only differs of Dahlhaus's algorithm by considering intervals of type M before interval of type E. In each overlap class there exist at least one interval of type

M to begin with at step 2, and by lemma 1, all rows in a given interval belong to the same overlap class.Thus, we have:

Proposition 1 ([5]). *Algorithm A computes all overlap classes of $\mathcal{R}$.*

Worst case complexity of Algorithm A. Algorithm A can be implemented to run in $O(|\mathcal{R}|)$, provided that for a given row r computing $\text{Max}(r)$ is $O(1)$ time (we refer the reader to [12] for details on this computation).

3   Swap Overlap Order

A swap overlap order is an order $r_{i_1} \ldots r_{i_k}$ on the rows of an overlap class such that, for all $2 \leq l \leq k$, at least one of the two following cases is true: either (a) r_{i_l} overlaps one $r_{i_g}, 1 \leq g < l$, or (b) $l < k$ and $r_{i_{l+1}}$ overlaps $r_{i_g}, 1 \leq g < l$, and r_{i_l} overlaps $r_{i_{l+1}}$. We now modify Algorithm A to output for each overlap class a swap overlap order.

Algorithm B: outputing a swap overlap order for all overlap classes

1. Initialize the counter $nc = 1$ to count the overlap class we are tagging; Initialize O_{nc} to the empty word ϵ,
2. Choose an arbitrary $l, 1 \leq l \leq m$ such that there exist at least on interval in $TI[l]$ of type M;
3. For all interval(s) $I = [r_{i_1} \ldots r_{i_k}]$ of type M in $TI[l]$,
   (a) remove all occurrences of I out of TI;
   (b) concatenate to O_{nc} successively the rows $r_{i_1}, r_{i_k}, r_{i_2}, .. , r_{i_{k-1}}$ in this order, adding a row only if $NC[i_j] = 0$. After adding a row, change $NC[i_j]$ to *no*.
   (c) recurse this algorithm from step 3 on all $i_j, 1 \leq j \leq k$, such that $TI[i_j]$ is not empty;
4. For all interval(s) $J = [r_{i_1} \ldots r_{i_k}]$ of type E in $TI[l]$,
   (a) remove all occurrences of J out of TI;
   (b) recurse step 3 on $TI[i_1]$;
   (c) concatenate to O_{nc} successively the rows $r_{i_2}, r_{i_3}, .. , r_{i_k}$ in this order, adding a row only if $NC[i_j] = 0$. After adding a row, change $NC[i_j]$ to *no*.
   (d) recurse this algorithm from step 3 on all $i_j, 1 < j \leq k$, such that $TI[i_j]$ is not empty;
   (e) end the recursive procedure;
5. Increment nc and apply step 2 while $TI[l]$ is not empty.

The main difference with Algorithm A in terms of recursive call is step 4.(b), where we first recurse on $\text{First}(J)$ when considering an interval of type E before processing the interval itself. A trace of the execution of Algorithm B can be found in [12]. For the largest overlap class of our current example, it returns the swap overlap order $O_1 = r_2 r_3 r_4 r_5 r_7 r_1 r_9 r_{11}$.

What is the idea behind algorithm B ? We begin an order by considering and interval of type M, say $I = [r_{i_1} \ldots r_{i_k}]$. By placing r_{i_1} and then $r_{i_k} = \text{Max}(r_{i_1})$ before all other rows in I, Lemma 1 guaranties that the following rows in I overlap either r_{i_1} or r_{i_k}.

Then, assume that there exits a row X between r_{i_1} and r_{i_k} in I. We recurse on X. If the line corresponding to X in TI contains and interval, say $I' = [r'_{i_1} \ldots r'_{i_{k'}}]$, it be of two types, M or E. **Case 1**. If I' is of type M then it will be process first before all type E intervals corresponding to X. Then, either X is the fist row of the interval, either not. Whatever, as X already appears in O_{nc} by interval I, then by concatenating the rows in the order $r'_{i_1} r'_{i_{k'}} \ldots$ if not already in O_{nc}, we guaranty that: (a) one of $r'_{i_{k'}} = \text{Max}(r'_{i_1})$ or r'_{i_1} overlaps X that is already placed in O_{nc} by Lemma 1, and (b) each following row in I', if any, either overlaps $r'_{i_{k'}}$ or r'_{i_1}, or already appears in O_{nc}. **Case 2**. If I' is of type E, then r'_{i_k} is not $\text{Max}(r'_{i_1})$. Thus there is not guaranty that $\text{Max}(r'_{i_1})$ (that has to exist since I' is an interval beginning in r'_{i_1}) has already been placed in O_{nc}. Thus we first recurse on r'_{i_1} (step 4-(a)) to guaranty that after some recursion the rows r'_{i_1} and $\text{Max}(r'_{i_1})$ appear somewhere in O_{nc} before processing I. Then, by lemma 1, each row following r'_{i_1} in I' overlaps either $\text{Max}(r'_{i_1})$ or r'_{i_1}. As both are already in O_{nc}, we simply concatenate them to O_{nc} in step 4-(c).

Thus, summarizing the 2 cases, when concatenating new rows to O_{nc}, we can insure that either (a) we add a couple $(X, \text{Max}(X))$, provided that at least one of those rows overlaps a row Y already placed in O_{nc} (note that if one of those rows is already in O_{nc}, then the result also holds), or (b) a row X that surely overlaps a row already in O_{nc}. Using this approach we identify each overlap class and in the same time we build a swap overlap order for each overlap class. **Complexity.** It is obvious that the time complexity is the same that Algorithm A, that is, $O(|\mathcal{R}|)$.

4   Partitioning Each Overlap Class

At this point, we built a swap overlap order for each non trivial overlap class. It remains to explain how to test C1P on each such class using this order.

We use a partitioning that is relatively similar to that of [10], except that instead of being driven by a spanning tree it uses a swap overlap order that is easier to build since it is in the direct continuation of Dahlhaus's approach for computing overlap classes. However, the important difference is that using a swap overlap order we can not certify that we *cut* each time the current partition when refined by a new row. Instead, we can certify that if the new row r_1 does not cut, the following row r_2 will, and r_1 will then cut r_2. We thus *swap* the two rows in the partitioning.

Let us enter details. We maintain an ordered set of sets, called *parts*, of columns of $\mathcal{C}$. When adding a row, a part C can only be cut in two parts $C'C''$ such that $C' \cup C'' = C$ and $C' \cap C'' = \emptyset$. In the partitioning, C is replaced by $C'C''$ or $C''C'$ depending the case, but the general order of the initial partition is maintained.

To begin the partitioning phase, we consider the first row r_{i_1} of the overlap order $O_{nc} = r_{i_1} r_{i_2} \ldots r_{i_k}$ of overlap class nc. We create a first part in our partition P_1 that is composed of the columns of r_{i_1}. We then refine this partition with r_2 by first marking all elements of r_2. Suppose first that r_2 overlaps (or *cuts*) r_1 and let $X = r_1 \cap r_2$. We partition P by r_2 in $P_2 = (r_1 \setminus X)(X)(r_2 \setminus X)$, thus we simply placed all common elements of r_1 and r_2 on a line in such a way that both r_1 and r_2 are intervals of P, which is the core of the C1P.

Let us now consider a new row r_{i_j}. We mark elements of r_{i_j} in P_{j-1}. Suppose again that r_{i_j} cuts a row already integrated to P_{j-1}. Let Y be the set of elements of r_{i_j} that already appear in P_j. Two cases may occur:

(a) if $Y = r_{i_j}$, we only try to group together the elements of r_3 in P_2. If we can, we only cut the parts accordingly to build P_j
(b) if $Y \neq r_{i_j}$, then we try to cluster the elements of Y on a border (left or right) of P_{j-1}. If we can, we cut the parts accordingly and add a new part $(r_{i_j} \setminus Y)$ before (resp. after) all parts of P_{j-1} if the border was the left (resp. right) one to eventually build P_j.

	Row	Columns	Partition
Example of partitioning on the first overlap class of our current data set with the order $r_2 r_3 r_4 r_5 r_7 r_3 r_1 r_9 r_{11}$. The main point of this approach is that if this process fails for a given row, the overlap call does not verify C1P.	r_2	$\{b, c, d\}$	(bcd)
	r_3	$\{c, d, e, f, g, h\}$	(b)(cd)(efgh)
	r_4	$\{d, e\}$	(b)(c)(d)(e)(fgh)
	r_5	$\{e, f, g, h\}$	(b)(c)(d)(e)(fgh)
	r_7	$\{b, h\}$	fail

Proposition 2 ([10]). *Let $r_{i_1} r_{i_2} \ldots r_{i_k}$ be a total order of the rows of a given overlap class nc such that each row $r_{i_j}, j > 2$, overlaps a previous row $r_{i_l}, 1 \leq l < j$. Then the above partitioning fails if and only if the overlap class nc does not verify C1P.*

In our approach, as we manipulate swap overlap orders, the partitioning phase must be slightly modified in the following way. Suppose that we want to refine the partition P_{j-1} with r_{i_j}. If r_{i_j} does not overlap any previous row use in the partitioning, that is if all columns of r_{i_j} either belong to the same part of P_{j-1} of to none, we swap r_{i_j} and $r_{i_{j+1}}$, refine the partition with $r_{i_{j+1}}$ and only then with r_{i_j}. The swap overlap order guaranties that $r_{i_{j+1}}$ will cut a previous row, and that r_{i_j} overlaps $r_{i_{j+1}}$. We call this partitioning a *swap partitioning*.

Theorem 1. *Let $r_{i_1} r_{i_2} \ldots r_{i_k}$ be a swap overlap order of the rows of a given overlap class nc. Then the above swap partitioning fails if and only if the overlap class nc does not verify C1P.*

Proof. By swapping the rows when necessary, we insure that the order of the $r_{i_1} r_{i_2} \ldots r_{i_k}$ rows in which we refine the partition verifies that each row $r_{i_j}, j > 2$, overlaps a previous row $r_{i_l}, 1 \leq l < j$, thus satisfying the conditions of proposition 2. □

Implementation issues. Let us now consider the time complexity of our partitioning. We show below how it might be implemented in time $O(|O_{nc}|)$ where $|O_{nc}|$ is the sum of the size of all rows belonging to the overlap class.
The data structure we need must allow us to

1. split a part C in $C'C''$ in the number of the elements of C touched;
2. add a new part to the left of to the right of the current partition in the number of the elements added;
3. test if the elements touched can be made consecutive;
4. test if a new row cut another one already embedded in the partition;

There might be many data structures implementation having these properties. We propose below a simple one. This structure can also replace that used in [3] for identifying all $\text{Max}(X)$ used by Dahlhaus's algorithm, and thus our whole algorithm only uses a single data structure.

We basically use an array of size $|\mathcal{C}|$ to store a stack which encodes a permutation of elements of C. Each cell of this array contains a column and a link to the part it belongs to. A part is coded as a pair of its beginning and ending positions in the array, relatively to the beginning of the array. A schematic representation of this data structure is given in Figure 2.

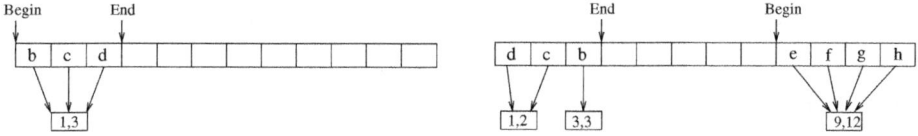

Fig. 2. Example continued: implementation of (bcd) and then $(efgh)(dc)(b)$ when refining $r_2 = \{b, c, d\}$ by $r_3 = \{c, d, e, f, g, h\}$

Using this data structure, refining a part C by one of its subset C'' can be easily done in $O(|C''|)$. Indeed, let $[i, j]$ be the bounds of C. We swap elements in the subtable $[i, j]$ to place all $s = |C''|$ elements of C'' at the end or at the beginning of this subtable as necessary. We then adjust the bounds of C to $[i, j - s]$ or $[i + s, j]$ depending of the case and create a new set $[j - s + 1, j]$ or $[i, i + s - 1]$ on which the s elements of C'' now point.

Adding a new part to the left of to the right of the current partition in the number of the elements added is easy since it suffices to create a new part and move the pointers of the beginning or ending modulo $|\mathcal{C}|$. An example of such operation is shown in Figure 2.

Assume that a new row r used for refining cut a class in the partition P, and let $Y \subset r$ be the elements of r that are already in the partition.

If $Y \neq r$, then, to verify C1P, all classes touched by Y must be placed at an extremity of P, all parts from this extremity must be fully touched except the last one of which all elements touched has to be placed on the side of the extremity we considered. All these requirements can easily be checked in the number of elements of r, and if they are verified, a new part containing $r \setminus Y$ is added to the extremity.

If $Y = r$, then to verify C1P there should be a left part that might not be fully touched followed by a series (that can be empty) of plenty touched parts and eventually a last part also not necessary fully touched. This is also not difficult to check in $O(|r|)$.

The novelty in our approach is that a new row r might not cut the current partition, which has to be tested efficiently. This can also easily be checked in $O(|r|)$ on our structure. Indeed, it suffices to test if r is included in a single part, in none, or contains all parts. We thus have:

Theorem 2. *Testing the C1P of the rows belonging to a same overlap class can be done in $O(|O_{nc}|)$ time provided a swap overlap order O_{nc} of it.*

And eventually:

Corollary 1. *Testing the C1P of a family $\mathcal{R}$ can be done in $O(|\mathcal{R}|)$ using a swap overlap order of each overlap class.*

Proof. It suffices to compute all overlap classes of $\mathcal{R}$ using Algorithm B that provides for each overlap class a swap overlap order. Then Theorem 2 insures that C1P can be tested on each overlap class in the number of rows belonging to this class. As overlap classes partition $\mathcal{R}$ and that $\mathcal{R}$ verifies C1P if an only if each overlap class verifies C1P, the whole test can be done in $O(|\mathcal{R}|)$ time. □

References

1. Blin, G., Rizzi, R., Vialette, S.: A faster algorithm for finding minimum tucker submatrices. In: Ferreira, F., Löwe, B., Mayordomo, E., Mendes Gomes, L. (eds.) CiE 2010. LNCS, vol. 6158, pp. 69–77. Springer, Heidelberg (2010)
2. Booth, K.S., Lueker, G.S.: Testing for the consecutive ones properties, interval graphs and graph planarity using pq-tree algorithm. J. Comput. Syst. Sci. 13, 335–379 (1976)
3. Charbit, P., Habib, M., Limouzy, V., de Montgolfier, F., Raffinot, M., Rao, M.: A note on computing set overlap classes. Information Processing Letters 108(4), 186–191 (2008)
4. Chauve, C., Tannier, E.: A methodological framework for the reconstruction of contiguous regions of ancestral genomes and its application to mammalian genomes. PLoS Comput. Biol. 4(11), 11 (2008)
5. Dahlhaus, E.: Parallel algorithms for hierarchical clustering and applications to split decomposition and parity graph recognition. J. Algorithms 36(2), 205–240 (2000)
6. Habib, M., McConnell, R., Paul, C., Viennot, L.: Lex-bfs and partition refinement, with applications to transitive orientation, interval graph recognition and consecutive ones testing. Theoretical Computer Science 234, 59–84 (2000)
7. Haeupler, B., Tarjan, R.E.: Planarity algorithms via pq-trees (extended abstract). Electronic Notes in Discrete Mathematics 31, 143–149 (2008)
8. Hsu, W.-L.: A simple test for the consecutive ones property. J. Algorithms 43(1), 1–16 (2002)

9. Hsu, W.-L., McConnell, R.M.: PC-trees and circular-ones arrangements. Theoretical Computer Science 296, 99–116 (2003)

10. McConnell, R.M.: A certifying algorithm for the consecutive-ones property. In: SODA, pp. 768–777 (2004)

11. Nishizeki, T., Rahman, M.S.: Planar Graph Drawing. World Scientific, Singapore (2004)

12. Raffinot, M.: Consecutive ones property testing: cut or swap. CoRR, abs/1008.3786 (2010)

13. Young, S.: Implementation of PQ-tree Algorithms. Master's thesis, University of Washington (1977)

Current Developments on Computational Modeling Using P Systems

Agustín Riscos-Núñez*

Research Group on Natural Computing, Department of Computer Science and
Artificial Intelligence, University of Sevilla, Avda. Reina Mercedes s/n, 41012, Sevilla,
Spain
ariscosn@us.es

Abstract. A P system based general framework for modeling ecosystems dynamics will be described. Roughly speaking, the idea is to use several regions (environments) that can be connected, each of them containing a probabilistic P system with active membranes (having identical skeleton for every environment).

Some real case studies will be displayed, discussing the usefulness of this tool in simulating complex ecosystems dynamics to aid managers, conservationists and policy-makers in making appropriate decisions for the improvement of management and conservation programs.

Membrane Computing is a quite active research field, initiated by Păun in 1998 [6]. It is a theoretical machine-oriented model, where the computational devices (known as *P systems*) are in some sense an abstraction of a living cell. There exist a large number of different definitions of P systems models, but most of them share some common features: a *membrane structure* (defining in a natural way a number of regions or compartments), and an alphabet of *objects* that are able to evolve and/or move within the membrane structure according to a *set of rules* (emulating the way substances undergo biochemical reactions in a cell).

Recently, P systems are being used as tools for modeling purposes, adapting their semantics in an appropriate way. Indeed, interesting achievements are being obtained, which show that membrane computing can be an alternative to classical modeling frameworks (e.g. those based on differential equations). Works on this direction rely on the development of associated simulation software, since P systems have not yet been implemented neither in hardware nor in biological means.

The talk will try to provide a general overview of modeling frameworks for systems biology and population dynamics based on P systems (see e.g. [1,7]), displaying some real case studies [2,3]. It is interesting to note that some biological parameters of the modeled processes can be obtained experimentally

* The author acknowledges the support of the projects TIN2008-04487-E and TIN2009-13192 of the *Ministerio de Ciencia e Innovación* of Spain, cofinanced by FEDER funds, and the support of the Project of Excellence with *Investigador de Reconocida Valía* of the *Junta de Andalucía*, grant P08-TIC-04200.

B. Löwe et al. (Eds.): CiE 2011, LNCS 6735, pp. 250–251, 2011.

by Biologists or Ecologists (and therefore they can be taken into account when designing the model), while some other relevant constants may be unknown. Software tools are thus necessary to enable virtual experimentation, as well as for the process of model validation [4,5].

References

1. Cardona, M., Colomer, M.A., Margalida, A., Palau, A., Pérez-Hurtado, I., Pérez-Jiménez, M.J., Sanuy, D.: A computational modeling for real ecosystems based on P systems. Natural Computing, online version doi: 10.1007/s11047-010-9191-3
2. Cardona, M., Colomer, M.A., Margalida, A., Pérez-Hurtado, I., Pérez-Jiménez, M.J., Sanuy, D.: A P system based model of an ecosystem of some scavenger birds. In: Păun, G., Pérez-Jiménez, M.J., Riscos-Núñez, A., Rozenberg, G., Salomaa, A. (eds.) WMC 2009. LNCS, vol. 5957, pp. 182–195. Springer, Heidelberg (2010)
3. Colomer, M.A., Margalida, A., Sanuy, D., Pérez-Jiménez, M.J.: A bio-inspired computing model as a new tool for modeling ecosystems: The avian scavengers as a case study. Ecological modelling 222(1), 33–47 (2011)
4. García-Quismondo, M., Gutiérrez-Escudero, R., Pérez-Hurtado, I., Pérez-Jiménez, M.J., Riscos-Núñez, A.: An Overview of P-Lingua 2.0. In: Păun, G., Pérez-Jiménez, M.J., Riscos-Núñez, A., Rozenberg, G., Salomaa, A. (eds.) WMC 2009. LNCS, vol. 5957, pp. 264–288. Springer, Heidelberg (2010)
5. Martínez-del-Amor, M.A., Pérez-Hurtado, I., Pérez-Jiménez, M.J., Riscos-Núñez, A., Sancho-Caparrini, F.: A simulation algorithm for multienvironment probabilistic P systems: A formal verification. Int. J. Foundations of Computer Science 22(1), 107–118 (2011)
6. Păun, G.: Computing with membranes. Journal of Computer and System Sciences 61(1), 108–143 (2000); Turku Center for Computer Science-TUCS Report No 208
7. Păun, G., Romero-Campero, F.J.: Membrane Computing as a Modeling Framework. Cellular Systems Case Studies. In: Bernardo, M., Degano, P., Tennenholtz, M. (eds.) SFM 2008. LNCS, vol. 5016, pp. 168–214. Springer, Heidelberg (2008)

Automata on Ordinals and Linear Orders

Philipp Schlicht[1] and Frank Stephan[2],[*]

[1] Mathematisches Institut, Rheinische Friedrich-Wilhelms-Universität Bonn,
Endenicher Allee 60, 53115 Bonn, Germany
schlicht@math.uni-bonn.de
[2] Department of Mathematics, National University of Singapore,
Block S17, 10 Lower Kent Ridge Road, Singapore 119076
fstephan@comp.nus.edu.sg

Abstract. We investigate structures recognizable by α-automata with running time a limit ordinal α. The domain of such a structure consists of finite α-words with gaps. An α-automaton resembles a finite automaton but has a limit rule which maps the set of states which appear cofinally often before the limit to a limit state. We determine the suprema of the α-automatic ordinals and the ranks of α-automatic linear orders. The power of α-automata increases with every power of ω.

1   Introduction

Various types of automata indexed by ordinals have been studied for instance by Büchi, Choueka [1], Wojciechowski [6]. Let us consider a finite automaton which has read an infinite input string. The run of the automaton can be extended to the ordinals by specifying the behaviour at limits via a transition function; this is the extra component that distinguishes automata on ordinals from finite automata. We consider the following type of automaton with running time a limit ordinal α.

Definition 1. Suppose α is a limit ordinal and Σ is a finite alphabet. An α-*automaton* consists of

- a finite set S of states,
- an initial state,
- a set of accepting states,
- a successor transition function $S \times (\Sigma \cup \{\diamond\}) \to S$, and
- a limit transition function $\mathcal{P}(S) \to S$.

The input of such an α-automaton is a word of length α in a given finite alphabet extended by an extra symbol $\diamond$. The automaton will run for α steps and read all letters of the α-word successively. At each limit $\lambda \leq \alpha$ the limit state is determined by a transition function which maps the set of states occurring cofinally often before λ to the state at λ. After running for α steps, the transition function is applied again and the input is accepted if the state at α is an

---

[*] This work is supported in part by NUS grant R252-000-420-112.

B. Löwe et al. (Eds.): CiE 2011, LNCS 6735, pp. 252–259, 2011.

accepting state. Note that in comparison, for a Büchi automaton there is only one limit transition rule at stage ω which checks if at least one of the accepting states occurred infinitely often. Thus a Büchi automaton can be simulated by a ω-automaton in the sense of this paper. Since we do not repeat the definition and the basic properties of finite automata, we refer to [3] for an introduction to the theory of automata and automatic structures.

Example 2. *Suppose an ω^n-automaton moves into state 0 in successor steps and into state $m + 1$ in limits, where m is the maximum of the states appearing cofinally often before the limit. This detects the limit type of the current step: in any step of the form $\omega^{n-1}n_{n-1} + \omega^{n-2}n_{n-2} + \ldots + \omega^m n_m$ with $n_m \neq 0$, the state is $m < n$.*

Any ω^n-automaton can be converted into an ω^n-automaton accepting the same words whose state varies depending on the limit type of the current step, by forming the product with an automaton detecting the limit type.

An α-*automatic presentation* of a relational structure M in a finite language is a structure $N \cong M$ whose domain consists of α-words together with α-automata for the domain and each relation in N accepting exactly the words in the domain or the respective relation of N. Functions will be replaced by their graphs, thus converting any structure into a relational structure. Let us say an α-word is *finite* if all but finitely many of its letters are equal to $\diamond$.

Definition 3. A structure is *finite word α-automatic*, or α-*automatic* for short, if it is has an α-automatic presentation whose domain consists of finite α-words.

Note that if necessary, we can mark the end of every finite string in the domain of a presentation by attaching an extra symbol, and thus obtain another α-automatic presentation of the same structure. It is not hard to prove that the finite word ω-automatic structures are exactly the automatic structures, i.e. structures with a presentation by finite automata which do not run for ω steps but halt at the end of the finite input word. Comparison of the current input letter with the limit type yields an ω^n-automatic presentation of the following set.

Example 4. *The set of finite ω^n-words whose letter at place $\omega^{n-1}n_{n-1} + \omega^{n-2} n_{n-2} + \ldots + \omega^m n_m$ where $n_m \neq 0$ is either n_m or $\diamond$, is ω^n-automatic.*

A natural question is: what is the supremum of ordinals β so that $(\beta, <)$ is α-automatic? Delhommé [2] proved that the supremum of the automatic ordinals is ω^ω and we first conjectured that the supremum of the α-automatic ordinals is ω^α, however this is false for any ϵ with $\omega^\epsilon = \epsilon$. Since $(\alpha, <)$ is easily α-automatic and any finite product of α-automatic structures is again α-automatic, the supremum is at least α^ω.

Example 5. *Let $(n_0, \ldots, n_k) <^* (m_0, \ldots, m_l)$ if $k < l$ and $n_i = m_i$ for all $i \leq k$, or there is $i \leq \min\{k, l\}$ with $n_i < m_i$ and $n_j = m_j$ for all $j < i$. This is a well ordering of type ω^ω. The representation of the finite sequence $(n_0, \ldots, n_k)$ by the* *finite* ω^2-*word* $0^{n_0}1\diamond^\omega \ldots 0^{n_k}1\diamond^\omega \diamond^{\omega^2}$ *leads to an* ω^2-*automatic presentation of* $<^*$.

Thus $(\omega^\omega, <)$ is ω^2-automatic and the supremum of the ω^2-automatic ordinals is at least ω^{ω^2}. A straightforward generalization of this example shows that ω^β is $\omega \cdot \beta$-automatic and the supremum of the $\omega \cdot \beta$-automatic ordinals is at least $\omega^{\beta \cdot \omega}$.

To prove that ω^{ω^2} is not ω^2-automatic, we will follow Delhommé's proof [2,3] that ω^ω is not automatic. In generalizing this proof, the step of splitting the domain of an automatic structure of order type ω^n into finitely many pieces is a problem. For automatic structures the pieces of the partition are parametrized by finite words of fixed length so that one of the pieces necessarily has order type ω^n. However, in this setting, the words have infinite length so that the domain is split into infinitely many pieces and the argument breaks down. We will find a piece of the partition of the same order type when ω^n is replaced by a sufficiently closed ordinal such as $\omega^{\omega \cdot n}$. To do this, note that there are only finitely many possibilities, or types, in which two pieces can be arranged relative to each other. Thus the ordinals occurring in α-automatic structures can be characterized via a commutative product of ordinals. This leads to the computation of the supremum of the α-automatic ordinals and the supremum of ranks of α-automatic linear orders for limit ordinals α.

2   Ordinals

Let us introduce a *finite type product* of ordinals which naturally occurs in automatic structures. We fix the following notation. When C is a set of ordinals, let $\mathrm{tp}(C)$ denote its order type and $C(\mu)$ its μth element. When C, D are sets of ordinals, let $\mathrm{tp}(C, D)$ denote the isomorphism type of $(C \cup D, <, C, D)$.

Definition 6. $\alpha \cdot_{\mathrm{fin}} \beta$ is the supremum of the ordinals γ such that there is $(C_\delta : \delta < \epsilon)$ with $\gamma = \bigcup_{\delta < \epsilon} C_\delta$ and

- $\forall \delta < \epsilon \ \mathrm{tp}(C_\delta) \leq \alpha$,
- there are only finitely many $\mathrm{tp}(C_\delta, C_\eta)$ for $\delta, \eta < \epsilon$, and
- $\mathrm{tp}(\mathrm{Tr}_\mu) \leq \beta$ for $\mathrm{Tr}_\mu = (C_\delta(\mu) : \delta < \epsilon)$ (the *trace* of μ) where $\mu < \alpha$.

To define $\mathrm{tp}(\mathrm{Tr}_\mu)$, we identify Tr_μ with its image. Note that the length ϵ of the sequence can be arbitrarily large. We will see that there is always such a sequence $(C_\delta : \delta < \epsilon)$ with $\alpha \cdot_{\mathrm{fin}} \beta = \bigcup_{\delta < \epsilon} C_\delta$. Let us call such a sequence an *instance* of $\alpha \cdot_{\mathrm{fin}} \beta$.

Remark 7. If $(C_\delta : \delta < \epsilon)$ is as above, then there are only finitely many $\mathrm{tp}(\mathrm{Tr}_\mu)$.

Proof. There are finitely many $\mathrm{tp}(C_\gamma, C_\delta)$ and hence there is a finite set F of ordinals such that $\mathrm{tp}(C_\gamma \upharpoonright F, C_\delta \upharpoonright F)$ determines $\mathrm{tp}(C_\gamma, C_\delta)$ for all γ, δ. To each η we associate the function f_η which maps each type $\mathrm{tp}(G, H)$ for $|G| = |H| = |F|$ to $0, 1$ or 2 depending on whether $C_\gamma(\eta) < C_\delta(\eta)$, $C_\gamma(\eta) = C_\delta(\eta)$, or $C_\gamma(\eta) >$

$C_\delta(\eta)$ for all γ, δ with $\mathrm{tp}(C_\gamma \restriction F, C_\delta \restriction F) = \mathrm{tp}(G, H)$. We partition α into finitely many pieces so that f is constant on each piece. Then $\mathrm{tp}(\mathrm{Tr}_\eta) = \mathrm{tp}(\mathrm{Tr}_\zeta)$ for η, ζ in the same piece.

The finite type product is the same as the well-known commutative product.

Definition 8. [5] Suppose $\alpha = \Sigma_i \omega^{\alpha_i}$ and $\beta = \Sigma_j \omega^{\beta_j}$ are in Cantor normal form. The *commutative sum* $\alpha \oplus \beta$ is defined as the sum of all ω^{α_i} and ω^{β_j} arranged in decreasing order. The *commutative product* $\alpha \otimes \beta$ is defined as the sum of all $\omega^{\alpha_i \oplus \beta_j}$ arranged in decreasing order.

Note that the commutative sum and product are strictly monotone in both coordinates.

Lemma 9. $\alpha \cdot_{\mathrm{fin}} \beta = \alpha \otimes \beta$ for all α, β.

Proof. We construct an instance of $\alpha \cdot_{\mathrm{fin}} \beta$ of order type $\alpha \otimes \beta$ such that $\mathrm{tp}(C_\delta) = \alpha$ and $\mathrm{tp}(\mathrm{Tr}_\mu) = \beta$ for all δ, μ. Suppose $\alpha = \omega^{\bar\alpha} = \omega^{\Sigma_i \omega^{\bar\alpha_i}}$ and $\beta = \omega^{\bar\beta} = \omega^{\Sigma_i \omega^{\bar\beta_i}}$. Let us suppose the least term in the Cantor normal form of $\bar\alpha \oplus \bar\beta$ is $\omega^{\bar\alpha_k}$ and suppose $(C_\alpha : \alpha < \epsilon)$ is an instance of $\omega^{\bar\alpha - \omega^{\bar\alpha_k}} \otimes \omega^{\bar\beta} = \omega^{(\bar\alpha \oplus \bar\beta) - \omega^{\bar\alpha_k}}$. We let $C_{\alpha,0} = C_\alpha$ and find sets $C_{\alpha,\gamma}$ for $0 < \gamma < \delta := \omega^{\omega^{\bar\alpha_k}}$ such that

- $\mathrm{tp}(C_{\alpha,\gamma}, C_{\beta,\gamma}) = \mathrm{tp}(C_\alpha, C_\beta)$ for all $\alpha, \beta < \epsilon$ and all $\gamma < \delta$, and
- $\sup(C_{\alpha,\gamma}) < \min(C_{\beta,\eta})$ for all $\alpha, \beta < \epsilon$ and all $\gamma < \eta < \delta$.

Let $C_\alpha := \bigcup_{\gamma < \delta} C_{\alpha,\gamma}$ for $\alpha < \epsilon$. Then $(C_\alpha : \alpha < \epsilon)$ is an instance of $\alpha \cdot_{\mathrm{fin}} \beta$ of order type $\alpha \otimes \beta$. If the least term in $\bar\alpha \oplus \bar\beta$ is $\omega^{\bar\beta_k}$, we choose instead an instance $(C_\alpha : \alpha < \epsilon)$ of $\omega^{\bar\alpha} \otimes \omega^{\bar\beta - \omega^{\bar\beta_k}} = \omega^{(\bar\alpha \oplus \bar\beta) - \omega^{\bar\beta_k}}$ and define $C_{\alpha,\gamma}$ as before. Then $(C_{\alpha,\gamma} : \alpha < \epsilon, \gamma < \delta)$ is an instance of $\alpha \cdot_{\mathrm{fin}} \beta$ of order type $\alpha \otimes \beta$.

Suppose $\alpha = \alpha_0 + \alpha_1$ with $\alpha_1 = \omega^\alpha$ and $\beta = \beta_0 + \beta_1$ with $\beta_1 = \omega^\beta$. We can choose (by inductive hypothesis) instances $(C_\delta^{i,j} : \delta < \epsilon_{i,j})$ of $\alpha_i \otimes \beta_j$ and there is freedom to choose the relative position of $C_\gamma^{i,j}$ and $C_\delta^{k,l}$ for $(i,j) \neq (k,l)$. We can assume that $\epsilon_{i,j} = \epsilon$ for all (i,j). It is not difficult to see from the definition of the commutative sum that the relative positions of $C_\delta^{i,j}$ can be arranged so that their union has order type $\alpha \otimes \beta = \Sigma_{i,j}^\oplus \alpha_i \otimes \beta_j$. Let $C_\delta = C_\delta^{0,0} \cup C_\delta^{1,0}$ for $\delta < \epsilon$ and $C_\delta = C_{\beta_0+\delta}^{1,0} \cup C_{\beta_0+\delta}^{1,1}$ for $\epsilon \leq \delta < \epsilon \cdot 2$. Then $(C_\delta : \delta < \epsilon \cdot 2)$ is an instance of $\alpha \otimes \beta$.

In the finite type product, each C_δ comes with its increasing enumeration. We define a variant of the finite type product by replacing the enumerations of the sets C_δ with arbitrary partial functions from α onto C_δ. Let us call this product $\alpha \cdot_p \beta$. We prove $\alpha \cdot_p \beta \leq \alpha \otimes \beta$ by induction and this implies $\alpha \cdot_{\mathrm{fin}} \beta = \alpha \cdot_p \beta = \alpha \otimes \beta$ for all α and β. It is sufficient to prove this for $\alpha = \omega^\alpha$ and $\beta = \omega^\beta$, since it can be extended to all α and β using the fact that $\alpha \otimes \beta$ is the maximal order type of sets which are unions of two sets of order types α and β. There exists $\theta < \mathrm{tp}(C_\delta)$ for every $\delta < \epsilon$ such that for all ζ, η, if $\sup C_\zeta < \sup C_\eta$ and $C_\eta(\theta)$ is defined, then $\sup C_\zeta < C_\eta(\theta)$. We can assume that $\sup C_\gamma < \sup C_\eta$ implies $\sup C_\gamma < \min C_\eta$ since this can be achieved by splitting C into finitely many pieces. We may also assume that $\mathrm{tp}(C_\gamma) = \mathrm{tp}(C_\delta)$ for all $\gamma, \delta < \epsilon$. If there are $\gamma \neq \delta$ such that $\sup C_\gamma \neq \sup C_\delta$, then every proper initial segment of $\bigcup_{\gamma < \epsilon} C_\gamma$ is bounded by $\alpha \cdot_{\mathrm{fin}} \bar{\beta} = \alpha \otimes \bar{\beta} < \alpha \otimes \beta$ for some $\bar{\beta} < \beta$. If $\sup C_\gamma = \sup C_\delta$ for all $\gamma, \delta < \epsilon$, then every proper initial segment is bounded by $\bar{\alpha} \cdot_{\mathrm{fin}} \beta = \bar{\alpha} \otimes \beta$ for some $\bar{\alpha} < \alpha = \mathrm{tp}(C_\gamma)$.

If α is a limit, then every $\alpha \cdot n$-automaton can be simulated by an α-automaton and thus every $\alpha \cdot n$-automatic structure is α-automatic. Although not every α-automatic structure is necessarily ω^γ-automatic for $\omega^\gamma \leq \alpha < \omega^{\gamma+1}$, the following arguments easily adapt to show that the supremum of the α-automatic ordinals is the same as that of the ω^γ-automatic ordinals. If x is an α-word, let $|x|$ denote the least $\beta \leq \alpha$ such that $x(\gamma) = \diamond$ for all $\beta \leq \gamma < \alpha$.

Proposition 10. *Suppose $\alpha = \omega \cdot \beta = \omega^\gamma$. Then $\omega^{\beta \cdot \omega}$ is the supremum of the α-automatic ordinals.*

Proof. The ordinals $\omega^{\beta \cdot n}$ are α-automatic as in Example 5. Let us suppose there is an α-automatic structure of order type at least $\omega^{\beta \cdot \omega}$. We follow the proof for automatic ordinals [3, Theorem 17]. We choose an element u_n of rank $\omega^{\beta \cdot n}$ for each $n \geq 1$ and write the set of its predecessors as

$$u_n \downarrow = Y \sqcup \bigsqcup_{|v|=|u_n|} X_v^{u_n},$$

where $X_v^u = \{vw : vw < u\}$ and $Y = \{x : |x| < |u_n| \& x < u_n\}$. Since $|u_n| < \alpha$, every $|u_n|$-automatic ordinal is strictly below ω^β by the induction hypothesis and $\gamma := \mathrm{tp}(Y) < \omega^\beta$. The number of types $\mathrm{tp}(X_v^u, X_w^u)$ is bounded by the product of the numbers of states of the automata on inputs (u_n, v) and (v, v), since the type depends only on the states reached after reading u_n and v. Hence the sets $X_v^{u_n}$ form a finite type product. Since each Tr_δ is ϵ-automatic for some $\epsilon < \alpha$, $\mathrm{tp}(\mathrm{Tr}_\delta) < \omega^\beta$. Let δ be the maximal $\mathrm{tp}(X_v^{u_n})$. Then $\omega^{\beta \cdot n} \leq \gamma \otimes \delta$ and thus $\delta = \omega^{\beta \cdot n}$. This implies that for any n there are u_n, v_n with $|u_n| = |v_n|$ and $\mathrm{tp}(X_{v_n}^{u_n}) = \omega^{\beta \cdot n}$, however there are only finitely many types.

In particular, the supremum of the ω^n-automatic ordinals is ω^{ω^n}.

3   Linear Orders

In a similar fashion as for ordinals, we analyze the ranks of α-automatic linear orders following [4]. Since there is no trace function for families of linear orders, let us define a finite type product of linear orders as follows. If $(f_n : \alpha \to C_n : n < m)$ is a tuple of partial onto functions and $C = \bigcup_{n<m} C_n$, let $\mathrm{tp}(f_0, ..., f_{m-1})$ denote the isomorphism type of the two-sorted structure $(C, \alpha, <, f_0, ..., f_{m-1})$.

Definition 11. A linear order C is a *finite type product* of linear orders A and B if there is are sequences $(C_\gamma : \gamma < \epsilon)$ of subsets of C and $(f_\gamma : \alpha \to C_\gamma : \gamma < \epsilon)$ of partial onto functions with

- $C = \bigcup_{\gamma < \epsilon} C_\gamma$,
- $C_\gamma \hookrightarrow A$ for all $\gamma < \epsilon$,
- for each n and $\gamma_i < \epsilon$, there are only finitely many $\mathrm{tp}(f_{\gamma_0}, ..., f_{\gamma_n})$, and
- $\mathrm{ran}(g_\mu) \hookrightarrow B$ for all $\mu < \alpha$, where $g_\mu(\gamma) = f_\gamma(\mu)$.

If L is a linear order and $(L_i : i \in L)$ is a family of linear orders, the L-sum of $(L_i : i \in L)$ is defined as the lexicographic order on pairs (i, j) where $i \in L$ and $j \in L_i$.

Definition 12. Suppose L is a linear order. Let $\mathrm{rk}(L) = 0$ if L is finite. Let $\mathrm{rk}(L) \leq \alpha$ if L is a $\mathbb{Z} \cdot n$-sum of linear orders of rank below α for some n.

The linear orders into which $(\mathbb{Q}, <)$ does not embed are called *scattered*. Easily every scattered linear order has a rank.

Lemma 13. *Suppose C is a scattered linear order and C is a finite type product of A and B. Then $\mathrm{rk}(C) \leq \mathrm{rk}(A) \oplus \mathrm{rk}(B)$.*

Proof. Suppose $(C_\gamma : \gamma < \epsilon)$ and $(f_\gamma : \alpha \to C_\gamma : \gamma < \epsilon)$ witness the finite type product. Let us split A and B into finitely many $\mathbb{Z}$-sums of linear orders of rank below $\mathrm{rk}(C)$ and accordingly C into finitely many finite type products. We can assume that $\sup C_\gamma < \sup C_\delta$ implies $C_\gamma < C_\delta$ and $\inf C_\gamma < \inf C_\delta$ implies $C_\gamma < C_\delta$ and that each f_γ is total, since we can achieve this by splitting C into finitely many pieces and the rank is closed under finite unions. Let $A = \Sigma_{z \in \mathbb{Z}} A_z$ and $B = \Sigma_{z \in \mathbb{Z}} B_z$ such that $\mathrm{rk}(A_z) < \mathrm{rk}(A)$ and $\mathrm{rk}(B_z) < \mathrm{rk}(B)$. We partition C into $C^0 < C^1 < C^2$ such that

- $\sup C_\gamma = \sup C_\delta$ and $C_\gamma \subseteq C^i$ imply $C_\delta \subseteq C^i$,
- if $\{\inf C_\gamma : C_\gamma \subseteq C^i\}$ has a minimum and $C_\gamma, C_\delta \subseteq C^i$, then $\inf C_\gamma = \inf C_\delta$, and
- if $\{\sup C_\gamma : C_\gamma \subseteq C^i\}$ has a maximum and $C_\gamma, C_\delta \subseteq C^i$, then $\sup C_\gamma = \sup C_\delta$.

If there exist $\gamma \neq \delta$ with $\sup C_\gamma \neq \sup C_\delta$, we may choose $C^0, C^2 \neq \emptyset$. It suffices to prove for all $i \leq 2$ and $c < d$ in C^i that $\mathrm{rk}([c, d]) < \mathrm{rk}(A) \oplus \mathrm{rk}(B)$. If $\{\inf C_\gamma : C_\gamma \subseteq C^0\}$ has no minimum, then $[c, d]$ is a subset of a finite type product of A and $\Sigma_{z \in F} B_z$ for some finite $F \subseteq \mathbb{Z}$ and hence $\mathrm{rk}([c, d]) \leq \mathrm{rk}(A) \oplus \mathrm{rk}(\Sigma_{z \in F} B_z) < \mathrm{rk}(A) \oplus \mathrm{rk}(B)$ by the induction hypothesis. If $C_\gamma, C_\delta \subseteq C^0$ implies $\inf C_\gamma = \inf C_\delta$ and hence $\sup C_\gamma = \sup C_\delta$, then $[c, d]$ is a subset of a finite type product of $\Sigma_{z \in F} A_z$ and B for some finite $F \subseteq \mathbb{Z}$ and hence $\mathrm{rk}([c, d]) \leq \mathrm{rk}(\Sigma_{z \in F} A_z) \oplus \mathrm{rk}(B) < \mathrm{rk}(A) \oplus \mathrm{rk}(B)$. The argument for $c, d \in C^2$ is similar. For all $c < d$ in C^1, $[c, d]$ is a subset of a finite type product of A and $\Sigma_{z \in F} B_z$ for some finite $F \subseteq \mathbb{Z}$ and hence $\mathrm{rk}([c, d]) < \mathrm{rk}(A) \oplus \mathrm{rk}(B)$.

If $\sup C_\gamma = \sup C_\delta$ and hence $\inf C_\gamma = \inf C_\delta$ for all γ, δ, let $C^0, C^2 = \emptyset$. For all $c < d$ in C^1, $[c, d]$ is a subset of a finite type product of $\Sigma_{z \in F} A_z$ and B for some finite $F \subseteq \mathbb{Z}$ and hence $\mathrm{rk}([c, d]) \leq \mathrm{rk}(\Sigma_{z \in F} A_z) \oplus \mathrm{rk}(B) < \mathrm{rk}(A) \oplus \mathrm{rk}(B)$.

This is sufficient to determine the supremum of ranks of α-automatic scattered linear orders, similar to ordinals.

Proposition 14. *Suppose* $\alpha = \omega \cdot \beta = \omega^{\gamma}$. *Then* $\beta \cdot \omega$ *is the supremum of ranks of scattered* α-*automatic linear orders.*

Proof. Instead of choosing an element u_n of rank $\omega^{\beta \cdot n}$ as we did for ordinals, we choose intervals $[u_n^0, u_n^1]$ of rank $\omega^{\beta \cdot n}$ for each n, and proceed as in the proof of Proposition 10.

The rank can be extended to non-scattered linear orders via a finite condensation function c_{FC} which amalgamates elements of a partial order with only finitely many elements in between. Let c_{FC}^{α} be the αth iterate of c_{FC} and $\mathrm{rk}(L)$ the least α so that $c_{FC}^{\alpha}(L)$ does not contain a convex suborder isomorphic to ω or ω^{*}. Note that this implies $c_{FC}^{\alpha+1}(L) = c_{FC}^{\alpha+2}(L)$.

Proposition 15. *Suppose* $\alpha = \omega \cdot \beta = \omega^{\gamma}$. *Then* $\beta \cdot \omega$ *is the supremum of ranks of* α-*automatic linear orders.*

Proof. Note that every linear order A is a dense sum of scattered linear orders. It is sufficient to prove that there is a uniform bound below $\beta \cdot \omega$ on the rank of scattered intervals. Suppose $[u_0, u_1]$ is a scattered interval in A and $\nu = \max\{|u_0|, |u_1|\}$. Let us write

$$[u_0, u_1] = Y \sqcup \bigsqcup_{|v|=\nu} X_v^{u_0, u_1},$$

where $X_v^{u_0, u_1} = \{vw \in A : u_0 < vw < u_1\}$ and $Y = \{x : u_0 < x < u_1 \,\&\, |x| < \nu\}$. Then Y is μ-automatic for some $\mu < \omega \cdot \beta$ and hence $\mathrm{rk}(Y) < \beta$ by the induction hypothesis. The number of isomorphism types of $X_v^{u_0, u_1}$ for $u_0, u_1, v \in A$ is bounded by the product of the numbers of states of the automata on inputs (u_0, v), (u_1, v), and (v, v). Hence there is m such that $\mathrm{rk}(X_v^{u_0, u_1}) \leq \beta \cdot m$ for all $u_0, u_1, v \in A$. Since the sets $X_v^{u_0, u_1}$ form a finite type product, $\mathrm{rk}(\bigsqcup_{|v|=\nu} X_v^{u_0, u_1}) \leq \beta \cdot m \oplus \beta$.

We leave open the following questions:

- Is the theory of any α-automatic structure decidable?
- Is the isomorphism problem for α-automatic ordinals decidable if α is computable?
- Is every countable infinite word α-automatic structure finite word α-automatic?

References

1. Choueka, Y.: Finite automata, definable sets, and regular expressions over ω^n-tapes. J. Comput. System Sci. 17(1), 81–97 (1978)
2. Delhommé, C.: Automaticité des ordinaux et des graphes homogènes. C. R. Math. Acad. Sci. Paris 339(1), 5–10 (2004)
3. Khoussainov, B., Minnes, M.: Three lectures on automatic structures. In: Logic Colloquium 2007. Lect. Notes Log., vol. 35, pp. 132–176. Assoc. Symbol. Logic, La Jolla, CA (2010)

4. Khoussainov, B., Rubin, S., Stephan, F.: Automatic linear orders and trees. ACM Trans. Comput. Log. 6(4), 675–700 (2005)
5. Levitz, H.: The Cartesian product of sets and the Hessenberg natural product of ordinals. Czechoslovak Math. J. 29(104)(3), 353–358 (1979)
6. Wojciechowski, J.: Finite automata on transfinite sequences and regular expressions. Fund. Inform. 8(3-4), 379–396 (1985)

A Fine Hierarchy of ω-Regular k-Partitions

Victor Selivanov[*]

A.P. Ershov Institute of Informatics Systems,
Siberian Division Russian Academy of Sciences,
Lavrentyev av. 6, 630090 Novosibirsk, Russia
vseliv@iis.nsk.su

Abstract. We develop the theory of ω-regular k-partitions (for arbitrary $k \geq 2$) that extends the theory around the Wagner hierarchy of regular ω-languages. In particular, we characterize the structure of Wadge degrees of ω-regular k-partitions, prove the decidability of any level of the corresponding hierarchy, establish coincidence of the reducibilities by continuous functions and by functions computed by finite automata on the ω-regular k-partitions.

1 Introduction

In [Wag79], Wagner gave in a sense the finest possible topological classification of regular ω-languages (i.e., of subsets of X^ω for a finite alphabet X recognized by finite automata). In particular, he characterized the quotient-poset $(\mathbb{R}; \leq_{\mathrm{CA}})$ of the preorder $(\mathcal{R}; \leq_{\mathrm{CA}})$ formed by the class $\mathcal{R}$ of regular subsets of X^ω and the reducibility $\leq_{\mathrm{CA}}$ by functions continuous in the Cantor topology on X^ω. In [Se98] the Wagner hierarchy was related to the Wadge hierarchy and to the author's fine hierarchy.

The aim of this paper is to generalize this theory from the case ω-regular languages to the case of regular k-partitions $\alpha : X^\omega \to k$ of X^ω, i.e. k-tuples $(A_0, \ldots, A_{k-1})$ of pairwise disjoint regular sets satisfying $A_0 \cup \cdots \cup A_{k-1} = X^\omega$. Note that the ω-languages are in a bijective correspondence with 2-partitions of X^ω. Motivation for this generalization comes from the fact that similar objects are quite important in descriptive set theory, computability theory and complexity theory, see also [Ko00, KW00]. In particular, we will extend to k-partitions the following facts from [Wag79]:

1. The structure $(\mathbb{R}; \leq_{\mathrm{CA}})$ is almost well-ordered with the order type ω^ω.
2. The CA-reducibility coincides on $\mathcal{R}$ with the reducibility by functions computed by deterministic asynchronous finite transducers.
3. Any level of the Wagner hierarchy is decidable.

We will characterize the structure $(\mathbb{R}_k; \leq_{\mathrm{CA}})$ of Wadge degrees of ω-regular k-partitions, prove the decidability of any level of the corresponding hierarchy,

---

[*] Supported by DFG-RFBR (grant No 436 RUS 113/1002/01, grant No 09-01-91334).

B. Löwe et al. (Eds.): CiE 2011, LNCS 6735, pp. 260–269, 2011.

and establish the coincidence of CA- and DA-reducibilities on the ω-regular k-partitions. Though for $k \geq 3$ the structure $(\mathbb{R}_k; \leq_{\mathrm{CA}})$ turns out to be much more complicated than for the case of sets (e.g., the first-order theory of $(\mathbb{R}_k; \leq_{\mathrm{CA}})$ is undecidable for $k \geq 3$ while it is decidable for $k = 2$) [Se07a], our proofs are not hard (modulo [Wag79, Se98, Se07a]). They are even shorter than the existing corresponding proofs for the particular case $k = 2$. This is because our definition of the fine hierarchy of k-partitions here is distinct (even for $k = 2$) from the definition in [Se98] though of course equivalent to it for $k = 2$). This enables us to separate automata-theoretic and topological arguments from combinatorial arguments related to ordinal arithmetic. For some related hierarchies of k-partitions see, e.g., [H93, Ko00, KW00, Se04, Se07, Se07a].

2 Preliminaries

Fix a finite alphabet X containing more than one symbol (for simplicity we may assume that $X = m = \{x \mid x < m\}$ for a natural number $m > 1$, so $0, 1 \in X$). Let X^* and X^ω denote respectively the sets of all words and of all ω-words (i.e. sequences $\alpha : \omega \to X$) over X. The empty word is denoted by ε. Let $X^{\leq \omega} = X^* \cup X^\omega$. We use some standard notation concerning words and ω-words. For $w \in X^*$ and $\xi \in X^{\leq \omega}$, $w \sqsubseteq \xi$ means that w is a substring of ξ, $w \cdot \xi = w\xi$ denote the concatenation, $l = |w|$ is the length of $w = w(0) \cdots w(l-1)$. For $w \in X^*, W \subseteq X^*$ and $A \subseteq X^{\leq \omega}$, let $w \cdot A = \{w\xi : \xi \in A\}$ and $W \cdot A = \{w\xi : w \in W, \xi \in A\}$. For $k, l < \omega$ and $\xi \in X^{\leq \omega}$, let $\xi[k, l] = \xi(k) \cdots \xi(l-1)$ and $\xi[k] = \xi[0, k)$.

The set X^ω carries the *Cantor topology* with the open sets $W \cdot X^\omega$, where $W \subseteq X^*$. Continuous functions in this topology are called also CA-*functions* [Wag79]. A CS-*function* is a function $f : X^\omega \to X^\omega$ satisfying $f(\xi)(n) = \phi(\xi[n+1])$ for some $\phi : X^* \to X$. Every CS-function is a CA-function. Both classes of functions are closed under composition.

By an *automaton* (over X) we mean a triple $\mathcal{A} = (Q, f, i)$ consisting of a finite non-empty set Q of states, a transition function $f : Q \times X \to Q$ and an initial state $i \in Q$. The transition function is naturally extended to the function $f : Q \times X^* \to Q$ defined by induction $f(q, \varepsilon) = q$ and $f(q, u \cdot x) = f(f(q, u), x)$, where $u \in X^*$ and $x \in X$. Similarly, we may define the function $f : Q \times X^\omega \to Q^\omega$ by $f(q, \xi)(n) = f(q, \xi[n])$. Relate to any automaton $\mathcal{A}$ the set of *cycles* $C_{\mathcal{A}} = \{f_{\mathcal{A}}(\xi) \mid \xi \in X^\omega\}$ where $f_{\mathcal{A}}(\xi)$ is the set of states which occur infinitely often in the sequence $f(i, \xi) \in Q^\omega$.

A *Muller acceptor* has the form $(\mathcal{A}, \mathcal{F})$ where $\mathcal{A}$ is an automaton and $\mathcal{F} \subseteq C_{\mathcal{A}}$; it recognizes the set $L(\mathcal{A}, \mathcal{F}) = \{\xi \in X^\omega \mid f_{\mathcal{A}}(\xi) \in \mathcal{F}\}$. It is well known that Muller acceptors recognize exactly the *regular ω-languages* called also regular sets in this paper. The class $\mathcal{R}$ of all regular ω-languages is a proper subclass of $\mathrm{BC}(\Sigma_2^0)$ that in turn is a proper subclass of Δ_3^0. As usual, $\Sigma_n^0, \Pi_n^0, \Delta_n^0$ denote levels of the Borel hierarchy [Ke94] and $\mathrm{BC}(\mathcal{C})$ the Boolean closure of a class of sets $\mathcal{C}$. The notion of a Muller acceptor is extended to the notion of a *Muller k-acceptor* as follows: it is a pair $(\mathcal{A}, c)$ where $\mathcal{A}$ is an automaton and $c : C_{\mathcal{A}} \to k$ is a k-partition of $C_{\mathcal{A}}$. Such a k-acceptor recognizes the k-partition $L(\mathcal{A}, c) = c \circ f_{\mathcal{A}}$ where $f_{\mathcal{A}} : X^{\omega} \to C_{\mathcal{A}}$ is the function defined above. We have the following simple characterization of the ω-regular k-partitions established in [Se07a].

Proposition 1. *A k-partition $L : X^{\omega} \to k$ is regular iff it is recognized by a Muller k-acceptor.*

Functions on ω-words computed by synchronous (resp. asynchronous) transducers are called DS-functions (resp. DA-functions) [Wag79]. As is well known, both of these classes of functions are closed under composition, and every DS-function (resp. DA-function) is a CS-function (resp., CA-function).

A k-partition A is said to be CA-*reducible* to a k-partition B (in symbols $A \leq_{\mathrm{CA}} B$), if $A = B \circ g$ for some CA-function $g : X^{\omega} \to X^{\omega}$. The relations $\leq_{\mathrm{DA}}$, $\leq_{\mathrm{CS}}$ and $\leq_{\mathrm{DS}}$ on $k^{X^{\omega}}$ are defined in the same way but using the other three classes of functions. The CA-reducibility is known in descriptive set theory as the Wadge reducibility, and CS-reducibility as the Lipschitz reducibility. We also need the more recent notion of $\mathbf{\Delta}_2^0$-reducibility [An06], i.e. the reducibility by functions f such that $f^{-1}(A) \in \mathbf{\Delta}_2^0$ whenever $A \in \mathbf{\Delta}_2^0$ (such functions are called $\mathbf{\Delta}_2^0$-*functions*). Since CA-reducibility coincides with $\mathbf{\Delta}_1^0$-reducibility, $A \leq_{\mathrm{CA}} B$ implies $A \leq_{\mathbf{\Delta}_2^0} B$. By taking functions from X^{ω} to Y^{ω} one can extend the above-mentioned reducibilities to compare k-partitions of X^{ω} and Y^{ω} for different alphabets X, Y. The introduced relations on $k^{X^{\omega}}$ are preorders.

By $\equiv_{\mathrm{CA}}$ we denote the induced equivalence relation which gives rise to the corresponding quotient partial order. Following a well established jargon, we call this partial order the *structure of* CA-*degrees*. For a set A and a class of sets $\mathcal{C}$, $\mathcal{C} \leq_{\mathrm{CA}} A$ means that any set from $\mathcal{C}$ is CA-reducible to A. The same applies to the other reducibilities and to partitions in place of sets. Note that the set $\mathcal{R}_k$ of regular k-partitions is closed downwards under DA- and DS-reducibilities, but is not closed under CA- and CS-reducibilities. Any level of the Borel hierarchy is closed under CA-reducibility (and thus under all four reducibilities). Every $\mathbf{\Sigma}$-level $\mathcal{C}$ (and also every $\mathbf{\Pi}$-level) of the Borel hierarchy has a CA-complete set C which means that $\mathcal{C} = \{A : A \leq_{\mathrm{CA}} C\}$.

The operation $A \oplus B$ on k-partitions of X^{ω}, defined by $(A \oplus B)(0\xi) = A(\xi)$ and $(A \oplus B)(i\xi) = B(\xi)$ for all $0 < i < m$ and $\xi \in X^{\omega}$ (recall that $X = m$), induces the operation of supremum in the structures of degrees under all reducibilities introduced above. The set $\mathcal{R}_k$ is closed under the operation $\oplus$. The structure of CA-degrees of regular k-partitions is denoted by $(\mathbb{R}_k; \leq_{\mathrm{CA}})$, and similarly for the other reducibilities.

3   Labeled Trees and Forests

In this and the next section we briefly describe an extension of the fine hierarchy (FH) of sets [Se98] to the FH of k-partitions. More details are given in [Se10].

Let $(P; \leq)$ be a poset. Any subset of P may be considered as a poset with the induced partial ordering. By a *forest* we mean a finite poset in which every upper cone $\uparrow x$ is a chain. A *tree* is a forest having the largest element (called the *root* of the tree).

Let $(Q; \leq)$ be a poset. A *Q-poset* is a triple $(P, \leq, c)$ consisting of a finite nonempty poset $(P; \leq)$, $P \subseteq \omega$, and a labeling $c : P \to Q$. Such labeled posets will be often shorter denoted just by P. By default, we denote the labeling in a Q-poset by c. A *morphism* $f : (P, \leq, c) \to (P', \leq', c')$ of Q-posets is a monotone function $f : (P; \leq) \to (P'; \leq')$ satisfying $\forall x \in P(c(x) \leq c'(f(x)))$. Let $\mathcal{P}_Q$, $\mathcal{F}_Q$, $\mathcal{T}_Q$ and $\mathcal{C}_Q$ denote the sets of all finite Q-posets, Q-forests and Q-trees, respectively.

The *h-preorder* $\leq_h$ on $\mathcal{P}_Q$ is defined as follows: $P \leq_h P'$, if there is a morphism from P to P'. The quotient-posets of $\mathcal{P}_Q$, $\mathcal{F}_Q$ and $\mathcal{T}_Q$ with the h-preorder are denoted $\mathbb{P}_Q$, $\mathbb{F}_Q$ and $\mathbb{T}_Q$, respectively. Note that for the particular case $Q = \bar{k}$ of the antichain with k elements we obtain the preorders $\mathcal{P}_k$, $\mathcal{F}_k$, $\mathcal{T}_k$ and $\mathcal{C}_k$. Some properties of $\mathcal{P}_k, \ldots$ from [Se04, Se07] are extended to $\mathcal{P}_Q, \ldots$ in a straightforward way. In particular, the next assertion is a straightforward extension of Lemma 1.2 in [Se04].

Proposition 2. *Let Q be a preorder. For any finite Q-poset $(P, \leq, c)$ there exist a finite Q-forest $(F, \leq, d)$ which is a largest element in $(\{G \in \mathcal{F}_Q \mid G \leq_h P\}; \leq_h)$ (F is obtained by a top-down unfolding of P to a forest).*

By a *minimal Q-forest* we mean a finite Q-forest not h-equivalent to a Q-forest of lesser cardinality. The next extension of Theorem 1.4 in [Se04] is straightforward.

Proposition 3.   1. *Any singleton Q-forest is minimal.*
   2. *A non-singleton Q-tree $(T, \leq, c)$ is minimal iff the Q-forest $(T \setminus T(1), c)$ is minimal, $\forall x \in T(1) \forall y \in T(2)(c(y) \not\leq c(x))$ and if $|T(2)| = 1$ then $\forall x \in T(1) \forall y \in T(2)(c(x)|c(y))$.*
   3. *A proper (i.e., not h-equivalent to a Q-tree) Q-forest is minimal iff all its Q-trees are minimal and pairwise incomparable under $\leq_h$.*

For posets P and Q we write $P \subseteq Q$ (resp. $P \sqsubseteq Q$) if P is a substructure of Q (resp. P is an initial segment of Q). The next assertion is obvious, and its analog holds for $\mathcal{P}, \mathcal{T}$ in place of $\mathcal{F}$: Let P, Q be arbitrary posets. Then $P \subseteq Q$ implies $\mathbb{F}_P \subseteq \mathbb{F}_Q$, and $P \sqsubseteq Q$ implies $\mathbb{F}_P \sqsubseteq \mathbb{F}_Q$. Now we can iterate the construction $Q \mapsto \mathbb{F}_Q$ starting with the antichain $\bar{k}$. Define the sequence $\{\mathcal{F}_k(n)\}_{n < \omega}$ of preorders by induction on n as follows: $\mathcal{F}_k(0) = \bar{k}$ and $\mathcal{F}_k(n+1) = \mathcal{F}_{\mathcal{F}_k(n)}$. Identifying the elements $i < k$ of $\bar{k}$ with the corresponding minimal elements $s(i)$ of $\mathcal{F}_k(1)$, we may think that $\mathcal{F}_k(0) \sqsubseteq \mathcal{F}_k(1)$. By the assertions above, $\mathcal{F}_k(n) \sqsubseteq \mathcal{F}_k(n+1)$ for each $n < \omega$, $\mathcal{F}_k(\omega) = \bigcup_{n < \omega} \mathcal{F}_k(n)$ is a wqo and $\mathcal{F}_{\mathcal{F}_k(\omega)}$ clearly coincides with $\mathcal{F}_k(\omega)$. For any $n \leq \omega$, let $\mathbb{F}_k(n)$ be the quotient-poset of $\mathcal{F}_k(n)$.

Of course, similar constructions can be done with $\mathcal{P}$ or $\mathcal{T}$ in place of $\mathcal{F}$. The introduced preorders, especially $\mathcal{T}_k(\omega)$ and the set $\mathcal{T}_k^{\sqcup}(\omega)$ of finite joins (i.e., disjoint unions denoted by $\sqcup$) of elements in $\mathcal{T}_k(\omega)$, play an important role in the study of the fine hierarchy. Note that $\mathcal{F}_k(1) = \mathcal{F}_k$ and $\mathcal{T}_k(1) = \mathcal{T}_k$.

Recall that a *well quasiorder* (wqo) is a preorder that has neither infinite descending chains nor infinite antichains Well-known results of the wqo-theory imply the following (non-obvious) assertions: If Q is a wqo then $(\mathcal{F}_Q; \leq_h)$ and $(\mathcal{T}_Q; \leq_h)$ are wqo's. Therefore, $\mathcal{F}_k(\omega)$ and $\mathcal{T}_k^{\sqcup}(\omega)$ are also wqo's.

Let us briefly introduce some operation of labeled forests and collect their properties used in the sequel (all these operations respect the h-equivalence). For any $i < k$ and $F \in \mathcal{F}_k(\omega)$, let $p_i(F)$ be the tree in $\mathcal{F}_k(\omega)$ obtained from F by adjoining a new largest element labeled by i. By a straightforward extension of [Se04, Theorem 2.2], $(\mathcal{F}_k(\omega); \sqcup, \leq_h, p_0, \ldots, p_{k-1})$ is a dc-semilattice. Recall that a dc-semilattice is a structure $(S; \leq, \cup, p_0, \ldots, p_{k-1})$ such that: $\leq$ is a preorder; $x \cup y$ is a supremum of x, y in $(S; \leq)$; any p_i is a closure operation on $(S; \leq)$ (i.e., $x \leq p_i(x)$, $x \leq y \rightarrow p_i(x) \leq p_i(y)$ and $p_i(p_i(x)) \leq p_i(x)$); for all distinct $i, j < k$, $p_i(x) \leq p_j(y) \rightarrow p_i(x) \leq y$; every $p_i(x)$ is join-irreducible, i.e. $p_i(x) \leq y \cup z \rightarrow (p_i(x) \leq y \vee p_i(x) \leq z)$.

Define the binary operation $+$ on $\mathcal{F}_k(\omega)$ as follows: $F + G$ is obtained by adjoining a copy of G below any leaf of F. One easily checks that $i + F = p_i(F)$ (here $i < k$ is identified with the singleton tree labeled by i), $F \leq_h F + G$, $G \leq_h F + G$, $F \leq_h F_1 \rightarrow F + G \leq_h F_1 \leq_h G$, $G \leq_h G_1 \rightarrow F + G \leq_h F \leq_h G_1$, $(F + G) + H \equiv_h F + (G + H)$.

Define the unary operation s on $\mathcal{F}_k(\omega)$ as follows: $s(F)$ is a singleton tree carrying the label F (we identify $s(i)$ with the tree i for each $i < k$). Then $x \in \mathbb{F}_k(n + 1) \setminus \mathbb{F}_k(n)$ implies $s(x) \in \mathbb{F}_k(n + 2) \setminus \mathbb{F}_k(n + 1)$ for each $n < \omega$, and s is an embedding of $\mathbb{F}_k(n)$ into $\mathbb{F}_k(n + 1)$. For the operation $F * G = s(F) + G$ one easily checks the following properties:

Lemma 1. *1. For any $x \in \mathbb{F}_k(\omega)$, $x \mapsto x * y$ is a closure operation on $\mathbb{F}_k(\omega)$.*
2. For all $x, y, x_1, y_1 \in \mathbb{F}_k(\omega)$, if $x * y \leq x_1 * y_1$ and $x \not\leq x_1$ then $x * y \leq y_1$.

We need also the following binary operation on $\mathcal{F}_k'(\omega) = \mathcal{F}_k \cup \{\emptyset\}$: $R(F, G) = \{b \in F \mid \exists a \geq b(c(a) \not\leq_h G)\}$, with the partial order and labeling induced from F. One easily checks the following property:

Lemma 2. *Let $X * Y$ be a minimal (in the sense of Proposition 3) forest in $\mathcal{F}_k'(\omega)$. Then $F \leq_h X * Y$ iff $R(F, X) \leq_h Y$.*

For this paper, the preorders $\mathcal{T}_k^{\sqcup}(2)$, $\mathcal{F}_k(2)$, $\mathcal{P}_k(2)$ are especially relevant. By Proposition 2, for any $P \in \mathcal{P}_k(2)$ there is an $\leq_h$-largest element $F \in \mathcal{F}_k(2)$ below P. It is constructed in two steps: first unfold P to a labeled forest F_1, then unfold any label of F_1. Note that if any label in P has a largest element then $F \in \mathcal{T}_k^{\sqcup}(2)$.

We conclude this section by a result about a categorical structure on $\mathcal{P}_k(2)$. For $P, Q \in \mathcal{P}_k(2)$, an *explicit morphism* $\varphi : P \to Q$ is a pair (φ_0, φ_1) where φ_0 is a morphism from P to Q and $\varphi_1 = \{\varphi_{1,p_0}\}_{p_0 \in P}$ is a family of morphisms from $c(p_0)$ to $c(\varphi_0(p_0))$ (this notion makes use of the convention that $i = s(i)$ for each $i < k$). Note that $P \leq_h Q$ iff there is an explicit morphism from P ro Q. By a *k-labeled 2-preorder* we mean a tuple $(S; \leq_0, \leq_1, d)$ where S is a finite nonempty set, $d : S \to k$ is a labeling of S, and $\leq_0, \leq_1$ are preorders on S such that $x \leq_1 y$ implies $x \equiv_0 y$. By a morphism of k-labeled 2-preorders S, U we mean a function $\psi : S \to U$ that respects the labelings and the preorders.

Proposition 4. *The category $\mathcal{P}_k(2)$ with the explicit morphisms is equivalent to the category $\mathcal{S}_2(k)$ of k-labeled n-preordrs.*

4   Fine Hierarchy of k-Partitions

Let M be a set and α a positive ordinal. By an α-*base* in M we mean a sequence $\mathcal{L} = \{\mathcal{L}_\beta\}_{\beta<\alpha}$ of subsets of $P(M)$ such that each $\mathcal{L}_\beta$ is closed under finite unions and intersections and $\mathcal{L}_\gamma \subseteq \mathcal{L}_\beta \cap co\text{-}\mathcal{L}_\beta$ for all $\gamma < \beta < \alpha$. By a *morphism* of α-bases $\mathcal{K}, \mathcal{L}$ in U, M resectively, we mean a function $f : U \to M$ such that $f^{-1}(A) \in \mathcal{K}_\beta$ for all $\beta < \alpha$ and $A \in \mathcal{L}_\beta$. An α-base $\mathcal{L}$ is *reducible* if any $\mathcal{L}_\beta$ has the reduction property [Ke94]. For any $n < \omega$ the $(n+1)$-bases are sequences $(\mathcal{L}_0, \ldots, \mathcal{L}_n)$ of bounded lattices with the corresponding inclusions. In this section we deal mostly with 2-bases (though all facts extend easily to ω-bases). In fact, for this paper the 2-bases $\Sigma = (\boldsymbol{\Sigma}_1^0, \boldsymbol{\Sigma}_2^0)$ and $\mathcal{R}\Sigma = (\mathcal{R}\cap\boldsymbol{\Sigma}_1^0, \mathcal{R}\cap \boldsymbol{\Sigma}_2^0)$ are the most important. These 2-bases are known to be reducible [Se98].

In [Se04, Se07a] we considered difference hierarchy of k-partitions over 1-bases (called there bases); the FH defined below extends the difference hierarchy. Let $\mathcal{L}$ be a 2-base in M and $P \in \mathcal{P}_k(2)$. By a P-*family over* $\mathcal{L}$ we mean a family $\{B_{p_0}, B_{p_0,p_1}\}$ where $p = (p_0, p_1) \in P^\circ$ (see the end of Section 3), $B_{p_0} \in \mathcal{L}_0, B_{p_0,p_1} \in \mathcal{L}_1$, and $\tilde{B}_{p_0} = \bigcup\{B_{p_0,p_1} \mid p_1 \in c(p_0)\}$. Here $\tilde{B}_{p_0} = B_{p_0} \setminus \bigcup\{B_r \mid r \leq_0 p\}$, and similarly $\tilde{B}_{p_0,p_1} = B_{p_0,p_1} \setminus \bigcup\{B_r \mid r \leq_1 p\}$. To simplify notation we often denote families just by $\{B_p\}$. Note that $d(p) = c(p_1)$ is always in $\mathcal{P}_k(0) = \{0, \ldots, k-1\}$, and $\tilde{B}_{p_0} = \bigcup\{\tilde{B}_{p_0,p_1} \mid p_1 \in c(p_0)\}$. Obviously, $\bigcup_{p_0} B_{p_0} = \bigcup_{p\in P^\circ} \tilde{B}_p$. We call a P-family $\{B_p\}$ over $\mathcal{L}$ *consistent* if $d(p) = d(q)$ whenever the components $\tilde{B}_p$ and $\tilde{B}_q$ have a nonempty intersection. Any such consistent P-family determines the k-partition $A : \bigcup_{p_0} B_{p_0} \to k$ where $A(x) = d(p)$ for some (equivalently, for any) $p \in P^\circ$ with $x \in \tilde{B}_p$; we also say in this case that A is *defined by* $\{B_p\}$.

Let $\mathcal{L}^Y(P)$ denote the set of k-partitions $A : Y \to k$ defined by some P-family over $\mathcal{L}$ with $Y = \bigcup_{p_0} B_{p_0}$. In case $Y = X$ we omit the superscript X and call $\{\mathcal{L}(P)\}_{P\in\mathcal{P}_k(\omega)}$ the FH of k-partitions over $\mathcal{L}$.

Lemma 3.   1. *Any $A \in \mathcal{L}^Y(P)$ is defined by a monotone P-family $\{C_p\}$ (monotonicity means that $C_{q_0} \subseteq C_{p_0}$ for $q_0 \leq p_0$ and $C_{p_0,q_1} \subseteq C_{p_0,p_1}$ for $q_1 \leq p_1$).*
2. *Let $f : U \to M$ be a morphism of 2-spaces $\mathcal{K}, \mathcal{L}$, and let $A \in \mathcal{L}^Y(P)$. Then $f^{-1}(A) \in \mathcal{K}^{f^{-1}(Y)}(P)$.*
3. *If $P \leq_h Q$ then $\mathcal{L}^Y(P) \subseteq \mathcal{L}^Y(Q)$.*
4. *$\mathcal{L}^Y(P) = \mathcal{L}^Y(F(P))$ where $F(P)$ is from Proposition 3.*

For $F \in \mathcal{F}_k(\omega)$, a *reduced F-family over* $\mathcal{L}$ is a monotone F-family $\{B_p\}$ over $\mathcal{L}$ such that $B_{p_0} \cap B_{q_0} = \emptyset$ for all incomparable $p_0, q_0 \in F$ and $B_{p_0,p_1} \cap B_{p_0,q_1} = \emptyset$ for all incomparable $p_1, q_1 \in c(p_0)$. Let $\mathcal{L}_r^Y(F)$ be the set of partial k-partitions defined by the reduced F-families $\{B_p\}$ over $\mathcal{L}$ such that $\bigcup_{p_0} B_{p_0} = Y$. The next result is a straightforward extension of [Se04, Theorem 3.1].

Proposition 5. *Let the ω-base $\mathcal{L}$ be reducible, $Y \in \mathcal{L}_0$ and $F \in \mathcal{F}_k(\omega)$. Then $\mathcal{L}^Y(F) = \mathcal{L}_r^Y(F)$.*

5   Operations on k-Partitions

Here we define and study some operations on k^{X^ω} which extend (and slightly modify) the corresponding operations on subsets of the Baire space ω^ω introduced in [Wad84]. First we recall some operations introduced in [Se04, Se07]. For all $i < k$ and $A \in k^{X^\omega}$, define the k-partition $p_i(A)$ by

$$[p_i(A)](\xi) = \begin{cases} i, & \text{if } \xi \notin 0^*1X^\omega, \\ A(\eta), & \text{if } \xi = 0^n 1\eta. \end{cases}$$

The first item below was established in [Se04, Se07], the second one is checked in the obvious way.

Lemma 4.   *1. The set $\mathcal{R}_k$ is closed under $p_0, \ldots, p_{k-1}$, and the structures $(k^{X^\omega}; \leq_{CA}, \oplus, p_0, \ldots, p_{k-1})$, $(k^{X^\omega}; \leq_{DA}, \oplus, p_0, \ldots, p_{k-1})$ are dc-semilattices.*
2. If $B \leq_{CA} p_i(A)$ (resp. $B \leq_{DA} p_i(A)$) then $B \leq_{CS} p_i(A)$ (resp. $B \leq_{DS} p_i(A)$).

Next we define unary operations $q_0, \ldots, q_{k-1}$ on k^{X^ω}. To simplify notation, we do this only for the particular case $X = \{0,1\}$ (the general case is treated similarly). Define a DA-function $f : 3^\omega \to 2^\omega$ by $f(x_0 x_1 \cdots) = \tilde{x}_0 \tilde{x}_1 \cdots$ where $x_0, x_1 \ldots < 3$ and $\tilde{0} = 110000, \tilde{1} = 110100, \tilde{2} = 110010$ (in the same way we may define $f : 3^* \to 2^*$). Obviously, $f(3^\omega) \in \mathcal{R} \cap \mathbf{\Pi}_1^0$ and there is a DA-function $f_1 : 2^\omega \to 3^\omega$ such that $f_1 \circ f = id_{3^\omega}$. For all $i < k$ and k-partition A of X^ω, define the k-partition $q_i(A)$ by

$$[q_i(A)](\xi) = \begin{cases} i, & \text{if } \xi \notin f(3^\omega) \vee \forall p \exists n \geq p(\xi[n, n+5] = \tilde{2}), \\ A(f_1(\xi)), & \text{if } \xi \in f(2^\omega), \\ A(\eta), & \text{if } \xi = f(\sigma 3\eta) \end{cases}$$

for some $\sigma \in 3^\omega$ and $\eta \in 2^\omega$. These operations extend and modify the operation $\#$ from [Wad84, An06]. From the properties of that operation one easily derives

Lemma 5.   *The class of regular k-partitions of X^ω is closed under $q_0, \ldots, q_{k-1}$, the structure $(k^{X^\omega}; \leq_{\mathbf{\Delta}_2^0}, \oplus, q_0, \ldots, q_{k-1})$ is a dc-semilattice, and $A \leq_{\mathbf{\Delta}_2^0} q_i(B)$ implies $A \leq_{CA} q_i(B)$.*

Finally, we define the binary operation $+$ on k^{X^ω} as follows. Define a DA-function $g : X^\omega \to X^\omega$ by $g(x_0 x_1 \cdots) = x_0 0 x_1 0 \cdots$ where $x_0, x_1, \ldots \in X$ (in the same way we may define $g : X^* \to X^*$). Obviously, $g(X^\omega) \in \mathcal{R} \cap \mathbf{\Pi}_1^0$ and there is a DA-function $g_1 : X^\omega \to X^\omega$ such that $g_1 \circ g = id_{X^\omega}$. For all k-partitions A, B of X^ω, we set

$$[A + B)](\xi) = \begin{cases} A(g_1(\xi)), & \text{if } \xi \in g(X^\omega), \\ B(\eta), & \text{if } \xi = g(\sigma)i\eta \end{cases}$$

for some $\sigma \in X^\omega, i \in X \setminus \{0\}$ and $\eta \in X^\omega$. The operation $+$ extends and modifies the corresponding operation from [Wad84, An06]. From the properties of that operation one easily derives

Lemma 6.   *1. $A, B \leq_{DA} A + B$.*

2. Modulo $\equiv_{DA}$, $+$ is associative (but not commutative).

3. With respect to any of $\leq_{DA}, \leq_{CA},$ operation $+$ is monotone in each argument.

4. For all $A \in k^{X^\omega}$ and $i < k$, $B \mapsto q_i(A) + B$ is a closure operation on $(k^{X^\omega}; \leq_{CA}))$ and $(k^{X^\omega}; \leq_{DA})$.

5. For all $A, A_1 B, B_1 \in k^{X^\omega}$ and $i, j < k$, if $q_i(A) + B \leq_{CA} q_j(A_1) + B_1$ and $q_i(A) \nleq_{CA} q_j(A_1))$ then $q_i(A) + B \leq_{CA} B_1$; the same holds for $\leq_{DA}$.

One could notice analogies of the properties of the operations on k-partitions to those on labeled forests in Section 3. To make this precise, let us relate to any minimal $F \in \mathcal{T}_k^{\sqcup}(2)$ the k-partition $r(F)$ by induction on the rank of $[F]$ in $\mathbb{T}_k^{\sqcup}(2)$ as follows (to see that the cases are exhaustive cf. Proposition 3):

1. if $F = T_0 \sqcup \cdots \sqcup T_n$ for some $n \geq 1$ and $T_i \in \mathcal{T}_k(2)$ then $r(F) = r(T_0) \oplus \cdots \oplus r(T_n)$;
2. if $F \in \mathcal{T}_k(1)$ then represent F by a variable-free term of signature $\{0, \ldots, k-1, \sqcup, p_0, \ldots, p_{k-1}\}$ and let $r(F)$ be the value of this term when $i < k$ is interpreted as $\lambda \xi.i$, $\sqcup$ as $\oplus$ and p_i as the corresponding operation on k-partitions;
3. if $F = s(T)$ where $T \in \mathcal{T}_k(1) \setminus \{0, \ldots, k-1\}$ then $r(F)$ is defined as in the previous item with T in place of F and q_i in place of p_i;
4. if $F = s(T) + K$ is not in $\mathcal{T}_k^{\sqcup}(1)$, where $T \in \mathcal{T}_k(1)$ and $K \in \mathcal{T}_k^{\sqcup}(2)$, then set $r(F) = r(s(T)) + r(K)$.

The next crucial fact, proved by induction on the rank of $[F]$, extends the corresponding facts in [Se04, Se07a].

Theorem 1. *The function r induces isomorphic embeddings of $(\mathbb{T}_k^{\sqcup}(2); \leq_h)$ into $(\mathbb{R}_k; \leq_{CA}))$ and $(\mathbb{R}_k; \leq_{DA})$.*

The next result shows a close relation of the embedding r to the fine hierarchies over Σ and $\mathcal{R}\Sigma$.

Theorem 2. *For any $F \in \mathcal{T}_k^{\sqcup}(2)$, $r(F)$ is CA-complete in $\Sigma(F)$ and DA-complete in $\mathcal{R}\Sigma(F)$. Moreover, if $F \in \mathcal{T}_k(2)$ then $r(F)$ is CS-complete in $\Sigma(F)$ and DS-complete in $\mathcal{R}\Sigma(F)$*

6   Main Results

Define the preorder $\leq_0$ and the partial order $\leq_1$ on the set of cycles $C_{\mathcal{A}}$ of an automaton $\mathcal{A}$ as follows (we in fact use the reverse orderings compared with [Wag79]): $D \leq_0 E$ iff some state in D is reachable in the graph of the automaton $\mathcal{A}$ from some state in E; $D \leq_1 E$ iff $D \subseteq E$. For any $D \in C_{\mathcal{A}}$ the $\equiv_0$-class of D contains the largest element w.r.t. $\subseteq$. Let $\mathcal{A} = (A, c)$ be a Muller k-acceptor. Then $(C_{\mathcal{A}}; \leq_0, \leq_1, c)$ is a k-labeled 2-preorder, hence $P_{\mathcal{A}} = (C_{\mathcal{A}}; \leq_0, \leq_1)^+$ is in $\mathcal{P}_k(2)$ (see the end of Section 4). By the previous paragraph, any label of $P_{\mathcal{A}}$ has a largest element. By remarks before Proposition 4, $F_{\mathcal{A}} = F(P_{\mathcal{A}})$ is in $\mathcal{T}_k^{\sqcup}(2)$.

Theorem 3. *For any Muller k-acceptor $\mathcal{A}$, $L(\mathcal{A})$ is CA-complete in $\Sigma(F_\mathcal{A})$ and DA-complete in $\mathcal{R}\Sigma(F_\mathcal{A})$. Moreover, if $F_\mathcal{A} \in \mathcal{T}_k(2)$ then $L(\mathcal{A})$ is CS-complete in $\Sigma(F_\mathcal{A})$ and DS-complete in $\mathcal{R}\Sigma(F_\mathcal{A})$.*

Proof Sketch. For the first assertion, it suffices to show that $L(\mathcal{A}) \in \mathcal{R}\Sigma(F_\mathcal{A})$, $\mathcal{R}\Sigma(F_\mathcal{A}) \leq_{\mathrm{DA}} L(\mathcal{A})$ and $\Sigma(F_\mathcal{A}) \leq_{\mathrm{CA}} L(\mathcal{A})$. Relate to $\mathcal{A}$ the 2-base $\mathcal{L} = (\mathcal{L}_0, \mathcal{L}_1)$ in $C_\mathcal{A}$ where $\mathcal{L}_i$ is the set of initial segments (including the empty segment) of $(C_\mathcal{A}; \leq_i)$ (cf. [Se98, Proposition 5.2]). The k-partition $c : C_\mathcal{A} \to k$ is defined by the $P_\mathcal{A}$-family $\{B_{p_0}, B_{p_0,p_1}\}$ over $\mathcal{L}$ where $B_{p_0} = \{D \mid D \leq_0 C\}$ for each $p_0 = [C]_0 \in P$ and $B_{p_0,p_1} = \{D \mid D \leq_1 p_1\}$ for each $p_1 \in c(p_0) = [C]_0$. Therefore, $c \in \mathcal{L}(P_\mathcal{A})$. By the proof of Theorem 7.3 in [Se98], the function $f_\mathcal{A} : X^\omega \to C_\mathcal{A}$ is a morphism of 2-bases $\mathcal{R}\Sigma$ and $\mathcal{L}$, hence $L(\mathcal{A}) = c \circ f_\mathcal{A} \in \mathcal{R}\Sigma(P_\mathcal{A})$ by item 2 of Lemma 3. By item 4 of Lemma 3, $L(\mathcal{A}) \in \mathcal{R}\Sigma(F_\mathcal{A})$.

Let now $U \in \mathcal{R}\Sigma(F_\mathcal{A})$ (resp. $U \in \Sigma(F_\mathcal{A})$). Since $\mathcal{R}\Sigma$ and Σ are reducible, by Proposition 5 U is defined by a reduced $F_\mathcal{A}$-family $\{B_{p_0}, B_{p_0,p_1}\}$ over $\mathcal{R}\Sigma$ (resp. over Σ). Repeating the corresponding reductions in [Wag79, Se98], we obtain a DA-reduction (resp. a CA-reduction) of U to $L(\mathcal{A})$.

The second assertion follows from the previous paragraph and item 2 of Lemma 4 because it is easy to see that $L(\mathcal{A}) \equiv_{\mathrm{DS}} p_i(L(\mathcal{A}))$ for some $i < k$. $\square$

We are ready to state the main result of this paper which extends the properties of the Wagner hierarchy mentioned in the introduction, as well as a property established in [Se98].

Theorem 4.
1. *The structures $(\mathbb{T}_k^\sqcup(2); \leq_h)$, $(\mathbb{R}_k; \leq_{\mathrm{CA}})$ and $(\mathbb{R}_k; \leq_{\mathrm{DA}})$ are isomorphic.*
2. *The relations $\leq_{\mathrm{CA}}$ and $\leq_{\mathrm{DA}}$ coincide on $\mathcal{R}_k$.*
3. *The relations $L(\mathcal{A}) \in \mathcal{R}\Sigma(F)$ and $L(\mathcal{A}) \leq_{\mathrm{DA}} L(\mathcal{B})$ are decidable (here $\mathcal{A}, \mathcal{B}$ are Muller k-acceptors and $F \in \mathcal{T}_k^\sqcup(2)$).*
4. *For any $F \in \mathcal{T}_k^\sqcup(2)$, $\mathcal{R}\Sigma(F) = \mathcal{R}_k \cap \Sigma(F)$.*

Proof. 1. By Theorem 1, it suffices to show that for any $A \in \mathcal{R}_k$ there is $F \in \mathcal{T}_k^\sqcup(2)$ such that $A \equiv_{\mathrm{DA}} r(F)$. Let $\mathcal{A}$ be a Muller k-acceptor that recognizes A, i.e. $A = L(\mathcal{A})$. Since by Theorems 2 and 3, $L(\mathcal{A}) \equiv_{\mathrm{DA}} r(F_\mathcal{A})$, we can take $F = F_\mathcal{A}$.

2. We have to show that, for all $A, B \in \mathcal{R}_k$, $A \leq_{\mathrm{CA}} B$ implies $A \leq_{\mathrm{DA}} B$. Let $\mathcal{A}$ and $\mathcal{B}$ be Muller k-acceptors that recognize A and B, respectively. Assume first that $F_\mathcal{B} \in \mathcal{T}_k(2)$. By Theorem 3, $A \leq_{\mathrm{CS}} B$. As shown in [Se07a], the relations $\leq_{\mathrm{CS}}$ and $\leq_{\mathrm{DS}}$ coincide on $\mathcal{R}_k$. Thus, $A \leq_{\mathrm{DS}} B$ and hence $A \leq_{\mathrm{DA}} B$.

Now consider the general case when $F_\mathcal{A} = T_1 \sqcup \cdots \sqcup T_l$, $F_\mathcal{B} = S_1 \sqcup \cdots \sqcup S_n$ for some $l, n \geq 0$ and $T_i, S_j \in \mathcal{T}_k(2)$. Using Theorems 1, 2 and 3, and the arguments of preceding paragraph and item 1, we successively deduce from $A \leq_{\mathrm{CA}} B$ that $r(F_\mathcal{A}) \leq_{\mathrm{CA}} r(F_\mathcal{B})$, $F_\mathcal{A} \leq_h F_\mathcal{B}$, $\forall i \leq l \exists j \leq n(T_i \leq_h S_j)$, $\forall i \leq l \exists j \leq n(r(T_i) \leq_{\mathrm{DA}} r(S_j))$, $\forall i \leq l \, n(r(T_i) \leq_{\mathrm{DA}} B)$, $A \leq_{\mathrm{DA}} B$.

3. We check that $L(\mathcal{A}) \in \mathcal{R}\Sigma(F)$ iff $F_\mathcal{A} \leq_h F$ (this implies the first assertion because $F_\mathcal{A}$ is computable from $\mathcal{A}$ and the structure $(\mathcal{T}_k^\sqcup(2); \leq_h)$ is computable). Implication from right to left follows from item 3 of Lemma 3 and Theorem 3.

Conversely, let $L(\mathcal{A}) \in \mathcal{R}\Sigma(F)$. By Theorems 3 and 1, $r(F_{\mathcal{A}}) \leq_h r(F)$, hence $F_{\mathcal{A}} \leq_h F$. The second assertion follows from the equivalence: $L(\mathcal{A}) \leq_{\mathrm{DA}} L(\mathcal{B})$ iff $F_{\mathcal{A}} \leq_h F_{\mathcal{A}}$ that follows from Theorems 3 and 1.

4. The inclusion from left to right is obvious. Conversely, let A be in $\mathcal{R}\Sigma(F)$. By Theorem 1, $A \leq_{\mathrm{DA}} r(F)$, hence $A \in \mathcal{R}\Sigma(F)$ by item 2 of Lemma 3. $\square$

Notice that item 1 above really subsumes item 1 in Introduction, because it is not hard to show that the structure $(\mathbb{T}_2^{\sqcup}(2); \leq_h)$ is isomorphic to the structure of levels of the Wagner hierarchy (see [Se10] for details).

References

[An06]   Andretta, A.: More on Wadge determinacy. Annals of Pure and Applied Logic 144, 2–32 (2006)

[H93]   Hertling, P.: Topologische Komplexitätsgrade von Funktionen mit endlichem Bild. Informatik-Berichte 152, 34 pages, Fernuniversität Hagen (December 1993)

[Ke94]   Kechris, A.S.: Classical Descriptive Set Theory. Springer, New York (1994)

[Ko00]   Kosub, S.: On NP-partitions over posets with an application to reducing the set of solutions of NP problems. In: Nielsen, M., Rovan, B. (eds.) MFCS 2000. LNCS, vol. 1893, pp. 467–476. Springer, Heidelberg (2000)

[KW00]   Kosub, S., Wagner, K.W.: The Boolean Hierarchy of NP-Partitions. In: Reichel, H., Tison, S. (eds.) STACS 2000. LNCS, vol. 1770, pp. 157–168. Springer, Heidelberg (2000)

[Se98]   Selivanov, V.L.: Fine hierarchy of regular ω-languages. Theoretical Computer Science 191, 37–59 (1998)

[Se04]   Selivanov, V.L.: Boolean hierarchy of partitions over reducible bases. Algebra and Logic 43(1), 77–109 (2004); (Russian, there is an English translation)

[Se07]   Selivanov, V.L.: The quotient algebra of labeled forests modulo h-equivalence. Algebra and Logic 46(2), 120–133 (2007)

[Se07a] Selivanov, V.L.: Classifying omega-regular partitions. In: PreProceedings of LATA-2007, Universitat Rovira i Virgili Report Series, 35/07, pp. 529–540 (2007)

[Se10]   Selivanov, V.L.: Fine hierarchies via Priestley duality (submitted)

[Wad84] Wadge, W.: Reducibility and determinateness in the Baire space. PhD thesis, University of California, Berkely (1984)

[Wag79] Wagner, K.: On ω-regular sets. Information and Control 43, 123–177 (1979)

On a Relative Computability Notion for Real Functions

Dimiter Skordev[1] and Ivan Georgiev[2]

[1] Sofia University, Faculty of Mathematics and Informatics, Sofia, Bulgaria
skordev@fmi.uni-sofia.bg
[2] Burgas Prof. Assen Zlatarov University, Faculty of Natural Sciences, Burgas,
Bulgaria
ivandg@yahoo.com

Abstract. For any class of total functions in the set of natural numbers, we define what it means for a real function to be conditionally computable with respect to this class. This notion extends a notion of relative uniform computability of real functions introduced in a previous paper co-authored by Andreas Weiermann. If the given class consists of recursive functions then the conditionally computable real functions are computable in the usual sense. Under certain weak assumptions about the class in question, we show that conditional computability is preserved by substitution, that all conditionally computable real functions are locally uniformly computable, and that the ones with compact domains are uniformly computable. All elementary functions of calculus turn out to be conditionally computable with respect to one of the small subrecursive classes.

1 Introduction

In the paper [1], a notion of relative computability of real functions was introduced, namely, when a class $\mathcal{F}$ of total functions in $\mathbb{N}$ is given, certain real functions were called uniformly $\mathcal{F}$-computable. It was shown that the elementary functions of calculus are uniformly $\mathcal{M}^2$-computable if we consider them on compact subsets of their domains[1]. In the present paper, we introduce a wider notion of relative computability called conditional $\mathcal{F}$-computability[2]. This notion is close to uniform computability in a certain sense, but nevertheless all elementary functions of calculus, considered on their whole domains, turn out to be conditionally $\mathcal{M}^2$-computable. The supplementary feature of conditional $\mathcal{F}$-computability in comparison to the uniform one can be informally described

---

[1] The class $\mathcal{M}^2$ consists of the argumentless constants $0, 1, 2, \dots$ and all functions in $\mathbb{N}$ which can be obtained from the successor function, the function $\lambda xy.x \doteq y$, the multiplication function and projection functions by finitely many applications of substitution and bounded least number operation.

[2] As in [1], we are interested mainly in the case when $\mathcal{F}$ is some of the small subrecursive classes (and especially in the case $\mathcal{F} = \mathcal{M}^2$).

B. Löwe et al. (Eds.): CiE 2011, LNCS 6735, pp. 270–279, 2011.

as follows. It is now allowed the approximation process to depend on an additional natural parameter. Its value at any point of the domain of the considered real function can be determined by means of a search until a certain term vanishes. The term in question must be constructed by using function symbols for functions in $\mathcal{F}$ and for the functions approximating the coordinates of the point[3].

We will use many definitions and results from [1] without explanation. Although not obligatory for the reading of the present paper, a more detailed acquaintance with the paper [1] could help for a better comprehension of the present one.

2 $\mathcal{F}$-Substitutional Mappings

For any $m \in \mathbb{N}$, we will denote by $\mathbb{T}_m$ the set of all m-argument total functions in $\mathbb{N}$. Let $\mathcal{F} \subseteq \bigcup_{m \in \mathbb{N}} \mathbb{T}_m$. For any $k, m \in \mathbb{N}$, certain mappings of $\mathbb{T}_1^k$ into $\mathbb{T}_m$ will be called $\mathcal{F}$-substitutional, as follows[4].

Definition 1. *We proceed by induction:*

1. *For any m-argument projection function h in $\mathbb{N}$ the mapping F defined by means of the equality $F(f_1, \ldots, f_k) = h$ is $\mathcal{F}$-substitutional.*
2. *For any $i \in \{1, \ldots, k\}$, if F_0 is a $\mathcal{F}$-substitutional mapping of $\mathbb{T}_1^k$ into $\mathbb{T}_m$ then so is the mapping F defined by means of the equality*

$$F(f_1, \ldots, f_k)(n_1, \ldots, n_m) = f_i(F_0(f_1, \ldots, f_k)(n_1, \ldots, n_m)).$$

3. *For any natural number r and any r-argument function f from $\mathcal{F}$, if $F_1, \ldots, F_r$ are $\mathcal{F}$-substitutional mappings of $\mathbb{T}_1^k$ into $\mathbb{T}_m$ then so is the mapping F defined by means of the equality*

$$F(f_1, \ldots, f_k)(n_1, \ldots, n_m) =$$
$$f(F_1(f_1, \ldots, f_k)(n_1, \ldots, n_m), \ldots, F_r(f_1, \ldots, f_k)(n_1, \ldots, n_m)).$$

Intuitively, a mapping F of $\mathbb{T}_1^k$ into $\mathbb{T}_m$ is $\mathcal{F}$-substitutional iff there is an expression for $F(f_1, \ldots, f_k)(n_1, \ldots, n_m)$ built from the variables $n_1, \ldots, n_m$ by using function symbols $f_1, \ldots, f_k$ and function symbols for functions from $\mathcal{F}$[5].

The following statements are straightforward generalizations of statements from [1], and can be proved in the same way (i.e. by induction on F).

---

[3] In the situation studied in [3], also a dependance of the approximation process on an additional parameter is admitted, but its description uses the distance to the complement of the domain of the function and makes no use of $\mathcal{F}$.

[4] Only the cases $m = 1$ and $m = 2$ will be actually needed (the first of them is considered, by using a slightly different terminology, also in [1]).

[5] If $k = 0$ then such a mapping can be identified with an m-argument function in $\mathbb{N}$ which is explicitly definable through functions from $\mathcal{F}$.

Proposition 1. *Let $F : \mathbb{T}_1^k \to \mathbb{T}_m$ and $G_1, \ldots, G_m : \mathbb{T}_1^k \to \mathbb{T}_l$ be $\mathcal{F}$-substitutional. Then so is the mapping $H : \mathbb{T}_1^k \to \mathbb{T}_l$ defined by*

$$H(f_1, \ldots, f_k)(\overline{n}) = F(f_1, \ldots, f_k)(G_1(f_1, \ldots, f_k)(\overline{n}), \ldots, G_m(f_1, \ldots, f_k)(\overline{n})),$$

where $\overline{n}$ is an abbreviation for $n_1, \ldots, n_l$.

Proposition 2. *Let $F : \mathbb{T}_1^k \to \mathbb{T}_m$ and $G_1, \ldots, G_k : \mathbb{T}_1^l \to \mathbb{T}_1$ be $\mathcal{F}$-substitutional. Then so is the mapping $H : \mathbb{T}_1^l \to \mathbb{T}_m$ defined by*

$$H(g_1, \ldots, g_l) = F(G_1(g_1, \ldots, g_l), \ldots, G_k(g_1, \ldots, g_l)).$$

In the sequel, we will sometimes need to consider an expression with values in $\mathbb{N}$ as a function of some variable in it ranging over $\mathbb{N}$. Instead of a λ-notation for the function defined in this way, we will use another one obtained by replacing the variable in question by the symbol $\bullet$ (for instance, we could write $\bullet^2 + \bullet + 1$ instead of $\lambda x.x^2 + x + 1$). Despite having much more restricted usability than λ-notation, this notation will be sufficient for most of our needs here.

Proposition 2 can be generalized as follows.

Proposition 3. *Let $F : \mathbb{T}_1^k \to \mathbb{T}_m$ and $G_1, \ldots, G_k : \mathbb{T}_1^l \to \mathbb{T}_{p+1}$ be $\mathcal{F}$-substitutional. Then so is the mapping $H : \mathbb{T}_1^l \to \mathbb{T}_{p+m}$ defined by the equality*

$$H(g_1, \ldots, g_l)(\overline{u}, \overline{n}) = F(G_1(g_1, \ldots, g_l)(\overline{u}, \bullet), \ldots, G_k(g_1, \ldots, g_l)(\overline{u}, \bullet))(\overline{n}),$$

where $\overline{u}$ and $\overline{n}$ are abbreviations for $u_1, \ldots, u_p$ and $n_1, \ldots, n_m$, respectively.

Proof. If F is $\mathcal{F}$-substitutional by clause 1 of Definition 1 then so is H. Suppose now F has the form from clause 2 of Definition 1, and the mapping F_0 has the considered property. Then

$$H(g_1, \ldots, g_l)(\overline{u}, \overline{n}) = G_i(g_1, \ldots, g_l)(\overline{u}, \bullet)(H_0(g_1, \ldots, g_l)(\overline{u}, \overline{n})) =$$
$$G_i(g_1, \ldots, g_l)(\overline{u}, H_0(g_1, \ldots, g_l)(\overline{u}, \overline{n})),$$

where

$$H_0(g_1, \ldots, g_l)(\overline{u}, \overline{n}) = F_0(G_1(g_1, \ldots, g_l)(\overline{u}, \bullet), \ldots, G_k(g_1, \ldots, g_l)(\overline{u}, \bullet))(\overline{n}).$$

Since G_i and H_0 are $\mathcal{F}$-substitutional, H is also $\mathcal{F}$-substitutional (by Proposition 1 and clause 1 of Definition 1). Finally suppose that F has the form from clause 3 of Definition 1, and the mappings $F_1, \ldots, F_r$ have the considered property. Then

$$H(g_1, \ldots, g_l)(\overline{u}, \overline{n}) = f(H_1(g_1, \ldots, g_l)(\overline{u}, \overline{n}), \ldots, H_r(g_1, \ldots, g_l)(\overline{u}, \overline{n})),$$

where

$$H_i(g_1, \ldots, g_l)(\overline{u}, \overline{n}) = F_i(G_1(g_1, \ldots, g_l)(\overline{u}, \bullet), \ldots, G_k(g_1, \ldots, g_l)(\overline{u}, \bullet))(\overline{n})$$

for $i = 1, \ldots, r$. Since $H_1, \ldots, H_r$ are $\mathcal{F}$-substitutional, H is also $\mathcal{F}$-substitutional (by clause 3 of Definition 1).

3 Conditional $\mathcal{F}$-Computability of Real Functions

As in [1], a triple $(f, g, h) \in \mathbb{T}_1^3$ is called to name a real number ξ if

$$\left| \frac{f(t) - g(t)}{h(t) + 1} - \xi \right| < \frac{1}{t + 1}$$

for all $t \in \mathbb{N}$.

Definition 2. *Let $N \in \mathbb{N}$ and $\theta : D \to \mathbb{R}$, where $D \subseteq \mathbb{R}^N$. The function θ will be called* conditionally computable with respect to $\mathcal{F}$ *or conditionally $\mathcal{F}$-computable, for short, if there exist $\mathcal{F}$-substitutional mappings $E : \mathbb{T}_1^{3N} \to \mathbb{T}_1$ and $F, G, H : \mathbb{T}_1^{3N} \to \mathbb{T}_2$ such that, whenever $(\xi_1, \ldots, \xi_N) \in D$ and $(f_1, g_1, h_1), \ldots, (f_N, g_N, h_N)$ are triples from $\mathbb{T}_1^3$ naming $\xi_1, \ldots, \xi_N$, respectively, the following holds:*

1. *There exists a natural number s satisfying the equality*

$$E(f_1, g_1, h_1, \ldots, f_N, g_N, h_N)(s) = 0. \tag{1}$$

2. *For any natural number s satisfying the equality (1), the number $\theta(\xi_1, \ldots, \xi_N)$ is named by the triple $(\tilde{f}, \tilde{g}, \tilde{h})$, where*

$$\tilde{f} = F(f_1, g_1, h_1, \ldots, f_N, g_N, h_N)(s, \bullet),$$
$$\tilde{g} = G(f_1, g_1, h_1, \ldots, f_N, g_N, h_N)(s, \bullet),$$
$$\tilde{h} = H(f_1, g_1, h_1, \ldots, f_N, g_N, h_N)(s, \bullet).$$

Example 1. If a real function is uniformly $\mathcal{F}$-computable in the sense of [1] then it is conditionally $\mathcal{F}$-computable. Indeed, let $(F^\circ, G^\circ, H^\circ)$ be an $\mathcal{F}$-substitutional $(3,3)$-computing system in the sense of [1] for the N-argument real function θ. Then we can satisfy the requirements of Definition 2 by setting

$$E(f_1, g_1, h_1, \ldots, f_N, g_N, h_N)(s) = s,$$
$$F(f_1, g_1, h_1, \ldots, f_N, g_N, h_N)(s, t) = F^\circ(f_1, g_1, h_1, \ldots, f_N, g_N, h_N)(t),$$
$$G(f_1, g_1, h_1, \ldots, f_N, g_N, h_N)(s, t) = G^\circ(f_1, g_1, h_1, \ldots, f_N, g_N, h_N)(t),$$
$$H(f_1, g_1, h_1, \ldots, f_N, g_N, h_N)(s, t) = H^\circ(f_1, g_1, h_1, \ldots, f_N, g_N, h_N)(t).$$

In the next three examples, we assume $\mathcal{F} = \mathcal{M}^2$.

Example 2. Despite not being uniformly $\mathcal{F}$-computable, the function $\theta(\xi) = \frac{1}{\xi}$ is conditionally $\mathcal{F}$-computable. To satisfy the requirements of Definition 2, we may set

$$E(f, g, h)(s) = (2h(s) + 3) \div (s + 1)|f(s) - g(s)|,$$
$$F(f, g, h)(s, t) = (h(u(s, t)) + 1) \operatorname{sg}(f(u(s, t)) \div g(u(s, t))),$$
$$G(f, g, h)(s, t) = (h(u(s, t)) + 1) \operatorname{sg}(g(u(s, t)) \div f(u(s, t))),$$
$$H(f, g, h)(s, t) = |f(u(s, t)) - g(u(s, t))| \div 1,$$

where $u(s, t) = s + (s + 1)^2 (t + 1)$.

Example 3. The function $\theta(\xi) = \exp(\xi)$ is not uniformly $\mathcal{F}$-computable, but it is conditionally $\mathcal{F}$-computable. To show the conditional $\mathcal{F}$-computability of θ, we may use Theorem 7 of [1]. According to it, $\min(\exp(\xi), \eta)$ is a uniformly $\mathcal{M}^2$-computable function of ξ and η. Let $(F^\circ, G^\circ, H^\circ)$ be an $\mathcal{M}^2$-substitutional (3,3)-computing system for this function. To satisfy the requirements of Definition 2, we may set

$$
\begin{aligned}
E(f, g, h)(s) &= (f(0) + h(0) + 1) \div ((s + 1)_1(h(0) + 1) + g(0)), \\
F(f, g, h)(s, t) &= F^\circ(f, g, h, \lambda x.s + 1, \lambda x.0, \lambda x.0)(t), \\
G(f, g, h)(s, t) &= G^\circ(f, g, h, \lambda x.s + 1, \lambda x.0, \lambda x.0)(t), \\
H(f, g, h)(s, t) &= H^\circ(f, g, h, \lambda x.s + 1, \lambda x.0, \lambda x.0)(t),
\end{aligned}
$$

where $(s + 1)_1$ is the exponent of the prime number 3 in $s + 1$.

Example 4. Any partial recursive function in $\mathbb{N}$ regarded as a function in $\mathbb{R}$ is conditionally $\mathcal{F}$-computable. Indeed, let θ be an N-argument partial recursive function. Then θ has a representation of the form

$$
\theta(x_1, \ldots, x_N) = U(\mu y[T(x_1, \ldots, x_N, y) = 0]),
$$

where $T, U \in \mathcal{F}$. To satisfy the requirements of Definition 2, we may set

$$
\begin{aligned}
E(f_1, g_1, h_1, \ldots, f_N, g_N, h_N)(s) &= T(x_1, \ldots, x_N, s) + \max_{y < s} \overline{\mathrm{sg}}\, T(x_1, \ldots, x_N, y), \\
F(f_1, g_1, h_1, \ldots, f_N, g_N, h_N)(s, t) &= U(s), \\
G(f_1, g_1, h_1, \ldots, f_N, g_N, h_N)(s, t) &= 0, \\
H(f_1, g_1, h_1, \ldots, f_N, g_N, h_N)(s, t) &= 0,
\end{aligned}
$$

where

$$
x_i = \left\lfloor \frac{f_i(1) \div g_i(1)}{h_i(1) + 1} + \frac{1}{2} \right\rfloor, \quad i = 1, \ldots, N.
$$

It is easy to see that all conditionally $\mathcal{F}$-computable real functions are computable in the usual sense, whenever all functions in $\mathcal{F}$ are recursive ones.

4   Substitution in Conditionally $\mathcal{F}$-Computable Real Functions

Theorem 1. *Let the class $\mathcal{F}$ contain the addition function and one-argument functions L and R such that $\{(L(s), R(s)) \mid s \in \mathbb{N}\} = \mathbb{N}^2$. Then the substitution operation on real functions preserves conditional $\mathcal{F}$-computability.*

Proof. To avoid writing excessively long expressions, we will restrict ourselves to the case of one-argument functions. Let θ_0 and θ_1 be conditionally $\mathcal{F}$-computable one-argument real functions. We will show the conditional $\mathcal{F}$-computability of the function θ defined by $\theta(\xi) = \theta_0(\theta_1(\xi))$. For $i = 0, 1$, let E_i, F_i, G_i, H_i be $\mathcal{F}$-substitutional mappings such that $\exists s(E_i(f, g, h)(s) = 0)$ and

$$\forall s(E_i(f, g, h)(s) = 0 \Rightarrow$$
$$(F_i(f, g, h)(s, \bullet), G_i(f, g, h)(s, \bullet), H_i(f, g, h)(s, \bullet)) \text{ names } \theta_i(\xi))$$

for any $\xi \in \mathrm{dom}(\theta_i)$ and any triple (f, g, h) naming ξ. We will show that the requirements of Definition 2 for the function θ are satisfied through the mappings E, F, G, H defined as follows:

$$E(f, g, h)(s) = E_1(f, g, h)(R(s)) +$$
$$E_0(F_1(f, g, h)(R(s), \bullet), G_1(f, g, h)(R(s), \bullet), H_1(f, g, h)(R(s), \bullet))(L(s)),$$
$$F(f, g, h)(s, t) =$$
$$F_0(F_1(f, g, h)(R(s), \bullet), G_1(f, g, h)(R(s), \bullet), H_1(f, g, h)(R(s), \bullet))(L(s), t),$$
$$G(f, g, h)(s, t) =$$
$$G_0(F_1(f, g, h)(R(s), \bullet), G_1(f, g, h)(R(s), \bullet), H_1(f, g, h)(R(s), \bullet))(L(s), t),$$
$$H(f, g, h)(s, t) =$$
$$H_0(F_1(f, g, h)(R(s), \bullet), G_1(f, g, h)(R(s), \bullet), H_1(f, g, h)(R(s), \bullet))(L(s), t)$$

for all $s, t \in \mathbb{N}$. By Propositions 1 and 3, these mappings are also $\mathcal{F}$-substitutional. Suppose now $\xi \in \mathrm{dom}(\theta)$ and (f, g, h) is a triple naming ξ. By the conditional $\mathcal{F}$-computability of θ_1, there exists $s_1 \in \mathbb{N}$ such that

$$E_1(f, g, h)(s_1) = 0, \tag{2}$$

and if we choose such an s_1 then the number $\theta_1(\xi)$ is named by the triple (f_1, g_1, h_1), where

$$f_1 = F_1(f, g, h)(s_1, \bullet), \quad g_1 = G_1(f, g, h)(s_1, \bullet), \quad h_1 = H_1(f, g, h)(s_1, \bullet). \tag{3}$$

By the conditional $\mathcal{F}$-computability of θ_0, there exists $s_0 \in \mathbb{N}$ such that

$$E_0(f_1, g_1, h_1)(s_0) = 0. \tag{4}$$

If s is a natural number such that $L(s) = s_0$, $R(s) = s_1$, then $E(f, g, h)(s) = 0$. Consider now any natural number s such that $E(f, g, h)(s) = 0$. Let $s_0 = L(s)$, $s_1 = R(s)$. The equality $E(f, g, h)(s) = 0$ implies the equality (2), as well as the equality (4) for the functions f_1, g_1, h_1 defined by means of the equalities (3). It follows from the equality (2) that (f_1, g_1, h_1) names $\theta_1(\xi)$, and, together with the equality (4), this fact implies that $\theta(\xi)$ is named by the triple

$$(F(f, g, h)(s, \bullet), G(f, g, h)(s, \bullet), H(f, g, h)(s, \bullet)).$$

Example 5. The function $\theta(\xi) = \ln \xi$ is conditionally $\mathcal{M}^2$-computable, although it is not uniformly $\mathcal{M}^2$-computable . To prove the conditional $\mathcal{M}^2$-computability of θ, let us consider the function θ° having domain $\{(\xi_1, \xi_2) \in \mathbb{R}^2 \mid \xi_1 > 0, \xi_1 \xi_2 \geq 1\}$

and defined by $\theta^\circ(\xi_1, \xi_2) = \ln \xi_1$. This function is uniformly $\mathcal{M}^2$-computable by Theorem 6 of [1], hence (by Example 1) it is conditionally $\mathcal{M}^2$-computable. Since $\theta(\xi) = \theta^\circ(\xi, 1/\xi)$ for all $\xi \in \mathrm{dom}(\theta)$, the conditional $\mathcal{M}^2$-computability of θ follows from here by Theorem 1 and Example 2.

Corollary 1. *All elementary functions of calculus are conditionally computable with respect to the class* $\mathcal{M}^2$.

Proof. Any elementary function of calculus can be obtained by means of substitution from some functions shown in [1] to be uniformly $\mathcal{M}^2$-computable and the functions shown in Examples 2, 3 and 5 to be conditionally $\mathcal{M}^2$-computable. By Theorem 1, this implies the conditional $\mathcal{M}^2$-computability of all elementary functions of calculus.

5   Local Uniform $\mathcal{F}$-Computability of the Conditionally $\mathcal{F}$-Computable Functions

Definition 3. *Let* $N \in \mathbb{N}$ *and* $\theta : D \to \mathbb{R}$, *where* $D \subseteq \mathbb{R}^N$. *The function* θ *will be called* locally uniformly $\mathcal{F}$-computable *if any point of* D *has some neighbourhood* U *such that the restriction of* θ *to* $D \cap U$ *is uniformly* $\mathcal{F}$-computable.

Theorem 2. *Let for any* $a, b \in \mathbb{N}$ *the class* $\mathcal{F}$ *contain the two-argument function whose value at* (x, y) *is* b *or* y *depending on whether or not* $x = a$. *Let also all one-argument constant functions in* $\mathbb{N}$ *belong to* $\mathcal{F}$. *Then all conditionally* $\mathcal{F}$-computable real functions are locally uniformly $\mathcal{F}$-computable.

Proof. Let θ be a conditionally $\mathcal{F}$-computable real function, and ξ_0 be a point of its domain (for the sake of simplicity, we assume additionally that θ is a one-argument function). Let E, F, G, H be such $\mathcal{F}$-substitutional mappings as in Definition 2 (with $N = 1$). Let (f_0, g_0, h_0) be a triple naming ξ_0, and let s_0 be a natural number satisfying the equality $E(f_0, g_0, h_0)(s_0) = 0$. There exists a finite set A of natural numbers such that $E(f, g, h)(s_0) = 0$, whenever f, g, h are functions from $\mathbb{T}_1$ coinciding, respectively, with f_0, g_0, h_0 on A. The assumptions imposed on $\mathcal{F}$ imply the existence of $\mathcal{F}$-substitutional mappings P, Q, R of $\mathbb{T}_1$ into itself, such that, for any $f, g, h \in \mathbb{T}_1$, the functions $P(f), Q(g), R(h)$ coincide, respectively, with the functions f_0, g_0, h_0 on A and with the functions f, g, h on $\mathbb{N} \smallsetminus A$. Let

$$U = \left(\max_{i \in A} \left(\frac{f_0(i) - g_0(i)}{h_0(i) + 1} - \frac{1}{i + 1} \right), \min_{i \in A} \left(\frac{f_0(i) - g_0(i)}{h_0(i) + 1} + \frac{1}{i + 1} \right) \right).$$

Then $\xi_0 \in U$, and, whenever a triple (f, g, h) names a number belonging to U, the triple $(P(f), Q(g), R(h))$ also names this number. Let us set

$$F'(f, g, h) = F(P(f), Q(g), R(h))(s_0, \bullet),$$
$$G'(f, g, h) = G(P(f), Q(g), R(h))(s_0, \bullet),$$
$$H'(f, g, h) = H(P(f), Q(g), R(h))(s_0, \bullet).$$

The mappings F', G', H' are also $\mathcal{F}$-substitutional. Since

$$E(P(f), Q(g), R(h))(s_0) = 0$$

for any $f, g, h \in \mathbb{T}_1$, it is clear that, whenever (f, g, h) names a number $\xi \in U$ that belongs to the domain of θ, the triple $(F'(f, g, h), G'(f, g, h), H'(f, g, h))$ names $\theta(\xi)$. Thus (F', G', H') is an $\mathcal{F}$-substitutional 3,3-computing system for the restriction of θ to $D \cap U$.

Theorem 2 and the Characterization Theorem from [2] imply that, under the assumptions about $\mathcal{F}$ in them, if θ is a conditionally $\mathcal{F}$-computable function then each point of $\mathrm{dom}(\theta)$ has some neighbourhood U such that θ is uniformly continuous in $\mathrm{dom}(\theta) \cap U$.[6] Since these assumptions are satisfied when $\mathcal{F} = \bigcup_{m \in \mathbb{N}} \mathbb{T}_m$, it follows that there exist computable real functions, which are not conditionally $\mathcal{F}$-computable, whatever be the class $\mathcal{F}$, e.g. the function $\theta(\xi) = \sum_{k=1}^{\infty} \frac{1}{2^k} \sigma\left(\xi - \frac{1}{k}\right)$, where σ is the restriction of the sign function to $\mathbb{R} \setminus \{0\}$.

6 Uniform $\mathcal{F}$-Computability of the Locally Uniformly $\mathcal{F}$-Computable Functions with Compact Domains

The conclusion of the next theorem is natural to be expected under some assumptions about the class $\mathcal{F}$. The premise of the theorem contains a choice of such ones which allows a proof along more or less usual lines.

Theorem 3. *Let the class $\mathcal{F}$ be closed under substitution, and let $\mathcal{F}$ contain the projection functions, the successor function, the addition function, the function $\lambda xy.x \dot{-} y$ and the function $\lambda xy.x(1 \dot{-} y)$. Then all locally uniformly $\mathcal{F}$-computable real functions with compact domains are uniformly $\mathcal{F}$-computable.*

Proof. The assumptions of the theorem imply that any constant function in $\mathbb{N}$ with a non-zero number of arguments belongs to $\mathcal{F}$. Since $1 \dot{-} y = \overline{\mathrm{sg}}\, y$, and $\mathrm{sg}\, y = \overline{\mathrm{sg}}\; \overline{\mathrm{sg}}\, y$, the class $\mathcal{F}$ contains also the functions $\lambda xy.x\,\overline{\mathrm{sg}}\, y$ and $\lambda xy.x\,\mathrm{sg}\, y$. Suppose now $N \in \mathbb{N}$, $\theta : D \to \mathbb{R}$, where D is a compact subset of $\mathbb{R}^N$, and θ is locally uniformly $\mathcal{F}$-computable. Then there exist a natural number n, rational numbers a_{ij} $(i = 1, \ldots, n, \; j = 1, \ldots, N)$ and positive rational numbers $d_1, \ldots, d_n$ such that $D \subseteq U_1 \cup \ldots \cup U_n$, where, for $i = 1, \ldots, n$,

$$U_i = \left\{ (\xi_1, \ldots, \xi_N) \in \mathbb{R}^N \;\middle|\; |\xi_1 - a_{i1}| < d_i, \; \ldots, \; |\xi_N - a_{iN}| < d_i \right\}$$

and the restriction of θ to $D \cap U_i$ is uniformly $\mathcal{F}$-computable. We will prove that θ is also uniformly $\mathcal{F}$-computable. In order to do this, we consider the continuous function $\delta(\xi_1, \ldots, \xi_N) = \max_{i=1,\ldots,n} \min_{j=1,\ldots,N} (d_i - |\xi_j - a_{ij}|)$. Since $\delta(\overline{\xi}) > 0$ for all $\overline{\xi} \in D$, there exists a natural number k, such that $\delta(\overline{\xi}) \geq \frac{2}{k+1}$

---

[6] The above-mentioned theorem from [2] characterizes the uniformly $\mathcal{F}$-computable real functions by essentially the same property which is required on the last line of Definition 3.1 in [3].

for any $\bar{\xi} \in D$. For such a k, as it is easy to see, whenever $(\xi_1, \ldots, \xi_N) \in D$, and $x_1, \ldots, x_N$ are rational numbers satisfying the inequalities $|x_j - \xi_j| < \frac{1}{k+1}$ $(j = 1, \ldots, N)$, at least one of the numbers $r_1 = \min_{j=1,\ldots,N}(d_1 - |x_j - a_{1j}|)$, $\ldots$, $r_n = \min_{j=1,\ldots,N}(d_n - |x_j - a_{nj}|)$ will be greater than $\frac{1}{k+1}$, and $(\xi_1, \ldots, \xi_N)$ will belong to U_i for any $i \in \{1, \ldots, n\}$ such that $r_i > \frac{1}{k+1}$. In particular, that will be the case, whenever $(\xi_1, \ldots, \xi_N) \in D$, (f_1, g_1, h_1), $\ldots$, (f_N, g_N, h_N) are triples naming $\xi_1, \ldots, \xi_N$, respectively, and

$$x_j = \frac{f_j(k) - g_j(k)}{h_j(k) + 1}$$

for $j = 1, \ldots, N$. Let, for $i = 1, \ldots, n$, the triple (F_i, G_i, H_i) be a $\mathcal{F}$-substitutional 3,3-computing system for the restriction of θ to $D \cap U_i$. We define mappings $F, G, H : \mathbb{T}_1^{3N} \to \mathbb{T}_1$ by setting

$$F(f_1, g_1, h_1, \ldots, f_N, g_N, h_N)(t) = F_l(f_1, g_1, h_1, \ldots, f_N, g_N, h_N)(t),$$
$$G(f_1, g_1, h_1, \ldots, f_N, g_N, h_N)(t) = G_l(f_1, g_1, h_1, \ldots, f_N, g_N, h_N)(t),$$
$$H(f_1, g_1, h_1, \ldots, f_N, g_N, h_N)(t) = H_l(f_1, g_1, h_1, \ldots, f_N, g_N, h_N)(t),$$

where l is the least of the numbers $i \in \{1, \ldots, n\}$ satisfying the inequality

$$\min_{j=1,\ldots,N} \left(d_i - \left| \frac{f_j(k) - g_j(k)}{h_j(k) + 1} - a_{ij} \right| \right) > \frac{1}{k+1}, \tag{5}$$

if there exists such an i, and

$$F(f_1, g_1, h_1, \ldots, f_N, g_N, h_N)(t) = G(f_1, g_1, h_1, \ldots, f_N, g_N, h_N)(t) =$$
$$H(f_1, g_1, h_1, \ldots, f_N, g_N, h_N)(t) = 0$$

otherwise. The triple (F, G, H) is a 3,3-computing system for θ. This triple is $\mathcal{F}$-substitutional. To show this, we note that the inequality (5) is equivalent to the conjunction of the inequalities

$$d_i - \left| \frac{f_j(k) - g_j(k)}{h_j(k) + 1} - a_{ij} \right| > \frac{1}{k+1}, \quad j = 1, \ldots, N. \tag{6}$$

For $i = 1, \ldots, n$, $j = 1, \ldots, N$, let

$$a_{ij} = \frac{p_{ij} - q_{ij}}{r_{ij} + 1},$$

where p_{ij}, q_{ij}, r_{ij} are natural numbers. Let m be a positive integer divisible by $k+1$, by all numbers $r_{ij} + 1$ and by the denominators of all d_i. Then the numbers

$$e = \frac{m}{k+1}, \quad e_i = md_i, \quad e_{ij} = \frac{m}{r_{ij} + 1}$$

are positive integers, and the inequality (6) is equivalent to the inequality

$$e_i(h_j(k) + 1) >$$
$$e(h_j(k) + 1) + |(mf_j(k) + q_{ij}e_{ij}(h_j(k) + 1)) - (mg_j(k) + p_{ij}e_{ij}(h_j(k) + 1))|.$$

In its turn, this inequality is equivalent to the equality $K_{ij}(f_j, g_j, h_j) = 0$, where

$$K_{ij}(f_j, g_j, h_j) =$$
$$(e(h_j(k) + 1) + |(mf_j(k) + q_{ij}e_{ij}(h_j(k) + 1)) - (mg_j(k) + p_{ij}e_{ij}(h_j(k) + 1))| + 1)$$
$$\dot{-}\, e_i(h_j(k) + 1).$$

Therefore the inequality (5) is equivalent to the equality $\sum_{j=1}^{N} K_{ij}(f_j, g_j, h_j) = 0$. Let us define mappings $F_1', \ldots, F_n' : \mathbb{T}_1^{3N} \to \mathbb{T}_1$ as follows:

$$F_i'(f_1, g_1, h_1, \ldots, f_N, g_N, h_N)(t) =$$
$$F_i(f_1, g_1, h_1, \ldots, f_N, g_N, h_N)(t) \, \overline{sg} \sum_{j=1}^{N} K_{ij}(f_j, g_j, h_j) \prod_{i'=1}^{i-1} sg \sum_{j=1}^{N} K_{i'j}(f_j, g_j, h_j).$$

Then we have the equality

$$F(f_1, g_1, h_1, \ldots, f_N, g_N, h_N)(t) = \sum_{i=1}^{n} F_i'(f_1, g_1, h_1, \ldots, f_N, g_N, h_N)(t),$$

and it implies that F is a $\mathcal{F}$-substitutional mapping. It is seen in a similar way that G and H are also $\mathcal{F}$-substitutional mappings.

Corollary 2. *If the class $\mathcal{F}$ satisfies the assumptions of Theorem 3 then all conditionally $\mathcal{F}$-computable real functions with compact domains are uniformly $\mathcal{F}$-computable.*

Proof. The assumptions of Theorem 3 imply the assumptions of Theorem 2.

Notes and Comments. The conditional $\mathcal{F}$-computability of real functions has some similarity in its spirit with the notion of a real function in $\mathcal{F}$ introduced in [3] (under some restrictions on the class $\mathcal{F}$) for functions whose domains are open sets. However, there are many essential differences between the two notions. For instance, if $\mathcal{F}$ is the class of the lower elementary functions then the class of the real functions in $\mathcal{F}$ is not closed under substitution, it is not true that it contains all elementary functions of calculus, and there are real functions in $\mathcal{F}$ which are not computable in the usual sense.

Acknowledgement. We thank the anonymous referees for many useful suggestions.

References

1. Skordev, D., Weiermann, A., Georgiev, I.: $\mathcal{M}^2$-computable real numbers. J. Logic Comput. (Advance Access published September 21, 2010), doi:10.1093/logcom/exq050
2. Skordev, D.: Uniform computability of real functions. In: Collection of Summaries of Talks Delivered at the Scientific Session on the Occasion of the 120th Anniversary of FMI, Sofia, October 24 (2009) (to appear)
3. Tent, K., Ziegler, M.: Computable functions of reals. Münster J. Math. 3, 43–66 (2010)

Towards an Axiomatic System for Kolmogorov Complexity

Antoine Taveneaux

Laboratoire d'Informatique Algorithmique: Fondements et Applications (LIAFA),
CNRS & Université Paris Diderot—Paris 7, 75205 Paris Cedex 13, France

Abstract. In [6], it is shown that four of its basic functional properties are enough to characterize plain Kolmogorov complexity, hence obtaining an axiomatic characterization of this notion. In this paper, we try to extend this work, both by looking at alternative axiomatic systems for plain complexity and by considering potential axiomatic systems for other types of complexity. First we show that the axiomatic system given by Shen cannot be weakened (at least in any natural way). We then give an analogue of Shen's axiomatic system for conditional complexity. In the second part of the paper, we look at prefix-free complexity and try to construct an axiomatic system for it. We show however that the natural analogues of Shen's axiomatic systems fail to characterize prefix-free complexity.

1 Introduction

The concept of Kolmogorov complexity was introduced independently by Kolmogorov (in [3]) and Chaitin (in [1]). The aim of Kolmogorov complexity is to quantify the amount of "information" contained in finite objects, such as binary strings. This idea can be used to give an answer to the philosophical question, "What does it mean for a single object to be random?"

The usual definition of Kolmogorov complexity uses the existence of an optimal Turing machine. However, it is not immediate from that definition that Kolmogorov complexity is satisfactory as a measure of information. One is only convinced after deriving certain fundamental facts about it, such as: most strings have maximal complexity, the complexity of a pair (x, y) is not (much) greater than the sum of the complexities of x and y, etc. Therefore, a natural question is to ask whether there exists an axiomatic system characterizing Kolmogorov complexity uniquely via some of its functional properties. And of course, as with any axiomatic system, we want the axiomatic system to be minimal, i.e. to contain no superfluous axiom.

Such a characterization was given by Shen in [6] for plain complexity. In Section 2 we recall this characterization and adapt it to provide an axiomatic system for conditional complexity. We then study whether we can weaken the hypotheses of this characterization of plain complexity in a natural way and show that it is indeed not possible. In particular, one of the hypotheses in the theorem states that applying a partial computable function to a string does not increase

B. Löwe et al. (Eds.): CiE 2011, LNCS 6735, pp. 280–289, 2011.

its Kolmogorov complexity (up to an additive constant), and we show that this hypothesis cannot be restricted to total computable functions. To show that we need the power of partial computable functions to characterize plain complexity, we introduce a notion of complexity for functions that are total on a initial segment of the integers; this notion of complexity is robust under the application of total computable functions, but differs from Kolmogorov complexity.

A second natural question would be, "Is there a similar axiomatic system for prefix-free Kolmogorov complexity?" Unlike the plain complexity case, we show that the classical properties of prefix-free complexity are not sufficient to characterize it. Since prefix-free complexity is greater than plain complexity, we have to choose a larger upper bound and a tighter lower bound to characterize K (where $K(x)$ is prefix-free complexity of the string x, see below). Actually, all basic upper bounds on prefix-free complexity fail to characterize it. To show that our the classical properties of prefix-free complexity do not characterize it, we construct a counter-example defined by $A = K + f$ with f a very slow growing function. To build such a slow function, we define an operator that slows down sub-linear non-decreasing functions while preserving their computational properties (computability or semicomputability).

Throughout the paper we will identify natural numbers and finite strings in a natural way (the set of finite strings is denoted by $2^{<\omega}$). We denote by $\log(x)$ the discrete binary logarithm of x. We fix an effective enumeration of the Turing machines and we denote by ψ_e the function computed by the eth machine. For each machine T and string x, the complexity of x relatively to T is:

$$C_T(x) = \min\{n \mid \exists y \in 2^{<\omega} \text{ such that } |y| = n \text{ and } T(y) = x\}$$

and throughout the paper we fix an optimal machine $\mathbb{U}$ (i.e. a machine such that for all machines T we have $C_{\mathbb{U}} \leq C_T + O(1)$, see [5] for a existence proof of a such a machine) and set $C = C_{\mathbb{U}}$. In the same way, we fix an optimal prefix-free machine $\mathbb{U}'$ (i.e. a machine with prefix-free domain) and set $K = C_{\mathbb{U}'}$. $C(x)$ and $K(x)$ denote the plain complexity and prefix-free complexity of x, respectively.

Conditional Kolmogorov complexity is an extension of the above notions which quantifies the information of a string x *relative* to another string y. More precisely, the complexity of x given y, relative to the machine T, is:

$$C_T(x|y) = \min\{n \mid \exists z \in 2^{<\omega} \text{ such that } |z| = n \text{ and } T(\langle z, y \rangle) = x\}$$

As above we can define $C(.|.) = C_{\mathbb{U}}(.|.)$ and $K(.|.) = C_{\mathbb{U}'}(.|.)$.

2   Plain Complexity

As mentioned above, Shen showed in [6] that four basic properties are sufficient to fully characterize plain Kolmogorov complexity:

1. Upper-semicomputability: C is not computable but it is upper semicomputable (i.e. the predicate $C(x) \leq k$ is uniformly computably enumerable in x and k).

2. Stability: a recursive function cannot increase the complexity of a string by more than an additive constant.
3. Explicit description: the length of the smallest description of a string (i.e. its plain complexity) is not much bigger than the string itself.
4. Counting: no more than 2^n of the strings have a complexity less than n.

Formally, Shen's theorem states the following.

Theorem 1. *[6] Let* $A : 2^{<\omega} \to \mathbb{N}$ *be some function. Suppose* A *satisfies the following four properties:*

1. *A is upper semi-computable.*
2. *For every partial computable function $f : 2^{<\omega} \to 2^{<\omega}$ there exists a constant c_f such that for each $A(f(x)) \leq A(x) + c_f$ for each $x \in 2^{<\omega}$.*
3. *$A(x) \leq |x| + O(1)$ for all $x \in 2^{<\omega}$.*
4. *$|\{x | A(x) \leq n\}| = O(2^n)$.*

Then $A(x) = C(x) + O(1)$.

Proof. We give a quick sketch of the proof.

To show $A \leq C + O(1)$, let x^* denote the shortest description of x (for the complexity C). By hypotheses 2 and 3 we have:

$$A(x) = A(\mathbb{U}(x^*)) \leq A(x^*) + O(1) \leq |x^*| + O(1) = C(x) + O(1).$$

To show $C \leq A + O(1)$, we consider y and n such that $A(y) = n$. Since A is upper semi-computable, the set $\{x | A(x) \leq n\}$ is uniformly computably enumerable. Since there exists a uniform d such that $|\{x | A(x) \leq n\}| = 2^{n+d}$, we can describe y with only $n + d$ bits (this description $\overline{y}$ such that $|\overline{y}| = n + d$ represents the rank of y in an enumeration of $\{x | A(x) \leq |\overline{y}| - d\}$). So, for all y we have $C(y) \leq n + d + O(1) = A(y) + O(1)$. $\quad\square$

Remark 1: The authors of [7] show that conditions 3 and 4 can be replaced by "There exists a constant c such that $|\{x | A(x) \leq n\}| \in [2^{n-c}, 2^{n+c}]$." We can also replace conditions 2 and 3 by "For every partial computable function f there exists a constant c_f such that $A(f(x)) \leq |x| + c_f$ for each $x \in 2^{<\omega}$." Finally condition 4 can be replaced by the stronger version "$|\{x | A(x) \leq n - k\}| = O(2^{n-k})$."

Remark 2: With essentially the same proof, one can show a similar result for conditional plain complexity. The following system characterizes of conditional plain complexity:

- Uniformly in $x, y \in 2^{<\omega}$, $B(x|y)$ is computable from above.
- For all $x, y \in 2^{<\omega}$, $B(x|y) \leq |x| + O(1)$.
- For each $y \in 2^{<\omega}$ we have $|\{x | B(x|y) \leq n\}| = O(2^n)$ (such that $O(2^n)$ do not depend of y).
- For all y and for every partial computable function f from $2^{<\omega}$ to $2^{<\omega}$ there exists a constant c_f such that for each $x \in 2^{<\omega}$:

$$B(f(x)|y) \leq B(x|y) + c_f.$$

To characterize the conditional aspect, we add to the four previous items the hypothesis "$B(\langle x, y \rangle | y) \leq B(x|y) + O(1)$". Note however that replacing this last condition by $B(x|x) = O(1)$ would not be sufficient.

2.1   Weakening the Hypotheses

Shen's theorem raises a natural question: Are all 4 conditions actually needed? In this subsection we discuss this question. First, it is not hard to see that none of the hypotheses can be removed.

- We need the hypothesis 3 because the function 2C satisfies the three others hypotheses.
- The hypothesis 4 is necessary because the function 0 satisfies the three others hypotheses.
- The hypothesis 1 cannot be removed since $C^{\emptyset'}$ (plain Kolmogorov complexity relativised to the halting problem oracle) satisfies each of three others hypotheses (and clearly differs from the unrelativized version C).
- The hypothesis 2 cannot be removed because the length function satisfies the three others hypotheses.

It could however be the case that hypothesis 2 be replaced by the weaker "for all total computable functions f there exists c_f such that $A(f(x)) \leq A(x) + c_f$". Our first main result is that this is not case.

Theorem 2. *There exists a function* $A : 2^{<\omega} \to \mathbb{N}$ *satisfying hypotheses 1, 3, 4 (of Theorem 1) and:*

- *For every total computable function f from $2^{<\omega}$ to $2^{<\omega}$ there exists a constant c_f such that $A(f(x)) \leq A(x) + c_f$ for each $x \in 2^{<\omega}$.*
- $|A(x) - C(x)|$ *is not bounded.*

Proof. In [4], the authors define a notion of total conditional complexity $\overline{C}(x|y)$ as the smallest length of a program for a total function f code such that $f(y) = x$. Of course, $\overline{C}$ is stable over all total computable functions (i.e. $\overline{C}(f(x)|y) \leq \overline{C}(x|y) + c_f$ for all total computable functions f) and the authors show that $\overline{C}$ significantly differs from the plain conditional complexity. However, the function $\overline{C}$ is not quite suitable for our purposes, for two reasons. First, it is not upper semi-computable and second, its non-conditional version $\overline{C}(x|\lambda)$ is equal to C up to a constant.

   In order to construct our counter-example, we first define a way to encode compositions of partial computable functions by a set of strings having the prefix-free property. This encoding is not at all optimal, which is precisely what will make our proof work. We define:

$$P = \{1^{p_1} 0001^{p_2} 000 \ldots 1^{p_k} 01 | \ \forall k, \ p_k > 0\}.$$

Notice that P is a prefix-free set. For $\tau = 1^{p_1} 0001^{p_2} 000 \ldots 1^{p_k} 01 \in P$ we now denote by φ_τ the function $\varphi_\tau = \psi_{p_1} \circ \psi_{p_2} \circ \cdots \circ \psi_{p_i}$ (recall that (ψ_i) is a standard enumeration of partial computable functions). We now define a function V, as follows. For all $x \in 2^{<\omega}$ and $\tau \in P$, set

$$V(\tau x) = \begin{cases} \varphi_\tau(x) & \text{if for all } y \text{ such that } |y| \le |x| \text{ we have } \varphi_\tau(y) \downarrow \\ \uparrow & \text{otherwise} \end{cases}$$

P is prefix-free, so V is defined without ambiguity and clearly V is a (partial) computable function. If φ_τ is not a total function, there are only a finite number of strings x such that $V(\tau x) \downarrow$. We shall prove that $A = C_V$ satisfies the conditions of the theorem. First, C_V is upper semicomputable and satisfies the counting condition, as it is just the Kolmogorov complexity function associated to the machine V. Moreover, let i be an index for the identity function (i.e. $\psi_i = id$). By definition of V, one has $V(1^i 01x) = x$, hence $A(x) \le |x| + (i + 2)$. To see that A is stable over all total computable functions, let f be a total computable function and let e be an index for f. Now, for any string x, let τy be such the shortest description of x for V with $\tau \in P$. By definition of V, this means that $\varphi_\tau(z) \downarrow$ for all $|z| \le |y|$. And since $f = \psi_e$ is total, we also know that $\psi_e \circ \varphi_\tau(z) \downarrow$ for all $|z| \le |y|$. Therefore $\sigma = 1^e 000\tau$ is a description of x for V. We have proven that $C_V(f(x)) \le C_V(x) + e + 3$ for all x.

It remains to prove that C_V differs from C, i.e. that $C_V - C$ takes arbitrarily large values. We prove this by contradiction: Suppose that $|A(x) - C(x)|$ is bounded by a constant. For $x \in 2^{<\omega}$, we denote by $\hat{x}$ the smallest description of x for V (by definition this means that $C_V(x) = |\hat{x}|$).

Let x be a string. Let us first write

$$\hat{x} = 1^{p_1} 0001^{p_2} 0000 \ldots 1^{p_k} 01y.$$

It is easy to see that

$$C(\hat{x}) \le 2\log(p_1) + 2\log(p_2) + \cdots + 2\log(p_k) + 2k + |y| + O(1),$$

and since x can be computed from $\hat{x}$, this implies a fortiori:

$$C(x) \le 2\log(p_1) + 2\log(p_2) + \cdots + 2\log(p_k) + 2k + |y| + O(1).$$

Moreover, by definition of V,

$$C_V(x) = |\hat{x}| = p_1 + p_2 + \cdots + p_k + 3(k - 1) + 2 + |y|.$$

Thus, since we have assumed that $C_V(x) - C(x)$ bounded, this shows two things:

- the (p_i) appearing in the $\hat{x}$'s are bounded, and
- the number of p_i's used in each $\hat{x}$ is bounded.

Formally, we have proven that $\{\tau \in P \mid \exists x \in 2^{<\omega} \text{ such that } \hat{x} = \tau y\}$ is a finite set and that for each τ in this set, either φ_τ is a total function or for y large enough, τy is not in the domain of V. Thus, this τ appears only in a finite number of $\hat{x}$.

Finally for $|\widehat{x}|$ large enough (and hence for $|x|$ large enough because $\{x|A(x) \leq n\}$ is finite for all n), $\widehat{x} = \tau x$ with $\tau \in P$, and φ_τ is a total computable function. So

$$Q = \{\tau \in P | \exists^\infty x \in 2^{<\omega} \text{ such that } \widehat{x} = \tau y\}.$$

is a finite set of codes of total functions and thus there is only a finite number of $\tau \in P$ in the prefixes of $\widehat{x}$'s.

Therefore, for x large enough, $\widehat{x}$ is of the form τy with $\tau \in Q$ and hence:

$$A(x) = \min\{|\tau y| \mid \tau \in Q \text{ and } \varphi_\tau(y) = x\}.$$

Since Q is finite and all $(\varphi_\tau)_{\tau \in Q}$ are total, this makes A computable, contradicting $A = C + O(1)$ because no non-trivial lower-bound of C is computable.

$\square$

3   An Axiomatic System for Prefix Complexity

As we have seen in the last section, there exists a minimal set of simple properties that characterize plain complexity. One may ask whether it is possible to obtain a similar characterization of prefix-free complexity K .

It is natural to keep the hypotheses 1 and 2, but the other two hypotheses need to be adapted. Indeed, hypothesis 3 fails to hold for K (i.e. $K(x) \not\leq |x| + O(1)$), and the sharpest classical upper bound is $K(x) \leq |x| + K(|x|) + O(1)$ (see [2]).

Accordingly, the hypothesis 4 (i.e. $|\{x|K(x) \leq n\}| = O(2^n)$) is too weak. The analogue of that counting argument for K is the classical

$$|\{x | |x| = n \text{ and } K(x) \leq n + K(n) - k\}| = O(2^{n-k}).$$

Another property of K that is very often used is $\sum_{x \in 2^{<\omega}} 2^{-K(x)} < \infty$ (in fact, any upper semi-computable function A satisfying $\sum_{x \in 2^{<\omega}} 2^{-A(x)} < \infty$ is such that $K \leq A + O(1)$). Perhaps surprisingly, this set of properties alone is not enough to characterize K.

Theorem 3. *There exists a function A satisfying the following:*

1. *A is upper semi-computable.*
2. *For every partial computable function f from $2^{<\omega}$ to $2^{<\omega}$ there exists a constant c_f such that for each $A(f(x)) \leq A(x) + c_f$ for each $x \in 2^{<\omega}$.*
3. $\sum_{x \in 2^{<\omega}} 2^{-A(x)} < \infty.$
4. $A(x) \leq |x| + A(|x|) + O(1)$ *for each $x \in 2^{<\omega}$.*
5. $|\{x \in 2^n \mid A(x) \leq |n| + A(n) - b\}| \leq O(2^{n-b}).$
6. $|A - K|$ *is not bounded.*

Remark: Since hypothesis 3 guarantees the inequality $K \leq A + O(1)$, it would be sufficient, in order to obtain a full characterization of K, to add the property: "For every f partial computable prefix-free function there exists c_f such that

$A(f(x)) \leq |x| + c_f$". Indeed, for all x if we denote by x^* a shortest string such that $\mathbb{U}'(x^*) = x$ then $A(x) = A(\mathbb{U}'(x*)) \leq |x^*| + c_{\mathbb{U}'} = K(x) + c_{\mathbb{U}'}$. However, such a system would not be very satisfactory because it uses the prefix-freeness of functions and thus is mostly a rewording of the definition of K.

Proof. We will construct A by taking $A = K + \beta$ with β an unbounded function with certain nice properties. The function β will be upper semicomputable, non-decreasing, unbounded, such that

$$\beta(x) = \beta(|x|) + O(1),$$

and such that for f partial computable function, there is c_f such that

$$\beta(f(n)) \leq \beta(n) + c_f. \tag{1}$$

Simple considerations show that β has to have a very low growth speed. First let us define Solovay's α-function:

Definition 1. *The Solovay's α-function is defined by:*

$$\alpha(n) = \min\{K(i)|i > n\}.$$

We call order a total, non-decreasing and unbounded (not necessarily computable) function $f : \mathbb{N} \to \mathbb{N}$.

Equivalently $\alpha(n)$ is the length of the shortest string τ such that $\mathbb{U}'(\tau) > n$ since $\mathbb{U}'$ is the optimal optimal prefix-free machine chosen to define K.

α is an order with a very low rate of growth, and actually one can show that it grows more slowly than any computable order.

Lemma 1. *For each h computable order:*

(i) for all n, $\alpha(h(n)) = \alpha(n) + O(1)$
(ii) for all n, $\alpha(n) \leq h(n) + O(1)$

Proof. To prove this lemma, we need the following list of trivial facts.

- By the definition of α, there exists $j > n$ such that $\alpha(n) = K(j)$.
- There exists c_h such that for all n, $K(h(n)) \leq K(n) + c_h$.
- $K(n) \geq \alpha(n)$ for all n.
- Since h and α are order functions, $h(j) \geq h(n)$ and $\alpha(h(j)) \geq \alpha(h(n))$.

Now, one can apply these facts in order to get:

$$\alpha(n) = K(j) \geq K(h(j)) - c_f \geq \alpha(h(j)) - c_f \geq \alpha(h(n)) - c_f.$$

To prove $\alpha(n) \leq \alpha(h(n)) + O(1)$, it suffices to consider an inverse order $\widehat{h}$ of the order function h defined by: $\widehat{h}(n) = \max\{i|h(i) \leq n\}$. Since $\widehat{h}$ is a computable order we have:

$$\alpha(n) \leq \alpha(\widehat{f}(h(n))) \leq \alpha(h(n)) + c.$$

To show that $\alpha(n) \leq h(n) + O(1)$, notice that $K(n) \leq n + O(1)$ and so there exists c such that $\alpha(n) \leq n + c$. Finally, by the previous point, we have:

$$\alpha(n) \leq \alpha(h(n)) + c_h \leq h(n) + d.$$

$\square$

We can show that α satisfies 1 for each total computable function, but there exists some partial computable functions such that α does not satisfy 1. In the same way we can show that $K + \alpha$ does not satisfy condition 2 in the statement of the theorem. However, we have a weaker version for partial functions:

Lemma 2. *For each partial computable function* $f : \mathbb{N} \to \mathbb{N}$ *there exists* c_f *such that for all* n

$$\alpha(\alpha(f(n))) \leq \alpha(n) + c_f.$$

Proof. This follows from the following simple fact. For each f there exists c_f such that:

$$\alpha(f(n)) \leq \mathrm{K}(f(n)) \leq \mathrm{K}(n) + c_f \leq n + c_f.$$

Since α is a sub-linear order:

$$\alpha(\alpha(f(n))) \leq \alpha(n + c_f) \leq \alpha(n) + c_f + O(1).$$

$\square$

As stated above, the partial computable functions can increase too quickly to satisfy the second condition of the theorem. For this reason we introduce a general operator to slow down sub-linear and upper semi-computable orders:

Definition 2 (Star-operator). *Let* f *be a sub-linear (i.e.* $f(n) = o(n)$*) order function. If we set*

$$p_f = \max\{n | f(n) \geq n\}$$

which is well-defined by sub-linearity of f*, then,* f^* *is defined by:*

$$f^*(n) = \min\{k | f^{(k)}(n) \leq p_f\}.$$

This operator is a generalization of the so-called log*, which is precisely the function one gets by taking $f = \log$ in our definition of f^*.

Remark: A simpler definition could be $f^*(n) = \min\{k | f^{(k)}(n) = f^{(k+1)}(n)\}$ but for small values of n and for some function f (for functions with more than one fixed point, for example) this definition is not exactly the same and is in fact less natural. This star operator will suit our purposes because it possesses some nice properties.

Lemma 3. *Let* f *a sub-linear order function. The following properties hold:*

1. f^* *is a sub-linear order function.*
2. *If* f *is a computable function then so is* f^**.*
3. *If* f *is a upper semi-computable function then so is* f^**.*
4. $0 \leq f^*(n) - f^*(f^{(i)}(n)) \leq i$*.*

Proof. (1) The last claim will ensure sub-linearity. To see that f^* is non-decreasing, if $x \leq y$ then $f^{(i)}(x) \leq f^{(i)}(y)$ for all i because f is non-decreasing. Finally, f^* is unbounded, for if f^* had a finite limit d, then $f^{(d)}$ would be bounded. But this is not possible because $\hat{f}$ tends to infinity.

(2) If f is computable then to determine $f^*(n)$ we can compute the sequence $f^{(1)}(n), f^{(2)}(n), \ldots, f^{(k)}(n), \ldots$ until we find the first j such that $f^{(j)}(n) \leq p_f$ and we return j.

(3) If f is upper semi-computable then we compute in parallel the approximations of $f^{(k)}(n)$ for all k, and we return the least k such that $f^{(k)}(n) \leq p_f$.

(4) If $f^*(n) \leq i$ then $f^*(f^{(i)})(n) = 0$ because necessarily, $f^{(i)}(n) \leq p_f$. So $f^*(n) \leq f^*(f^{(i)}(n)) + i$.

If $f^*(n) > i$ then $f^*(n) = f^*(f^{(i)})(n) + i$ by definition of the star-operator. In both cases $f^*(n) \geq f^*(f^{(i)}(n))$.

We shall use Solovay's α function transformed by the star-operator. We will show that the function $A(x) = K(x) + \alpha^*(x)$ has all the necessary properties to prove the theorem.

By Lemma 3 the function α^* is upper semicomputable, and thus $K + \alpha^*$ is as well.

By Lemma 2 we have that for each partial computable function $f : \mathbb{N} \to \mathbb{N}$ there exists c_f such that for all n

$$\alpha(\alpha(f(n))) \leq \alpha(n) + c_f$$

and by Lemma 3 (claim 4), if we apply α^* on each term of the previous inequality we have (since α^* is sub-linear)

$$\alpha^*(f(n)) - 2 \leq \alpha^*(\alpha(n) + c_f) \leq \alpha^*(n) + O(1).$$

This proves the second condition of the theorem.

The property 3 is clear because we have $\sum_{x \in 2^{<\omega}} 2^{-K(x)} < \infty$ and $A(x) \geq K(x)$ (because $\alpha^*(x) \geq 0$).

Finally, since $|.|$ is a computable order function, by Lemma 1 we have $\alpha(x) = \alpha(|x|) + O(1)$. By condition 4 of Lemma 3, the equality

$$\alpha^*(x) = \alpha^*(|x|) + O(1)$$

holds. This equality shows that A satisfies hypotheses 4 and 5.

□

It is interesting to notice that the counter-example we produced also invalidates several similar attempts for an axiomatization. For example, one could add the condition:

$$K(xy) \leq K(x, y) \leq K(x) + K(y) + O(1).$$

But $K + \alpha^*$ also satisfies this. One could then ask whether the more precise inequality $K(x, y) \leq K(x) + K(y|x) + O(1)$ could help characterizing conditional prefix-free Kolmogorov complexity, but then again, defining $\alpha(x|y)$ by

$$\alpha(n|m) = \min\{K(i|m)|i > n\}$$

and then $\alpha^*(.|y)$ for each y, we get a counter-example by taking $A(.|.) = K(.|.) + \alpha^*(.|.)$.

This, together with Theorem 3, shows that the situation is more subtle in the prefix-free complexity case than in the plain complexity case. Finding a natural characteristic set of properties for K is left as an open question.

Acknowledgements

I would like to express my gratitude to Laurent Bienvenu without whom this paper would never had existed. Thanks also to Serge Grigorieff for our numerous discussions during which he helped me progress on this work. Finally, thanks to the Chris Porter and three anonymous reviewers for their help in preparing the final version of this paper.

References

1. Gregory, J.: Chaitin On the Length of Programs for Computing Finite Binary Sequences. J. ACM 13(4), 547–569 (1966)
2. Rod, G.: Downey and Denis Hirschfeldt. In: Algorithmic Randomness and Complexity. Springer, Heidelberg (2010)
3. Andreï, N.: Kolmogorov. Three approaches to the definition of the concept "quantity of information". In: Problemy Peredači Informacii, pp. 3–11 (1965)
4. Andrej, A.: Muchnik, Ilya Mezhirov, Alexander Shen, and Nikolay Vereshchagin. Game interpretation of Kolmogorov complexity. Draft version (2010)
5. Nies, A.: Computability and Randomness. Oxford University Press, Oxford (2009)
6. Shen, A.: Axiomatic description of the entropy notion for finite objects. In: VIII All-USSR Conference on Logika i metodologija nauki, Vilnjus (1982) Paper in Russian
7. Uspensky, V.A., Shen, A., Vereshchagin, N.K.: Kolmogorov complexity and randomness. Book draft version (2010)

A New Optimum-Time Firing Squad Synchronization Algorithm for Two-Dimensional Rectangle Arrays: One-Sided Recursive Halving Based

H. Umeo, K. Nishide, and T. Yamawaki

Division of Information and Computer Sciences,
Osaka Electro-Communication University,
Neyagawa-shi, Hastu-cho, 18-8, Osaka, 572-8530, Japan
umeo@cyt.osakac.ac.jp

Abstract. The firing squad synchronization problem on cellular automata has been studied extensively for more than fifty years, and a rich variety of synchronization algorithms have been proposed for not only one-dimensional arrays but two-dimensional arrays. In the present paper, we propose a new optimum-time synchronization algorithm that can synchronize any two-dimensional rectangle arrays of size $m \times n$ with a general at one corner in $m + n + \max(m, n) - 3$ steps. The algorithm is based on a simple recursive halving marking schema which helps synchronization operations on two-dimensional arrays. A proposed computer-assisted implementation of the algorithm gives a description of a two-dimensional cellular automaton in terms of a finite 384-state set and a local 112690-rule set.

1 Introduction

We study a synchronization problem that gives a finite-state protocol for synchronizing a large scale of cellular automata. The synchronization in cellular automata has been known as a firing squad synchronization problem (FSSP) since its development, in which it was originally proposed by J. Myhill in Moore [1964] to synchronize all parts of self-reproducing cellular automata. The problem has been studied extensively for more than 50 years [1-18].

In the present paper, we propose a new time-optimum synchronization algorithm for rectangle cellular automata. The algorithm can synchronize any rectangle arrays of size $m \times n$ with a general at one corner in $m + n + \max(m, n) - 3$ steps. The algorithm is based on a simple marking schema which prints a special mark in a given cellular space. We also give a computer-assisted implementation of the algorithm operating on a two-dimensional cellular automaton with 384-state and 112690-rule. Our computer simulation shows that the state and rule sets presented is valid for the synchronization on any rectangle arrays of size $m \times n$ such that $2 \leq m, n \leq 253$.

B. Löwe et al. (Eds.): CiE 2011, LNCS 6735, pp. 290–299, 2011.

2   Firing Squad Synchronization Problem on Two-Dimensional Arrays

Figure 1 shows a finite two-dimensional (2-D) cellular array consisting of $m \times n$ cells, that is, m rows and n columns, each denoted by $C_{ij}, 1 \leq i \leq m, 1 \leq j \leq n$. Each cell is an identical (except the border cells) finite-state automaton. The array operates in lock-step mode in such a way that the next state of each cell (except border cells) is determined by both its own present state and the present states of its north, south, east and west neighbors. Thus we assume the well-known *von Neumann neighborhood*. All cells (*soldiers*), except the north-west corner cell (*general*), are initially in the quiescent state at time $t = 0$ with the property that the next state of a quiescent cell with quiescent neighbors is the quiescent state again. At time $t = 0$, the north-west corner cell C_{11} is in the *fire-when-ready* state, which is the initiation signal of the synchronization for the array. The firing squad synchronization problem (FSSP) is to determine a description (a state set $\mathcal{Q}$ and a next-state function such that $\delta : \mathcal{Q}^5 \to \mathcal{Q}$) for cells that ensures all cells enter the *fire* state at exactly the same time and for the first time. The tricky part of the problem is that the same kind of soldier having a fixed number of states must be synchronized, regardless of the size $m \times n$ of the array. The set of states and next state function must be independent of m and n.

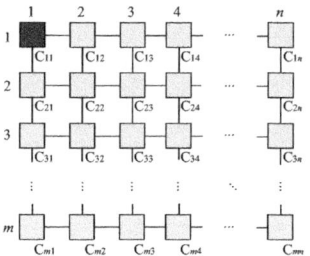

Fig. 1. A two-dimensional cellular automaton

Several synchronization algorithms on 2-D arrays have been proposed by Beyer [1969], Grasselli [1975], Shinahr [1974], Szwerinski [1982], Umeo, Maeda, Hisaoka and Teraoka [2006], and Umeo and Uchino [2008]. It has been shown by Beyer [1969] and Shinahr [1974] independently that there exists no two-dimensional cellular automaton that can synchronize any 2-D array of size $m \times n$ in less than $m + n + \max(m, n) - 3$ steps. In addition they first proposed an optimum-time synchronization algorithm that can synchronize any 2-D array of size $m \times n$ in optimum $m + n + \max(m, n) - 3$ steps. Shinahr [1974] gave a 28-state implementation. Umeo, Hisaoka and Akiguchi [2005] presented a new 12-state synchronization algorithm operating in optimum-step, realizing the smallest, in the number of states, solution to the rectangle synchronization problem at present.

3   One-Sided Recursive Halving Mark

In this section, we develop a marking schema for one-dimensional arrays, which is referred to as *one-sided recursive halving mark*. The schema prints a special mark on cells in a given cellular space, which is defined by one-sided recursive halving. The marking itself is based on a well-known optimum-time one-dimensional synchronization algorithm.

Let S be a one-dimensional cellular space consisting of cells C_i, C_{i+1}, ..., C_j, denoted by $[i...j]$, where $j > i$. Let $|S|$ denote the number of cells in S, that is, $|S| = j - i + 1$. The center cell(s) C_x of S is defined by

$$
x = \begin{cases} (i+j)/2 & |S|\text{: odd,} \\ (i+j-1)/2, (i+j+1)/2 & |S|\text{: even.} \end{cases} \tag{1}
$$

Note that we have two center cells when $|S|$ is even. The one-sided recursive halving mark for a given cellular space $[1...n]$ is defined as follows:

One-sided recursive halving mark ────────────────────────────

```
    begin
        S := [1...n];
        while |S| ≥ 2 do
            if |S| is odd then
                mark a center cell C_x in S
                S := [x...n];
            else
                mark center cells C_x and C_{x+1} in S
                S := [x + 1...n];
    end
```

──

We call the schema one-sided recursive halving mark. For example, we consider a cellular space $S = [1...15]$ consisting of 15 cells. The first center cell is C_8, then the second one is C_{11} and C_{12}, and the last one is C_{13} and C_{14}, respectively. In case $S = [1...17]$, we get C_9, C_{13}, C_{15}, and C_{16} after four iterations.

The one-sided recursive halving marking schema can be realized on a cellular automaton. Figure 2 (left) shows a space-time diagram for the marking. At time $t = 0$, the leftmost cell C_1 emits two signals simultaneously, each propagating in the right direction at 1/1- and 1/3-speed. The 1/1-speed signal arrives at C_n at time $t = n - 1$. Then, the rightmost cell C_n generates an infinite set of signals $w_1, w_2, ..., w_k, ..$, each propagating in the left direction at $1/(2^k - 1)$ speed, where $k = 1, 2, 3, ...,$. The readers can find that each crossing, made by two signals of the first right-going 1/3-speed signal and with each $w_1, w_2, ..., w_k$, shown in Fig. 2 (left), enables the marking at middle points defined by the one-sided recursive halving. A finite state realization for generating the infinite set of signals above

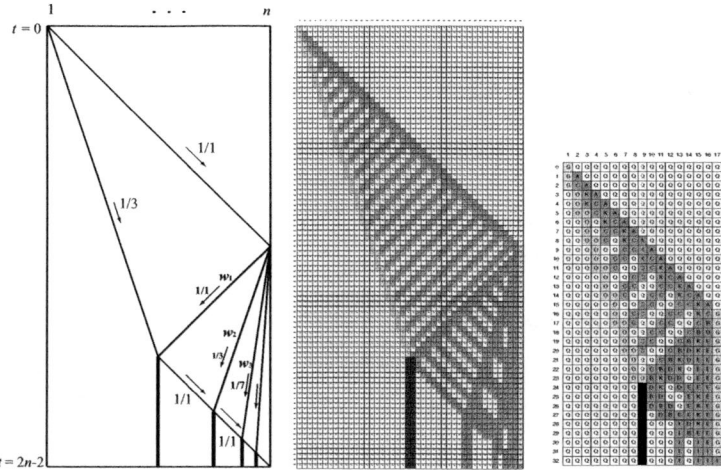

Fig. 2. Space-time diagram for one-sided recursive halving on one-dimensional array of length n (left) and some snapshots on 44 (middle) and 17 (right) cells, respectively, for the marking implemented on a 13-state cellular automaton

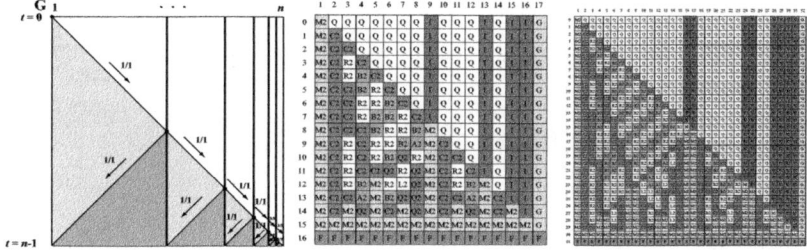

Fig. 3. Space-time diagram for synchronizing a cellular space with the one-sided recursive halving mark (left) and some snapshots for the synchronization on 17 (middle) and 32 (right) cells, respectively

is a well-known technique used for the optimum-time synchronization algorithms on one-dimensional arrays in Balzer [1967], Gerken [1987], and Waksman [1966].

We have developed a simple implementation of the one-sided recursive halving marking on a 13-state cellular automaton. In Fig. 2 (middle and right) we present several snapshots for the marking on 44 and 17 cells, respectively. Thus we have:

Lemma 1. There exists a 13-state cellular automaton that can print the one-sided recursive halving mark in any cellular space of length n in $2n - 2$ steps.

It can be seen that any cellular space of length n initially set up with a recursive halving mark and a general at left end can be synchronized in optimum $n - 1$ steps. Figure 3 (left) shows the space-time diagram for synchronizing a cellular space with the one-sided recursive halving mark. Note that a general is located at left end of the cellular space with the marking. Several snapshots for the synchronization on 17 and 32 cells are shown in Fig. 3 (middle and right), respectively.

Lemma 2. Any cellular space of length n with the one-sided recursive halving mark can be synchronized in $n - 1$ steps.

4   A New Two-Dimensional Optimum-Time Synchronization Algorithm

4.1   Overview of the Algorithm

In this section we give an overview of the synchronization algorithm $\mathcal{A}$ proposed for two-dimensional arrays. First we consider the algorithm which can synchronize any rectangle arrays wider than long, that is, $m \leq n$. A more generalized case will be discussed later.

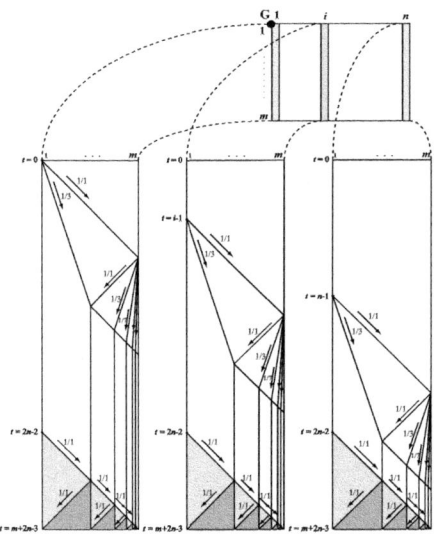

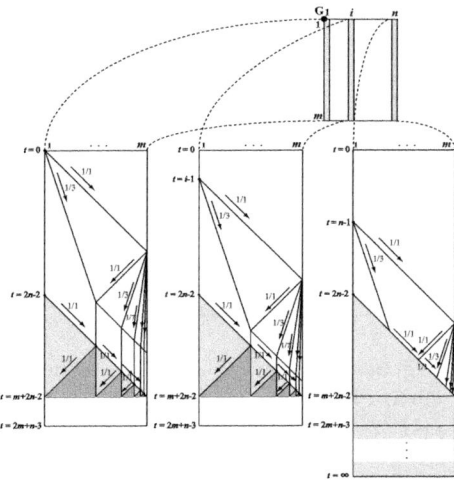

Fig. 4. Space-time diagrams for the wider-than-long column synchronization algorithm on the 1st, ith, and nth columns, respectively

Fig. 5. Space-time diagram of the wider-than-long column synchronization algorithm, being applied to a rectangle longer than wide

An overview of the algorithm is as follows:

1. At time $t = 0$ an initial general on the north west corner emits a wake-up signal along the first row to synchronize the first row with a tentative firing state. The first row will be synchronized at time $t = 2n - 2$. The wake-up signal on the first row also acts as an activation signal for the one-sided

recursive halving marking which prints special marks on each column. The wake-up signal reaches the ith column at time $t = i - 1$ and it begins the marking operation for the column. The recursive-halving making for the ith column will be finished at time $t = i - 1 + 2m - 2$. The computed time is due to Lemma 1.

2. Once the first row could be synchronized with the tentative firing state, then the cell C_{1i} in the tentative firing state initiates the synchronization for the ith column for each i such that $1 \leq i \leq n$. Using Lemma 2, the ith column will be synchronized at time $t = (2n - 2) + (m - 1) = m + 2n - 3 = m + n + \max(m, n) - 3$ in the case where $m \leq n$.

For any i such that $1 \leq i \leq n$, the first center cell(s) of the ith column is marked at time $t_1 = i - 2 + \lceil 3m/2 \rceil$ and the wake-up signal for the synchronization arrives at the center cell(s) of the ith column at time $t_2 = 2n - 2 + \lceil m/2 \rceil$. The first center cell(s) has been marked before the arrival of the wake-up signal in the case where $m \leq n$, since $t_2 - t_1 = 2n - m - i \geq 0$, for any i such that $1 \leq i \leq n$. Note that the signal propagation for the halving and the wake-up signal for the column synchronization are made at the same $1/1$ speed. Thus, the synchronization can be performed successfully for each column and the array can be synchronized in optimum steps. We call the synchronization *wider-than-long column synchronization*. Figure 4 illustrates a space-time diagram for the wider-than-long column synchronization algorithm on the 1st, ith, and nth column, respectively. One can see that each marking operation has been finished before the arrival of the first synchronization signal on each column.

4.2 Applying the Wider-Than-Long Column Synchronization Algorithm to Rectangles Longer than Wide

Note that the above synchronization scheme doesn't work for the rectangle longer than wide, because the pre-synchronization on the first row is finished too early. Consider a longer-than-wide rectangle of size $2n \times n$. The synchronization on the first column is successfully done, however the last nth column fails to synchronize, since the wake-up signal for the synchronization continues to look for the markings.

In the case of rectangles longer than wide, each horizontal segment, that is a shorter one in this case, falls into a firing state at time $t = 2n + m - 3$, which is smaller than the optimum-time $t = 2m + n - 3$. Moreover at time $t = 2n + m - 3$ some cells on some vertical segments are still in quiescent state. Thus the algorithm presented above fails to synchronize rectangles longer than wide. See Fig. 5. The figure illustrates a space-time diagram for those unsuccessful synchronization operations on the 1st, ith, and nth column when applying the wider-than-long column synchronization algorithm to a rectangle longer than wide.

4.3 Synchronization of Rectangle Longer than Wide

The synchronization for the rectangles longer than wide can be done by interchanging the roles of row and column operations described above. By the similar way, the rectangle can be synchronized at time $t = n + 2m - 3 = m + n + \max(m, n) - 3$ in the case where $m \geq n$. We call the synchronization *longer-than-wide row synchronization*. By symmetry, the longer-than-wide row synchronization algorithm doesn't work for the rectangle wider than long.

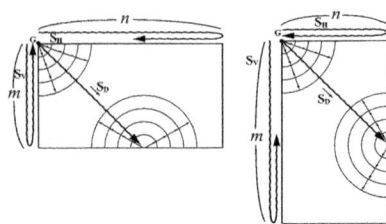

Fig. 6. A trajectory of inhibition signals generated by a general on a two-dimensional wider than long (left) and longer than wide (right) array of size $m \times n$. Three signals s_V, s_H, and s_D (denoted by bold arrows in the figure) for stopping undesirable synchronization operations are also illustrated.

4.4   Stopping of Undesirable Synchronization Operations

The initial general has no a priori knowledge on the side length of rectangle arrays. To synchronize a given rectangle in optimum steps, the array, at time $t = 0$, begins to prepare both two synchronization operations, one for wider than long and the other for longer than wide rectangles. In order to check the type of the given rectangle, the array generates a zigzag signal s_D at time $t = 0$ which propagates at a unit speed along the principal diagonal. If the signal meets the lower (south) boundary, then the rectangle is wider than long and if it meets the right (east) boundary, then the rectangle is longer than wide. In addition, at time $t = 0$, the array also generates two signals, s_V and s_H, each propagates in the vertical and horizontal direction at $1/1$-speed, respectively.

The general on C_{11} gets the side length information at time $t = 2m - 2$ or $t = 2n - 2$ by the first arrival of the return of the signal s_V or s_H. See Fig. 6. The array has to inhibit the operations of the column synchronization in the case it is longer than wide and the row synchronization operations in the case wider than long. Otherwise some cells fall into a firing state before the final firing time. To inhibit the column synchronization, the diagonal signal generates a $1/1$ speed signal when it hits the right boundary at time $t = 2n - 2$. The signal propagates horizontally to the left direction on the nth row and prints a special inhibition mark on each cell. The inhibition signal arrives at the left end at time $t = 3n - 3$. On the other hand, the wake-up signal for the column synchronization arrives at the nth row at time $t = 3n - 3$, thus all of the wake-up signal can be stopped on that row and inhibit the column synchronization. As for the row synchronization, a similar technique can be employed.

An alternative way for the inhibition is to generate a signal at the cell C_{11} and C_{mm} (or C_{nn}), where the s_D-signal hits depending on the side length. The signal spreads at $1/1$ speed in every direction like a circular wave, inhibiting

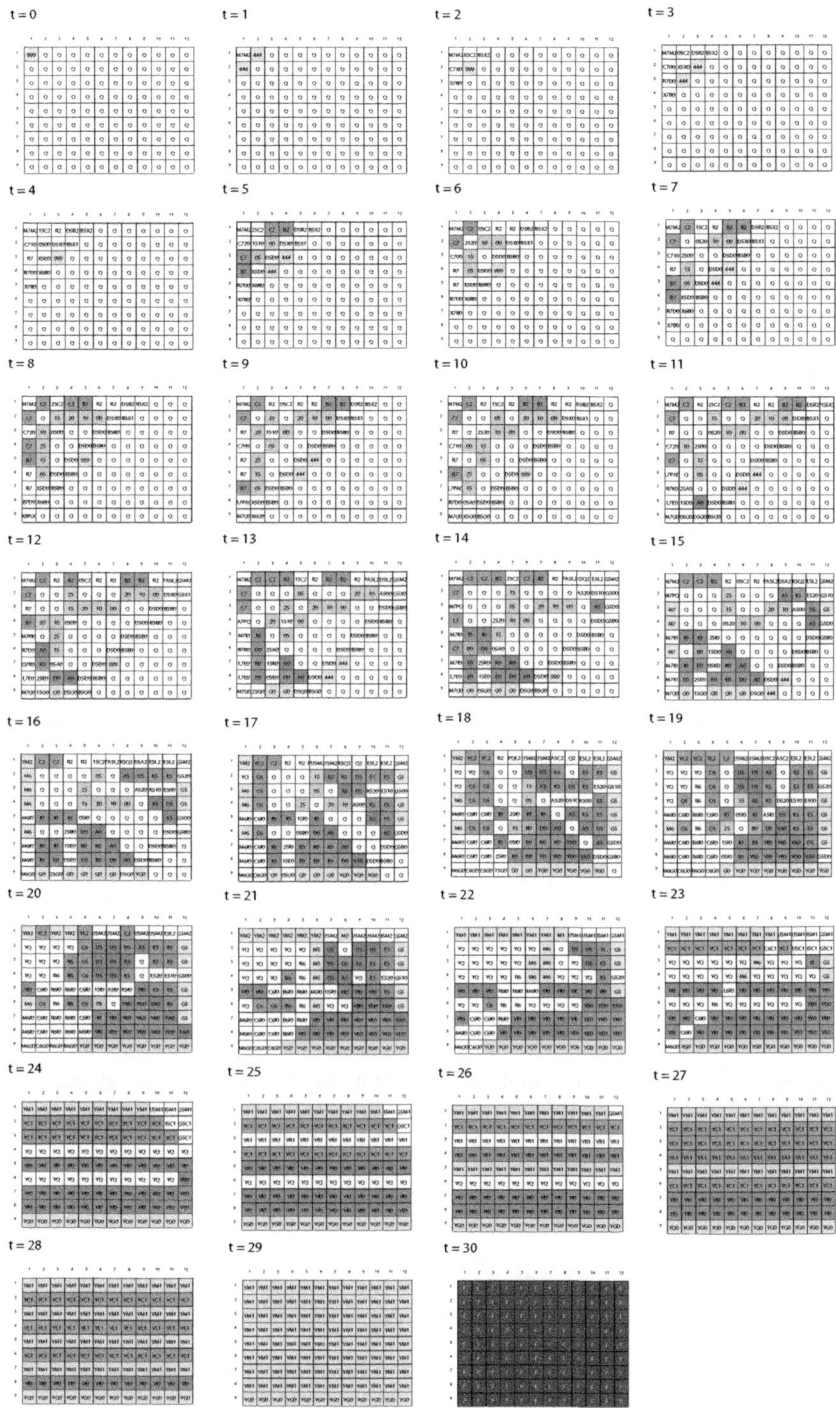

Fig. 7. Snapshots of the final synchronization algorithm on a 9×12 array

undesirable synchronization operations. Figure 6 shows how the signal spreads over the array.

4.5   Final Algorithm

Now we can establish the following theorem. We have also implemented the algorithm on a two-dimensional cellular automaton having 384 states and 112690 local rules. The rule set is generated by a computer program based on the algorithm $\mathcal{A}$. Our computer simulation shows that the rule set generated is valid for the synchronization on any rectangle of size $m \times n$ such that $2 \leq m, n \leq 253$. In Figure 7, we present some snapshots of the synchronization processes of the implementation, operating on a 9×12 array.

Theorem 3. The synchronization algorithm $\mathcal{A}$ can synchronize any $m \times n$ rectangular array in optimum $m + n + \max(m, n) - 3$ steps.

5   Conclusions

We have proposed a new optimum-time rectangle synchronization algorithm based on the one-sided recursive halving mark. It can synchronize any two-dimensional rectangle array of size $m \times n$ in optimum $m + n + \max(m, n) - 3$ steps. We have also implemented the algorithm on a two-dimensional cellular automaton having 384 states and 112690 local rules. A smaller-state implementation would not be so complicated.

Acknowledgements

The authors would like to express their thanks to reviewers for useful comments. A part of this work is supported by Grant-in-Aid for Scientific Research (C) 21500023.

References

1. Balzer, R.: An 8-state minimal time solution to the firing squad synchronization problem. Information and Control 10, 22–42 (1967)
2. Beyer, W.T.: Recognition of topological invariants by iterative arrays. Ph.D. Thesis, MIT, pp. 144 (September 1969)
3. Gerken, H.D.: Über Synchronisationsprobleme bei Zellularautomaten. Diplomarbeit, Institut für Theoretische Informatik, Technische Universität Braunschweig, pp. 50 (1987)
4. Goto, E.: A minimal time solution of the firing squad problem. In: Dittoed course notes for Applied Mathematics, vol. 298, pp. 52–59. Harvard University, Cambridge (1962)
5. Mazoyer, J.: A six-state minimal time solution to the firing squad synchronization problem. Theoretical Computer Science 50, 183–238 (1987)
6. Moore, E.F.: The firing squad synchronization problem. In: Moore, E.F. (ed.) Sequential Machines, Selected Papers, pp. 213–214. Addison-Wesley, Reading (1964)

7. Schmid, H.: Synchronisationsprobleme für zelluläre Automaten mit mehreren Generälen. Diplomarbeit, Universität Karsruhe (2003)
8. Shinahr, I.: Two- and three-dimensional firing squad synchronization problems. Information and Control 24, 163–180 (1974)
9. Szwerinski, H.: Time-optimum solution of the firing-squad-synchronizationproblem for n-dimensional rectangles with the general at an arbitrary position. Theoretical Computer Science 19, 305–320 (1982)
10. Umeo, H.: A simple design of time-efficient firing squad synchronization algorithms with fault-tolerance. IEICE Trans. on Information and Systems E87-D(3), 733–739 (2004)
11. Umeo, H.: Firing squad synchronization problem in cellular automata. In: Meyers, R.A. (ed.) Encyclopedia of Complexity and System Science, vol. 4, pp. 3537–3574. Springer, Heidelberg (2009)
12. Umeo, H., Hisaoka, M., Akiguchi, S.: A twelve-state optimum-time synchronization algorithm for two-dimensional rectangular cellular arrays. In: Calude, C.S., Dinneen, M.J., Păun, G., Jesús Pérez-Jímenez, M., Rozenberg, G. (eds.) UC 2005. LNCS, vol. 3699, pp. 214–223. Springer, Heidelberg (2005)
13. Umeo, H., Hisaoka, M., Sogabe, T.: A survey on optimum-time firing squad synchronization algorithms for one-dimensional cellular automata. Intern. J. of Unconventional Computing 1, 403–426 (2005)
14. Umeo, H., Hisaoka, M., Teraoka, M., Maeda, M.: Several new generalized linear- and optimum-time synchronization algorithms for two-dimensional rectangular arrays. In: Margenstern, M. (ed.) MCU 2004. LNCS, vol. 3354, pp. 223–232. Springer, Heidelberg (2005)
15. Umeo, H., Maeda, M., Hisaoka, M., Teraoka, M.: A state-efficient mapping scheme for designing two-dimensional firing squad synchronization algorithms. Fundamenta Informaticae 74, 603–623 (2006)
16. Umeo, H., Uchino, H.: A new time-optimum synchronization algorithm for rectangle arrays. Fundamenta Informaticae 87(2), 155–164 (2008)
17. Umeo, H., Yamawaki, T., Nishide, K.: An optimum-time firing squad synchronization algorithm for two-dimensional rectangle arrays—freezing-thawing technique based. In: Proceedings of the 2010 International Conference on High Performance Computing & Simulation (HPCS 2010), pp. 575–581 (2010)
18. Waksman, A.: An optimum solution to the firing squad synchronization problem. Information and Control 9, 66–78 (1966)

Conservative Extensions of Abstract Structures

Stefan Vatev⋆

Sofia University, Faculty of Mathematics and Informatics,
5 James Bourchier blvd., 1164 Sofia, Bulgaria
stefan.vatev@gmail.com

Abstract. In the present paper we investigate a relation, called conservative extension, between abstract structures $\mathfrak{A}$ and $\mathfrak{B}$, possibly with different signatures and $|\mathfrak{A}| \subseteq |\mathfrak{B}|$. We give a characterisation of this relation in terms of computable Σ_n formulae and we show that in some sense it provides a finer complexity measure than the one given by degree spectra of structures. As an application, we show that the n-th jump of a structure and its Marker's extension are conservative extensions of the original structure.

1 Introduction

We shall work with abstract structures of the form $\mathfrak{A} = (A; P_1, \ldots, P_s)$, where A is countable and infinite, $P_i \subseteq A^{n_i}$ and the equality is among $P_1, \ldots, P_s$. We shall use the letters $\mathfrak{A}$, $\mathfrak{B}$ to denote structures and the letters A, B to denote their domains.

Our initial motivation was to investigate the common features between the structures built in [8], namely the jump structure and the Marker's extension of a structure. It turns out that both structures relate to the initial structure in a similar way. In our terminology, the jump structure of $\mathfrak{A}$ is $(1,0)$-conservative extension of $\mathfrak{A}$ and the Marker's extension of $\mathfrak{A}$ is $(0,1)$-conservative extension of $\mathfrak{A}$. Our main results are Theorem 2 and Theorem 3 which show that a conservative extension of a structure preserves some families of sets definable with computable Σ formulae.

The main tool in our research is the enumeration of a structure. The pair $\alpha = (f_\alpha, R_\alpha)$ is called an *enumeration* of $\mathfrak{A}$ if R_α is a subset of natural numbers, f_α is a partial one-to-one mapping of $\mathbb{N}$ onto A and $f_\alpha^{-1}(\mathfrak{A})$ is computable in R_α, where
$$f_\alpha^{-1}(P_i) = \{\langle x_1, \ldots, x_{n_i}\rangle \mid x_1, \ldots, x_{n_i} \in Dom(f_\alpha) \ \& \ (f_\alpha(x_1), \ldots, f_\alpha(x_{n_i})) \in P_i\}$$
and $f_\alpha^{-1}(\mathfrak{A}) = f_\alpha^{-1}(P_1) \oplus \cdots \oplus f_\alpha^{-1}(P_s)$. For an enumeration $\alpha = (f_\alpha, R_\alpha)$ of $\mathfrak{A}$ we denote $\alpha^{(n)} = (f_\alpha, R_\alpha^{(n)})$, where $R_\alpha^{(n)}$ is the n-th Turing jump of the set R_α. Given a set $X \subseteq A$, by $X \le \alpha$ we shall denote that $f_\alpha^{-1}(X)$ is c.e. in R_α and by $\mathfrak{A} \le \alpha$ we shall denote that α is an enumeration of $\mathfrak{A}$.

⋆ This work was supported by the European Social Fund through the Human Resource Development Operational Programme 2007–2013 under contract BG051PO001-3.3.04/27/28.08.2009 and by the Bulgarian National Science Fund under contract D002-258/18.12.08

B. Löwe et al. (Eds.): CiE 2011, LNCS 6735, pp. 300–309, 2011.

We shall give an informal definition of the set of the computably infinitary Σ_n formulae in the language of $\mathfrak{A}$, denoted by Σ_n^c. The Σ_0^c and Π_0^c formulae are the finitary quantifier free formulae. A Σ_{n+1}^c formula $\varphi(\overline{x})$ is a disjunction of a c.e. set of formulae of the form $\exists \overline{y} \psi$, where ψ is a Π_n^c formula and $\overline{y}$ includes the variables of ψ which are not in $\overline{x}$. The Π_{n+1}^c formulae are the negations of the Σ_{n+1}^c formulae. We refer the reader to [1] for more background information on computably infinitary formulae.

A set $X \subseteq A$ is Σ_n^c definable in the structure $\mathfrak{A}$ if there is a Σ_n^c formula $\psi(x, \overline{y})$ and a finite number of parameters $\overline{a}$ in A such that $b \in X \leftrightarrow \mathfrak{A} \models \psi(b, \overline{a})$. We denote by $\Sigma_n^c(\mathfrak{A})$ the family of all sets Σ_n^c definable in $\mathfrak{A}$. A subset X of A is said to be *relatively intrinsically* Σ_{n+1}^0 in $\mathfrak{A}$ if for every enumeration α of $\mathfrak{A}$, $f_\alpha^{-1}(X)$ is Σ_{n+1}^0 relative to $f_\alpha^{-1}(\mathfrak{A})$ or equivalently, $f_\alpha^{-1}(X)$ is c.e. relative to $f_\alpha^{-1}(\mathfrak{A})^{(n)}$. In [2] and [3], it is shown that the relatively intrinsically Σ_{n+1}^0 in $\mathfrak{A}$ sets are exactly the Σ_{n+1}^c definable sets in $\mathfrak{A}$. We shall use this result in the following form.

Theorem 1 (Ash-Knight-Manasse-Slaman [2], Chisholm [3]). *Let $\mathfrak{A}$ be a countable structure. For every set $X \subseteq A$,*

$$X \in \Sigma_{n+1}^c(\mathfrak{A}) \; \text{iff} \; (\forall \alpha)[\mathfrak{A} \leq \alpha \to X \leq \alpha^{(n)}].$$

2 Conservative Extensions

Let $\alpha = (f_\alpha, R_\alpha)$ and $\beta = (f_\beta, R_\beta)$ be enumerations of the countable structures $\mathfrak{A}$ and $\mathfrak{B}$ respectively. We write $\alpha \leq \beta$ if

(i) $R_\alpha \leq_T R_\beta$ and
(ii) the set $E(f_\alpha, f_\beta) = \{(x, y) \mid x \in Dom(f_\alpha) \; \& \; y \in Dom(f_\beta) \; \& \; f_\alpha(x) = f_\beta(y)\}$ is c.e. in R_β.

Definition 1. *Let $\mathfrak{A}$ and $\mathfrak{B}$ be countable structures, possibly with different signatures and $A \subseteq B$.*

(i) $\mathfrak{A} \Rightarrow_n^k \mathfrak{B}$ *if for every enumeration β of $\mathfrak{B}$ there exists an enumeration α of $\mathfrak{A}$ such that $\alpha^{(k)} \leq \beta^{(n)}$.*

(ii) $\mathfrak{A} \Leftarrow_n^k \mathfrak{B}$ *if for every enumeration α of $\mathfrak{A}$ there exists an enumeration β of $\mathfrak{B}$ such that $\beta^{(k)} \leq \alpha^{(n)}$.*

(iii) $\mathfrak{A} \Leftrightarrow_n^k \mathfrak{B}$ *if $\mathfrak{A} \Rightarrow_n^k \mathfrak{B}$ and $\mathfrak{A} \Leftarrow_k^n \mathfrak{B}$. We shall say that $\mathfrak{B}$ is a (k, n)-conservative extension of $\mathfrak{A}$.*

The reader should be aware that the relation $\Leftrightarrow_n^k$ is not symmetric. The following theorem motivates the use of the term conservative extension, i.e. if $\mathfrak{B}$ is a (k, n)-conservative extension of $\mathfrak{A}$ then all Σ_{k+1}^c definable sets in $\mathfrak{A}$ are preserved as Σ_{n+1}^c definable sets in $\mathfrak{B}$.

Theorem 2. *Let $\mathfrak{A}$ and $\mathfrak{B}$ be countable structures with $A \subseteq B$. For all $k, n \in \omega$,*

(i) if $\mathfrak{A} \Rightarrow_n^k \mathfrak{B}$ then $(\forall X \subseteq A)[X \in \Sigma_{k+1}^c(\mathfrak{A}) \to X \in \Sigma_{n+1}^c(\mathfrak{B})]$;

(ii) if $\mathfrak{A} \Leftarrow^n_k \mathfrak{B}$ then $(\forall X \subseteq A)[X \in \Sigma^c_{n+1}(\mathfrak{B}) \to X \in \Sigma^c_{k+1}(\mathfrak{A})]$;
(iii) if $\mathfrak{A} \Leftrightarrow^k_n \mathfrak{B}$ then $(\forall X \subseteq A)[X \in \Sigma^c_{k+1}(\mathfrak{A}) \leftrightarrow X \in \Sigma^c_{n+1}(\mathfrak{B})]$.

Proof. *(i)* We have that for every enumeration β of $\mathfrak{B}$, there exists an enumeration α of $\mathfrak{A}$ such that $\alpha^{(k)} \leq \beta^{(n)}$. Let X be a subset of A such that $X \in \Sigma^c_{k+1}(\mathfrak{A})$. According to Theorem 1 it is equivalent to $(\forall\alpha)[\mathfrak{A} \leq \alpha \to X \leq \alpha^{(k)}]$. We wish to show $(\forall\beta)[\mathfrak{B} \leq \beta \to X \leq \beta^{(n)}]$. Let us take an arbitrary enumeration β of $\mathfrak{B}$. Since $\mathfrak{A} \Rightarrow^k_n \mathfrak{B}$, for some enumeration α of $\mathfrak{A}$, $\alpha^{(k)} \leq \beta^{(n)}$. It gives us that $R^{(k)}_\alpha$ is computable in $R^{(n)}_\beta$ and $E(f_\alpha, f_\beta)$ is c.e. in $R^{(n)}_\beta$. Moreover, $X \leq \alpha^{(k)}$ and then $f^{-1}_\alpha(X)$ is c.e. in $R^{(n)}_\beta$. From the equivalence

$$x \in f^{-1}_\beta(X) \leftrightarrow (\exists y)[(x, y) \in E(f_\alpha, f_\beta) \; \& \; y \in f^{-1}_\alpha(X)],$$

it follows that $f^{-1}_\beta(X)$ is c.e. in $R^{(n)}_\beta$ and then $X \leq \beta^{(n)}$ which is what we wanted to show. The proof of *(ii)* is similar to that of *(i)*. $\square$

Remark 1. Notice that we do not have the other directions in Theorem 2. Assume $A \subseteq B$ and if $(\forall X \subseteq A)[X \in \Sigma^c_{n+1}(\mathfrak{A}) \to X \in \Sigma^c_{k+1}(\mathfrak{B})]$ then $\mathfrak{A} \Rightarrow^n_k \mathfrak{B}$. We can give a simple counterexample. Let $\mathscr{O}_A = (A; =)$ and take $\mathfrak{A} = \mathfrak{B} = \mathscr{O}_A$. It is easy to see that for every natural number n, $X \subseteq A$ is Σ^c_n-definable in $\mathscr{O}_A$ iff X is a finite or co-finite subset of A. Therefore $\Sigma^c_1(\mathscr{O}_A) = \Sigma^c_n(\mathscr{O}_A)$ and then $(\forall n)(\forall X \subseteq A)[X \in \Sigma^c_{n+1}(\mathscr{O}_A) \to X \in \Sigma^c_1(\mathscr{O}_A)]$. We conclude that $(\forall n)[\mathscr{O}_A \Rightarrow^n_0 \mathscr{O}_A]$, which is evidently not true.

We shall proceed with the investigation of under what conditions we have the other directions in Theorem 2. For this purpose we shall firstly introduce some coding machinery and then the sets $K^{\mathfrak{A}}_n$.

2.1   Moschovakis' Extension

Following Moschovakis [6], we define the least acceptable extension $\mathfrak{A}^\star$ of $\mathfrak{A}$. Let 0 be an object which does not belong to A and Π be a pairing operation chosen so that neither 0 nor any element of A is an ordered pair. Let $A^\star$ be the least set containing all elements of $A_0 = A \cup \{0\}$ and closed under Π.

We associate an element $n^\star$ of $A^\star$ with each $n \in \omega$ by induction. Let $0^\star = 0$ and $(n + 1)^\star = \Pi(0, n^\star)$. We denote by $\mathbb{N}^\star$ the set of all elements $n^\star$. Let L and R be the functions on $A^\star$ satisfying the following conditions:

$$L(0) = R(0) = 0;$$
$$(\forall t \in A)[L(t) = R(t) = 1^\star];$$
$$(\forall s, t \in A^\star)[L(\Pi(s, t)) = s \; \& \; R(\Pi(s, t)) = t].$$

The pairing function allows us to code finite sequences of elements. Let $\Pi_1(t_1) = t_1$ and $\Pi_{n+1}(t_1, \ldots, t_{n+1}) = \Pi(t_1, \Pi_n(t_2, \ldots, t_{n+1}))$ for every $t_1, \ldots, t_{n+1} \in A^\star$. For each predicate P_i of the structure $\mathfrak{A}$ define the respective predicate $P^\star_i$ on $A^\star$ by $P^\star_i(t) \leftrightarrow (\exists a_1, \ldots, a_{n_i} \in A)[t = \Pi_{n_i}(a_1, \ldots, a_n) \& P_i(a_1, \ldots, a_n)]$.

Definition 2. *Moschovakis' extension of* $\mathfrak{A}$ *is the structure*

$$\mathfrak{A}^\star = (A^\star; A_0, P_1^\star, \ldots, P_s^\star, G_\Pi, G_L, R_R, =),$$

where G_Π, G_L *and* G_R *are the graphs of* Π, L *and* R *respectively.*

Proposition 1. *For every two structures* $\mathfrak{A}$, $\mathfrak{B}$ *with* $A \subseteq B$ *and* $n, k \in \omega$, $\mathfrak{A} \leftrightarrow_n^k \mathfrak{B}$ *iff* $\mathfrak{A}^\star \leftrightarrow_n^k \mathfrak{B}^\star$. *Moreover,* $\mathfrak{A} \leftrightarrow_n^n \mathfrak{A}^\star$.

2.2 The set $K_n^{\mathfrak{A}}$

Let $\alpha = (f_\alpha, R_\alpha)$ be an enumeration of $\mathfrak{A}$. For every $e, x \in \omega$ and every $n \in \omega$, we define the modelling relations $\models_n$ in the following way:

$$f_\alpha \models_0 F_e(x) \leftrightarrow x \in W_e^{f_\alpha^{-1}(\mathfrak{A})}$$

$$f_\alpha \models_{n+1} F_e(x) \leftrightarrow x \in W_e^{f^{-1}(\mathfrak{A})^{(n+1)}}$$

$$f_\alpha \models_n \neg F_e(x) \leftrightarrow f_\alpha \not\models_n F_e(x)$$

Following the modelling relation, we shall define a forcing relation with conditions all finite injective mappings from $\mathbb{N}$ into the domain A of $\mathfrak{A}$. We call them *finite parts* and we shall use the letters τ, ρ, δ to denote them. Let $\Delta(A)$ be the set of all finite parts and let Fin_2 be the set of all finite functions on the natural numbers taking values in $\{0, 1\}$. Given a finite part τ and a relation $R \subseteq A^n$, we define the finite function $\tau^{-1}(R)$ in Fin_2 as follows:

$$\tau^{-1}(R)(u) \downarrow = 1 \leftrightarrow (\exists x_1, \ldots, x_n \in Dom(\tau))[u = \langle x_1, \ldots, x_n \rangle \ \&$$
$$(\tau(x_1), \ldots, \tau(x_n)) \in R],$$

$$\tau^{-1}(R)(u) \downarrow = 0 \leftrightarrow (\exists x_1, \ldots, x_n \in Dom(\tau))[u = \langle x_1, \ldots, x_n \rangle \ \&$$
$$(\tau(x_1), \ldots, \tau(x_n)) \notin R].$$

By $\tau^{-1}(\mathfrak{A})$ we shall denote the finite function $\tau^{-1}(R_1) \oplus \cdots \oplus \tau^{-1}(R_s)$.

If φ is a partial function and $e \in \omega$, then by W_e^φ we shall denote the set of all x such that the computation $\{e\}^\varphi(x)$ halts successfully. We shall assume that if during a computation the oracle φ is called with an argument outside of its domain, then the computation halts unsuccessfully.

For every $e, x, n \in \omega$ and for every finite part τ, we define the forcing relations in the following way:

$$\tau \Vdash_0 F_e(x) \leftrightarrow x \in W_e^{\tau^{-1}(\mathfrak{A})},$$

$$\tau \Vdash_{n+1} F_e(x) \leftrightarrow (\exists \delta \in Fin_2)[x \in W_e^\delta \ \& \ (\forall z \in Dom(\delta))[$$
$$(\delta(z) = 1 \ \& \ \tau \Vdash_n F_z(z)) \vee (\delta(z) = 0 \ \& \ \tau \Vdash_n \neg F_z(z))]],$$

$$\tau \Vdash_n \neg F_e(x) \leftrightarrow (\forall \rho \in \Delta(A))[\tau \subseteq \rho \rightarrow \rho \not\Vdash_n F_e(x)].$$

An enumeration α of $\mathfrak{A}$ is called *n-generic* if for every $e, x \in \omega$ and every $j < n$, $(\exists \tau \subseteq f_\alpha)[\tau \Vdash_j F_e(x) \vee \tau \Vdash_j \neg F_e(x)]$.

Lemma 1 (Truth Lemma)

(i) *For every* $n, e, x \in \omega$ *and every finite parts* $\tau \subseteq \rho$,

$$\tau \Vdash_n (\neg) F_e(x) \to \rho \Vdash_n (\neg) F_e(x).$$

(ii) *For every* n-*generic enumeration* α *of* $\mathfrak{A}$ *and all* $e, x \in \omega$,

$$f_\alpha \models_n F_e(x) \leftrightarrow (\exists \tau \subseteq f_\alpha)[\tau \Vdash_n F_e(x)].$$

(iii) *For every* $(n+1)$-*generic enumeration* α *of* $\mathfrak{A}$ *and all* $e, x \in \omega$,

$$f_\alpha \models_n \neg F_e(x) \leftrightarrow (\exists \tau \subseteq f_\alpha)[\tau \Vdash_n \neg F_e(x)].$$

For each finite part $\tau \neq \emptyset$ with $Dom(\tau) = \{x_1, \dots, x_n\}$ and $\tau(x_i) = s_i$, we associate the element $\tau^\star = \Pi_n(\Pi(x_1^\star, s_1), \dots, \Pi(x_n^\star, s_n))$ of $A^\star$. For $\tau = \emptyset$, let $\tau^\star = 0$. We define for every $n \in \omega$ the set

$$K_n^{\mathfrak{A}} = \{\Pi_3(\delta^\star, e^\star, x^\star) \mid (\exists \tau \in \Delta(A))[\delta \subseteq \tau \,\&\, \tau \Vdash_n F_e(x)] \,\&\, e^\star, x^\star \in \mathbb{N}^\star\}.$$

Proposition 2. *For every countable structure* $\mathfrak{A}$ *and every* $n \in \omega$, *we have* $K_n^{\mathfrak{A}} \in \Sigma_{n+1}^c(\mathfrak{A}^\star)$ *and* $A^\star \setminus K_n^{\mathfrak{A}} \in \Sigma_{n+2}^c(\mathfrak{A}^\star)$.

Theorem 3. *Let* $\mathfrak{A}$ *and* $\mathfrak{B}$ *be countable structures with* $A^\star \subseteq B$ *and* $k, n \in \omega$. *Suppose that* $(\forall X \subseteq A^\star)[X \in \Sigma_{k+1}^c(\mathfrak{A}^\star) \to X \in \Sigma_{n+1}^c(\mathfrak{B})]$. *Then* $\mathfrak{A} \Rightarrow_n^k \mathfrak{B}$.

Proof. Let us fix an enumeration $\beta = (f_\beta, R_\beta)$ of $\mathfrak{B}$. We shall show that there exists an enumeration $\gamma = (f_\gamma, f_\gamma^{-1}(\mathfrak{A}))$ of $\mathfrak{A}$ such that $\gamma^{(k)} \leq \beta^{(n)}$.

Firstly, let $k = 0$. Since $A \in \Sigma_1^c(\mathfrak{A}^\star)$, $A \in \Sigma_{n+1}^c(\mathfrak{B})$ and then by Theorem 1, $f_\beta^{-1}(A)$ is c.e. in $R_\beta^{(n)}$. We can take a total enumeration f_γ of A defined as $f_\gamma = f_\beta \circ \mu$, where $\mu : \mathbb{N} \to f_\beta^{-1}(A)$ is a computable in $R_\beta^{(n)}$ bijection. Such μ exists because $f_\beta^{-1}(A)$ is c.e. in $R_\beta^{(n)}$. Clearly the set $E(f_\gamma, f_\beta)$ is c.e. in $R_\beta^{(n)}$. We have for all $P_i^{\mathfrak{A}}$ of $\mathfrak{A}$, $P_i^{\mathfrak{A}} \in \Sigma_{n+1}^c(\mathfrak{B})$ and $A^{n_i} \setminus P_i^{\mathfrak{A}} \in \Sigma_{n+1}^c(\mathfrak{B})$. Thus $f_\beta^{-1}(P_i^{\mathfrak{A}})$ and $f_\beta^{-1}(A^{n_i} \setminus P_i^{\mathfrak{A}})$ are c.e. in $R_\beta^{(n)}$. $f_\gamma^{-1}(P_i^{\mathfrak{A}})$ is c.e. in $R_\beta^{(n)}$ and since f_γ is total, $\mathbb{N} \setminus f_\gamma^{-1}(P_i^{\mathfrak{A}})$ is c.e. in $R_\beta^{(n)}$. Therefore, $f_\gamma^{-1}(\mathfrak{A}) \leq_T R_\beta^{(n)}$ and hence $\gamma \leq \beta^{(n)}$.

Let $k > 0$. We shall build a k-generic enumeration $\gamma = (f_\gamma, f_\gamma^{-1}(\mathfrak{A}))$ of $\mathfrak{A}$ such that $f_\gamma^{-1}(\mathfrak{A})^{(k)} \leq_T R_\beta^{(n)}$ and $E(f_\gamma, f_\beta)$ is c.e. in $R_\beta^{(n)}$. Before proceeding with its construction, we shall describe a way to encode finite parts $\tau \in \Delta(A)$ as natural numbers. We define a coding scheme for finite sequences of natural numbers belonging to $f_\beta^{-1}(A^\star)$ in the following way:

$$J(x, y) = f_\beta^{-1}(\Pi(f_\beta(x), f_\beta(y)));$$
$$J_1(x) = x, \quad J_{n+1}(x_1, \dots, x_{n+1}) = J(x_1, J_n(x_2, \dots, x_n, x_{n+1})).$$

For every natural number n, we denote $n^\sharp = f_\beta^{-1}(n^\star)$ and $\mathbb{N}^\sharp = f_\beta^{-1}(\mathbb{N}^\star)$. For finite parts $\tau \in \Delta(A)$, we associate with $\tau^\star$ the natural number $\tau^\sharp = f_\beta^{-1}(\tau^\star)$.

That is, if $\tau^\star = \Pi_n(\Pi(x_1^\star, y_1), \ldots, \Pi(x_n^\star, y_n))$ then $\tau^\sharp = J_n(J(x_1^\sharp, f_\beta^{-1}(y_1)), \ldots,$
$J(x_n^\sharp, f_\beta^{-1}(y_n)))$. Therefore, the set $\Delta^\sharp(A) = \{\tau^\sharp \mid \tau \in \Delta(A)\}$ is c.e. in $R_\beta^{(n)}$. Let
$Dom(\tau^\sharp) = \{x_1^\sharp, \ldots, x_n^\sharp\}$ and $\tau^\sharp(x_i^\sharp) = f_\beta^{-1}(y_i)$. We shall assume that $Dom(\tau^\sharp) = \emptyset$ if $\tau^\sharp = 0$. Notice that $Dom(\tau^\sharp) = \{x^\sharp \mid x \in Dom(\tau)\}$ and $f_\beta(\tau^\sharp(x^\sharp)) = \tau(x)$
for all $x \in Dom(\tau)$. There exists a partial computable in $R_\beta^{(n)}$ predicate P such
that for $\tau, \delta \in \Delta(A)$, $P(\tau^\sharp, \delta^\sharp) \downarrow = 1$ iff $\tau \subseteq \delta$. We shall write $\tau^\sharp \subseteq \delta^\sharp$ instead
of $P(\tau^\sharp, \delta^\sharp) \downarrow = 1$. From Proposition 2 we know that $K_{k-1}^{\mathfrak{A}}$ and $A^\star \setminus K_{k-1}^{\mathfrak{A}}$ are
Σ_{k+1}^c definable in $\mathfrak{A}^\star$. This means that $K_{k-1}^{\mathfrak{A}}$ and $A^\star \setminus K_{k-1}^{\mathfrak{A}}$ are Σ_{n+1}^c definable
in $\mathfrak{B}$. Thus $f_\beta^{-1}(K_{k-1}^{\mathfrak{A}})$ and $f_\beta^{-1}(A^\star \setminus K_{k-1}^{\mathfrak{A}})$ are both c.e. in $R_\beta^{(n)}$. It is not hard
to see that there exists a computable function χ such that for every $\tau \in \Delta(A)$,

$$\tau \Vdash_{k-1} F_e(x) \leftrightarrow x \in W_{\chi(\tau^\sharp, e)}^{R_\beta^{(n)}}.$$

Claim. There exists a k-generic enumeration γ of $\mathfrak{A}$ such that $f_\gamma^\sharp$ is partial computable in $R_\beta^{(n)}$, where $f_\gamma^\sharp : \mathbb{N}^\sharp \to f_\beta^{-1}(A)$ is defined as $f_\gamma^\sharp(x^\sharp) = f_\beta^{-1}(f_\gamma(x))$.

Proof. Since the set A is Σ_{k+1}^c definable in $\mathfrak{A}$, $f_\beta^{-1}(A)$ is c.e. in $R_\beta^{(n)}$. Let us fix a
computable in $R_\beta^{(n)}$ bijection $\mu : \mathbb{N} \to f_\beta^{-1}(A)$. We shall describe a construction
in which at each stage s we shall define a finite part $\tau_s \subseteq \tau_{s+1}$. In the end, the
k-generic enumeration of $\mathfrak{A}$ will be defined as $f_\gamma = \bigcup_s \tau_s$ and $R_\gamma = f_\gamma^{-1}(\mathfrak{A})$. Let
$\tau_0 = \emptyset$ and suppose we have already defined τ_s.

a) Case $s = 2r$. We make sure that f_γ is one-to-one and onto A. Let $x^\sharp$ be
the least natural number not in $Dom(\tau_s^\sharp)$. Find the least p such that $\mu(p) \notin Ran(\tau_s^\sharp)$. Set $\tau_{s+1}(x) = f_\beta(\mu(p))$ and $\tau_{s+1}(z) = \tau_s(z)$ for every $z \neq x$ and
$z \in Dom(\tau_s)$. Leave $\tau_{s+1}(z)$ undefined for any other z.

b) Case $s = 2\langle e, x \rangle + 1$. We satisfy the requirement that f_γ is k-generic. Check
whether there exists an extension δ of τ_s such that $\delta \Vdash_{k-1} F_e(x)$. This is
equivalent to asking whether $J_3(\tau_s^\sharp, e^\sharp, x^\sharp) \in f_\beta^{-1}(K_{k-1}^{\mathfrak{A}})$ or $J_3(\tau_s^\sharp, e^\sharp, x^\sharp) \in$
$f_\beta^{-1}(A^\star \setminus K_{k-1}^{\mathfrak{A}})$. We can do this effectively using the oracle $R_\beta^{(n)}$.
If $J_3(\tau_s^\sharp, e^\sharp, x^\sharp) \in f_\beta^{-1}(A^\star \setminus K_{k-1}^{\mathfrak{A}})$, then $\tau_s \Vdash_{k-1} \neg F_e(x)$ and we set $\tau_{s+1} = \tau_s$.
If $J_3(\tau_s^\sharp, e^\sharp, x^\sharp) \in f_\beta^{-1}(K_{k-1}^{\mathfrak{A}})$, we search for $\delta^\sharp \in \Delta^\sharp(A)$ such that $\tau_s^\sharp \subseteq \delta^\sharp$ and
$x \in W_{\chi(\delta^\sharp, e)}^{R_\beta^{(n)}}$. Since $J_3(\tau_s^\sharp, e^\sharp, x^\sharp) \in f_\beta^{-1}(K_{k-1}^{\mathfrak{A}})$ we know that such $\delta^\sharp$ exists
and we can find it effectively in $R_\beta^{(n)}$. Set $\tau_{s+1} = \delta$.

End of construction

It follows from the construction that $f_\gamma^\sharp$ is partial computable in $R_\beta^{(n)}$.   □

The equivalence $f_\gamma(x) = f_\beta(y) \leftrightarrow f_\gamma^\sharp(x^\sharp) = y$ and the fact that the graph of $f_\gamma^\sharp$
is c.e. in $R_\beta^{(n)}$ implies that the set $E(f_\gamma, f_\beta)$ is c.e. in $R_\beta^{(n)}$. Since f_γ is k-generic,
we have the equivalences

$$x \in f_\gamma^{-1}(\mathfrak{A})^{(k)} \leftrightarrow f_\gamma \models_{k-1} F_x(x) \leftrightarrow (\exists \tau \subseteq f_\gamma)[\tau \Vdash_{k-1} F_x(x)]$$

$$\leftrightarrow (\exists \tau^\sharp \subseteq f_\gamma^\sharp)[x \in W_{\chi(\tau^\sharp, x)}^{R_\beta^{(n)}}].$$

$$x \notin f_\gamma^{-1}(\mathfrak{A})^{(k)} \leftrightarrow f_\gamma \models_{k-1} \neg F_x(x) \leftrightarrow (\exists \tau \subseteq f_\gamma)[\tau \Vdash_{k-1} \neg F_x(x)]$$

$$\leftrightarrow (\exists \tau^\sharp \subseteq f_\gamma^\sharp)[J_3(\tau^\sharp, x^\sharp, x^\sharp) \in f_\beta^{-1}(A^\star \setminus K_{k-1}^\mathfrak{A})].$$

Since $f_\beta(\tau^\sharp(x^\sharp)) = \tau(x)$, we have the equivalence:

$$\tau^\sharp \subseteq f_\gamma^\sharp \leftrightarrow (\forall x^\sharp \in Dom(\tau^\sharp))(\exists y)[(x, y) \in E(f_\gamma, f_\beta) \& \langle \tau^\sharp(x^\sharp), y \rangle \in f_\beta^{-1}(=^{\mathfrak{A}^\star})].$$

It means that the relation $\tau^\sharp \subseteq f_\gamma^\sharp$ is c.e. in $R_\beta^{(n)}$. It follows that $f_\gamma^{-1}(\mathfrak{A})^{(k)}$ is computable in $R_\beta^{(n)}$. We conclude that for the enumeration $\gamma = (f_\gamma, f_\gamma^{-1}(\mathfrak{A}))$ of $\mathfrak{A}$, $\gamma^{(k)} \le \beta^{(n)}$ and hence $\mathfrak{A} \Rightarrow_n^k \mathfrak{B}$. □

Corollary 1. *For any two countable structures $\mathfrak{A}$, $\mathfrak{B}$ with $A \subseteq B$ and $n, k \in \omega$,*

$$\mathfrak{A} \Rightarrow_n^k \mathfrak{B} \leftrightarrow (\forall X \subseteq A^\star)[X \in \Sigma_{k+1}^c(\mathfrak{A}^\star) \to X \in \Sigma_{n+1}^c(\mathfrak{B}^\star)].$$

3 Applications

3.1 Degree Spectra of Structures

In [7], Richter initiates the study of the notion of the degree spectrum of a countable structure. Here we define the degree spectrum following [9].

Definition 3. *The Turing degree spectrum of $\mathfrak{A}$ is the set $DS(\mathfrak{A}) = \{d_T(R_\alpha) \mid \mathfrak{A} \le \alpha\}$. The k-th jump Turing degree spectrum of $\mathfrak{A}$ is the set $DS_k(\mathfrak{A}) = \{d_T(R_\alpha^{(k)}) \mid \mathfrak{A} \le \alpha\}$.*

Here by $d_T(X)$ we denote the Turing degree of the set X. A set of Turing degrees $\mathscr{A}$ is *closed upwards* if for all Turing degrees $\mathbf{a}$ and $\mathbf{b}$, $\mathbf{a} \in \mathscr{A}$ & $\mathbf{a} \le \mathbf{b} \to \mathbf{b} \in \mathscr{A}$. It is clear that for every structure $\mathfrak{A}$, its degree spectrum $DS(\mathfrak{A})$ is closed upwards.

Remark 2. Richter's definition of degree spectrum is slightly different. She defines the degree spectrum as the set of all Turing degrees $d_T(f^{-1}(\mathfrak{A}))$, where f is a *total* enumeration of the domain of $\mathfrak{A}$. Both definitions produce the same sets of Turing degrees for automorphically non-trivial structures.

Proposition 3. *Let $\mathfrak{A}$ and $\mathfrak{B}$ be countable structures with $A \subseteq B$.*

(i) *If $\mathfrak{A} \Rightarrow_n^k \mathfrak{B}$ then $DS_n(\mathfrak{B}) \subseteq DS_k(\mathfrak{A})$;*
(ii) *If $\mathfrak{A} \Leftarrow_k^n \mathfrak{B}$ then $DS_k(\mathfrak{A}) \subseteq DS_n(\mathfrak{B})$;*
(iii) *If $\mathfrak{A} \Leftrightarrow_n^k \mathfrak{B}$ then $DS_k(\mathfrak{A}) = DS_n(\mathfrak{B})$.*

Proof. We shall prove only *(i)* since the others are similar. Let $\mathfrak{A} \Rightarrow_n^k \mathfrak{B}$ and $\mathbf{b} \in DS_n(\mathfrak{B})$. We wish to show that $\mathbf{b} \in DS_k(\mathfrak{A})$. Since $DS_k(\mathfrak{A})$ is closed upwards, it is enough to prove that there exists a Turing degree $\mathbf{a} \in DS_k(\mathfrak{A})$ and $\mathbf{a} \le \mathbf{b}$. Let β be an enumeration of $\mathfrak{B}$ and $d_T(R_\beta^{(n)}) = \mathbf{b}$. $\mathfrak{A} \Rightarrow_n^k \mathfrak{B}$ gives us an enumeration α of $\mathfrak{A}$ such that $\alpha^{(k)} \le \beta^{(n)}$. For $\mathbf{a} = d_T(R_\alpha^{(k)})$ we have $\mathbf{a} \in DS_k(\mathfrak{A})$ and $\mathbf{a} \le \mathbf{b}$. □

Remark 3. We should note that we do not have the other directions in Proposition 3. Let us define the structures $\mathscr{O}_\mathbb{N} = (\mathbb{N}; =)$ and $\mathscr{S} = (\mathbb{N}; G_{Succ}, =)$, where G_{Succ} is the graph of the successor function. It is easy to see that $DS(\mathscr{O}_\mathbb{N}) = DS(\mathscr{S})$ whereas it follows easily from Theorem 2 that $\mathscr{S} \not\approx^0_0 \mathscr{O}_\mathbb{N}$.

3.2   Jumps of Structures

In [8], the jump of the structure $\mathfrak{A}$ is defined as $\mathfrak{A}' = (\mathfrak{A}^\star, K_0^\mathfrak{A})$. It is natural to ask whether we can extend it for $n > 0$.

Definition 4. *Let $\mathfrak{A}$ be a countable structure. For every natural number n, we define the n-th jump of $\mathfrak{A}$ in the following way.*

$$\mathfrak{A}^{(0)} = \mathfrak{A} \ and \ \mathfrak{A}^{(n+1)} = (\mathfrak{A}^\star, K_n^\mathfrak{A}).$$

Actually, the results in [8] are enough to produce a definition of the n-th jump of $\mathfrak{A}$, just let $\mathfrak{A}^{(n+1)} = (\mathfrak{A}^{(n)})'$. The difficulty with it is that we add a new relation symbol and a new layer of coding to the structure for each jump.

Using the enumeration built in Lemma 7 of [8], we can easily obtain the following useful result.

Proposition 4. *Let $\mathfrak{A}$ be a countable structure.*

(i) *For every enumeration α of $\mathfrak{A}$ there exists an enumeration α_0 of $\mathfrak{A}^{(n)}$ such that $\alpha_0 \leq \alpha^{(n)}$.*

(ii) *For every n-generic enumeration γ of $\mathfrak{A}$ there exists an enumeration $\gamma^\star = (f_{\gamma^\star}, f_{\gamma^\star}^{-1}(\mathfrak{A}^\star))$ of $\mathfrak{A}^\star$ such that $f_\gamma^{-1}(\mathfrak{A})^{(n)} \equiv_T f_{\gamma^\star}^{-1}(\mathfrak{A}^\star)^{(n)} \equiv_T f_{\gamma^\star}^{-1}(\mathfrak{A}^{(n)})$.*

Proposition 5. *For any countable structure $\mathfrak{A}$, we have*

(i) *For every $n \in \omega$, $K_n^\mathfrak{A} \notin \Sigma_n^c(\mathfrak{A}^\star)$.*

(ii) *For every $n, k \in \omega$ with $k > 0$, $K_{k+n}^\mathfrak{A} \in \Sigma_{n+1}^c(\mathfrak{A}^{(k)})$ and $K_{k+n}^\mathfrak{A} \notin \Sigma_n^c(\mathfrak{A}^{(k)})$.*

Proof. (i) Assume $K_n^\mathfrak{A} \in \Sigma_n^c(\mathfrak{A}^\star)$. If $n = 0$ then $K_0^\mathfrak{A}$ is definable in $\mathfrak{A}^\star$ by a finitary open formula. This means that for every enumeration α of $\mathfrak{A}^\star$, $f_\alpha^{-1}(K_0^A)$ is computable in $f_\alpha^{-1}(\mathfrak{A}^\star)$ and then $f_\alpha^{-1}(\mathfrak{A}')$ is computable in $f_\alpha^{-1}(\mathfrak{A}^\star)$. Take a 1-generic enumeration γ of $\mathfrak{A}$. Then $\gamma^\star$, as in (ii) of Proposition 4, is an enumeration of $\mathfrak{A}^\star$ and $f_\gamma^{-1}(\mathfrak{A})' \equiv_T f_{\gamma^\star}^{-1}(\mathfrak{A}') \leq_T f_\gamma^{-1}(\mathfrak{A})$. This is clearly a contradiction.

Let $n > 0$. Theorem 1 tells us that for every enumeration α of $\mathfrak{A}^\star$, $f_\alpha^{-1}(K_n^\mathfrak{A})$ is c.e. in $R_\alpha^{(n-1)}$ and therefore $f_\alpha^{-1}(\mathfrak{A}^{(n+1)})$ is computable in $R_\alpha^{(n)}$. Let γ be an $(n+1)$-generic enumeration of $\mathfrak{A}$ and $\gamma^\star$ be as in (ii) of Proposition 4. Since $\gamma^\star$ is an enumeration of $\mathfrak{A}^\star$, $f_{\gamma^\star}^{-1}(\mathfrak{A}^{(n+1)})$ is computable in $f_{\gamma^\star}^{-1}(\mathfrak{A}^\star)^{(n)}$. But we also have $f_{\gamma^\star}^{-1}(\mathfrak{A}^\star)^{(n+1)} \leq_T f_{\gamma^\star}^{-1}(\mathfrak{A}^{(n+1)})$. Thus we reach a contradiction.

The proof of the first part of (ii) uses Theorem 1 and follows by induction on k. For the second part, if we assume $K_{k+n}^\mathfrak{A} \in \Sigma_n^c(\mathfrak{A}^{(k)})$ then by taking an $(n+k)$-generic enumeration of $\mathfrak{A}$, we argue as above to reach a contradiction. $\square$

Proposition 6. *For every countable structure $\mathfrak{A}$ and natural number n,*

(i) $\mathfrak{A} \Leftrightarrow_0^n \mathfrak{A}^{(n)}$;
(ii) $\mathfrak{A}^{(n)} \Rightarrow_0^0 \mathfrak{A}^{(n+1)}$ *and* $\mathfrak{A}^{(n)} \not\Leftarrow_0^0 \mathfrak{A}^{(n+1)}$.

Proof. *(i)* Let $n > 0$ since it is obvious for $n = 0$. $\mathfrak{A} \Rightarrow_0^n \mathfrak{A}^{(n)}$ is a direct application of Theorem 3. Now we wish to show $\mathfrak{A} \Leftarrow_n^0 \mathfrak{A}^{(n)}$. Let us take an enumeration α of $\mathfrak{A}$. From *(i)* of Proposition 4, there is an enumeration α_0 of $\mathfrak{A}^{(n)}$ such that $\alpha_0 \leq \alpha^{(n)}$.

(ii) Let $n = 0$. Clearly $\mathfrak{A} \Rightarrow_0^0 \mathfrak{A}'$. Assume $\mathfrak{A} \Leftarrow_0^0 \mathfrak{A}'$. Let $\gamma = (f_\gamma, f_\gamma^{-1}(\mathfrak{A}))$ be a 1-generic enumeration of $\mathfrak{A}$ and $\beta = (f_\beta, R_\beta)$ be an enumeration of $\mathfrak{A}'$ such that $\beta \leq \gamma$. As in the proof of Theorem 3, we use β to define a coding scheme J_n and prove that the relation $\tau^\sharp \subseteq f_\gamma^\sharp$ is c.e. in $f_\gamma^{-1}(\mathfrak{A})$. From the equivalences $x \in f_\gamma^{-1}(\mathfrak{A})' \leftrightarrow (\exists \tau^\sharp \subseteq f_\gamma^\sharp)[x \in W_{\chi(\tau^\sharp, x)}^{f_\gamma^{-1}(\mathfrak{A})}]$ and $x \notin f_\gamma^{-1}(\mathfrak{A})' \leftrightarrow (\exists \tau^\sharp \subseteq f_\gamma^\sharp)[J_3(\tau^\sharp, x^\sharp, x^\sharp) \in f_\beta^{-1}(A^\star \setminus K_0^{\mathfrak{A}})]$, it follows that $f_\gamma^{-1}(\mathfrak{A})'$ is computable in $f_\gamma^{-1}(\mathfrak{A})$. Thus we reach a contradiction.

Let $n > 0$. It is clear that $\mathfrak{A}^{(n)} \Rightarrow_0^0 \mathfrak{A}^{(n+1)}$. Assume $\mathfrak{A}^{(n)} \Leftrightarrow_0^0 \mathfrak{A}^{(n+1)}$. Theorem 2 gives us $K_{n+1}^{\mathfrak{A}} \in \Sigma_1^c(\mathfrak{A}^{(n+1)})$ iff $K_{n+1}^{\mathfrak{A}} \in \Sigma_1^c(\mathfrak{A}^{(n)})$. On the other hand, *(ii)* of Proposition 5 tells us that $K_{n+1}^{\mathfrak{A}} \in \Sigma_1^c(\mathfrak{A}^{(n+1)})$ and $K_{n+1}^{\mathfrak{A}} \notin \Sigma_1^c(\mathfrak{A}^{(n)})$. Thus our assumption is absurd and hence $\mathfrak{A}^{(n)} \not\Leftarrow_0^0 \mathfrak{A}^{(n+1)}$. □

Since $\mathfrak{A} \Leftrightarrow_n^k \mathfrak{B}$ implies $DS_k(\mathfrak{A}) = DS_n(\mathfrak{B})$, we get the following.

Corollary 2. *For every countable structure $\mathfrak{A}$, $DS(\mathfrak{A}^{(n)}) = DS_n(\mathfrak{A})$.*

Proposition 7. *For all countable structures $\mathfrak{A}$, $\mathfrak{B}$ with $A \subseteq B$ and $n, k \in \omega$,*

$$\mathfrak{A} \Leftrightarrow_n^k \mathfrak{B} \ iff \ \mathfrak{A}^{(k)} \Leftrightarrow_0^0 \mathfrak{B}^{(n)}.$$

Proposition 8. *Let $\mathfrak{A}$ be a countable structure, $n, k \in \omega$ and $k > 0$.*

(i) $(\forall X \subseteq A^\star)[X \in \Sigma_{n+2}^c(\mathfrak{A}^\star) \leftrightarrow X \in \Sigma_{n+1}^c(\mathfrak{A}')]$;
(ii) $(\forall X \subseteq A^\star)[X \in \Sigma_{n+2}^c(\mathfrak{A}^{(k)}) \leftrightarrow X \in \Sigma_{n+1}^c(\mathfrak{A}^{(k+1)})]$;

3.3 Marker's Extensions

In [4], Goncharov and Khoussainov adapted Marker's construction from [5] to prove that for any natural number $n \geq 1$, there exists an $\aleph_1$-categorical theory T with a computable model of a finite language whose theories are Turing equivalent to $\emptyset^{(n)}$. Building on their results, A. Soskova and I. Soskov proved a theorem for the degree spectrum of structures resembling a jump inversion theorem, namely the following theorem.

Theorem 4 (A. Soskova - I. Soskov [8]). *Let $\mathfrak{A}$ and $\mathfrak{C}$ be countable structures and $DS(\mathfrak{A}) \subseteq DS_1(\mathfrak{C})$. There exists a structure $\mathfrak{B}$ such that $DS(\mathfrak{A}) = DS_1(\mathfrak{B})$ and $DS(\mathfrak{B}) \subseteq DS(\mathfrak{C})$.*

In [10], Stukachev proves an analogue of this theorem for the semilattices of Σ-degrees of structures with arbitrary cardinalities.

Theorem 5 (Stukachev [10]). *Let* $\mathfrak{A}$ *be a structure such that* $\mathbf{0}' \leq_{\Sigma} \mathfrak{A}$. *There exists a structure* $\mathfrak{B}$ *such that* $\mathfrak{A} \equiv_{\Sigma} \mathfrak{B}'$.

We can prove a result similar to Stukachev's.

Proposition 9. *Let* $\mathfrak{A}$ *be a countable structure and* $\mathscr{O}_A \Rightarrow_0^k \mathfrak{A}$ *for some* $k \in \omega$. *There exists a countable structure* $\mathfrak{B}$ *such that* $\mathfrak{A} \Leftrightarrow_0^0 \mathfrak{B}^{(k)}$.

Proof. We give a brief sketch of the proof for the case $k = 1$. The proof is easily generalized for $k > 1$. Following [4], let $\mathfrak{A}^\exists$ and $\mathfrak{A}^\forall$ be the respective Marker's $\exists$ and $\forall$ extensions of the structure $\mathfrak{A}$ and define $\mathfrak{B} = (\mathfrak{A}^\exists)^\forall$. With almost trivial modifications of the proof of Theorem 4 from above, we can show that $\mathfrak{A} \Leftrightarrow_1^0 \mathfrak{B}$. From Proposition 7 it follows that $\mathfrak{A} \Leftrightarrow_0^0 \mathfrak{B}'$. □

Proposition 10. *Let* $\mathfrak{A}$ *be a countable structure and* $\mathscr{O}_A \Rightarrow_0^k \mathfrak{A}$ *for some* $k \in \omega$. *There exists a countable structure* $\mathfrak{B}$ *such that for every* $n \in \omega$, $\mathfrak{A} \Leftrightarrow_k^n \mathfrak{B}^{(n)}$.

Combining Proposition 10 with Theorem 2, we get the following corollary.

Corollary 3. *Let* $\mathfrak{A}$ *be a countable structure and* $\mathscr{O}_A \Rightarrow_0^k \mathfrak{A}$ *for some* $k \in \omega$. *There exists a countable structure* $\mathfrak{B}$ *such that*

$$(\forall n \in \omega)(\forall X \subseteq A)[X \in \Sigma_{n+1}^c(\mathfrak{A}) \leftrightarrow X \in \Sigma_{k+1}^c(\mathfrak{B}^{(n)})].$$

References

1. Ash, C., Knight, J.: Computable Structures and the Hyperarithmetical Hierarchy. Elsevier Science, Amsterdam (2000)
2. Ash, C., Knight, J., Manasse, M., Slaman, T.: Generic Copies of Countable Structures. Annals of Pure and Applied Logic 42, 195–205 (1989)
3. Chisholm, J.: Effective Model Theory vs. Recursive Model Theory. The Journal of Symbolic Logic 55(3), 1168–1191 (1990)
4. Goncharov, S.S., Khoussainov, B.: Complexity of categorical theories with computable models. Algebra and Logic 43(6), 365–373 (2004)
5. Marker, D.: Non Σ_n Axiomatizable Almost Strongly Minimal Theories. The Journal of Symbolic Logic 54, 921–927 (1989)
6. Moschovakis, Y.: Elementary Induction on Abstract Structures. North - Holland, Amsterdam (1974)
7. Richter, L.: Degrees of Structures. The Journal of Symbolic Logic 46(4), 723–731 (1981)
8. Soskova, A., Soskov, I.: A Jump Inversion Theorem for the Degree Spectra. Journal of Logic and Computation 19, 199–215 (2009)
9. Soskov, I.: Degree Spectra and Co-spectra of structures. Ann, Univ. Sofia 96, 45–68 (2004)
10. Stukachev, A.: A Jump Inversion Theorem for the Semilattices of Sigma-degrees. Siberian Advances in Mathematics 20(1), 68–74 (2010)

On a Hierarchy of Plus-Cupping Degrees

Shenling Wang and Guohua Wu*

Division of Mathematical Sciences, School of Physical and Mathematical Sciences,
Nanyang Technological University, 21 Nanyang Link, Singapore 637371
wang0362@e.ntu.edu.sg, guohua@ntu.edu.sg

Abstract. Greenberg, Ng and Wu proved in a recent paper the existence of a cuppable degree $\mathbf{a}$ that can be cupped to $\mathbf{0}'$ by high degrees only. A corollary of this result shows that such a degree $\mathbf{a}$ can be high, and hence bounds noncuppable degrees. In this paper, we prove the existence of a plus-cupping degree which can only be cupped to $\mathbf{0}'$ by high degrees. This refutes Li-Wang's claim that every plus-cupping degree is 3-plus-cupping, where a nonzero c.e. degree $\mathbf{a}$ is n-plus-cupping if for every c.e. degree $\mathbf{x}$ with $\mathbf{0} < \mathbf{x} \leq \mathbf{a}$, there is a low_n c.e. degree $\mathbf{y}$ such that $\mathbf{x} \vee \mathbf{y} = \mathbf{0}'$.

1 Introduction

A c.e. degree $\mathbf{a}$ is called *cuppable* if there is an incomplete c.e. degree $\mathbf{b}$ such that $\mathbf{a} \vee \mathbf{b} = \mathbf{0}'$, and *noncuppable* otherwise. Sacks' splitting theorem implies the existence of an incomplete cuppable degree, and Yates (unpublished) and Cooper [1] proved the existence of noncuppable degrees. Harrington [3] extended the latter result by showing that every high c.e. degree bounds a high noncuppable degree.

Extending the cupping property, Harrington [6] (see Fejer and Soare [9]) proposed a much stronger notion: plus-cupping degrees. A nonzero c.e. degree $\mathbf{a}$ is called *plus-cupping* if every nonzero c.e. degree below $\mathbf{a}$ is cuppable, i.e. if $\mathbf{0} < \mathbf{b} \leq \mathbf{a}$, then there is an incomplete c.e. degree $\mathbf{c}$ cupping $\mathbf{b}$ to $\mathbf{0}'$. (Harrington's original version of plus-cupping degrees is actually even stronger: he proved the existence a degree $\mathbf{a}$ such that for every $\mathbf{b} < \mathbf{a}$ and every $\mathbf{d} > \mathbf{a}$, there is some $\mathbf{e} < \mathbf{d}$ such that $\mathbf{b} \vee \mathbf{e} = \mathbf{d}$.)

Based on the high/low hiearchy, Li, Wu and Zhang [7] proposed a hierarchy of cuppable c.e. degrees $\mathbf{LC}_1 \subseteq \mathbf{LC}_2 \subseteq \mathbf{LC}_3 \subseteq \cdots$, where for each $n > 0$, $\mathbf{LC}_n$ denotes the class of low_n-cuppable degrees. Here, a c.e. degree $\mathbf{a}$ is called low_n-*cuppable* if there is a low_n c.e. degree $\mathbf{b}$ such that $\mathbf{a} \vee \mathbf{b} = \mathbf{0}'$. Li, Wu and Zhang showed that $\mathbf{LC}_1$ is a proper subset of $\mathbf{LC}_2$. Recently, Greenberg, Ng and Wu proved in [4] that there is an incomplete cuppable degree which can only be cupped to $\mathbf{0}'$ by high degrees. This shows that $\bigcup_n \mathbf{LC}_n$ does not exhaust all of the cuppable degrees. This refutes a claim of Li [8] that all cuppable degrees are low_3-cuppable.

---

* The second author is partially supported by NTU grant RG37/09, M52110101.

B. Löwe et al. (Eds.): CiE 2011, LNCS 6735, pp. 310–318, 2011.

Extending low_n-cuppability, Li and Wang [5] defined a notion of n-plus-cupping degrees. A c. e. degree $\mathbf{a}$ is called *n-plus-cupping* if every c. e. degree $\mathbf{x}$ with $\mathbf{0} < \mathbf{x} \leq \mathbf{a}$ is low_n-cuppable, and the class of n-plus-cupping degrees is denoted by $\mathbf{PC}_n$. This gives rise to a hierarchy for plus-cupping degrees

$$\mathbf{PC}_1 \subseteq \mathbf{PC}_2 \subseteq \mathbf{PC}_3 \subseteq \cdots.$$

Note that $\mathbf{PC}_1 = \emptyset$, as low-cuppable degrees are noncappable, while plus-cupping degrees are cappable. Li and Wang proved the existence of a 2-plus-cupping degree, and hence $\mathbf{PC}_1$ is a proper subset of $\mathbf{PC}_2$. Li and Wang [5] also claimed that all plus-cupping degrees are 3-plus-cupping, and therefore $\mathbf{PC}_3 = \mathbf{PC}$, where $\mathbf{PC}$ is the class of plus-cupping degrees.

In this paper, we extend the result of Greenberg, Ng and Wu, by showing that there is a plus-cupping degree which is cupped to $\mathbf{0}'$ by high degrees only. This implies the existence of a plus-cupping degree $\mathbf{a}$ such that any nonzero c. e. degree $\mathbf{b} \leq \mathbf{a}$ cannot be cupped to $\mathbf{0}'$ by a low_n c. e. degree for any n. Hence, this refutes Li-Wang's claim mentioned above.

Theorem 1. *There is a plus-cupping degree $\mathbf{a}$ such that for any c. e. degree $\mathbf{w}$, if $\mathbf{a} \vee \mathbf{w} = \mathbf{0}'$, then $\mathbf{w}$ is high.*

We call the plus-cupping degrees which can only be cupped to $\mathbf{0}'$ by high degrees *only-high-plus-cupping degrees*. Let $\mathbf{HPC}$ denote the class of only-high-plus-cupping degrees. It is easy to see that this class of degrees is downwards closed, and hence we can consider the ideal generated by $\mathbf{HPC}$. It is not hard to show that $\mathbf{HPC}$ itself is not an ideal, as we can show that it is not closed under join: there are two only-high-plus-cupping degrees $\mathbf{a}$ and $\mathbf{b}$ such that the join $\mathbf{a} \vee \mathbf{b}$ is high.

Note that the join of any noncuppable degree and any only-high-cuppable degree is again an only-high-cuppable degree. This implies the existence of high cuppable degrees which can be cupped to $\mathbf{0}'$ by high degrees only.

Our notation and terminology are quite standard and generally follow Soare [10]. A parameter p is defined to be fresh at a stage s if $p > s$ and p is the least number not mentioned so far in the construction.

2   Proof of Theorem 1

We will construct a c. e. set A and an auxiliary c. e. set P satisfying the following requirements:

$\mathcal{P}_e : A \neq \varphi_e$,

$\mathcal{R}_e : W_e = \Phi_e^A \Rightarrow W_e$ is computable or there are C_e, Γ_e such that C_e is an incomplete c. e. set, Γ_e is a partial computable functional, and $K = \Gamma_e^{C_e, W_e}$,

$\mathcal{Q}_e : P = \Phi_e^{A, V_e} \Rightarrow$ there exists a partial computable functional Δ_e such that for every i,

$$Tot(i) = \lim_x \Delta_e^{V_e}(i, x).$$

Here $\{(W_e, V_e) : e \in \omega\}$ is a fixed effective list of pairs of c. e. sets. For each e, Γ_e, Δ_e are partial computable functionals built by us and $Tot = \{i : \varphi_i \text{ is total}\}$ is a Π_2^0-complete set.

Let $\mathbf{a}$ be the degree of A. By the $\mathcal{P}$-requirements, $\mathbf{a}$ is nonzero. By the $\mathcal{R}$-requirements, $\mathbf{a}$ is a plus-cupping degree. The $\mathcal{Q}$-requirements ensure that for any c. e. set V_e, if $K \leq_T A \oplus V_e$ then V_e has high degree, as in this case, the $\mathcal{Q}$-requirements ensure that $Tot \leq_T V_e'$. Therefore, satisfying all the requirements will be enough to prove Theorem 1.

2.1   Strategy for $\mathcal{P}_e$

A $\mathcal{P}_e$-strategy α is a standard diagonalization strategy. That is, we choose a witness x and then wait for a stage at which $\varphi_e(x) \downarrow= 0$. If there is no such stage, then $\mathcal{P}_e$ is satisfied obviously. Otherwise, we put x into A and so $A(x) = 1 \neq 0 = \varphi_e(x)$, $\mathcal{P}_e$ is also satisfied.

2.2   Strategy for $\mathcal{R}_e$

Assume that a strategy β works on an $\mathcal{R}_e$-requirement. We define the length of agreement function $l(\beta, s)$ at stage s as

$$l(\beta, s) = \max\{x < s : (\forall y < x)[W_e(y)[s] = \Phi_e^A(y)[s]]\},$$

and the maximum length of agreement function at stage s as

$$m(\beta, s) = \max\{l(\beta, t) : t < s \text{ and t is a } \beta\text{-stage}\}.$$

Say that a stage s is a β-expansionary stage if $s = 0$ or s is a β-stage with $l(\beta, s) > m(\beta, s)$. Here, we say that a stage s is a β-stage, we mean that β is visited at stage s.

If $W_e = \Phi_e^A$, then there are infinitely many β-expansionary stages. At β-expansionary stages, we will construct an incomplete c. e. set C_e and a p.c. functional Γ_e such that we can either show that $K = \Gamma_e^{C_e, W_e}$, or show that W_e is computable.

For convenience, we omit the subscripts in the following discussion. The functional Γ will be built as follows.

1. (Rectification) If $\Gamma^{C,W}(x) \downarrow= 0 \neq 1 = K(x)$, then we put $\gamma(x)$ into C to rectify the definition of Γ.
2. (Extension) Let k be the least x such that $\Gamma^{C,W}(x) \uparrow$, then we define $\Gamma^{C,W}(k) \downarrow= K(k)$ with use $\gamma(k)$ a fresh number.

We will ensure that if $x < y$, then $\gamma(x) \leq \gamma(y)$ and thus, we have: if $\Gamma^{C,W}(x)$ becomes undefined, then for all $y \geq x$, $\Gamma^{C,W}(y)$ will also become undefined automatically.

β has two outcomes $\infty <_L f$, where ∞ denotes that β has infinitely many expansionary stages, and f denotes that β has only finitely many expansionary stages. Below outcome ∞, we will construct a c.e. set C_e and ensure that it is incomplete by constructing an auxiliary c.e. set E to satisfy the following subrequirements

$$\mathcal{S}_{e,i} : E \neq \Phi_i^{C_e}.$$

2.3 Strategy for $\mathcal{S}_{e,i}$

Let η be an $\mathcal{S}_{e,i}$-strategy below β's outcome ∞. A single η-strategy is again a standard Friedberg-Muchnik strategy, which works as follows:

(1) Pick x as a fresh number.
(2) Wait for a stage s such that $\Phi_i^{C_e}(x)[s] \downarrow= 0$.
(3) Put x into E and preserve $C_e \restriction \varphi_{i,s}(x)$.

But note that, after η performs the diagonalization, to rectify the definition of Γ, β needs to enumerate γ-uses into C_e, and this enumeration may injure the computation $\Phi_i^{C_e}(x) \downarrow= 0$. To ensure that a computation $\Phi_i^{C_e}(x)$ is clear of the γ-uses, we apply the gap-cogap argument to ensure that if η fails to protect a computation $\Phi_i^{C_e}(x)$, then the corresponding W_e is computable.

We first fix a number $k(\eta)$, the *threshold* of η. In the construction, if $y < k(\eta)$ enters K, we just put $\gamma(y)$ into C_e to rectify Γ, and we will *reset* η by undefining the associated parameters, but with $k(\eta)$ unchanged. Clearly, as $k(\eta)$ is kept unchanged, η can be reset at most $k(\eta)$ many times.

In general, when we see $\Phi_i^{C_e}(x) \downarrow= 0$ at a stage s, we do not perform the diagonalization immediately, we open a gap for A to change (and hence expecting a W_e-change, which can be small enough for us to lift the γ-uses) and create a link between β and η. At the next β-expansionary stage s', we check whether W_e changes below $\gamma(k(\eta))[s]$ between stage s and s'. We define a partial computable function h_η towards showing that W_e is computable if η opens infinitely many gaps, and none of these can be closed successfully during the whole construction.

Case a. (Close the gap successfully)

> If $W_{e,s'} \restriction (\gamma(k(\eta))[s] + 1) \neq W_{e,s} \restriction (\gamma(k(\eta))[s] + 1)$, then we travel the link and put x into E to satisfy η. Cancel the link.
>
> In this case, this W_e-change can lift $\gamma(y)$ use, for all $y \geq k(\eta)$, to big numbers, and hence the computation $\Phi_i^{C_e}(x) \downarrow= 0$ is clear of the γ-uses, and can be preserved forever.

Case b. (Close the gap unsuccessfully)

> If $W_{e,s'} \restriction (\gamma(k(\eta))[s] + 1) = W_{e,s} \restriction (\gamma(k(\eta))[s] + 1)$, then we put $\gamma(k(\eta))[s]$ into C_e to lift $\gamma(k(\eta))$ to a big number. Cancel the link.
>
> In this case, for all $x \leq \gamma(k(\eta))[s]$, if $h_\eta(x) \uparrow$, then we define $h_\eta(x) = W_{e,s}(x)$. After stage s', we prevent A from changing to preserve $W_{e,s} \restriction (\gamma(k(\eta))[s] + 1)$ till the next β-expansionary stage at which η opens a gap again.

The η-strategy has three outcomes $g <_L w <_L d$, the outcome g denotes η opens gaps infinitely often, the outcome w denotes that η waits for $\Phi_i^{C_e}(x) \downarrow = 0$ forever, and the outcome d denotes that η successfully performs diagonalization. Note that if η has outcome g, then η shows that W_e is computable as $W_e = h_\eta$, β is satisfied at η, and there is no need to satisfy β's substrategies at the nodes extending $\eta^\frown g$. In this case, the use $\gamma(k(\eta))$ approaches to infinity. However this is okay since we meet β via the computability of W_e.

We now consider the interaction between more $\mathcal{R}$-strategies. Suppose that we also have an $\mathcal{R}_{e'}$-strategy β' and its substrategy η' with $\beta^\frown\langle\infty\rangle \subseteq \beta'^\frown\langle\infty\rangle \subseteq \eta^\frown\langle g\rangle \subseteq \eta'$, where η is a substrategy of β. Then a link is created between β' and η' (η' opens a gap) only when η opens a gap (a link is created between β and η). These two links are crossing which needs to be avoided in a gap-cogap argument (we can see the necessity of this concern in the construction of a high$_2$ nonbounding degree in Downey, Lempp and Shore's paper [2]). With this in mind, on the construction tree, when η has outcome g, we will say that β' becomes inactive, or β' is injured by η, and we need to arrange a back-up strategy β' below this outcome.

We assume that readers have some basic ideas of the framework of $0'''$ argument. The construction tree will be labelled in a way where a single $\mathcal{R}$-requirement might be labelled several times along a single path, corresponding to the "injuries" mentioned above.

In general, for a given $\mathcal{R}$-requirement, only finitely many $\mathcal{R}$-requirements can have higher priority, and only substrategies of these $\mathcal{R}$-strategies with higher priority can injure a strategy of the given $\mathcal{R}$-requirement. By induction, we can ensure that on any path of the construction tree, for each $\mathcal{R}$-requirement, there is an $\mathcal{R}$-strategy on the path such that no $\mathcal{S}$-strategy of other $\mathcal{R}$-strategies can injure it.

In particular, on the true path, the longest $\mathcal{R}$-strategy, β say, can either have infinitely many substrategies (Γ is contructed such that $K = \Gamma_\beta^{C_{e(\beta)} \oplus W_{e(\beta)}}$), or have only finitely many substrategies (in this case, the associated c. e. set $W_{e(\beta)}$ is computable).

2.4   Strategy for $\mathcal{Q}_e$

Let ξ be a $\mathcal{Q}_e$-strategy. We define the length of agreement function $l(\xi, s)$ at stage s as

$$l(\xi, s) = \max\{x < s : (\forall y < x)[P(y)[s] = \Phi_e^{A, V_e}(y)[s]]\},$$

and the maximum length of agreement function at stage s as

$$m(\xi, s) = \max\{l(\xi, t) : t < s \text{ and t is a } \xi\text{-stage}\}.$$

Say that a stage s is a ξ-expansionary stage if $s = 0$ or s is a ξ-stage with $l(\xi, s) > m(\xi, s)$.

If $P = \Phi_e^{A,V_e}$, then there are infinitely many ξ-expansionary stages. At ξ-expansionary stages, we will construct a p.c. functional Δ_e to ensure that for any i,

$$Tot(i) = \lim_x \Delta_e^{V_e}(i, x).$$

The strategy of ξ has two outcomes $\infty <_L f$, where ∞ denotes that ξ has infinitely many expansionary stages and f denotes that ξ has only finitely many expansionary stages. Below the ∞ outcome of ξ, we will ensure that Δ_e satisfies the following subrequirements

$$\mathcal{T}_{e,i} : Tot(i) = \lim_x \Delta_e^{V_e}(i, x).$$

2.5 Strategy for $\mathcal{T}_{e,i}$

Let ζ be a $\mathcal{T}_{e,i}$-strategy below the ∞ outcome of ξ. We define the length of convergence function $l(\zeta, s)$ at stage s as

$$l(\zeta, s) = \max\{x < s : (\forall y < x)[\varphi_i(y)[s] \downarrow]\},$$

and the maximum length of convergence function at stage s as

$$m(\zeta, s) = \max\{l(\zeta, t) : t < s \text{ and } t \text{ is a } \zeta\text{-stage}\}.$$

Say that a stage s is a ζ-expansionary stage if $s = 0$ or s is a ζ-stage with $l(\zeta, s) > m(\zeta, s)$. ζ has two outcomes $\infty <_L f$, where ∞ denotes the outcome that φ_i is total, in which we will define $\Delta_e^{V_e}(i, x) = 1$ for almost all x, and f denotes the outcome that φ_i is not total, in which case we will define $\Delta_e^{V_e}(i, x) = 0$ for almost all x.

Suppose that we define $\Delta_e^{V_e}(i, x) = 0$ under the outcome f at a previous stage, and now ζ changes its outcome to ∞, so we want to redefine $\Delta_e^{V_e}(i, x) = 1$, which requires V_e to have a corresponding change to first make $\Delta_e^{V_e}(i, x)$ undefined. For this purpose, before we define $\Delta_e^{V_e}(i, x)$ as 0, we first pick a big number p_ζ which is not in P, and wait for a ξ-expansionary stage, s_0 say, such that $l(\xi, s_0) > p_\zeta$, i.e. we see $\Phi_e^{A,V_e}(p_\zeta) \downarrow = 0$ at stage s_0, then we define $\Delta_e^{V_e}(i, x)[s_0] = 0$ with use $\delta_e(i, x)[s_0] > \varphi_e(p_\zeta)[s_0]$. At the next ζ-stage $s_1 > s_0$, if s_1 is a ζ-expansionary stage, i.e. ζ has outcome ∞, we first put p_ζ into P to force V_e to have a change below $\varphi_e(p_\zeta)[s_0]$ while restrain A on $\varphi_e(p_\zeta)[s_0]$ (if there is no such change, then P and $\Phi_e^{A,V_e}(p_\zeta)$ will differ at p_ζ, a global win for ξ), and hence below $\delta_e(i, x)[s_0]$. This change undefines $\Delta_e^{V_e}(i, x)$, and as a consequence, we can redefine it.

We now consider the interactions between two $\mathcal{T}$-strategies. We describe a potential problem: if there are two or more $\mathcal{T}$-strategies working below the ∞ outcome of ξ, ζ_1, ζ_2, with $\xi^\frown \langle \infty \rangle \subseteq \zeta_1^\frown \langle \infty \rangle \subseteq \zeta_2$. Then the action of enumerating p_{ζ_1} into P for ζ_1 as above can always force V_e to have a change to undefine $\Delta_e^{V_e}(i(\zeta_1), x)$, but this change can also undefine $\Delta_e^{V_e}(i(\zeta_2), x)$. Since ζ_2 assumes that ζ_1 has outcome ∞, it knows that p_{ζ_1} will be enumerated into P, and will be updated infinitely often. As a consequence, if ζ_2 has outcome f, then the action of ζ_1 may lead $\Delta_e^{V_e}(i(\zeta_2), x)$ to diverge eventually.

To avoid this, if $\Delta_e^{V_e}(i(\zeta_2), x)$ is undefined by a V_e-change, at the next ζ_2-stage, if ζ_2 has outcome f, we redefine $\Delta_e^{V_e}(i(\zeta_2), x) = 0$, but with use the same as before if the computation $\Phi_e^{A,V_e}(p_{\zeta_2})$ is the same as before. Of course, if the computation $\Phi_e^{A,V_e}(p_{\zeta_2})$ has been changed, then we define $\Delta_e^{V_e}(i(\zeta_2), x) = 0$ with use bigger than the current use $\varphi_e(p_{\zeta_2})$. Therefore, if $P = \Phi_e^{A,V_e}$, and ζ_2 has outcome f, then the parameter p_{ζ_2} will be updated finitely often and so it has a final value, and as a consequence, the $\varphi_e(p_{\zeta_2})$ use will have a fixed value, which ensures that $\delta_e(i(\zeta_2), x)$ use will be lifted only finitely often. That is, in this case, $\Delta_e^{V_e}(i(\zeta_2), x) \downarrow = 0$ for almost all x, and ζ_1 has no effect on ζ_2.

We now consider the interaction between one $\mathcal{P}_{e'}$-strategy α and one $\mathcal{Q}_e$-strategy ξ with $\xi^\frown\langle\infty\rangle \subset \alpha$. Fix a witness y for α, when we see $\varphi_{e'}(y) \downarrow = 0$, we want to put y into A directly to satisfy $\mathcal{P}_{e'}$, but this action may injure ξ, which has higher priority.

Consider the following scenario: Let ζ be a $\mathcal{T}_{e,i}$-strategy between ξ and α with $\xi^\frown\langle\infty\rangle \subseteq \zeta^\frown\langle f\rangle \subseteq \alpha$ and ζ defines $\Delta_e^{V_e}(i, x) = 0$ with use $\delta_e(i, x) > \varphi_e(p_\zeta)$ under the outcome f, but now α puts y into A, this enumeration can change the computation $\Phi_e^{A,V_e}(p_\zeta)$ and lead to a new use $\varphi_e(p_\zeta)$ bigger than $\delta_e(i, x)$. Now if later, at a ζ-expansionary stage, we put p_ζ into P, V_e may change below the new use $\varphi_e(p_\zeta)$, but not below $\delta_e(i, x)$, and this V_e-change cannot make $\Delta_e^{V_e}(i, x)$ undefined as needed.

To avoid this problem, when α selects y as a number for diagonalization, α also selects a number $a_{\alpha,\xi}$. When we want to put a number y into A, we first put $a_{\alpha,\xi}$ into P to force V_e-change to undefine $\Delta_e^{V_e}(i, x)$ defined by ζ, and we put y into A only after we see such a V_e-change. Such a process delays the diagonalization, but it does not affect the satisfaction of α, since once $\varphi_{e'}(y) \downarrow = 0$, it will converge to 0 forever.

The interactions between one $\mathcal{P}$-strategy and several $\mathcal{Q}$-strategies is a direct generalization of the simple case discussed above.

Without loss of generality, suppose that, above a $\mathcal{P}_e$-strategy α, there are n many $\mathcal{Q}$-strategies, $\xi_1, \xi_2, \cdots, \xi_n$ say, with $\xi_1^\frown\langle\infty\rangle \subset \xi_2^\frown\langle\infty\rangle \subset \cdots \subset \xi_n^\frown\langle\infty\rangle \subset \alpha$. We will associate α with parameters $a_{\alpha,\xi_1} < \cdots < a_{\alpha,\xi_n}$ for those $\mathcal{Q}$-strategies. After α sees that $\varphi_e(y) \downarrow = 0$ at some stage s_0, where y is a candidate for the $\mathcal{P}_e$-strategy α. It first creates a link between α and ξ_n and puts the number a_{α,ξ_n} into P simultaneously. Thus, at the next ξ_n-expansionary stage s_1, a $V_{e(\xi_n)}$-change appears, and this $V_{e(\xi_n)}$-change will undefine those $\Delta_{e(\xi_n)}^{V_{e(\xi_n)}}(i, x)$ defined by ζ-strategies between α and ξ_n. So if we now put y into A, this enumeration will not cause incorrectness of $\Delta_{e(\xi_n)}^{V_{e(\xi_n)}}$. At stage s_1, the previous link between α and ξ_n is cancelled, we create a new link between α and ξ_{n-1} and put the number $a_{\alpha,\xi_{n-1}}$ into P simultaneously (note that once $\varphi_e(y) \downarrow = 0$, it will converge to 0 forever). At the next ξ_{n-1}-expansionary stage s_2, a $V_{e(\xi_{n-1})}$-change appears, and this $V_{e(\xi_{n-1})}$-change will undefine those $\Delta_{e(\xi_{n-1})}^{V_{e(\xi_{n-1})}}(i, x)$ defined by ζ-strategies between α and ξ_{n-1}. We repeat this process for all the other such $\mathcal{Q}$-strategies. After we cancel the link between α and ξ_1, then α can perform the diagonalization by putting y into A, this enumeration will not cause incorrectness of those $\Delta_{e(\xi_i)}^{V_{e(\xi_i)}}$ for all $1 \leq i \leq n$.

2.6   More Interactions

We now consider the interaction between a single $\mathcal{R}$-, $\mathcal{Q}$- and $\mathcal{P}$-strategy.

Suppose that a $\mathcal{Q}_e$-strategy ξ works between an $\mathcal{R}_{e'}$-strategy β and an $\mathcal{S}_{e',i}$-strategy η with $\beta^\frown\langle\infty\rangle \subseteq \xi \subset \xi^\frown\langle\infty\rangle \subseteq \eta \subset \eta^\frown\langle g\rangle \subset \alpha$, where α is a $\mathcal{P}$-strategy and $0 \leq e' \leq e$. Then we may have that η opens a gap and creates a link between β and η and α creates a link between α and ξ at the same stage, s_0 say. That is, we have two crossed links (β, η) and (ξ, α) at stage s_0. At the next β-expansionary stage s_1, suppose η closes the gap unsuccessfully and cancels the link (β, η), so η will impose an A-restraint after stage s_1 till the next η-stage, s_2 say, at which η opens another gap. But, before stage s_2, i.e. during the cogap of η-strategy, we will travel the link (ξ, α) created at stage s_0 and may create another link (ξ', α) if there is a $\mathcal{Q}$-strategy ξ' with $\xi'^\frown\langle\infty\rangle \subset \xi$. In this case, α will get to perform diagonalization before stage s_2. However we cannot put a small number ($\leq s_1$) into A before stage s_2. To avoid this problem, we use the backup strategy to deal with this. That is, we will put a backup strategy $\hat{\xi}$ below $\eta^\frown\langle g\rangle$ to try to satisfy the $\mathcal{Q}_e$-requirement. In the construction, when a $\mathcal{P}$-strategy α wants to create a link between α and a $\mathcal{Q}_e$-strategy ξ with $\xi^\frown\langle\infty\rangle \subset \alpha$, it will create a link between α and the longest node $\hat{\xi} \subset \alpha$ assigned to a $\mathcal{Q}_e$-strategy. This will avoid having the crossing of links as above. That is, we may have two links (β, η) and $(\hat{\xi}, \alpha)$ at the same stage such that $\beta^\frown\langle\infty\rangle \subseteq \eta \subset \eta^\frown\langle g\rangle \subseteq \hat{\xi} \subset \hat{\xi}^\frown\langle\infty\rangle \subset \alpha$.

We may have two nested links (β, η) and (ξ'', α) at the same stage for some $\mathcal{Q}$-strategy ξ'' with higher priority such that $\xi''^\frown\langle\infty\rangle \subseteq \beta \subset \beta^\frown\langle\infty\rangle \subseteq \eta \subset \eta^\frown\langle g\rangle \subset \alpha$. Since the η-gap is never closed until the outer link (ξ'', α) is travelled, α will perform the diagonalization during the η-gap.

In the construction, we put a backup $\mathcal{Q}_e$-strategy below the $\mathcal{S}_{e',i}$-strategy η with g outcome for some $e' \leq e$. Thus, for a given $e' \leq e$, on any path of the priority tree, if we have $\beta^\frown\langle\infty\rangle \subseteq \eta \subset \eta^\frown\langle g\rangle$ on the path, no other $\mathcal{S}_{e',i'}$ strategies will be listed on the path below $\eta^\frown\langle g\rangle$. Therefore, for a fixed e, there are at most finitely many backup $\mathcal{Q}_e$-strategies on any path of the priority tree, and the longest node assigned a $\mathcal{Q}_e$-requirement is responsible for satisfying the requirement.

This completes the description of our strategies for Theorem 1, and the possible interactions between them. A full construction can proceed on a priority tree, and Theorem 1 is proved.

References

1. Cooper, S.B.: On a theorem of C. E. M. Yates. Handwritten Notes (1974)
2. Downey, R., Lempp, S., Shore, R.: Highness and bounding minimal pairs. Mathematical Logic Quarterly 39(1), 475–491 (1993)
3. Harrington, L.A.: On Cooper's proof of a theorem of Yates. Handwritten Notes (1976)

4. Greenberg, N., Ng, K.M., Wu, G.: Cuppable degrees and the high/low hierarchy (to appear)
5. Wang, Y., Li, A.: A hierarchy for the plus cupping Turing degrees. J. Symbolic Logic 68(3), 972–988 (2003)
6. Harrington, L.A.: Plus-cupping in the recursively enumerable degrees. Handwritten Notes (1978)
7. Li, A., Wu, G., Zhang, Z.: A hierarchy for cuppable degrees. Illinois J. Math. 44(3), 619–632 (2000)
8. Li, A.: A hierarchy characterisation of cuppable degrees. University of Leeds, Dept. of Pure Math 1, 21 (2001) Preprint series
9. Fejer, P.A., Soare, R.I.: The plus-cupping theorem for the recursively enumerable degrees. In: logic year 1979-80: University of Connecticut, pp. 49–62 (1981)
10. Soare, R.I.: Recursively Enumerable Sets and Degrees. Perspectives in Mathematical Logic. Springer, Heidelberg (1987)

Author Index